Nickel Metal Hydride Batteries 2017

Special Issue Editor
Kwo Young

MDPI • Basel • Beijing • Wuhan • Barcelona • Belgrade

MDPI

Special Issue Editor
Kwo Young
Wayne State University
USA

Editorial Office
MDPI AG
St. Alban-Anlage 66
Basel, Switzerland

This edition is a reprint of the Special Issue published online in the open access journal *Batteries* (ISSN 2313-0105) from 2016–2018 (available at: http://www.mdpi.com/journal/batteries/special_issues/ni_mh_batteries2017).

For citation purposes, cite each article independently as indicated on the article page online and as indicated below:

Lastname, F.M.; Lastname, F.M. Article title. *Journal Name* **Year**, *Article number, page range.*

First Edition 2018

ISBN 978-3-03842-328-2 (Pbk)
ISBN 978-3-03842-327-5 (PDF)

Table of Contents

About the Special Issue Editor

Kwo Young is a Chief Scientist in BASF-Ovonic located in Rochester Hills, Michigan, USA. He graduated from the National Taiwan University, Republic of China in 1982 with a BS in EE and from Princeton University, New Jersey in 1989 with a Ph.D. also in Electrical Engineering (EE). He has worked in the research field of Ni/MH batteries for over 25 years. He is one of the key inventors to have increased the power of NiMH battery technology, and successfully implemented the technology in EV and HEV with support from USABC. He has 44 US Patents in Ni/MH battery technology which form the basis of the licenses for the battery manufactures. Dr. Young also serves as a research professor at Wayne State University, Michigan, where he supervises Ph.D. students in electrochemical materials' research. In recent years, he has published over 125 technical papers in the field of metal hydrides for electrochemical applications.

Preface to "Nickel Metal Hydride Batteries 2017"

The Nickel/metal hydride (Ni/MH) battery continued to be an important energy storage source in 2017. New demands from small- and large-scale stationary and transportation applications pushed research and development work into multiple directions with targets of higher gravimetric energy, higher delivered power at low temperature, prolonged cycle life at high temperature, and lower material and manufacturing costs. In the second Special Issue of Batteries dedicated to Ni/MH batteries, reviews, current research, and future projection in the materials, fabrication methods, cell performance evaluation, failure analysis, and other subjects related to Ni/MH batteries are included.

Kwo Young
Special Issue Editor

batteries

MDPI

Editorial

Research in Nickel/Metal Hydride Batteries 2017

Kwo-Hsiung Young [1,2]

[1] BASF/Battery Materials—Ovonic, 2983 Waterview Drive, Rochester Hills, MI 48309, USA;
kwo.young@basf.com; Tel.: +1-248-293-7000

[2] Department of Chemical Engineering and Materials Science, Wayne State University, Detroit, MI 48202, USA

Academic Editor: Andreas Jossen
Received: 2 January 2018; Accepted: 26 January 2018; Published: 12 February 2018

Abstract: Continuing from a special issue in Batteries in 2016, nineteen new papers focusing on recent research activities in the field of nickel/metal hydride (Ni/MH) batteries have been selected for the 2017 Special Issue of Ni/MH Batteries. These papers summarize the international joint-efforts in Ni/MH battery research from BASF, Wayne State University, Michigan State University, FDK Corp. (Japan), Institute for Energy Technology (Norway), Central South University (China), University of Science and Technology Beijing (China), Zhengzhou University of Light Industry (China), Inner Mongolia University of Science and Technology (China), Shenzhen Highpower (China), and University of the Witwatersrand (South Africa) from 2016–2017 through reviews of AB_2 metal hydride alloys, Chinese and EU Patent Applications, as well as descriptions of research results in metal hydride alloys, nickel hydroxide, electrolyte, and new cell type, comparison work, and projections of future works.

Keywords: nickel-metal hydride battery; rechargeable alkaline battery; metal hydride alloy; electrochemistry; electrolyte; core-shell structure

1. Introduction

The Nickel/metal hydride (Ni/MH) battery continued to be an important energy storage source in 2017. Recent demonstrations of Ni/MH batteries in a few key applications, such as new hybrid electric vehicles manufactured in China [1], an integrated smart energy solution in Sweden [2], a Ni/MH battery system with a high robustness at high temperature in Middle East [3], fast charge (3–5 min) [4] and a wide temperature range (between −55 and 70 °C) [5] for bus transportation, the introduction of Mega Twicell for larger scale energy storage [6], and Cellect 600 for telecommunication backup power [7]. This progress pushed research and development work into multiple directions with targets of higher gravimetric energy, higher delivered power at low temperature, prolonged cycle life at high temperature, and lower material and manufacturing costs. Continuing from the work established in a United States Department of Energy funded program—Robust Affordable Next Generation Energy Storage System (RANGE) in 2015–2016 [8], further development, especially in the implementation of the advanced materials at the cell level, were accomplished by BASF and its collaborators. These accomplishments are reported in this special issue of Batteries.

2. Contributions

The selected papers presented in this Special Issue are highlighted in this section. They are mainly results obtained through international collaborations with other institutes, and can be divided into six general categories: reviews (three papers), metal hydride (MH) alloys used as negative electrode active materials (seven papers), nickel hydroxide as the positive electrode active materials (one paper), electrolyte (two papers), cell performance (five papers), and special analytic tools (one paper).

2.1. Reviews on Related Work

Three review papers are included in this Special Issue of Batteries: one on the C14-predominated AB$_2$ MH alloys [9] and two on Patent Applications related to Ni/MH batteries filed separately in China [10] and Europe [11]. AB$_2$ MH alloy has a 20% higher capacity than that from the conventional used misch-metal based AB$_5$ MH alloy, and is absent from the rare earth elements and immune from their price volatilities. However, the relatively lower performances in high-rate dischargeability and cycle stability in AB$_2$ require refinements in both the chemical composition and fabrication process, which are reviewed and discussed here. In the intellectual property area, related patents (or applications) are important to the researchers and companies in the field in addition to regular academic publications. Continuing from the reviews of patents from United States [12] and Japan [13] published last year, we focused on patent applications filed in the country that produces the most Ni/MH batteries—China—and the third largest consumer market—Europe. While the Chinese Patents focus more on the battery components and fabrication method, the European ones concentrate on applications, such as for the button cell and bipolar design.

2.2. Metal Hydride Alloys

There are studies of three families of MH alloys included in this Special Issue of Batteries, AB$_2$ (both C14 and C15), body-centered-cubic (BCC), and superlattice alloys. In the Laves phase-based AB$_2$ family, doping effects of Pd [14] and B [15] to a C14-predominated alloy were studied and a comparison between C14- and C15-MH alloys was also presented [16]. The preliminary conclusions are both Pd and a newly formed V$_3$B$_2$ phase improve the surface catalytic ability and C15 alloy is more suitable for high-rate application. In the BCC area, thermal annealing was found to be beneficial to the corrosion resistance of a Fe-containing alloy by introducing a new Ti-rich phase [17]. Lastly, the effects of annealing [18], addition of Fe [19], and alkaline etch were reported in superlattice based MH alloys [20]. Optimization of the annealing condition and Fe-content were obtained. Additionally, a superlattice MH alloy with high La-content treated by an alkaline etching was recommended for high-rate application.

2.3. Nickel Hydroxide

The discovery of a core-shell structured high-capacity Ni(OH)$_2$ used as active material in the positive electrode of Ni/MH batteries was a major accomplishment in the RANGE program [8]. This Special Issue includes a paper that further elaborates on the manufacture, properties, and half-cell results of the high-capacity α-β Ni(OH)$_2$ as compared α-Ni(OH)$_2$ fabricated by other means [21]. After the phase transformation step (initial cycling), a core (β-Ni(OH)$_2$)-shell (α-Ni(OH)$_2$) structured spherical powder with an excellent cycle stability in the flooded half-cell configuration was formed. The shell portion (higher Al-content) of the particle is composed of α-Ni(OH)$_2$ nano-crystals imbedded in a β-Ni(OH)$_2$ matrix, which helps to reduce the stress originating from the lattice expansion in the β-α transformation. A review of the research on α-Ni(OH)$_2$ is also included.

2.4. Electrolyte

Finding an alternative electrolyte with a less corrosion to the MH alloy and expanded voltage window was the top priority of the RANGE program [8]. In 2016, we reported that the reduction of the corrosion nature of the alkaline KOH electrolyte can be accomplished by selection of both adequate combination of alkaline species [22] and salt additives [23]. In this Special Issue, the effects of adding Cs$_2$CO$_3$ salt in the electrolyte was further investigated, and a newly formed surface amorphous oxide was credited for the reduction of surface oxidation of the MgNi-based MH alloy [24]. Another breakthrough was using the ionic liquid to replace the aqueous KOH solution as the electrolyte [8]. The non-aqueous ionic liquid enabled the use of ultrahigh-capacity Si anode (3635 mAh g^{-1}) [25] and expansion of the voltage window. A paper about the fundamental principle and selection of ionic liquid used in the proton-conducting MH battery is included in the Special Issue of Batteries [26].

2.5. Cell Performance Comparison

Sealed cells were built and tested to verify the results obtained from half-cell testing. In the C-size cell, performances of AB_2 (between C14 and C15) [27], a Fe-free [28] and a Fe-doped [29] superlattice MH alloys were measured, and that results can be summarized as follows: C15-based MH alloy was more suitable for high-rate application comparing to those from C14 alloy and confirm previous half-cell results [16], superlattice alloy showed better high-rate and low-temperature performances comparing to those of AB_5 MH alloy, and Fe in the superlattice alloy can extend the cycle life by preventing Al-leaching from the negative electrode. In the newly developed pouch type cell, the high-capacity core-shell $Ni(OH)_2$ was compared to the conventional single-phase $Ni(OH)_2$, and lower impedance and better charge retention were observed [30]. Lastly, the high-temperature storage characteristics of Ni/MH battery module based on superlattice MH alloy were compared to those from nickel-cadmium and valve-regulated lead-acid batteries, and favorable results were obtained [31].

2.6. Analytic Methodology

Continuing from previous reports on the crystallographic orientation alignments in phases in an AB_2-predominating alloy [32] and a BCC-C14 alloy [33], electron backscatter diffraction (EBSD) was applied to a series of La-Mg-Ni-based superlattice metal hydride alloys produced by a novel method of interacting a $LaNi_5$ alloy with Mg vapor [34]. Mg formed discrete grains and then diffused through the *ab*-phase of $LaNi_5$ and transformed it into AB_2, AB_3, and A_2B_7 phases. According to EBSD mapping, diffusion of Mg stops at the grain boundary of the host $LaNi_5$ alloy. A prefect alignment in the *c*-axis between the newly formed superlattice phases and $LaNi_5$ was observed. Understanding of Mg-$LaNi_5$ solid-state reaction contributes directly to development work for a low-cost fabrication method to produce high-value superlattice alloys for battery applications.

3. Conclusions

In this Special Issue of Batteries, the joint research efforts from BASF-Ovonic and their collaborators in 2016–2017 are highlighted by reviewing nineteen papers focused on the area of Ni/MH batteries. The majority of the works focused on the implementation of the advanced material/cell design developed from the previous year, including both the superlattice MH alloy and BCC-based multi-phase MH alloys used as negative electrode active materials, high-capacity core-shell β-α $Ni(OH)_2$ as positive electrode active materials, ionic liquid as electrolyte, and pouch cell design. Future research activities of the team aim to commercialize high-capacity Si-based negative electrode, development of high-capacity Mn-based positive electrode, thin separator to reduce the impedance of the ionic liquid electrolyte, and continuous improvement in the low-and high-temperature performance of Ni/MH batteries.

Acknowledgments: The Guest Editor (K.Y.) thanks both the colleagues who made impressive and important contributions to the articles and the editorial team at the publisher MDPI for providing precious guidance. K.Y. is also obliged to R.F. Jordan at the Rockefeller University for refinement of his writing.

Conflicts of Interest: The author declares no conflict of interest.

Abbreviations

The following abbreviations are used in this manuscript:

Ni/MH	Nickel/metal hydride
RANGE	Robust Affordable Next Generation Energy Storage System
MH	Metal hydride
BCC	Body-centered-cubic
EBSD	Electron backscatter diffraction

References

1. Ogawa, K. Toyota to Start Local Production of NiMH Battery Cells in CHINA by end of 2016. Available online: http://techon.nikkeibp.co.jp/atclen/news_en/15mk/030900433/ (accessed on 8 December 2017).
2. Ferroamp EnergyHub Powered by Nilar. Available online: http://www.nilar.com/wp-content/uploads/2017/09/Ferroamp-Customer-Case-NIL17.pdf (accessed on 8 December 2017).
3. Alpha Innovations, Arts Energy and Grolleau Showcase NiMH Solution for Telecom at GITEX, in Dubai. Available online: http://www.alphainnovations.eu/news/alpha-gitex-dubai (accessed on 8 December 2017).
4. Domestic Super Battery Debut: 3–5 Minutes is Full. Available online: http://www.bestchinanews.com/Science-Technology/4877.html (accessed on 8 December 2017).
5. NiMH Battery with an Extreme Temperature Range. Available online: http://www.sohu.com/a/150492105_117460 (accessed on 8 December 2017). (In Chinese)
6. Large Capacity "MEGA TWICELL" Ni-MH Battery Developed by FDK. Available online: http://www.fdk.com/whatsnew-e/release20170215-e.html (accessed on 8 December 2017).
7. Cellect^{TM} 600 Solves the Small Cell Power and Battery Dilemma. Available online: Https://www.alpha.ca/cellect (accessed on 8 December 2017).
8. Young, K.; Ng, K.Y.S.; Bendersky, L.A. A Technical Report of the Robust Affordable Next Generation Energy Storage System-BASF Program. *Batteries* **2016**, *2*, 2. [CrossRef]
9. Young, K.; Chang, S.; Lin, X. C14 Laves phase metal hydride alloys for Ni/MH batteries applications. *Batteries* **2017**, *3*, 27. [CrossRef]
10. Young, K.; Cai, X.; Chang, S. Reviews on the Chinese Patents regarding nickel/metal hydride battery. *Batteries* **2017**, *3*, 24. [CrossRef]
11. Chang, S.; Young, K.; Lien, Y. Reviews of European patents on nickel/metal hydride batteries. *Batteries* **2017**, *3*, 25. [CrossRef]
12. Chang, S.; Young, K.; Nei, J.; Fierro, C. Reviews on the U.S. Patents regarding nickel/metal hydride batteries. *Batteries* **2016**, *2*, 10. [CrossRef]
13. Ouchi, T.; Young, K.; Moghe, D. Reviews on the Japanese Patent Applications regarding nickel/metal hydride batteries. *Batteries* **2016**, *2*, 21. [CrossRef]
14. Young, K.; Ouchi, T.; Nei, J.; Chang, S. Increase in the surface catalytic ability by addition of palladium in C14 metal hydride alloy. *Batteries* **2017**, *3*, 26. [CrossRef]
15. Chang, S.; Young, K.; Ouchi, T.; Nei, J.; Wu, X. Effects of boron-incorporation in a V-containing Zr-based AB_2 metal hydride alloy. *Batteries* **2017**, *3*, 36. [CrossRef]
16. Young, K.; J. Nei, J.; Wan, C.; Denys, R.V.; Yartys, V.A. Comparison of C15- and C15-predominated AB_2 metal hydride alloys for electrochemical applications. *Batteries* **2017**, *3*, 22. [CrossRef]
17. Abdul, J.M.; Chown, L.H.; Odusote, J.K.; Nei, J.; Young, K.; Olayinka, W.T. Hydrogen storage characteristics and corrosion behavior of $Ti_{24}V_{40}Cr_{24}Fe_2$ alloy. *Batteries* **2017**, *3*, 19. [CrossRef]
18. Young, K.; Ouchi, T.; Nei, J.; Koch, J.M.; Lien, Y. Comparison among constituent phases in superlattice metal hydride alloys for battery applications. *Batteries* **2017**, *3*, 34. [CrossRef]
19. Young, K.; Ouchi, T.; Nei, J.; Yasuoka, S. Fe-substitution for Ni in misch metal-based superlattice hydrogen absorbing alloys—Part 1. Structural, hydrogen storage, and electrochemical properties. *Batteries* **2016**, *2*, 34. [CrossRef]
20. Meng, T.; Young, K.; Hu, C.; Reichman, B. Effects of alkaline pre-etching to metal hydride alloys. *Batteries* **2017**, *3*, 30. [CrossRef]
21. Young, K.; Wang, L.; Yan, S.; Liao, X.; Meng, T.; Shen, H.; Mays, W.C. Fabrications of high-capacity alpha-$Ni(OH)_2$. *Batteries* **2017**, *3*, 6. [CrossRef]
22. Nei, J.; Young, K.; Rotarov, D. Studies on MgNi-based metal hydride electrode with aqueous electrolytes composed of various hydroxides. *Batteries* **2016**, *2*, 27. [CrossRef]
23. Yan, S.; Young, K.; Ng, K.Y.S. Effects of salt additives to the KOH electrolyte used in Ni/MH batteries. *Batteries* **2015**, *1*, 54–73. [CrossRef]
24. Yan, S.; Nei, J.; Li, P.; Young, K.; Ng, K.Y.S. Effects of Cs_2CO_3 additives in KOH electrolyte used in Ni/MH batteries. *Batteries* **2017**, *3*, 41. [CrossRef]
25. Meng, T.; Young, K.; Beglau, D.; Yan, S.; Zeng, P.; Cheng, M.M.-C. Hydrogenated amorphous silicon thin film anode for proton conducting batteries. *J. Power Sources* **2016**, *302*, 31–38. [CrossRef]

4

26. Meng, T.; Young, K.; Wong, D.F.; Nei, J. Ionic liquid-based non-aqueous electrolytes for nickel/metal hydride batteries. *Batteries* **2017**, *3*, 4. [CrossRef]
27. Young, K.; Koch, J.M.; Wan, C.; Denys, R.V.; Yartys, V.A. Cell performance comparison between C14- and C15- predominated AB$_2$ metal hydride alloys. *Batteries* **2017**, *3*, 29. [CrossRef]
28. Koch, J.M.; Young, K.; Nei, J.; Hu, C.; Reichman, B. Performance comparison between AB$_5$ and superlattice metal hydride alloys in sealed cell. *Batteries* **2017**, *3*, 35. [CrossRef]
29. Meng, T.; Young, K.; Nei, J.; Koch, J.M.; Yasuoka, S. Fe-substitution for Ni in misch metal-based superlattice hydrogen absorbing alloys—Part 2. Ni/MH battery performance and failure mechanism. *Batteries* **2017**, *3*, 28. [CrossRef]
30. Yan, S.; Meng, T.; Young, K.; Nei, J. A Ni/MH pouch cell with high-capacity Ni(OH)$_2$. *Batteries* **2017**, *3*, 38. [CrossRef]
31. Zelinsky, M.A.; Koch, J.M.; Young, K. Performance comparison of rechargeable batteries for stationary applications (Ni/MH vs. Ni-Cd and VRLA). *Batteries* **2018**, *4*, 1. [CrossRef]
32. Liu, Y.; Young, K. Microstructure investigation on metal hydride alloys by electron backscatter diffraction technique. *Batteries* **2016**, *2*, 26. [CrossRef]
33. Shen, H.; Young, K.; Bendersky, L.A. Clean grain boundary observations in multi-phase metal hydride alloys. *Batteries* **2016**, *2*, 22. [CrossRef]
34. Yan, S.; Young, K.; Zhao, X.; Mei, Z.; Ng, K.Y.S. Electron backscatter diffraction studies on the formation of superlattice metal hydride alloys. *Batteries* **2017**, *3*, 40. [CrossRef]

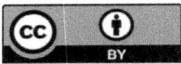

batteries

MDPI

Review

C14 Laves Phase Metal Hydride Alloys for Ni/MH Batteries Applications

Kwo-Hsiung Young [1,2,*], Shiuan Chang [2,3] and Xinting Lin [4]

1   Department of Chemical Engineering and Materials Science, Wayne State University, Detroit, MI 48202, USA
2   BASF/Battery Materials—Ovonic, 2983 Waterview Drive, Rochester Hills, MI 48309, USA; ShiuanC@wayne.edu
3   Department of Mechanical Engineering, Wayne State University, Detroit, MI 48202, USA
4   Department of Chemical Engineering and Materials Science, Michigan State University, East Lansing, MI 48824, USA; linxinti@msu.edu
*   Correspondence: kwo.young@basf.com; Tel.: +1-248-293-7000

Received: 6 June 2017; Accepted: 18 August 2017; Published: 14 September 2017

Abstract: C14 Laves phase alloys play a significant role in improving the performance of nickel/metal hydride batteries, which currently dominate the 1.2 V consumer-type rechargeable battery market and those for hybrid electric vehicles. In the current study, the properties of C14 Laves phase based metal hydride alloys are reviewed in relation to their electrochemical applications. Various preparation methods and failure mechanisms of the C14 Laves phase based metal hydride alloys, and the influence of all elements on the electrochemical performance, are discussed. The contributions of some commonly used constituting elements are compared to performance requirements. The importance of stoichiometry and its impact on electrochemical properties is also included. At the end, a discussion section addresses historical hurdles, previous trials, and future directions for implementing C14 Laves phase based metal hydride alloys in commercial nickel/metal hydride batteries.

Keywords: metal hydride; nickel metal hydride battery; Laves phase alloy; rare earth element; electrochemistry; pressure concentration isotherm

1. Introduction

Nickel/metal hydride (Ni/MH) batteries have a wide range of applications including portable consumer electronics [1], transportation [2], and stationary power sources [3,4] as it features benefits such as safe chemistry, sustainable life, a stable raw material supply, high performance, and reasonable cost [5]. The most common metal hydride (MH) alloy used as the active material in negative electrode of Ni/MH batteries is a misch-metal (combination of La, Ce, Pr, and Nd) based AB_5 alloy with a $CaCu_5$ crystal structure. Recently, a misch-metal and Mg in the superlattice alloy (A_2B_7 type) with its corresponding improvement in the gravimetric energy density has become more popular in consumer and stationary applications [5–9]. Meanwhile, the Laves phase AB_2 MH alloy has also been proposed to improve the energy density of Ni/MH batteries. Although the AB_2 MH alloy has the potential for relatively high capacity potential (440 mAh·g^{-1} [10,11] when compared to 330 mAh·g^{-1} from the AB_5 alloy), it suffers from a relatively slow electrochemical reaction rate and a less-desirable cycle life (see comparison in Table 1) [12]. A further comparison of the main battery performances between AB_2 and AB_5 MH alloys can be found in previous research (Table 1 in [13]). Numerous papers, including a few reviews [14–20], have been have examined how to improve the electrochemical properties of AB_2 MH alloys, however, an updated comprehensive review of this subject would be of use. The current work details the alloy preparation methods, performance requirements, failure mechanisms, constituent phases, selection of elements and stoichiometry, as well as the historical challenges and directions for future research.

Table 1. The performance comparison of hydrogen-absorbing alloys with different chemistries [12]. The numbers in parentheses are the nominal B/A ratios. The symbols are ++ (superb), + (good), 0 (acceptable), and − (poor).

Properties	AB_2 (2)	AB_3 (3)	A_2B_7 (3.5)	A_5B_{19} (3.8)	AB_5 (5)
AB_2 number of units	1	1	1	1	0
AB_5 number of units	0	1	2	3	1
Electrochemical capacity/weight	++	0	+	0	−
Electrochemical capacity/volume	0	0	+	0	−
Pulverization of alloy, oxidation (corrosion)	+	+	+	−	−
Reversibility of hydrogen absorption/release	−	0	+	0~+	++
Battery Life	−	0	++	0	+

C14, C15, and C36 Laves phases are composed of the same A_2B_4 unit-block, but with different stacking sequences; C14 has an *a-b-a-b* stacking sequence, which leads to a hexagonal structure (Figure 1a); C15 has an *a-b-c-a-b-c* stacking and a cubic structure (Figure 1b); and C36 exhibits *a-b-a-c* stacking and a rhombohedral (dihexagonal) structure (Figure 1c) [21–23]. The C14 and C15 phases are more commonly observed in MH alloys, whereas the C36 phase may exist between the C14 and C15 phases [24] but is difficult to identify using X-ray diffraction (XRD) [25,26]. Alloys with both C14 and C15 crystal structures serve as hydrogen-storage (H-storage) alloys and electrode materials. C15 alloys have better high-rate and low-temperature performance, however the C14 alloys were more widely used earlier when alloy composition design was driven by capacity and cycle stability. An AB_2 MH alloy with a C14-predominated structure was used in the first commercialized electric vehicle, EV1 by General Motor (Detroit, MI, USA). Similar alloys with the same structure were also used in the first-generation Ni/MH batteries made by Gold Peak Industries (Hong Kong, China) and Hitachi-Maxwell (Osaka, Japan). Recently, the electrochemical performances of a state-of-art C14-predominate MH alloy were compared to those from a recently proposed C15-based MH alloy together with a review of the development works done on the C15-based MH alloys [27]. In this paper, we will review only research focused on the C14-predominated MH alloys.

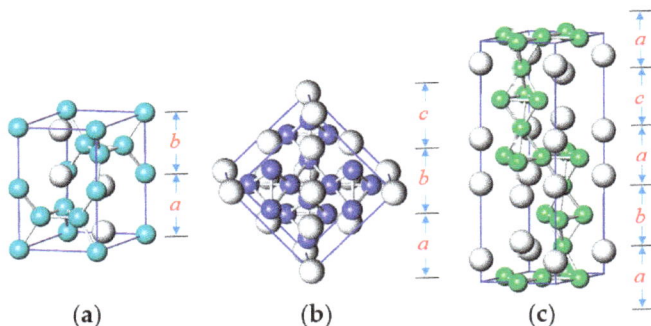

Figure 1. Schematic of (a) C14 (hexagonal), (b) C15 (cubic), (c) and C36 (rhombohedral) crystal structures [28].

2. Alloys Preparation

Conventionally, when used as the active material in negative electrodes of Ni/MH batteries, the MH alloys are initially prepared by melting and casting. Then, an annealing process is applied to the ingot to improve compositional uniformity and to reduce internal stress. To reach a final particle size of less than 200 mesh (75 μm), most ingots need to be crushed into powder through mechanical grinding or hydrogenation. The powder may require additional surface treatments to enhance its electrochemical properties. Negative electrode fabrication methods have been previously reviewed [13].

2.1. Melting and Casting

A few preparation methods for MH alloys are described in an introductory article [29] that covers conventional vacuum induction melting (VIM) [30], arc melting (AM) [31], centrifugal casting (CC), melt-spinning (MS) [32–34], gas atomization (GA) [35,36], and mechanical alloying (MA) [37–39]. A vacuum plasma spray (PS) has also been employed for research purposes [40]. While AM only produces small quantity of alloy (5 to 200 g) for laboratory use, VIM can produce ingots ranging from 1 kg to 1 ton (Figure 2). In terms of electrochemical property, no distinct difference is observed between AB_2 MH alloys prepared by VIM and those prepared by AM [41], except for a slightly higher capacity obtained from VIM due to a larger sample/chamber ratio [42]. Samples from different locations of a VIM-process ingot may have different microstructures, but the electrochemical capacities are very similar. For example, the microstructures of different pieces of ingot prepared by a 60-kg VIM furnace are shown in Figure 3. According to the analytic results from x-ray diffraction (*X'Pert Pro*, Philips, Amsterdam, The Netherlands) and scanning electron microscope (SEM) (*JEOL-JSM6320F*, JEOL, Tokyo, Japan) with the x-ray energy dispersive spectroscopy capability, this alloy is C14-predominant with C15, TiNi, and families of (Zr, Ni) secondary phases. We obtained this composition by continuing optimization and testing more than 400 alloys. The properties of C14-predominated alloys prepared by VIM and other techniques were also previously compared [43] and are summarized in Table 2. Because of the small sample size, use of expensive argon gas, and high utility costs from water chiller and arc power supply, AM is the most expensive method for making MH alloys. Photographs of equipment and different pouring processes are shown in Figures 4 and 5, respectively. SEM surface morphology and cross-section micrographs of products are compared in Figures 6 and 7, respectively. A review of Japanese Patent Applications shows 9 different ingot preparation methods for MH alloys [44]. For MH alloys that are sensitive to oxygen contamination from the reduction of ceramic refractory used as furnace crucible, a skull VIM melting can be used (Figure 8). MA in a shaker mill or an attritor is a popular method for making meta-stable/amorphous alloys not allowed in the phase diagram [45] (Figure 9).

(a)   (b)   (c)

Figure 2. Images of (**a**) a 200 g, (**b**) 60 kg, and (**c**) 1 ton vacuum induction melting (VIM) furnaces. While the first two are currently installed in BASF-Ovonic, the third one is operated by Rare-earth Ovonic Metal Hydride Alloy Company in Baotou, Inner Mongolia, China.

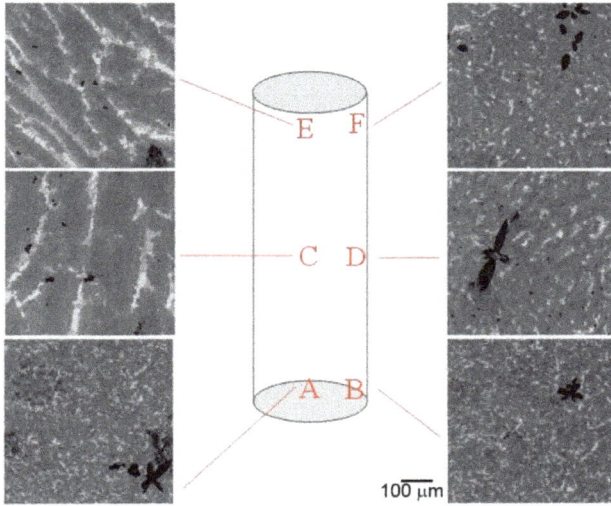

Figure 3. Scanning Electron Microscope (SEM) cross-section micrographs showing the microstructures from various locations of a 100-kg AB_2 ingot with a composition of $Ti_9Zr_{26.2}V_5Cr_{3.5}Mn_{15.6}$ $Co_{1.5}Ni_{38}Sn_{0.8}Al_{0.4}$ prepared by VIM. The electrochemical discharge capacities of ingots from locations A, B, C, D, E, and F are 390, 400, 391, 402, 389, and 397 $mAh \cdot g^{-1}$ with a discharge current density of 8 $mA \cdot g^{-1}$. The dark, white, and gray regions are ZrO_2 inclusions, secondary phases (TiNi, ZrNi, Zr_7Ni_{10}, and Zr_9Ni_{11}), and main C14 phases, respectively.

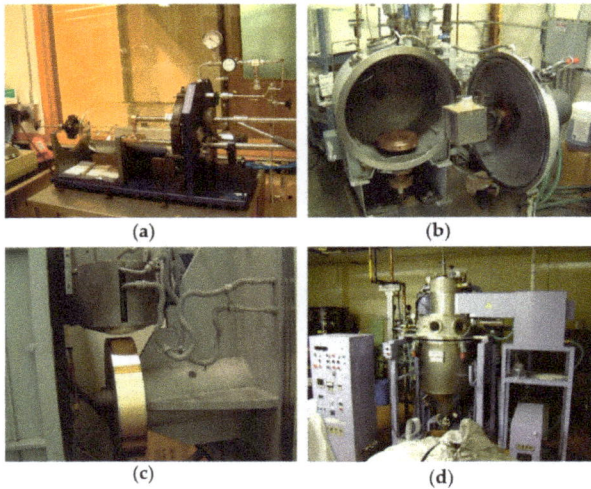

Figure 4. Images of (**a**) an arc melter, (**b**) a centrifugal casting VIM, (**c**) a melt-spinning VIM, and (**d**) a gas atomizer. The first three are operated in Ovonic-BASF and the fourth one is in Eutectix, Troy, MI, USA.

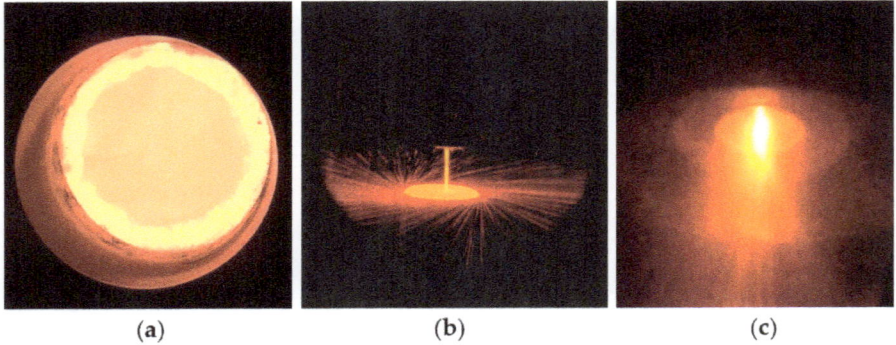

Figure 5. Images during operation of (**a**) a VIM, (**b**) a centrifugal casting (CC), (**c**) a gas atomization (GA) (Courtesy of Daido Steel, Nagoya, Japan).

Figure 6. SEM micrographs showing top morphologies of alloys prepared by (**a**) VIM with a mechanical crush-and-grind, (**b**) melt-spinning (MS), and (**c**) GA processes.

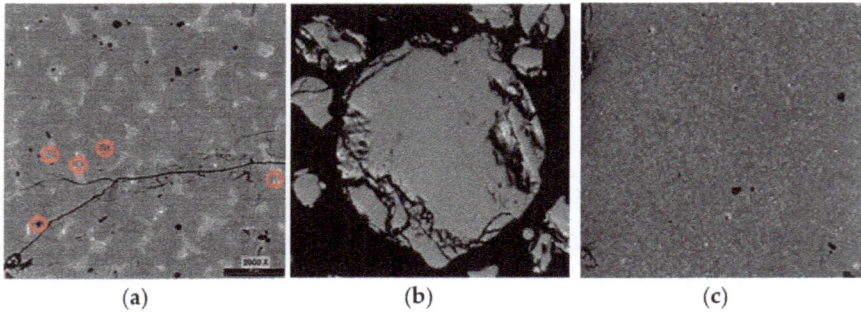

Figure 7. SEM cross-section micrographs from a C14-predominated alloy ($Ti_{12}Zr_{21.5}V_{10}Cr_{8.5}Mn_{13.6}Co_{1.5}Ni_{32.2}Sn_{0.3}Al_{0.4}$) prepared by (**a**) a conventional VIM, (**b**) GA, and (**c**) MS processes. Areas 1, 2, 3, 4, and 5 in (**a**) are C14, C15, TiNi, body-centered-cubic (bcc), and ZrO_2 phases, respectively, as determined by their compositions and crystal structures.

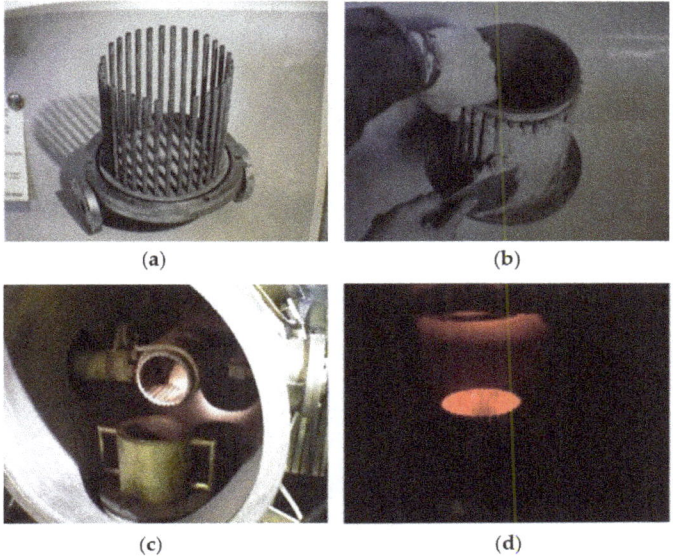

Figure 8. Images of skull melting: (**a**) water-cooling Cu rods, (**b**) application of refractory outside the crucible, (**c**) setup of a crucible and a mold in a vacuum chamber, and (**d**) during operation (Courtesy of Albany Research Center, Albany, OR, USA).

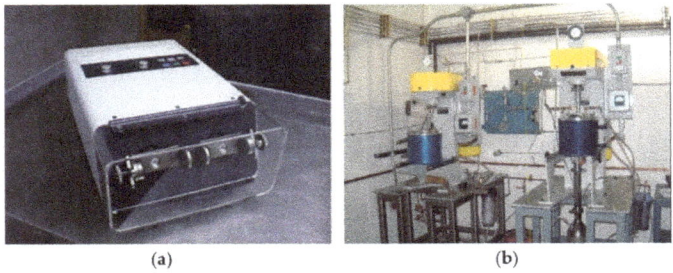

Figure 9. Images of (**a**) a shaker mill (Spex, Metuchen, NJ, USA) and (**b**) an attritor used in MA processes (Union Process, Akron, OH, USA).

Table 2. Comparison between alloy preparation methods. VIM, CC, MS, GA, MA, AM, and PS are short for vacuum induction melting, centrifugal casting, melt spinning, gas atomization, mechanical alloying, arc melting, and plasma spray, respectively.

Properties	VIM	CC	MS	GA	MA	AM	PS
Purpose	Production	Research/Production	Research/Production	Research/Production	Research	Research	Research
Batch Size	1–1000 kg	1–1000 kg	1–200 kg	1–1000 kg	1–1000 g	5–200 g	5–100 g
Equipment Cost	Medium	High	High	High	Low	Low	High
Production Cost	$3 kg^{-1}	$4 kg^{-1}	$4 kg^{-1}	$5 kg^{-1}	$1000 kg^{-1}	$1000–$5000 kg^{-1}	$2000 kg^{-1}
Cooling Rate	$100\ ^{\circ}C\ s^{-1}$	$1 \times 10^{3}\ ^{\circ}C\ s^{-1}$	$1 \times 10^{6}\ ^{\circ}C\ s^{-1}$	$1 \times 10^{4}\ ^{\circ}C\ s^{-1}$	$>1 \times 10^{6}\ ^{\circ}C\ s^{-1}$	$500\ ^{\circ}C\ s^{-1}$	$1 \times 10^{4}\ ^{\circ}C\ s^{-1}$
Micro-structure	Large crystallites	Medium crystallites	Nano-crystallites	Micro-crystallites	Amorphous	Large crystallites	Micro-crystallites
Alloy Discharge Capacity	Normal	Normal	Low	Low	High	Normal	Low
Alloy High-Rate Discharge-ability	Normal	Normal	High	Low	High	Normal	High
Alloy Cycle Stability (Anti-pulverization)	Normal	Better	Excellent	Excellent	Excellent	Normal	Excellent

2.2. Particle Size Reduction

Particle size influences capacity and high-rate dischargeability (HRD) behavior, but not activation properties [46]. As such, a proper particle size is required for battery operation. Unlike the conventional jar crusher used for AB_5 MH alloys, the high hardness of C14 MH alloys requires a hydrogenation process to create fissures and cracks for further grinding [47]. A stationary hydrogen reactor for laboratory use (2-kg load), a prototype rotational hydrogen reactor (50 kg), and a mass production stationary hydrogen reactor (5 tons) are shown in Figure 10. Operation details for the stationary and rotational hydrogen reactors can be found in the associated U.S. Patents [48,49]. Further reduction to a desirable size can be achieved by crushing [50], dry ball milling [51], wet ball milling [52], centrifugal grinding, or jet milling [53]. Three different powder fabrication methods—hydrogenation, dry ball milling, and wet ball milling—on a F-treated AB_2 MH alloy, were compared and the third showed the best electrochemical performance [54].

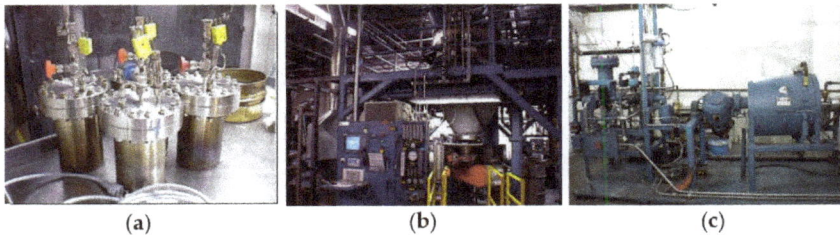

(a)   (b)   (c)

Figure 10. Images of (**a**) a 2-kg stationary, (**b**) a 3-ton stationary, and (**c**) a 20-kg rotational hydrogen reactors. While the first one was taken in BASF-Ovonic, the rest of two were taken in Eutectix.

2.3. Annealing

The main purpose of annealing is to improve uniformity in both composition and micro-structure. Annealing is a necessary step to prepare commercial grade AB_5 MH alloys with a long cycle life [55] (Figure 11a,b). In the case of AB_2 MH alloys, non-Laves secondary phases improve the electrochemical properties of an alloy [56] through synergetic effects [57] and are diminished by annealing treatments [58–61]. Therefore, annealing is not required for AB_2 MH alloys, except for those prepared by GA and with a surface oxide layer, which can be reduced to the metallic state by annealing in hydrogen [62] (Figure 11c). Annealing in Ar was also improved the capacity of GA- [63] and MS- [64] produced AB_2 MH alloys because of the ability to increase the surface crystallinity. The positive contribution of thermal annealing to electrochemical capacity was previously reported by Klein et al. [65].

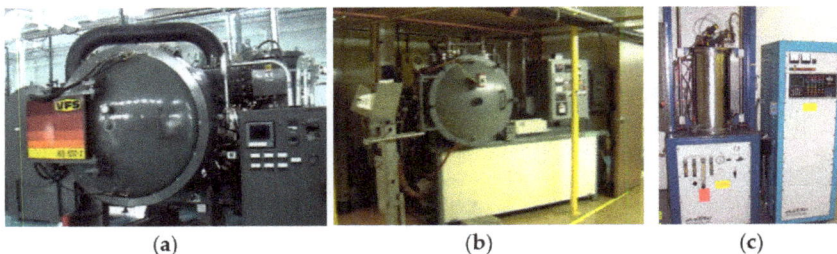

(a)   (b)   (c)

Figure 11. Images of (**a**) a 4-ton vacuum annealer, (**b**) a 200-kg vacuum annealer, and (**c**) a 10-kg flowing-hydrogen annealer. These pictures were taken in Rare-earth Ovonic Metal Hydride Company, Ovonic-BASF, and Eutectix.

2.4. Surface Treatment

Various surface treatments are available for AB_2 MH alloys. A fluorination process can improve the surface anti-corrosion capability [66–71] and HRD [72], while a hot alkaline bath helps remove the native oxide that forms during the processes and creates a less dense oxide with catalytic Ni, which further increases the surface's catalytic ability [51,66,73,74] and HRD [72]. Other methods include hot-charging performed at 80 °C [75], surface treatment in NH_4F and $NiCl_2$ solution [76], NiO coating by the sol-gel method [77], electroless Cu-coating [78], ball-milling with [79,80] or without Ni powder [81,82], ball-milling with La-containing AB_5 or A_2B_7 MH alloys [83], and surface reduction by KBH_4 and $NaBH_4$ [84].

3. Performance Criteria

The general requirements for an MH alloy that is suitable for use as the negative electrode in Ni/MH batteries include high electronic conductivity, a durable anti-corrosion surface, high H-storage capability, fast bulk hydrogen diffusibility, adequate metal-hydrogen (M-H) bond strength, acceptable cost, small lattice expansion during hydrogenation, and environmental friendliness [85,86]. As the applications of Ni/MH batteries become more versatile, the demands of MH alloys also vary. We classify the prerequisites of specific alloy properties in accordance with application in Table 3. For the AB_2 MH alloys, the surface catalytic ability is not as good as those in AB_5 and A_2B_7 MH alloys, because of its lower B/A ratio, which contributes to a lower level of metallic Ni on the surface [87]. But the corrosion product of AB_2 alloy is dissoluble in electrolyte, which is different from the highly packed oxide of rare earth elements (RE) in AB_5 alloys, and makes it a better candidate for high-temperature applications. The small hysteresis found in the pressure-concentration-temperature (PCT) analysis between absorption and desorption isotherm in AB_2 reduces the pulverization rate and helps improve cycle stability [88,89].

Table 3. Performance requirements of MH alloy in various type of applications. The symbols are ++ (highly desirable), + (important), 0 (not critical), and × (unrelated). EV and HEV represent electric vehicle and hybrid electric vehicle, respectively.

Alloy Requirements	High-Energy (EV)	High-Power (HEV)	Stationary General Purpose	Stationary at High Temperature	Stationary at Low Temperature
H-storage Capacity	++	0	+	+	+
H-diffusibility	+	++	+	+	++
Surface Catalysis	+	++	+	++	++
Anti-corrosion	+	0	+	++	+
Equilibrium Pressure	0	++	0	×	++
Pulverization	++	+	0	+	+
Cost	++	+	++	+	+

4. Failure Mechanism

While the failure mechanism of the AB_5 MH alloy has been studied extensively and reviewed in detail [90], similar studies on AB_2 MH alloys are rare [91,92]. Unlike the surface of AB_5 MH alloys, which are covered by an inert layer of $La(OH)_3$ after numerous cycles, the AB_2 MH alloy forms soluble complex ions, including $HZrO_3^-$, $HMnO_2^-$, and $AlO_2^{-\cdot}$, which migrate through the separator into the positive electrode, causing micro-shorts and particle pulverization at the electrodes, which leads to positive electrode failure. The surface passivation of AB_5 MH alloys causes stable capacity degradation, while the failure at the positive electrode with AB_2 MH alloys creates a pressure increase, cell venting, and a sudden capacity drop near the end of the cycle life (Figure 12). Three AB_2 MH alloy degradation modes (oxidation, pulverization, and amorphization) are discussed in this section.

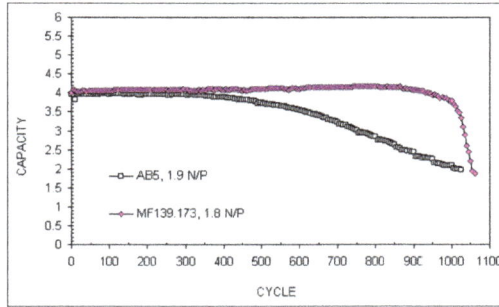

Figure 12. Comparison of cycle life in C-sized cell from AB_5 and AB_2 (MF139.173, $Ti_{12}Zr_{21.5}V_{10}Cr_{8.5}$ $Mn_{13.6}Co_{1.5}Ni_{32.2}Sn_{0.3}Al_{0.4}$) alloys as negative electrode active materials. N/P denotes the ratio of capacities of negative and positive electrodes and capacity in the unit of Ah was obtained with a C/2 charge rate and a C/2 discharge rate to 0.9 V cell voltage.

4.1. Oxidation

The activated (alkaline bath treated or cycled) AB_2 alloy surface consists of a top non-electrochemical reactive ZrO_2 patch, a surface oxide (~200 nm) with embedded metallic Ni/Co-clusters [10,93], and an amorphous buffer layer (~100 nm) underneath [94]. In one example, the Zr/Ti ratios are 8.2, 1.2, 0.8, and 1.9 for the top oxide, supporting oxide, buffer oxide, and bulk, respectively, which indicates a higher leaching rate for Zr compared to that of Ti [94]. The metallic inclusions on the surface of the AB_2 alloy were thought to be vital for the surface electrochemical reaction until it was determined that electrolyte-conducting channels coated with catalysts on the surface of the inner side of the alloy bulk can significantly improve low-temperature performance [95]. The leaching rates of constituent elements from a typical C14 MH alloy are calculated in Table 4 and follow the trend of Al > V > Zr > Ti > Mn. Oxidation and pulverization are the major capacity degradation mechanisms for AB_2 MH alloys [96,97].

Table 4. Calculation of leaching rate in 100 °C/30% KOH of a C14 MH alloy with a composition of $Ti_{12}Zr_{21.5}V_{9.5}Cr_{4.5}Mn_{13.6}Co_2Ni_{36.2}Sn_{0.3}Al_{0.4}$. Data are from [87].

Element	Ti	Zr	V	Al	Mn
Amount in alloy (at %)	12.0	21.5	9.5	0.4	13.6
Concentration in solution after 1 h etching (ppm)	0.6	15.6	11.4	7.8	0.4
Concentration in solution after 4 h etching (ppm)	1.0	48.9	33.8	28.1	0.5
Leaching rate after 1 h etching (ppm/at %)	0.05	0.73	1.2	19.5	0.03
Leaching rate after 4 h etching (ppm/at %)	0.08	2.27	3.6	70.2	0.04

4.2. Pulverization

Particle pulverization is a key failure mode for AB_2 MH alloys, because it increases the surface area of the electrode, which leads to more severe oxidation [98]. High internal stress exists in AB_2 MH alloy after hydrogenation, and cannot be removed by annealing because of the beneficial multi-phase structure. Thus, the initial pulverization due to the internal stress in AB_2 MH alloys is more severe than that in AB_5 MH alloys. But, because of the synergetic effect between the main storage phase and the catalytic secondary phase, the PCT hysteresis (the main reason for pulverization during hydride/dihydride cycling) in AB_2 is much smaller than that in AB_5 and, thus, a small degree of pulverization during cycling is expected from AB_2. Two SEM micrographs showing the pulverizations after activation (10 cycles) and near the end of cycle life (450 cycles) are presented in Figure 13.

Interestingly, a V-free AB_2 MH alloy also shows a higher pulverization rate than a V-containing alloy [92].

(a)   (b) $\overline{20\ \mu m}$

Figure 13. SEM cross-section micrographs of a C14 MH alloy at (**a**) 10th cycle, just after initial activation and (**b**) 450th cycle, at the end of cycle life.

4.3. Amorphization

Hydrogen-induced-amorphization (HIA) frequently occurs on AB_2 alloys [99,100], especially those with an atomic radius ratio between A- and B-site atoms (R_A/R_B) > 1.37 [101]. The amorphous alloy may have a lower electrochemical capacity [101,102] but better HRD [103,104] compared to its crystalline counterpart. An amorphous C14 MH alloy went through MA and its capacity decreased from 325 to 25 mAh·g^{-1} [105]. A similar degradation of capacity caused by amorphization from MS has been previously reported [64].

5. Non-Laves Secondary Phases

During cooling from the liquid, the C14 phase solidifies first, followed by the C15 phase, and then finally a TiNi phase. The TiNi phase further transforms into Zr_7Ni_{10}, Zr_9Ni_{11}, and ZrNi phases via solid-state reactions [106–109] (Figure 14). When other modifiers are present, additional secondary phases are also formed. These secondary phases may not store large amounts of hydrogen, but they may improve electrochemical performance through the synergetic effect [57].

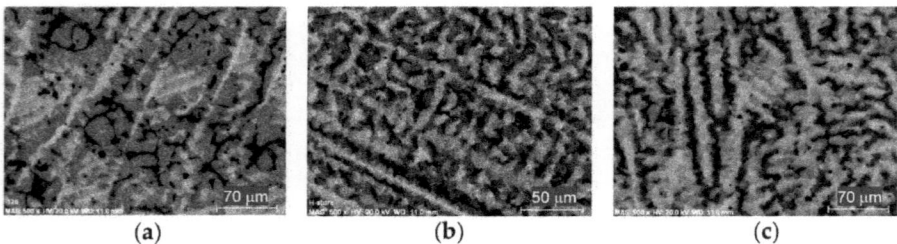

(a)   (b)   (c)

Figure 14. X-ray energy dispersive spectroscopy Cr-mapping from MH alloys with (**a**) 3.5 at % Cr and 28 wt % C14, (**b**) 5.5 at % Cr and 53 wt % C14, and (**c**) 8.5 at % Cr and 87 wt % C14. The three phases are C14, C15, and Zr_xNi_y in order of decreasing brightness.

5.1. TiNi

A TiNi phase is a commonly observed secondary phase in C14 MH alloys. TiNi has a large soluble Zr content [110]. Occasionally, the Zr-content in the TiNi phase can be larger than the

Ti-content [111]. A TiNi phase was found to increase the discharge capacity, and to improve the charge retention and cycle life while lowering HRD capability [56]. In addition, it is known to improve low-temperature kinetics [112,113]. The electrochemical performances of TiNi-based MH alloys were previously reported [114–117].

5.2. Zr_xNi_y

Three Zr-dominated secondary phases are commonly seen in C14 MH alloys, specifically ZrNi [30], Zr_9Ni_{11} [65,118], and Zr_7Ni_{10} [118]. ZrNi and Zr_7Ni_{10} phases are found to increase HRD, but sacrifice capacity and cycle stability, while Zr_9Ni_{11} does the opposite [56,119]. Zr_7Ni_{10} gradually shifted to Zr_9Ni_{11} as secondary phases in C14 MH alloys with increasing V-content [120]. The electrochemical properties of ZrNi [30,121–123], Zr_9Ni_{11} [121,124,125], Zr_7Ni_{10} [30,121,124–127], and other Zr_8Ni_{21} [30,121,124,125,128], Zr_2Ni_7 [30,121,129], and $ZrNi_5$ [130] based MH alloys are also available.

5.3. V-Based bcc Solid Solution

In the MH alloys with a high V-content, a V-based body-centered-cubic (bcc) is commonly observed [131–133]. This phase is unlikely to corrode and remains at the surface after activation [72]. As the V-content and the B/A ratio increases, the microstructure of the alloy will change from a predominated Laves phase into a Laves-phase-related bcc two-phase solid solution [134]. The electrochemical storage of the two-phase alloy is higher, but the HRD suffers in the Laves-phase alloy [135].

5.4. ZrO_2

ZrO_2 is formed during the melting stage and acts as an oxygen scavenger [136], which is ordinarily seen in C14 MH alloys with higher Zr/Ti ratios [111]. It forms a protective barrier against oxidation [72], but does not form hydride under normal operating conditions in batteries [137].

5.5. Other Secondary Phases

Other secondary phases can be detected using non-transition metal modifiers, such as RE. The solubility of RE in the C14 phase is extremely low and the addition of Y, La, Ce, Nd results in the formation of YNi [138], LaNi [139], CeNi [140], and NdNi [141] phases, respectively. These RE-Ni phases are detrimental to the electrochemical performance of MH alloys [112]. Like RE-Ni, the ScNi secondary phase increased discharge capacity, but weakened cycle stability [142]. A combination of hypo-stoichiometry and the addition of RE can boost low-temperature alloy performance [113]. High Mg-content secondary phases have been detected and facilitate the activation process [143,144]. A ZrC phase was formed with a C-addition and improves charge retention but deteriorates cycle life, HRD, and low-temperature performance. Zr_2Ni_2Sn [56,136] and Zr (Ni, Mn) $Sn_{0.35}$ [145] phases can be found in alloys with a relatively high Sn-content and are detrimental to the alloys' electrochemical performance. The structure and properties of Zr_2Ni_2Sn have been reported [146].

6. Selections of Element

AB_2 alloys have a greater compatibility of constituent elements, compared to that of AB_5 MH alloys, because of their superb solubility of components and great number of available phases. The compatibility of substituting elements makes AB_2 a better candidate to meet various application demands. Laves phase alloys are composed of A-site and B-site atoms (Figure 1). The A-site atom is usually larger than the B-site atom, but tends to shrink as the electronegativity of the B-site atom increases, resulting in electron transfer. The ideal R_A/R_B is 1.225 [147].

6.1. A-Site Element

For electrochemical applications, Ni, with an atomic radius of 1.377 Å in the Laves phase, is the major B-site element [148], which limits the choice of A-site element to those with an atomic radius of approximately 1.68 Å. Some physical properties of the A-site atoms are summarized in Table 5. Ti and Zr are the most frequently used A-site elements in Laves phase MH alloys because of their availability. Zr has the strongest M-H bond (lowest heat of hydride formation [149]) and the adjustment of Zr/Ti content is the most effective method to change M-H bond strength.

Table 5. Properties of A-site elements in AB_2 alloys. IMC denotes intermetallic compound.

Properties	Ti	Zr	Hf	Nb	Pd
Atomic Number	22	40	72	41	46
Atomic Radius in AB_2	1.614	1.771	1.743	1.625	1.521
Electronegativity	1.54	1.33	1.30	1.60	2.20
Earth Crust Abundance (%)	0.66	0.013	3.3×10^{-4}	1.7×10^{-3}	6×10^{-7}
Melting Temperature (°C)	1668	1855	2150	2468	1555
ΔH_h (kJ·mol H_2^{-1})	−136	−164	−161	−83	−41
Number of IMCs with Ni	3	8	8	3	0

6.1.1. Titanium

Ti results in weaker M-H bond strength, higher PCT equilibrium pressure [65], and improved HRD performance [41,150,151]. Ti forms an inert layer of TiO_2 and impedes activation [31,152], which improves cycle stability [153] and causes capacity degradation [154]. However, another group reported that Ti is beneficial for improving activation [155]. Ti-content impacts the discharge capacity in two competing ways: weaker M-H bonds lead to a lower capacity, but higher plateau pressure also improves the reversibility and increases the discharge capacity. Previous research has documented that increasing Ti-content has led to increasing [156–159], decreasing [155,158,160], and unchanged [150] electrochemical discharge capacities. A high Ti-content also contributes to a higher abundance of bcc phase [161] and a higher C14 abundance [162] because of its higher chemical potential [163].

6.1.2. Zirconium

To maintain stoichiometry in comparison to AB_2, the sum of Ti- and Zr-contents is approximately 33 at %. Therefore, Zr has the opposite effects of Ti. A higher Zr-content will stabilize the hydride [163] and lower HRD [150]. The bulk oxide of Zr is usually formed during solidification and has a higher solubility in KOH solution than TiO_2 [87]. A dense layer of ZrO_2 formed on the surface of the Zr particles during powder processing (grinding, sifting, and packaging), and impedes the activation process [152,164,165].

6.1.3. Hafnium

Not many studies have examined the electrochemical properties of Hf- substituted AB_2 MH alloys [42,160]. An XRD study confirms that Hf and Ti share the A-site and V and Ni are found in the B-site [166]. Hf-substitution yields a slightly smaller unit cell [167], which is unexpected because of its relatively large atomic radius (Table 5). The reduced lattice constant c/a ratio with the addition of Hf [167] predicts a lower pulverization rate. Partially substituting Hf for Zr results in a lower electrochemical capacity, but a higher HRD [160].

6.1.4. Niobium

Nb occupying the A-site was confirmed by a Rietveld analysis with XRD data from a substituted $ZrCr_2$ alloy [168]. Partial replacement of Zr by Nb leads to a smaller C14 unit cell volume and, consequently, a higher PCT plateau pressure [169]. It increases the C14 phase abundance and reduces lattice constants [167]. Like Hf, Nb also contributes to a reduced lattice constant c/a ratio [167] leading to a smaller pulverization rate. The NiNb secondary phase is a poor catalyst/H-storage material and, remarkably, reduces the surface exchange current [169]. Like Hf, Nb-substitution deteriorates capacity [42], but enhances HRD [160].

6.1.5. Palladium

Pd is smaller than both Ti and Zr, and when it enters the A-site, it reduces the unit cell of the Laves phases, which results in a decrease in discharge capacity. However, it also enhances the surface catalytic ability and improves both HRD and low-temperature performance for the alloy [170]. Improvement in HRD of C14 alloys was also reported by Yang, Ovshinsky, and their coworkers [171,172].

6.1.6. Scandium

Use of Sc, another expensive additive, in AB_2 MH alloy is scarce. The gaseous phase H-storages of C15/C36 ScM_2 (M = Fe, Co, and Ni) were reported [173] with an electrochemical study of a C15 $ScNiCo_{0.2}Mn_{0.5}Cr_{0.2}$ alloy [174]. The partial replacement by Zr and Y in Sc-based C15 showed cycle stability improvement with the corresponding sacrifice in discharge capacity [142]. The addition of Sc increased the abundance of the ZrNi secondary phase, improved the activation and capacity, but resulted in a trade-off in HRD [175].

6.2. B-Site Element

Ni is the most widely used B-site element in alkaline battery applications. However, Ti, Zr, and their mixture do not form Laves phase intermetallic compound (IMC) with Ni [176]. Therefore, B-site substitutions, mainly from the first row of transition metals—because of their availability and light weight—are essential for the stability of Laves phase IMC. Both the electron affinity and the atomic radius decrease as the atomic number of the modifier atom increases, which reduces the average metal-hydrogen bond strength, increases plateau pressure, and decreases H-storage capacity. With different solubility and chemical properties related to corrosion, these modifier elements are crucial for engineering various electrochemical properties and will be reviewed in the following sections. Various properties of the first-row transition metals are summarized in Table 6. The atomic radius of the element decreases as the atomic number increases until Ni and then increases from Cu and Zn. The number of IMC with Ti also increases roughly with increasing atomic number.

Table 6. Properties of first-row transition metal elements as B-site atoms in AB_2 alloys.

Properties	V	Cr	Mn	Fe	Co	Ni	Cu	Zn
Atomic Number	23	24	25	26	27	28	29	30
Atomic Radius in AB_2	1.491	1.423	1.428	1.411	1.385	1.377	1.413	1.538
Electronegativity	1.63	1.66	1.55	1.83	1.88	1.91	1.90	1.65
Earth Crust Abundance (%)	0.019	0.014	0.11	6.3	0.003	0.009	0.007	0.008
Melting Temperature (°C)	1890	1857	1245	1535	1495	1453	1083	420
ΔH_h (kJ·mol H_2^{-1})	−35	−8	−8	10	15	−3	20	8
Number of IMCs with Ti	0	1	1	2	4	3	5	7

6.2.1. Vanadium

V is the only B-site element that brings high H-storage capacity [17]. It results in a more stable hydride, more disorder, but also decreases peak power and charge retention [177]. The high leaching rate of V is the main cause of poor charge retention in conventional V-containing MH alloys [87,178]. The high electronic conductivity of V also improves the activation property of alloys [179]. A balance of V-content is needed to optimize battery performance [180]. V-free MH C14 MH alloys were designed to address the charge retention issue, but did so at the expense of cycle stability [144,178]. The large atomic radius of V enlarges the alloy unit cell and reduces the volume expansion during hydrogenation, which in turn reduces PCT hysteresis and pulverization tendency [181], which results in a shorter cycle life [42]. Reports of V lowering capacity are also available [42].

6.2.2. Chromium

The addition of Cr in C14 MH alloys improves cycle stability [11,131,182–185] and charge retention [11,184,186] but decreases the HRD [11,186,187], capacity [131], and activation tendency [11]. Cr can suppress segregation of Ti, Zr, and V on the surface of MH alloys [188], retarding the oxidation of V by forming a V-based bcc solid solution secondary phase [131,189], and slowing down pulverization [190]. Retarding HRD by adding Cr was further attributed to poor bulk hydrogen diffusibility and surface exchange current [133]. Compared to Co, Cr promotes a dendritic grain structure that is beneficial for cycle stability [118]. Its capability of reducing PCT hysteresis also contributes to a longer cycle life [11].

6.2.3. Manganese

Partial replacement of Ni by Mn increases capacity [17,186], facilitating the formation process [186], but deteriorating the cycle life of the battery because of its poor oxidation resistance [186,191] and results in micro short-circuits [192]. A high Mn-content in an alloy reduces the cycle life and charge retention because of poor oxidation resistance in the KOH electrolyte [11]. A low Mn-content provides improves activation, capacity, and HRD [193]. Therefore, a careful balance between the content of Mn- and other B-elements is necessary to optimize battery performance [11,42]. In a separate report, Mn was found to increase the discharge capacity and exchange current density while hindering HRD and cycle stability [194]. Mn facilitates homogenizing the chemical composition of different constituent phases in a multi-phase alloy system [11].

6.2.4. Iron

In V-containing C14 alloys, the addition of Fe facilitates activation, increases the discharge capacity and surface reaction area, decreases HRD [131] and hydrogen diffusibility, and impairs low-temperature performance [195]. However, reports also suggest that Fe contributes to a low capacity [42] and higher cycle life [196,197]. In a V-free C14 alloy, the addition of Fe improves low-temperature performance, but hinders cycle life and charge retention [144]. The solubility of Fe in Zr_7Ni_{10} and Zr_9Ni_{11} phases is lower than that in TiNi phases, which explains the promotion of TiNi phase by Fe-addition [195].

6.2.5. Cobalt

Co promotes an equiaxial grain structure that improves surface reaction kinetics, activation, and capacity, but deteriorates cycle stability [196]. An optimal Co-content in a C14 MH alloy of 1.5 at % provides the best performance in terms of formation, cycle life, and charge retention, but exhibits worse specific power and low-temperature performance compared to other compositions [111]. Compared to other modifiers, such as Fe, Cu, Mo, and Al, Co-substituted C14 MH alloys show a relatively higher discharge capacity [120,198,199], good cycle performance [120,193,199], and poor HRD [152]. Reports on Co-substituted C14 MH alloys with increased plateau pressure [200], a reduced capacity [200], and a lower cycle life [42] are also available.

6.2.6. Nickel

Ni is the most efficient catalytic B-site element because of the formation of metallic clusters embedded in the surface oxide after activation [87,201,202]. The introduction of Ni greatly improves electrochemical behavior [203], HRD [11], low-temperature performance [11,202], surface exchange current [204], and cycle stability [186], but impedes charge retention [11].

6.2.7. Copper

In a typical C14 MH alloy, the addition of Cu, with is high degree of pulverization, facilitates activation processes and improves discharge capacity and low-temperature performance, but decreases HRD because of lower hydrogen diffusibility [205]. Cu has also been reported to impair cycle stability [206] and lower capacity [42].

6.2.8. Zinc

Zn has an extremely low melting point compared to its neighbors' in the periodic table (Table 6) and is unable to remain in the C14 alloy at a melting point above 1350 °C. Therefore, ZnNi [207] and ZnCu [208] IMCs were used as raw materials to reduce Zn loss from evaporation. 1 at % Zn results in an increase in discharge capacities, a slight decrease in HRD, an easier activation process, and a satisfactory low-temperature performance [207].

6.2.9. Second Row Transition Metals (Mo)

Mo is the only element in the second row of transition metals that is used as a modifier in the C14 MH alloys, as it is accessible and not very toxic. According to our recent study, partial substitution of Co by Mo in a C14 MH alloy increases both the bulk diffusibility of hydrogen and surface catalytic ability, which leads to improved HRD and low-temperature performance [209]. Mo also improves charge retention and cycle stability, but yields a slightly lower capacity [209]. The cycle stability improvement of C14 MH alloy using Mo was also reported [42]. However, reports on the negative impact of Mo on the capacity, cycle life, and activation of C14 MH alloy are also available, presumably due to lack of Cr or Co in the composition [197,199]. Partially replacing Cr with Mo increases the gaseous phase H-storage capacity [210,211], ease of activation, and HRD [212].

6.2.10. Third Row Transition Metals (W, Pt)

W and Pt are the only two elements in the third row of the transition metals that are used because of concerns with availability and weight. W improves corrosion resistance and self-discharge of C14 MH alloys [213]. The addition of Pt in a C14 ZrCrNi alloy with an annealing treatment improved HRD and capacity because of the high catalytic activity of Pt for hydrogen electrosorption processes [214].

6.2.11. Group 13 Elements (B, Al)

B exhibits excellent solubility in V-free C14 MH alloys and contributes to a lower capacity, a high HRD, and improved superior low-temperature performance [144]. In V-containing alloys, B increased cycle stability [215] and bulk diffusion [216], however it also results in decreased capacity and HRD [216]. Reports on the positive contributions of B to HRD and low-temperature performance in V-containing C14 alloys are also available [217]. Ball milling C14 alloys with B show improved cycle stability at the cost of a lowered capacity [218].

Al is the most frequently used non-transition metal in Laves phase MH alloys. Al use has been shown to strongly correlate with a high capacity and a good HRD performance [219]. Al additives, together with Co, positively contribute to activation, charge retention, HRD, and low-temperature performances [136,220]. In the gaseous phase, Al was also shown to reduce PCT hysteresis, which in turn retards pulverization [200]. This also contributes to a lower flammability, which enhances safety in powder handling [221]. Even with the highest leaching out rate observed in Table 4, Al has been reported to improve the corrosion resistance of alloys, leading to better charge retention performance [164,213,222].

6.2.12. Group 14 Elements (C, Si, Ge, Sn)

In V-free C14 alloys, the addition of C promotes the formation of a ZrC secondary phase and improves charge retention, but also deteriorates low-temperature performance, cycle life, and HRD [144]. In a Ti-only C14 alloy, adding C promoted the formation of a TiC secondary phase, which facilitated the activation and decreased the gaseous phase H-storage capacity [223]. In one Zr-based C14 MH alloy, C improved cycle stability [17].

The addition of Si to C14 MH alloys improved the self-discharge [17], high-temperature discharge performances [17,224,225], low-temperature performance [226], and cycle stability [17,226] with trade-offs for lower capacity caused by an inert layer of SiO_2 on the surface of the alloy [224,225]. Si-incorporated C15 MH alloys' improved low-temperature performance is attributed to a layer of Ni_2O_3 that formed on the metal surface and can be detected by transmission electron microscopy (TEM, CM200/FEG, Philips, Amsterdam, The Netherlands) [227]. Si also contributes to a lower flammability that enhances safety in powder handling [221]. Both Si and Ge additives to $ZrCr_2$ MH alloy reduce gaseous phase H-storage capacity [228].

Zr is widely used in nuclear power plants because of its stability from a low neutron scattering cross-section. The costly separation process from Hf (neutron absorber) makes pure Zr metal expensive [229]. Future cost reduction of Zr is expected as the demand of a Hf-containing Zr for battery applications (Hf impurities do not matter since Hf is also a hydride former) increases. Currently, a large amount of Zircaloy (a Sn-Zr alloy [230]) scrap is available from demolished nuclear power plants and the addition of Sn is essential for cost reduction in raw materials. According to our own study, a small amount of Sn (0.1 at %) can improve activation, HRD, charge retention, and low-temperature performances at the expense of capacity and cycle life loss [219,231]. Sn also promotes the formation of detrimental secondary phases. Thus, the content of Sn should be limited [145].

6.2.13. Group 16 Elements (O, S, Se)

The only report on the addition of O, S, and Se to C14 MH alloys compares gaseous phase characteristics and shows the same order of effectiveness (S > Se > O) in raising plateau pressure and PCT hysteresis, increasing storage capacity, and reducing the PCT slope factor [232].

6.3. *Secondary Phase Promoter*

The addition of elements with large atomic radii, but extremely low solubility in Ti-Zr based AB_2 MH alloys, in the Laves phase will form secondary phases either in an element form or a Ni-containing alloy form. These secondary phases may play an important role in the electrochemical performance of the host alloy.

6.3.1. Group 1 Elements (Li, K)

Li has an extremely low melting point and very high vapor pressure at liquid temperatures for C14 MH alloys, therefore adding Li at the last stage and in the mold results in failure. However, Li-incorporated AB_5 MH alloys with their lower melting temperature, were successfully prepared by AM [233], VIM [234], and a diffusion method [235]. Li did promote unidentified secondary phases, based on XRD analysis [235]. KBH4 was added in a Ti_2Ni based MH alloy as a source of K and showed improved cycle stability [215].

6.3.2. Group 2 Elements (Mg)

Solubility of Mg in the Laves phase is extremely low (0.1–0.3 at % [143]). In V-containing C14 alloys, it forms a cubic secondary phase, which facilitated the formation process but did not improve other electrochemical properties [143]. In V-free C14 alloys, the addition of Mg formed a Mg_2Ni secondary phase and improved low-temperature performance and charge retention, but hampered cycle life and HRD [144]. Ball milling with Mg_2Ni can also introduce Mg into AB_2 MH alloys [236]. In recent years, Mg and La were introduced to AB_2 MH alloys by either remelting [237] or MA [238] with an AB_2/superlattice alloy mixture where superlattice alloys contain both La and Mg.

6.3.3. Rare Earth (RE) Metals

A great number of studies have focused on how to solve the slow activation of RE-incorporated C14MH alloys [67,165,184,239–243]. Low solubility of RE in Zr, Ti-based C14 IMC promotes a RE-Ni (AB) secondary phase and facilitates activation processes [244]. Properties of commonly used REs as modifiers in the AB$_2$ MH alloy are summarized in Table 7. La, Ce, and Sm are the least expensive. In general, the ionic radius, ease of oxidation, and MH-bond strength become smaller when the atomic number increases. Previous studies in a comparison series of RE-doped C14 MH alloys [112,113,138–141] showed that the RE-Ni phase is beneficial for activation, but detrimental to HRD and low-temperature performance, and that a careful stoichiometric design is needed to balance the abundance of the RE-Ni and TiNi phases [113]. The incorporation of La provides the greatest low-temperature performance, but a contradictory microstructure (an increase in the detrimental LaNi phase and corresponding decrease in the beneficial TiNi phase). Further TEM studies indicated that the Ni-Cr alloy channeled and catalyzed by a single-crystal sheet contributes to improvements in the surface catalytic ability of La-incorporated alloys [95]. Gd remains as a metallic inclusion when melted with C14 MH alloys and improves low-temperature performance, but deteriorates HRD, cycle stability, and charge retention [144]. Instead of using RE, which are nearly insoluble in Laves phase alloys, MA was used and achieved similar effects of capacity and activation [245]. In addition to adding RE in elemental form, RE-containing AB$_5$ [246,247] and A$_2$B$_7$ [248] can be used for similar results.

Table 7. Properties of some REs used as additives in AB$_2$ MH alloys.

Properties	Y	La	Ce	Pr	Nd	Sm	Gd	Yb
Atomic Number	39	57	58	59	60	61	55	70
Price (US$/kg) [249]	35	7	7	85	60	7	55	95
Ionic Radius in Laves (Å) [148]	1.990	3.335	2.017	2.013	2.013	1.990	1.992	1.990
Electronegativity	1.22	1.10	1.12	1.13	1.14	1.17	1.20	1.24
Melting Temperature (°C)	1522	918	798	931	1021	1072	1313	819
Oxidation potential (V)	−2.372	−2.379	−2.335	−2.353	−2.323	−2.304	−2.279	−2.19
Heat of Hydride Formation (kJ·mol H$_2$$^{-1}$) [250]	−114	−97	−103	−104	−106	−100	−98	−91

6.4. Summary of Modifier Studies

Unlike AB$_5$, AB$_2$ MH alloys are compatible with a remarkable number of elements. The degree of impact for commonly used elements on MH alloy properties are summarized in Table 8. Proper selection of elements can be based on the alloy requirements after checking each of the seven rows. Meanwhile, the pros and cons of each element can be easily evaluated in the corresponding column.

Table 8. Influence of commonly used modifiers to C14 alloy performance. The symbols are ++ (highly beneficial), + (beneficial), 0 (no significant effect), and − (detrimental). Degree of pulverization is the combination of actual measurement [88] and estimation from PCT hysteresis and C14 phase lattice constant ratio a/c. Equilibrium pressure changes are indicated with up or down arrows.

Alloy Requirements	Ti	Zr	V	Cr	Mn	Fe	Co	Ni	Cu	Zn	Al	Si	La
H-storage Capacity	−	++	+	0	0/+	0	+	0	0	0	+	0	0
H-diffusibility	+	−	−	0	0	−	0	−	+	+	+	+	+
Surface Catalysis	−	+	+	−	−	−	0	++	+	+	+	+	++
Anti-corrosion	+	−	−	++	−	0	0	++	0	0	0	0	0
Equilibrium Pressure	↑↑	↓↓↓	↓↓	↓	↓	0	↑	↑	0	0	↑	0	0
Anti-pulverization	+	−	0	−	0	0	0	0	−	−	+	−	0
Cost	0	0	−	+	++	++	−	0	++	++	++	++	+

7. Stoichiometry

An IMC with a fixed stoichiometry comes with a large negative heat formation [45]. The different energy levels of anti-site and vacancy defects in a specific IMC may change its stoichiometry range [251–253]. If the energy for such defects is relatively low, a wide range of compositions (off-stoichiometry) are expected, among which AB$_2$ Laves phase alloys make up the largest portion in

IMC [21]. Compositions of a few IMCs based on Group 4 elements (Ti, Zr, and Hf) as A-site atoms are listed in Table 9 and plotted in Figure 15 as a function of R_A/R_B. IMCs with a C14 structure have a wider composition range (more likely to suffer from point defects) and the phase solubility of IMCs with a C15 structure becomes larger as the radii ratio increases. An AB_2 alloy with a B/A ratio below 2.0 is a hypo-stoichiometric IMC. The reason that more hypo-stoichiometric IMCs tend to form as the radii ratio increases is that a relatively small B-atom more rapidly forms a double B in A-site (dumbbell model [254]) and further increases the B/A ratio. The characteristics of a few off-stoichiometric C14 MH alloys were previously compared [45,255,256] and important results are summarized in the next three sections.

Table 9. Solubility (range of A-site atom concentration in at %) of Laves phase IMCs (radius from [148] and solubility from [257]).

IMC	Structure	R_A/R_B	Solubility (at%)
ZrV_2	C15	1.19	33.3
HfV_2	C15	1.17	33.5–34.5
$TiCr_2$	C15	1.13	35–37
$ZrCr_2$	C15	1.24	31–36
$HfCr_2$	C15	1.22	33–35
$TiMn_2$	C14	1.13	30–40
$ZrMn_2$	C14	1.24	20.8–40
$HfMn_2$	C14	1.22	25.5–38
$TiMn_2$	C14	1.14	27.5–35.5
$ZrMn_2$	C15	1.26	27.1–34
$HfMn_2$	C15	1.24	32–33.5
$TiCo_2$	C15	1.17	33–33.5
$ZrCo_2$	C15	1.28	27–35
$HfCo_2$	C15	1.26	27–36

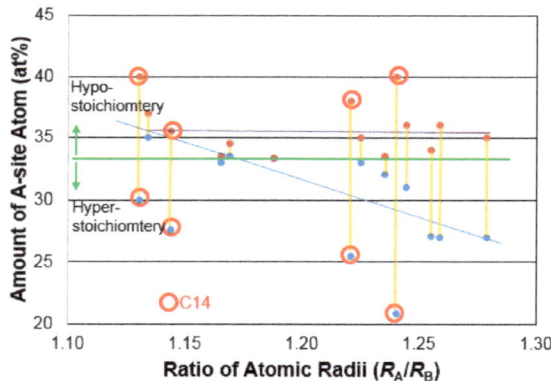

Figure 15. Plot of solubility (composition range) vs. ratio of atomic radii between A-site and B-site atoms from Ti, Zr, Hf-based Laves phase intermetallic compounds (IMCs). Red and blue dots are the upper and lower bounds of the AB_2 soluble range (yellow vertical line) in the composition. Data points in red circles are from C14 phases and other unmarked data points are from C15 phases.

7.1. Stoichiometric Alloy

The stoichiometric C14 MH alloys with a B/A ratio close to 2.0 have the highest electrochemical capacity and surface exchange current, and the smallest PCT hysteresis (least pulverization) [256] and is the most studied alloy stoichiometry in the field of battery applications.

7.2. Hypo-Stoichiometry

A C14 alloy with a B/A ratio below 2.0 is a hypo-stoichiometric IMC. The relatively low B/A ratio promotes the occurrence of AB phase (typically TiNi with a B2 structure) and improves activation behavior [256]. In a MS + annealing sample, the additional Ti (hypo-stoichiometry) increases both the capacity and cycle life of the electrode [64]. In a gaseous phase study of a TiCr based C14 alloy, a hypo-stoichiometry ratio (B/A of 1.8) gave the highest H-storage capacity [211].

7.3. Hyper-Stoichiometry

A hyper-stoichiometric C14 alloy has a B/A ratio higher than 2.0 and shows a high HRD [151,256], a high open-circuit voltage (from a higher PCT hydrogen equilibrium pressure) [256], and a flat PCT plateau [256]. The hyper-stoichiometric alloy also shows an improved cycle life, due to increased mechanical stability during cycling [185,258]. As x increases in a series of hyper-stoichiometric alloys AB_x (x = 2–6), the bcc phase abundance increases, and results in improvements in capacity, kinetics [259], and cycle stability [260].

8. Discussions

The Ovonic Battery Company (Troy, MI, USA) focused on research about AB_2 MH alloys for Ni/MH battery applications for more than three decades before being acquired by BASF (Ludwigshafen, Germany) in 2012. Throughout the years, we witnessed changing market demands and the corresponding reaction from the research communities. Before the successful debut of the commercialized Ni/MH battery in 1988, two main hurdles of the AB_2 MH alloy were slow activation and poor cycle stability. While the former was addressed by fluorination and hot alkaline baths, the latter was improved substantially by composition modifications, mainly the introduction of Cr and consequent bcc secondary phase. The challenge to the first generation AB_2-made Ni/MH battery was the self-discharge and was solved by the combination of V-free alloys and the use of the sulfonated separator. At that stage, AB_2 MH alloys were successfully used in products made by Hitachi-Maxell (Tokyo, Japan), Gold-peak (Hong Kong, China), and General Motors (Detroit, MI, USA). With the cheap misch-metal (main raw material for the rival AB_5 MH alloy) available from China, AB_2 MH alloys started to phase out from the consumer market at the turn of the century. Later, the upset of RE prices in 2010 created opportunities for AB_2 MH alloys. However, the bar to enter the market has not been raised in favor of AB_2, mostly in the high-rate dischargeability requirement for HEV and power tools. The superlattice alloy used by FDK (Tokyo, Japan), Panasonic (Osaka, Japan), and Kawasaki Heavy Industry (Tokyo, Japan) became a strong competitor with a capacity between AB_2 and AB_5, but better high-rate performance. Mg, an indispensable element in the superlattice alloy, makes it more vulnerable to KOH electrolyte attack, but was mitigated using surfactant additives in the negative electrode paste [44]. The structure of the superlattice allows the removal of Mn and Co from the composition and improves the charge-retention characteristics tremendously [261]. At the current stage, AB_2 has a hard time competing with AB_5 in term of price and superlattice alloys in terms of performance.

However, new opportunities started to appear when the conventional Ni/MH battery based on the RE-based AB_5 or A_2B_7 alloy lost the battle against the rival Li-ion technology. New applications require greater performance improvement in certain areas. For example, a solid-state battery with a thin solid separator [262] and ultra-high-power application with a very thin separator require a spherical MH alloy shape, which can be produced using GA techniques. Superlattice alloy with Mg inside cannot tolerate the Mg-vaporization because of the large surface area of the powder produced in the GA process, whereas GA with an associated annealing process already been successfully developed for AB_2 MH alloy [62]. The battery/fuel cell combination requires a MH alloy operated at intermediate temperature range (200–250 °C) [263]. The other opportunity is in stationary applications that require tolerance in the environment of above 50 °C [3,4,264]. While the RE-containing AB_5 and

A$_2$B$_7$ MH alloys form an impeccable surface oxide in the high-temperature environment, the transition metal-based AB$_2$ MH alloys leach out at a higher speed and are totally controllable by the composition re-adjustment. In the case of the new start-stop automobile application, a low-temperature cranking power is desperately needed [265]. There are a few elements contributing to the low-temperature performance of AB$_2$ MH alloys through different mechanisms, for example, Y for surface area increase, La, Pd and Mo for surface catalytic ability improvement, and Fe for the increase of beneficial TiNi secondary phase. The pouch cell design (like commercially available Li-ion battery for cell phones) of Ni/MH battery operating under a flooded electrolyte configuration requires new MH alloys with a very low plateau pressure and a highly electrochemically catalytic surface [266,267]. The last example is the need for new high-capacity MH alloys that fully utilize the wide voltage window of the newly developed ionic liquid electrolyte [268]. The new chemical environment (electrolyte interface) and lifting of voltage constraint (from competition of hydrogen evolution) present new challenges (i.e., opportunities) for the development of new MH alloys. Although the function of each modifying element in specific areas has been studied and reported in the current work, combinations and the accompanying synergetic effects among these elements have not been reported. Therefore, a review like the current work only provides a guideline for future research. Readers are encouraged to use the results of this study to pursue further performance improvements in C14 Laves phase based MH alloys.

9. Conclusions

The main advantages of C14 Laves phase based metal hydride alloys are higher capacities and flexibilities in composition, stoichiometry, and constituent phases, which allow fine tailoring in the electrochemical performance to meet the demands of different applications. Decades of research in this area have already solved historical hurdles in C14-based alloys, such as slow activation, low cycle life, high self-discharge, and poor low-temperature performance. To face the new challenges from future applications, we have compiled comprehensive comparisons of preparation methods, alloy property requirements, the pros and cons of each constituting/modifying element, and choice of stoichiometry in the current study. Based on these previous works, new directions for additional improvement in the electrochemical performance should focus on the combination of modifier elements with different functions.

Acknowledgments: The authors would like to thank the following individuals from BASF-Ovonic making contributions to the ongoing research in AB$_2$ MH alloys: Michael A. Fetcenko, Taihei Ouchi, Jean Nei, Diana F. Wong, Benjamin Reichman, Sherry Hu, John Koch, Suiling Chen, Cheryl Setterington, David Pawlik, Allen Chan, and Ryan J. Blankenship and former employees of Ovonics: Marvin Siskind, Su Cronogue, Jun Im, Joseph Tribu, Timothy Gizinski, David Smith, Xiaorong Tang, Crystal Yeh, Aereon Bisel, Anna Lewis, Lixin Wang, Tiejun Meng, Baoquan Huang, Feng Li, Melanie Reinhout, Nan Yu, Betty Cai, and Benjamin Chao for their technical assistance and valuable contributions. Kwo-Hsiung Young is obligated to his long-time partners in research: Ronald Tsang from Gold Peak, Koichi Morii from Daido Steel, Shigekazu Yasuoka from FDK Co., Leonid A. Bendersky from the National Institute of Standards and Technology, and Simon Ng from Wayne State University.

Author Contributions: Kwo-Hsiung Young prepared the main text and photographs. Shiuan Chang and Xinting Lin helped in the preparation of manuscript.

Conflicts of Interest: The authors declare no conflict of interest.

Abbreviations

The following abbreviations are used in this manuscript:

Ni/MH	Nickel/metal hydride
MH	Metal hydride
H-storage	Hydrogen-strorage
VIM	Vacuum induction melting
AM	Arc melting

CC   Centrifugal casting
MS   Melt spinning
GA   Gas atomization
MA   Mechanical alloying
PS   Plasma spray
SEM   Scanning electron microscope
bcc   Bady-centered cubic
HRD   High-rate dischargeability
M-H   metal-hydrogen
RE   Rare-earth elements
PCT   Pressure-concomposition-temperature
EV   Electric vehicle
HEV   Hybrid electric vehicle
N/P   Negative/Positive
HIA   Hydrogen-induced amorphization
IMC   Intermettalic compound
TEM   Transmission electron microscope

References

1. Cai, X.; Young, K.; Chang, S. Reviews on the Chinese Patent regarding nickel/metal hydride battery. *Batteries* **2017**, *3*, 24. [CrossRef]
2. Ogawa, K. Toyota to Start Local Production of NiMH Battery Cells in China by End of 2016. Available online: http://techon.nikkeibp.co.jp/atclen/news_en/15mk/030900433/ (accessed on 2 November 2016).
3. Zelinsky, M.; Koch, J.; Fetcenko, M. Heat Tolerant NiMH Batteries for Stationary Power. Available online: www.battcon.com/PapersFinal2010/ZelinskyPaper2010Final_12.pdf (accessed on 28 March 2016).
4. Zelinsky, M.; Koch, J. Batteries and Heat—A Recipe for Success? Available online: www.battcon.com/PapersFinal2013/16-Mike%20Zelinsky%20-%20Batteries%20and%20Heat.pdf (accessed on 28 March 2016).
5. Teraoka, H. Development of Ni-MH ESS with Lifetime and Performance Estimation Technology. In Proceedings of the 34th International Battery Seminar & Exhibit, Fort Lauderdale, FL, USA, 20–23 March 2017.
6. Teraoka, H. Ni-MH Stationary Energy Storage: Extreme Temperature & Long Life Developments. In Proceedings of the 33th International Battery Seminar & Exhibit, Fort Lauderdale, FL, USA, 21–24 March 2016.
7. Teraoka, H. Development of Highly Durable and Long Life Ni-MH Batteries for Energy Storage Systems. In Proceedings of the 32th International Battery Seminar & Exhibit, Fort Lauderdale, FL, USA, 9–12 March 2015.
8. Young, K.; Ouchi, T.; Nei, J.; Yasuoka, S. Fe-substitution for Ni in misch metal-based superlattice hydrogen absorbing alloys—Part 1. Structural, hydrogen storage, and electrochemical properties. *Batteries* **2016**, *2*, 34. [CrossRef]
9. Kai, T.; Ishida, J.; Yasuoka, S.; Takeno, K. The Effect of Nickel-Metal Hydride Battery's Characteristics with Structure of the Alloy. In Proceedings of the 54th Battery Symposium, Osaka, Japan, 7–9 October 2013; p. 210.
10. Fetcenko, M.A.; Ovshinsky, S.A.; Young, K.; Reichman, B.; Fierro, C.; Koch, J.; Martin, F.; Mays, W.; Ouchi, T.; Sommers, B.; et al. High catalytic activity disordered VTiZrNiCrCoMnAlSn hydrogen storage alloys for nickel–metal hydride batteries. *J. Alloys Compd.* **2002**, *330*, 752–759. [CrossRef]
11. Young, K.; Ouchi, T.; Koch, J.; Fetcenko, M.A. The role of Mn in C14 Laves phase multi-component alloys for NiMH battery application. *J. Alloys Compd.* **2009**, *477*, 749–758. [CrossRef]
12. Young, K.; Yasuoka, S. Past, Present, and Future of Metal Hydride Alloys in Nickel-Metal Hydride Batteries. In Proceedings of the MH2014 Conference, Salford, UK, 20–25 July 2014.
13. Chang, S.; Young, K.; Nei, J.; Fierro, C. Reviews on the U.S. Patents regarding nickel/metal hydride batteries. *Batteries* **2016**, *2*, 10. [CrossRef]
14. Young, K.; Nei, J. The current status of hydrogen storage alloy development for electrochemical applications. *Materials* **2013**, *6*, 4574–4608. [CrossRef] [PubMed]
15. Feng, F.; Geng, M.; Northwood, D.O. Electrochemical behaviour of intermetallic-based metal hydrides used in Ni/metal hydride (MH) batteries: A review. *Int. J. Hydrogen Energy* **2001**, *26*, 725–734. [CrossRef]

16. Kleperis, J.; Wójcik, G.; Czerwinski, A.; Skowronski, J.; Kopxzyk, M.; Beltowska-Brzezinska, M. Electrochemical behavior of metal hydrides. *J. Solid State Electrochem.* **2001**, *5*, 229–249. [CrossRef]

17. Han, S.; Zhao, M.; Wu, L.; Zheng, Y. Effect of additive elements on electrochemical properties of AB$_2$-type Laves phase alloys. *Chem. J. Chin. Univ.* **2003**, *24*, 2256–2259. (In Chinese)

18. Wang, J.; Yu, R.; Liu, Q. Effects of alloying side B on Ti-based AB$_2$ hydrogen storage alloys. *J. Harbin Inst. Technol. (New Ser.)* **2004**, *11*, 485–492.

19. Zhao, X.; Ma, L. Recent progress in hydrogen storage alloys for nickel/metal hydride secondary batteries. *Int. J. Hydrogen Energy* **2009**, *34*, 4788–4796. [CrossRef]

20. Ouyang, L.; Huang, J.; Wang, H.; Liu, J.; Zhu, M. Progress of hydrogen storage alloys for Ni-MH rechargeable power batteries in electric vehicles: A review. *Mater. Chem. Phys.* **2017**, *200*, 164–178. [CrossRef]

21. Stein, F.; Palm, M.; Sauthoff, G. Structure and stability of Laves phases. Part I. Critical assessment of factors controlling Laves phase stability. *Intermetallics* **2011**, *12*, 713–720. [CrossRef]

22. Kumar, K.S.; Hazzledine, P.M. Polytypic transformations in Laves phases. *Intermetallics* **2004**, *12*, 763–770. [CrossRef]

23. Chisholm, M.F.; Kumar, S.; Hazzledine, P. Dislocations in complex materials. *Sciences* **2005**, *307*, 701–703. [CrossRef] [PubMed]

24. Baumann, W.; Leineweber, A.; Mittemeijer, E.J. The kinetics of a polytypic Laves phase transformation in TiCr$_2$. *Intermetallics* **2011**, *19*, 526–535. [CrossRef]

25. Scudino, S.; Donnadieu, P.; Surreddi, K.B.; Nikolowski, K.; Stoica, M.; Eckert, J. Microstructure and mechanical properties of Laves phase-reinforced Fe–Zr–Cr alloys. *Intermetallics* **2009**, *17*, 532. [CrossRef]

26. Abraham, D.P.; Richardson, J.W., Jr.; McDeavitt, S.M. Laves intermetallics in stainless steel–zirconium alloys. *Mater. Sci. Eng. A* **1997**, *239–240*, 658–664. [CrossRef]

27. Young, K.; Nei, J.; Wan, C.; Yartys, V. Comparison of C14- and C15-predominated AB$_2$ metal hydride alloys for electrochemical applications. *Batteries* **2017**, *3*, 22. [CrossRef]

28. Laves Crystal Structure of Laves Phase. Available online: http://www.geocities.jp/ohba_lab_ob_page/structure5.html (accessed on 26 April 2017).

29. Young, K. Metal Hydride. In *Elsevier Reference Module in Chemistry, Molecular Sciences and Chemical Engineering*; Reedijk, J., Ed.; Elsevier: Waltham, MA, USA, 2013.

30. Joubert, J.-M.; Latroche, M.; Percheron-Guégan, A.; Bouet, J. Improvement of the electrochemical activity of Zr–Ni–Cr Laves phase hydride electrodes by secondary phase precipitation. *J. Alloys Compd.* **1996**, *240*, 219–228. [CrossRef]

31. Lee, H.-H.; Lee, K.-Y.; Lee, J.-Y. The hydrogenation characteristics of Ti-Zr-V-Mn-Ni C14 type Laves phase alloys for metal hydride electrodes. *J. Alloys Compd.* **1997**, *253*, 601–604. [CrossRef]

32. Shu, K.Y.; Lei, Y.Q.; Yang, X.G.; Zhang, S.K.; Chen, L.S.; Lu, G.L.; Wang, Q.D. A comparative study on the electrochemical performance of rapidly solidified and conventionally cast hydride electrode alloy Zr(NiMnM)$_{2.1}$. *J. Alloys Compd.* **1999**, *290*, 124–128. [CrossRef]

33. Chen, L.; Wu, F.; Tong, M.; Chen, D.M.; Long, R.B.; Shang, Z.Q.; Liu, H.; Sun, W.S.; Yang, Y.; Wang, L.B.; et al. Advanced nanocrystalline Zr-based AB$_2$ hydrogen storage electrode materials for NiMH EV batteries. *J. Alloys Compd.* **1999**, *293*, 508–520. [CrossRef]

34. Zhu, Y.; Liu, Y.; Hua, F.; Li, L. Effect of rapid solidification on the structural and electrochemical properties of the Ti–V-based hydrogen storage electrode alloy. *J. Alloys Compd.* **2008**, *463*, 528–532. [CrossRef]

35. Huang, S.S.; Chuang, H.J.; Chan, S.L.I. Effects of fluorination on the hydriding and electrochemical properties of a gas-atomized Zr-based hydrogen storage alloy. *J. Alloys Compd.* **2002**, *330*, 617–621. [CrossRef]

36. Kim, J.H.; Lee, H.; Hwang, K.T.; Han, J.S. Hydriding behavior in Zr-based AB$_2$ alloy by gas atomization process. *Int. J. Hydrogen Energy* **2009**, *34*, 9424–9430. [CrossRef]

37. Jung, C.B.; Lee, K.S. Electrode characteristics of metal hydride electrodes prepared by mechanical alloying. *J. Alloys Compd.* **1997**, *253*, 605–608. [CrossRef]

38. Qiu, S.J.; Chu, H.L.; Zhang, Y.; Sun, L.X.; Xu, F.; Cao, Z. The electrochemical performances of Ti–V-based hydrogen storage composite electrodes prepared by ball milling method. *Int. J. Hydrogen Energy* **2008**, *33*, 7471–7478. [CrossRef]

39. Kazemipour, M.; Salimijazi, H.; Saidi, A.; Saatchi, A. Hydrogen storage properties of Ti$_{0.72}$Zr$_{0.28}$Mn$_{1.6}$V$_{0.4}$ alloy prepared by mechanical alloying and copper boat induction melting. *Int. J. Hydrogen Energy* **2014**, *39*, 12784–12788. [CrossRef]

40. Kazemipour, M.; Salimijazi, H.; Saidi, A.; Saatchi, A.; Mostaghimi, J.; Pershin, L. The electrochemical hydrogen storage properties of $Ti_{0.72}Zr_{0.28}Mn_{1.6}V_{0.4}$ alloy synthesized by vacuum plasma spraying and vacuum copper boat induction melting: A comparative study. *Int. J. Hydrogen Energy* **2015**, *40*, 15569–15577. [CrossRef]

41. Young, K.; Fetcenko, M.A.; Li, F.; Ouchi, T. Structural, thermodynamic, and electrochemical properties of $Ti_xZr_{1-x}(VNiCrMnCoAl)_2$ C14 Laves phase alloys. *J. Alloys Compd.* **2008**, *464*, 238–247. [CrossRef]

42. Izumi, Y.; Moriwaki, Y.; Yamashita, K.; Tokuhiro, T. Nickel-Metal Hydride Storage Battery and Alloy for Configuring Negative Electrode of the Same. U.S. Patent 5,962,156, 5 October 1999.

43. Young, K.; Koch, J.; Ouchi, T.; Banik, A.; Fetcenko, M.A. Study of AB_2 alloy electrodes for Ni/MH battery prepared by centrifugal casting and gas atomization. *J. Alloys Compd.* **2010**, *496*, 669–677. [CrossRef]

44. Ouchi, T.; Young, K.; Moghe, D. Reviews on the Japanese Patent Applications regarding nickel/metal hydride batteries. *Batteries* **2016**, *2*, 21. [CrossRef]

45. Young, K. Stoichiometry in inter-metallic compounds for hydrogen storage applications. In *Stoichiometry and Materials Science—When Numbers Matter*; Innocenti, A., Kamarulzaman, N., Eds.; InTech: Rijeka, Crotia, 2012.

46. Humana, R.M.; Thomas, J.E.; Ruiz, F.; Real, S.G.; Castro, E.B.; Visintin, A. Electrochemical behavior of metal hydride electrode with different particle size. *Int. J. Hydrogen Energy* **2012**, *37*, 14966–14971. [CrossRef]

47. Matsuoka, M.; Tamura, K. Effects of mechanical modification on electrochemical performance of Zr–Ti-based Laves-phase alloy electrode. *J. Electrochem. Soc.* **2007**, *154*, A119–A122. [CrossRef]

48. Fetcenko, M.A.; Kaatz, T.; Sumner, S.P.; LaRocca, J. Hydride Reactor Apparatus for Hydrogen Comminution of Metal Hydride Hydrogen Storage Materials. U.S. Patent 4,893,756, 16 January 1990.

49. Young, K.; Fetcenko, M.A. Method for Powder Formation of a Hydrogen Storage Alloy. U.S. Patent 6,120,936, 19 September 2000.

50. Kim, D.M.; Lee, H.; Cho, K.; Lee, J.Y. Effect of Cu powder as an additive material on the inner pressure of a sealed-type Ni–MH rechargeable battery using a Zr-based alloy as an anode. *J. Alloys Compd.* **1999**, *282*, 261–267. [CrossRef]

51. Tan, S.; Shen, Y.; Şahin, E.O.; Noréus, D.; Öztürk, T. Activation behavior of an AB_2 type metal hydride alloy for NiMH batteries. *Int. J. Hydrogen Energy* **2016**, *41*, 9948–9953. [CrossRef]

52. Liu, B.H.; Li, Z.P.; Okutsu, A.; Suda, S. A wet ball milling treatment of Zr-based AB_2 alloys as negative electrode materials. *J. Alloys Compd.* **2000**, *296*, 148–151. [CrossRef]

53. Great Powder Website. Available online: http://www.greatpowder.com/yingwen/ (accessed on 10 May 2017).

54. Higuchi, E.; Li, Z.P.; Suda, S.; Nohara, S.; Inoue, H.; Iwakura, C. Structural and electrochemical characterization of fluorinated AB_2-type Laves phase alloys obtained by different pulverization methods. *J. Alloys Compd.* **2002**, *335*, 241–245. [CrossRef]

55. Hu, W.K.; Kim, D.M.; Jeon, S.W.; Lee, J.Y. Effect of annealing treatment on electrochemical properties of Mm-based hydrogen storage alloys for Ni/MH batteries. *J. Alloys Compd.* **1998**, *270*, 255–264. [CrossRef]

56. Young, K.; Nei, J.; Ouchi, T.; Fetcenko, M.A. Phase abundances in AB_2 metal hydride alloys and their correlations to various properties. *J. Alloys Compd.* **2011**, *509*, 2277–2284. [CrossRef]

57. Young, K.; Ouchi, T.; Meng, T.; Wong, D.F. Studies on the synergetic effects in multi-phase metal hydride alloys. *Batteries* **2016**, *2*, 15. [CrossRef]

58. Züttel, A.; Meli, F.; Chartouni, D.; Schlapbach, L.; Lichtenberg, F.; Friedrich, B. Properties of $Zr(V_{0.25}Ni_{0.75})_2$ metal hydride as active electrode material. *J. Alloys Compd.* **1996**, *239*, 175–282. [CrossRef]

59. Zhang, Q.A.; Lei, Y.Q.; Yang, X.G.; Ren, K.; Wang, Q.D. Annealing treatment of AB_2-type hydrogen storage alloys: I. crystal structures. *J. Alloys Compd.* **1999**, *292*, 236–240. [CrossRef]

60. Zhang, Q.A.; Lei, Y.Q.; Yang, X.G.; Du, Y.L.; Wang, Q.D. Effects of annealing treatment on phase structures, hydrogen absorption–desorption characteristics and electrochemical properties of a $V_3TiNi_{0.56}Hf_{0.24}Mn_{0.15}Cr_{0.1}$ alloy. *J. Alloys Compd.* **2000**, *305*, 125–129. [CrossRef]

61. Young, K.; Ouchi, T.; Huang, B.; Chao, B.; Fetcenko, M.A.; Bendersky, L.A.; Wang, K.; Chiu, C. The correlation of C14/C15 phase abundance and electrochemical properties in the AB_2 alloys. *J. Alloys Compd.* **2010**, *506*, 841–848. [CrossRef]

62. Young, K.; Ouchi, T.; Banik, A.; Koch, J.; Fetcenko, M.A. Improvement in the electrochemical properties of gas atomized AB_2 metal hydride alloys by hydrogen annealing. *Int. J. Hydrogen Energy* **2011**, *36*, 3547–3555. [CrossRef]

63. Chuang, H.J.; Huang, S.S.; Ma, C.Y.; Chan, S.L.I. Effect of annealing heat treatment on an atomized AB_2 hydrogen storage alloy. *J. Alloys Compd.* **1999**, *285*, 284–291. [CrossRef]

64. Liu, H.; Li, R. Effect of variation of constituent on microstructure and electrochemical properties of AB_2 hydrogen storage alloys. *Foundry Technol.* **2007**, *29*, 179–183. (In Chinese)

65. Klein, B.; Simon, N.; Klyamkine, S.; Latroche, M.; Percheron-Guégan, A. Improvement of the thermodynamical and electrochemical properties of multicomponent Laves phase hydrides by thermal annealing. *J. Alloys Compd.* **1998**, *280*, 284–289. [CrossRef]

66. Züttel, A.; Meli, F.; Schlapbach, L. Effects of pretreatment on the activation behavior of $Zr(V_{0.25}Ni_{0.75})_2$ metal hydride electrodes in alkaline solution. *J. Alloys Compd.* **1994**, *209*, 99–105. [CrossRef]

67. Liu, F.J.; Kitayama, K.; Suda, S. La and Ce-incorporation effects on the surface properties of the fluorinated $(Ti,Xr)(Mn,Cr,Ci)_2$ hydriding alloys. *Vacuum* **1996**, *47*, 903–906. [CrossRef]

68. Liu, B.H.; Li, Z.P.; Higuchi, E.; Suda, S. Improvement of the electrochemical properties of Zr-based AB_2 alloys by an advanced fluorination technique. *J. Alloys Compd.* **1999**, *293*, 702–706. [CrossRef]

69. Li, Z.P.; Higuchi, E.; Liu, B.H.; Suda, S. Effects of fluorination temperature on surface structure and electrochemical properties of AB_2 electrode alloys. *Electrochim. Acta* **2000**, *45*, 1773–1779. [CrossRef]

70. Park, H.Y.; Cho, W.I.; Cho, B.W.; Lee, S.R.; Yun, K.S. Effect of fluorination on the lanthanum-doped AB_2-type metal hydride electrodes. *J. Power Sources* **2001**, *92*, 149–156. [CrossRef]

71. Li, Z.P.; Liu, B.H.; Hitaka, K.; Suda, S. Effects of surface structure of fluorinated AB_2 alloys on their electrodes and battery performances. *J. Alloys Compd.* **2002**, *330*, 776–781. [CrossRef]

72. Young, K.H.; Fetcenko, M.A.; Ovshinsky, S.R.; Ouchi, T.; Reichman, B.; Mays, W.C. Improved surface catalysis of Zr-based Laves phase alloys for NiMH batteries. In *Hydrogen at Surface and Interfaces*; Jerkiewicz, G., Feliu, J.M., Popov, B.N., Eds.; Electrochemical Society: Pennington, NJ, USA, 2000.

73. Reichman, B.; Venkatesan, S.; Fetcenko, M.A.; Jeffries, K.; Stahl, S.; Bennett, C. Activated Rechargeable Hydrogen Storage Electrode and Method. U.S. Patent 4,716,088, 29 December 1987.

74. Jung, J.H.; Liu, B.H.; Lee, J.Y. Activation behavior of $Zr_{0.7}Ti_{0.3}Cr_{0.3}Mn_{0.3}V_{0.4}Ni$ alloy electrode modified by the hot-charging treatment. *J. Alloys Compd.* **1998**, *264*, 306–310. [CrossRef]

75. Liu, B.; Jung, J.; Lee, H.; Lee, K.; Lee, J. Improved electrochemical performance of AB_2-type metal hydride electrode activated by the hot-charging process. *J. Alloys Compd.* **1996**, *245*, 132–141. [CrossRef]

76. Cao, J.; Gao, X.; Lin, D.; Zhou, X.; Yuan, H.; Song, D.; Shen, P. Activation behavior of the Zr-based Laves phase alloy electrode. *J. Power Sources* **2011**, *93*, 141–144.

77. Şahin, E.O. Development of Rare Earth-Free Negative Electrode Materials for NiMH Batteries. In Proceeding of the 18th International Metallurgy & Materials Congress, Istanbul, Turkey, 29 September–1 October 2016; pp. 770–773.

78. Jung, J.H.; Lee, H.H.; Kim, D.M.; Jang, K.J.; Lee, J.Y. Degradation behavior of Cu-coated Ti–Zr–V–Mn–Ni metal hydride electrodes. *J. Alloys Compd.* **1998**, *266*, 266–270. [CrossRef]

79. Sun, D.; Latroche, M.; Percheron-Guégan, A. Activation behaviour of mechanically Ni-coated Zr-based laves phase hydride electrode. *J. Alloys Compd.* **1997**, *257*, 302–305. [CrossRef]

80. Jurczyk, M.; Rajewski, W.; Majchrzycki, W.; Wojcik, G. Synthesis and electrochemical properties of high-energy ball-milled Laves phase $(Zr,Ti)(V,Mn,Cr)_2$ alloys with nickel powder. *J. Alloys Compd.* **1998**, *274*, 299–302. [CrossRef]

81. Lee, S.; Lee, H.; Yu, J.S.; Fateev, G.A.; Lee, J.Y. The activation characteristics of a Zr-based hydrogen storage alloy electrode surface-modified by ball-milling process. *J. Alloys Compd.* **1999**, *292*, 258–265. [CrossRef]

82. Li, S.; Zhao, M.; Wang, Y.; Zhai, J. Structure and electrochemical property of ball-milled $Ti_{0.26}Zr_{0.07}Mn_{0.1}Ni_{0.33}V_{0.24}$ alloy. *Mater. Chem. Phys.* **2009**, *118*, 51–56. [CrossRef]

83. Chu, H.L.; Zhang, Y.; Sun, L.X.; Qiu, S.J.; Qi, Y.N.; Xu, F.; Yuan, H. Structure and electrochemical properties of composite electrodes synthesized by mechanical milling Ni-free $TiMn_2$-based alloy with La-based alloy. *J. Alloys Compd.* **2007**, *446–447*, 614–619. [CrossRef]

84. Li, P.; Wang, X.; Wu, J.; Zhao, W.; Li, R.; Ma, N. Present status on research of AB_2-type Laves phase hydrogen storage electrode materials. *Met. Funct. Mater.* **2000**, *7*, 7–12. (In Chinese)

85. Young, K. Electrochemical applications of metal hydride. In *Compendium of Hydrogen Energy Volume 3: Hydrogen Energy Conversion*; Barbir, F., Basile, A., Veziroglu, T.N., Eds.; Woodhead Publishing: Cambridge, UK, 2016.

86. Hong, K. The development of hydrogen storage alloys and the progress of nickel hydride batteries. *J. Alloys Compd.* **2001**, *321*, 307–313. [CrossRef]
87. Young, K.; Huang, B.; Regmi, R.K.; Lawes, G.; Liu, Y. Comparisons of metallic clusters imbedded in the surface oxide of AB_2, AB_5, and A_2B_7 alloys. *J. Alloys Compd.* **2010**, *506*, 831–840. [CrossRef]
88. Young, K.; Ouchi, T.; Fetcenko, M.A. Pressure–composition–temperature hysteresis in C14 Laves phase alloys: Part 1. Simple ternary alloys. *J. Alloys Compd.* **2009**, *480*, 428–433. [CrossRef]
89. Young, K.; Ouchi, T.; Mays, W.; Reichman, B.; Fetcenko, M.A. Pressure–composition–temperature hysteresis in C14 Laves phase alloys: Part 2. Applications in NiMH batteries. *J. Alloys Compd.* **2009**, *480*, 434–439. [CrossRef]
90. Young, K.; Yasuoka, S. Capacity degradation mechanisms in nickel/metal hydride batteries. *Batteries* **2016**, *2*, 3. [CrossRef]
91. Knosp, B.; Vallet, L.; Blanchard, P. Performance of an AB_2 alloy in sealed Ni–MH batteries for electric vehicles: Quantification of corrosion rate and consequences on the battery performance. *J. Alloys Compd.* **1999**, *293*, 770–774. [CrossRef]
92. Young, K.; Wong, D.F.; Yasuoka, S.; Ishida, J.; Nei, J.; Koch, J. Different failure modes for V-containing and V-free AB_2 metal hydride alloys. *J. Power Sources* **2014**, *251*, 170–177. [CrossRef]
93. Fetcenko, M.A.; Ovshinsky, S.R.; Reichman, B.; Young, K.; Fierro, C.; Koch, J.; Zallen, A.; Mays, W.; Ouchi, T. Recent advances in NiMH battery technology. *J. Power Sources* **2007**, *165*, 544–551. [CrossRef]
94. Young, K.; Chao, B.; Liu, Y.; Nei, J. Microstructures of the oxides on the activated AB_2 and AB_5 metal hydride alloys surface. *J. Alloys Compd.* **2014**, *606*, 97–104. [CrossRef]
95. Young, K.; Chao, B.; Pawlik, D.; Shen, H.T. Transmission electron microscope studies in the surface oxide on the La-containing AB_2 metal hydride alloy. *J. Alloys Compd.* **2016**, *672*, 356–365. [CrossRef]
96. Zhu, Y.; Pan, H.; Gao, M.; Liu, Y.; Li, R.; Lei, Y.; Wang, Q. Degradation mechanisms of Ti–V-based multiphase hydrogen storage alloy electrode. *Int. J. Hydrogen Energy* **2004**, *29*, 313–318. [CrossRef]
97. Liu, B.H.; Li, Z.P.; Matsuyama, Y.; Kitani, R.; Suda, S. Corrosion and degradation behavior of Zr-based AB_2 alloy electrodes during electrochemical cycling. *J. Alloys Compd.* **2000**, *296*, 201–208. [CrossRef]
98. Chuang, H.J.; Chan, S.L.I. Study of the performance of Ti–Zr based hydrogen storage alloys. *J. Power Sources* **1999**, *77*, 159–163. [CrossRef]
99. Aoki, K.; Yamamoto, T.; Masumoto, T. Hydrogen induced amorphization in RNi_2 laves phases. *Scr. Metall.* **1987**, *21*, 27–31. [CrossRef]
100. Aoki, K. Amorphous phase formation by hydrogen absorption. *Mater. Sci. Eng. A* **2001**, *304–306*, 45–53. [CrossRef]
101. Aoki, K.; Masumoto, T. Hydrogen-induced amorphization of intermetallics. *J. Alloys Compd.* **1995**, *231*, 20–28. [CrossRef]
102. Züttel, A.; Chartouni, D.; Nützenadel, C.; Gross, K.; Schlapbach, L. Bulk and surface properties of crystalline and amorphous $Zr_{36}(V_{0.33}Ni_{0.66})_{64}$ alloy as active electrode material. *J. Alloys Compd.* **1998**, *266*, 321–326. [CrossRef]
103. Hu, W.K.; Zhang, Y.S.; Song, D.Y.; Shen, P.W. Electrochemical hydrogen storage properties of amorphous and crystalline MI-Ni alloy films. *Int. J. Hydrogen Energy* **1996**, *21*, 651–656. [CrossRef]
104. Zhao, X.; Ma, L.; Gou, Y.; Shen, X. Structure, morphology and hydrogen desorption characteristics of amorphous and crystalline Ti-Ni alloys. *Mater. Sci. Eng. A* **2009**, *516*, 50–53. [CrossRef]
105. Yang, X.; Lei, Y.; Wang, C.; Zhu, G.; Zhang, W.; Wang, Q. Influence of amorphization on electrode performances of AB_2 type hydrogen storage alloys. *J. Alloys Compd.* **1998**, *265*, 264–268. [CrossRef]
106. Boettinger, W.J.; Newbury, D.E.; Wang, K.; Bendersky, L.A.; Chiu, C.; Kattner, U.R.; Young, K.; Chao, B. Examination of multiphase $(Zr,Ti)(V,Cr,Mn,Ni)_2$ Ni-MH electrode alloys: Part I. Dendritic solidification structure. *Metall. Mater. Trans. A* **2010**, *41*, 2033–2047. [CrossRef]
107. Bendersky, L.A.; Wang, K.; Boettinger, W.J.; Newbury, D.E.; Young, K.; Chao, B. Examination of multiphase $(Zr,Ti)(V,Cr,Mn,Ni)_2$ Ni-MH electrode alloys: Part II. Solid-state transformation of the interdendritic B2 phase. *Metall. Mater. Trans. A* **2010**, *41*, 1891–1906. [CrossRef]
108. Liu, Y.; Young, K. Microstructure investigation on metal hydride alloys by electron backscatter Diffraction Technique. *Batteries* **2016**, *2*, 26. [CrossRef]

109. Bendersky, L.A.; Wang, K.; Levin, I.; Newbury, D.; Young, K.; Chao, B.; Creuziger, A. $Ti_{12.5}Zr_{21}V_{10}Cr_{8.5}$ $Mn_xCo_{1.5}Ni_{46.5-x}$ AB$_2$-type metal hydride alloys for electrochemical storage application: Part 1. Structural characteristics. *J. Power Sources* **2012**, *218*, 474–486. [CrossRef]

110. Tokunaga, T.; Motsumoto, S.; Ohtani, H.; Hasebe, M. Thermodynamic calculation of phase equilibria in the Nb-Ni-Ti-Zr quaternary system. *Mater. Trans.* **2007**, *48*, 89–96. [CrossRef]

111. Young, K.; Ouchi, T.; Reichman, B.; Mays, W.; Regmi, R.; Lawes, G.; Fetcenko, M.A.; Wu, A. Optimization of Co-content in C14 Laves phase multi-component alloys for NiMH battery application. *J. Alloys Compd.* **2010**, *489*, 202–210. [CrossRef]

112. Young, K.; Reichman, B.; Fetcenko, M.A. Electrochemical performance of AB$_2$ metal hydride alloys measured at −40 °C. *J. Alloys Compd.* **2013**, *580*, S349–S352. [CrossRef]

113. Young, K.; Wong, D.F.; Nei, J.; Reichman, B. Electrochemical properties of hypo-stoichiometric Y-doped AB$_2$ metal hydride alloys at ultra-low temperature. *J. Alloys Compd.* **2015**, *643*, 17–27. [CrossRef]

114. Zhang, Q.A.; Lei, Y.Q.; Wang, C.S.; Wang, F.S.; Wang, Q.D. Structure of the secondary phase and its effects on hydrogen-storage properties in a $Ti_{0.7}Zr_{0.2}V_{0.1}Ni$ alloy. *J. Power Sources* **1998**, *75*, 288–291. [CrossRef]

115. Saldan, I.; Burtovyy, R.; Becker, H.W.; Ader, V.; Wöll, C. Ti–Ni alloys as MH electrodes in Ni–MH accumulators. *Int. J. Hydrog. Energy* **2008**, *33*, 7177–7184. [CrossRef]

116. Nei, J.; Young, K. Gaseous phase and electrochemical hydrogen storage properties of $Ti_{50}Zr_1Ni_{44}X_5$ (X = Ni, Cr, Mn, Fe, Co, or Cu) for nickel metal hydride battery applications. *Batteries* **2016**, *2*, 24. [CrossRef]

117. Emami, H.; Cuevas, F.; Latroche, M. Ti(Ni,Cu) pseudobinary compounds as efficient negative electrodes for Ni–MH batteries. *J. Power Sources* **2014**, *265*, 182–191. [CrossRef]

118. Shu, K.Y.; Yang, X.G.; Zhang, S.K.; Lü, G.L.; Lei, Y.Q.; Wang, Q.D. Effect of Cr and Co additives on microstructure and electrochemical performance of $Zr(NiVMn)_2M_{0.1}$ alloys. *J. Alloys Compd.* **2000**, *306*, 122–126. [CrossRef]

119. Young, K.; Chao, B.; Bendersky, L.A.; Wang, K. $Ti_{12.5}Zr_{21}V_{10}Cr_{8.5}Mn_xCo_{1.5}Ni_{46.5-x}$ AB$_2$-type metal hydride alloys for electrochemical storage application: Part 2. Hydrogen storage and electrochemical properties. *J. Power Sources* **2012**, *218*, 487–494. [CrossRef]

120. Ji, S.; Li, S.; Sun, J. Effect of alloys with Mn, V on phase structures and electrochemical properties of Zr-Cr-Ni based AB$_2$ hydrogen storage electrode alloys. *Chin. J. Rare Met.* **2004**, *28*, 657–661. (In Chinese)

121. Nei, J.; Young, K.; Regmi, R.; Lawes, G.; Salley, S.O.; Ng, K.Y.S. Gaseous phase hydrogen storage and electrochemical properties of Zr_8Ni_{21}, Zr_7Ni_{10}, Zr_9Ni_{11}, and ZrNi metal hydride alloys. *Int. J. Hydrogen Energy* **2012**, *37*, 16042–16055. [CrossRef]

122. Matsuyama, A.; Mizutani, H.; Kozuka, T.; Inoue, H. Effect of Ti substitution on electrochemical properties of ZrNi alloy electrode for use in nickel-metal hydride batteries. *Int. J. Hydrogen Energy* **2017**. [CrossRef]

123. Matsuyama, A.; Mizutani, H.; Kozuka, T.; Inoue, H. Crystal structure and hydrogen absorption-desorption properties of $Zr_{1-x}Ti_xNi$ ($0.05 \leq x \leq 0.5$) alloys. *J. Alloys Compd.* **2017**, *714*, 467–475. [CrossRef]

124. Ruiz, F.C.; Castro, E.B.; Real, S.G.; Peretti, H.A.; Visintin, A.; Triaca, W.E. Electrochemical characterization of AB$_2$ alloys used for negative electrodes in Ni/MH batteries. *Int. J. Hydrogen Energy* **2008**, *33*, 3576–3580. [CrossRef]

125. Ruiz, F.C.; Castro, E.B.; Peretti, H.A.; Visintin, A. Study of the different Zr_xNi_y phases of Zr-based AB$_2$ materials. *Int. J. Hydrogen Energy* **2010**, *35*, 9879–9887. [CrossRef]

126. Young, K.; Ouchi, T.; Liu, Y.; Reichman, B.; Mays, W.; Fetcenko, M.A. Structural and electrochemical properties of $Ti_xZr_{7-x}Ni_{10}$. *J. Alloys Compd.* **2009**, *480*, 521–528. [CrossRef]

127. Young, K.; Ouchi, T.; Fetcenko, M.A.; Mays, W.; Reichman, B. Structural and electrochemical properties of $Ti_{1.5}Zr_{5.5}V_xNi_{10-x}$. *Int. J. Hydrogen Energy* **2009**, *34*, 8695–8706. [CrossRef]

128. Nei, J.; Young, K.; Salley, S.O.; Ng, K.Y.S. Effects of annealing on $Zr_8Ni_{19}X_2$ (X = Ni, Mg, Al, Sc, V, Mn, Co, Sn, La, and Hf): Hydrogen storage and electrochemical properties. *Int. J. Hydrogen Energy* **2012**, *37*, 8418–8427. [CrossRef]

129. Young, M.; Chang, S.; Young, K.; Nei, J. Hydrogen storage properties of $ZrV_xNi_{3.5-x}$ (x = 0.0–0.9) metal hydride alloys. *J. Alloys Compd.* **2013**, *580*, S171–S174. [CrossRef]

130. Young, K.; Young, M.; Chang, S.; Huang, B. Synergetic effects in electrochemical properties of $ZrV_xNi_{4.5-x}$ (x = 0.0, 0.1, 0.2, 0.3, 0.4, and 0.5) metal hydride alloys. *J. Alloys Compd.* **2013**, *560*, 33–41. [CrossRef]

131. Zhu, Y.F.; Pan, H.G.; Wang, G.Y.; Gao, M.X.; Ma, J.X.; Chen, C.P.; Wang, Q.D. Phase structure, crystallography and electrochemical properties of Laves phase compounds $Ti_{0.8}Zr_{0.2}V_{1.6}Mn_{0.8-x}M_xNi_{0.6}$ (M = Fe, Al, Cr, Co). *Int. J. Hydrogen Energy* **2001**, *26*, 807–816. [CrossRef]

132. Du, Y.L.; Yang, X.G.; Lei, Y.Q.; Zhang, M.S. Hydrogen storage properties of $Zr_{0.8}Ti_{0.2}(Ni_{0.6}Mn_{0.3-x}V_{0.1+x}Cr_{0.05})_2$ (x = 0.0, 0.05, 0.15, 0.2) alloys. *Int. J. Hydrogen Energy* **2002**, *27*, 695–697. [CrossRef]

133. Li, S.; Zhao, M.; Zhai, J.; Wang, F. Effect of substitution of chromium for nickel on structure and electrochemical characteristics of $Ti_{0.26}Zr_{0.07}V_{0.24}Mn_{0.1}Ni_{0.33}$ multi-phase hydrogen storage alloy. *Mater. Chem. Phys.* **2009**, *113*, 96–102.

134. Young, K.; Nei, J.; Wong, D.F.; Wang, L. Structural, hydrogen storage, and electrochemical properties of Laves phase-related body-centered-cubic solid solution metal hydride alloys. *Int. J. Hydrogen Energy* **2014**, *39*, 21489–21499. [CrossRef]

135. Young, K.; Ng, K.Y.S.; Bendersky, L.A. A technical report of the robust affordable next generation energy storage system-BASF program. *Batteries* **2016**, *2*, 2. [CrossRef]

136. Young, K.; Regmi, R.; Lawes, G.; Ouchi, T.; Reichman, B.; Fetcenko, M.A.; Wu, A. Effects of aluminum substitution in C14-rich multi-component alloys for NiMH battery application. *J. Alloys Compd.* **2010**, *490*, 282–292. [CrossRef]

137. Iba, H.; Akiba, E. The relation between microstructure and hydrogen absorbing property in Laves phase-solid solution multiphase alloys. *J. Alloys Compd.* **1995**, *231*, 508–512. [CrossRef]

138. Young, K.; Young, M.; Ouchi, T.; Reichman, B.; Fetcenko, M.A. Improvement in high-rate dischargeability, activation, and low-temperature performance in multi-phase AB_2 alloys by partial substitution of Zr with Y. *J. Power Sources* **2012**, *215*, 279–287. [CrossRef]

139. Young, K.; Wong, D.F.; Ouchi, T.; Huang, B.; Reichman, B. Effects of La-addition to the structure, hydrogen storage, and electrochemical properties of C14 metal hydride alloys. *Electrochim. Acta* **2015**, *174*, 815–825. [CrossRef]

140. Young, K.; Ouchi, T.; Nei, J.; Moghe, D. The importance of rare-earth additions in Zr-based AB_2 metal hydride alloys. *Batteries* **2016**, *2*, 25. [CrossRef]

141. Wong, D.F.; Young, K.; Nei, J.; Wang, L.; Ng, K.Y.S. Effects of Nd-addition on the structural, hydrogen storage, and electrochemical properties of C14 metal hydride alloys. *J. Alloys Compd.* **2015**, *647*, 507–518. [CrossRef]

142. Yoshida, M.; Ishibashi, H.; Susa, K.; Ogura, T.; Akiba, E. The multiphase effect on crystal structure, hydrogen absorbing properties and electrode performance of Sc-Zr based Laves phase alloys. *J. Alloys Compd.* **1995**, *230*, 100–108. [CrossRef]

143. Chang, S.; Young, K.; Ouchi, T.; Meng, T.; Nei, J.; Wu, X. Studies on incorporation of Mg in Zr-based AB_2 metal hydride alloys. *Batteries* **2016**, *2*, 11. [CrossRef]

144. Young, K.; Ouchi, T.; Huang, B.; Fetcenko, M.A. Effects of B, Fe, Gd, Mg, and C on the structure, hydrogen storage, and electrochemical properties of vanadium-free AB_2 metal hydride alloy. *J. Alloys Compd.* **2012**, *511*, 242–250. [CrossRef]

145. Liu, H.; Li, R. Effect of Sn content on properties of AB_2 hydrogen storage alloy. *Foundry Technol.* **2006**, *27*, 503–505. (In Chinese)

146. Pottgen, R.; Dronskowski, R. Structure and properties of Zr_2Ni_2In and Zr_2Ni_2Sn. *J. Solid State Chem.* **1997**, *128*, 289–294. [CrossRef]

147. Pearson, W.B. *The Crystal Chemistry and Physics of Metals and Alloys*; Wiley—Interscience: New York, NY, USA, 1972; p. 59.

148. Nihon, K.G. *Hi Kagaku Ryouronteki Kinzokukan Kagoubutsu*; Maruzen: Tokyo, Japan, 1975; p. 296. (In Japanese)

149. Yukawa, H.; Nakatsuka, K.; Morinaga, M. Design of hydrogen storage alloys in view of chemical bond between atoms. *Sol. Energy Mater. Sol. Cells* **2000**, *62*, 75–80. [CrossRef]

150. Liu, S.; Sun, D. The current research and the development trend of hydrogen storage alloy. *Rare Met. Cem. Carbides* **2005**, *33*, 46–51. (In Chinese)

151. Kim, D.M.; Jeon, S.W.; Lee, J.Y. A study of the development of a high capacity and high performance Zr–Ti–Mn–V–Ni hydrogen storage alloy for Ni–MH rechargeable batteries. *J. Alloys Compd.* **1998**, *279*, 209–214. [CrossRef]

152. Kim, S.R.; Lee, J.Y.; Park, H.H. A study of the activation behaviour of Zr-Cr-Ni-La metal hydride electrodes in alkaline solution. *J. Alloys Compd.* **1994**, *205*, 225–229. [CrossRef]

153. Shu, K.; Zhang, S.; Lei, Y.; Lü, G.; Wang, Q. Effect of Ti on the structure and electrochemical performance of Zr-based AB$_2$ alloys for nickel-metal rechargeable batteries. *J. Alloys Compd.* **2003**, *349*, 237–241. [CrossRef]
154. Lee, H.H.; Lee, K.Y.; Lee, J.Y. Degradation mechanism of Ti-Zr-V-Mn-Ni metal hydride electrodes. *J. Alloys Compd.* **1997**, *260*, 201–207. [CrossRef]
155. Züttel, A.; Meli, F.; Schlapbach, L. Surface and bulk properties of the Ti$_y$Zr$_{1-y}$(V$_x$Ni$_{1-x}$)$_2$ alloy system as active electrode material in alkaline electrolyte. *J. Alloys Compd.* **1995**, *231*, 645–649. [CrossRef]
156. Visintin, A.; Peretti, H.A.; Tori, C.A.; Triaca, W.E. Hydrogen absorption characteristics and electrochemical properties of Ti substituted Zr-based AB$_2$ alloys. *Int. J. Hydrogen Energy* **2001**, *26*, 683–689. [CrossRef]
157. Sun, J.C.; Li, S.; Ji, S.J. Phase composition and electrochemical performances of the Zr$_{1-x}$Ti$_x$Cr$_{0.4}$Mn$_{0.2}$V$_{0.1}$Ni$_{1.3}$ alloys with $0.1 \leq x \leq 0.3$. *J. Alloys Compd.* **2005**, *404*, 687–690. [CrossRef]
158. Zhu, Y.F.; Pan, H.G.; Gao, M.X.; Ma, J.X.; Li, S.Q.; Wang, Q.D. The effect of Zr substitution for Ti on the microstructures and electrochemical properties of electrode alloys Ti$_{1-x}$Zr$_x$V$_{1.6}$Mn$_{0.32}$Cr$_{0.48}$Ni$_{0.6}$. *Int. J. Hydrogen Energy* **2002**, *27*, 287–293. [CrossRef]
159. Hariprakash, B.; Martha, S.K.; Shukla, A.K. Effect of copper additive on Zr$_{0.9}$Ti$_{0.1}$V$_{0.2}$Mn$_{0.6}$Cr$_{0.05}$Co$_{0.05}$Ni$_{1.2}$ alloy anode for nickel-metal hydride batteries. *J. Appl. Electrochem.* **2003**, *33*, 497–504. [CrossRef]
160. Huot, J.; Akiba, E.; Ogura, T.; Ishido, Y. Crystal structure, phase abundance and electrode performance of Laves phase compounds (Zr, A)V$_{0.5}$Ni$_{1.1}$Mn$_{0.2}$Fe$_{0.2}$ (A≡Ti, Nb or Hf). *J. Alloys Compd.* **1995**, *218*, 101–109. [CrossRef]
161. Huot, J.; Akiba, E.; Ishido, Y. Crystal structure of multiphase alloys (Zr,Ti)(Mn,V)$_2$. *J. Alloys Compd.* **1995**, *231*, 85–89. [CrossRef]
162. Nei, J.; Young, K.; Salley, S.O.; Ng, K.Y.S. Determination of C14/C15 phase abundance in Laves phase alloys. *Mater. Chem. Phys.* **2012**, *136*, 520–527. [CrossRef]
163. Kandavel, M.; Bhat, V.V.; Rougier, A.; Aymard, L.; Nazri, G.A.; Tarascon, J.M. Improvement of hydrogen storage properties of the AB$_2$ Laves phase alloys for automotive application. *Int. J. Hydrogen Energy* **2008**, *33*, 3754–3761. [CrossRef]
164. Wakao, S.; Sawa, H.; Furukawa, J. Effects of partial substitution and anodic oxidation treatment of Zr–V–Ni alloys on electrochemical properties. *J. Less Common Met.* **1991**, *172*, 1219–1226. [CrossRef]
165. Kim, S.; Lee, J. Activation behaviour of ZrCrNiM$_{0.05}$ metal hydride electrode (M = La, Mm (misch metal), Nd). *J. Alloys Compd.* **1992**, *185*, L1–L4. [CrossRef]
166. Rönnebro, E.; Noréus, D.; Sakai, T.; Tsukahara, M. Structural studies of a new Laves phase alloy (Hf,Ti)(Ni,V)$_2$ and its very stable hydride. *J. Alloys Compd.* **1995**, *231*, 90–94. [CrossRef]
167. Huot, J.; Akiba, E.; Orura, O.; Ishido, Y. Crystal structure of electrode materials (Zr,A)V$_{0.5}$Ni$_{1.1}$Mn$_{0.2}$Fe$_{0.2}$ (A=Ti, Nb and Hf) for Ni-hydrogen rechargeable batteries. *Trans. Mater. Res. Soc. Jpn.* **1994**, *18B*, 1197–1200.
168. Zhou, O.; Yao, Q.; Sun, X.; Gu, Q.; Sun, J. Lattice-substitution of alloying elements and its influences on mechanical properties of ZrCr$_2$ Laves phase. *Chin. J. Nonferr. Met.* **2006**, *16*, 1603–1607. (In Chinese)
169. Santos, A.R.; Ambrosio, R.C.; Ticianelli, E.A. Electrochemical and structural studies on nonstoichiometric AB$_2$-type metal hydride alloys. *Int. J. Hydrogen Energy* **2004**, *29*, 1253–1261.
170. Young, K.; Ouchi, T.; Nei, J.; Chang, S. Increase in the Surface Catalytic Ability by Addition of Palladium in C14 Metal Hydride Alloy. *Batteries* **2017**, *3*, 26. [CrossRef]
171. Yang, X.G.; Zhang, W.K.; Lei, Y.Q.; Wang, Q.D. Electrochemical properties of Zr-V-Ni system hydrogen storage alloys. *J. Electrochem. Soc.* **1999**, *146*, 1245–1250. [CrossRef]
172. Ovshinsky, S.R.; Young, R. High Power Nickel-Metal Hydride Batteries and High Power Alloys/Electrodes for Use Therein. U.S. Patent 6,413,670 B1, 2 July 2002.
173. Yoshida, M.; Akiba, E. Hydrogen absorbing properties of ScM$_2$ Laves phase alloys (M = Fe, Co and Ni). *J. Alloys Compd.* **1995**, *226*, 75–80. [CrossRef]
174. Yoshida, M.; Ishibashi, H.; Susa, K.; Ogura, T.; Akiba, E. Crystal structure, hydrogen absorbing properties and electrode performances of Sc-based Laves phase alloys. *J. Alloys Compd.* **1995**, *226*, 161–165. [CrossRef]
175. Li, K.; Luo, Y.; Wang, W.; Qiu, J.; Kang, L. Effects of scandium on hydrogen storage and electrochemical properties of AB$_2$-type Zr$_{1-x}$Sc$_x$Mn$_{0.6}$V$_{0.2}$Ni$_{1.2}$Co$_{0.1}$ (x = 0~1) alloys. *J. Chin. Rare Earth Soc.* **2013**, *31*, 442–449. (In Chinese)
176. Liu, X.J.; Yang, S.Y.; Huang, Y.X.; Zhang, J.B.; Wang, C.P. Experimental investigation of isothermal sections (100, 1200 °C) in the Ni-Ti-Zr system. *J. Ph. Equilib. Diffus.* **2015**, *36*, 414–421. [CrossRef]

177. Young, K.; Fetcenko, M.A.; Li, F.; Ouchi, T.; Koch, J. Effect of vanadium substitution in C14 Laves phase alloys for NiMH battery application. *J. Alloys Compd.* **2009**, *468*, 482–492. [CrossRef]

178. Young, K.; Ouchi, T.; Koch, J.; Fetcenko, M.A. Compositional optimization of vanadium-free hypo-stoichiometric AB$_2$ metal hydride alloy for Ni/MH battery application. *J. Alloys Compd.* **2012**, *510*, 97–106. [CrossRef]

179. Wang, J.; Yu, R.; Liu, Q. Relation of element substitution to discharge capacity and activation of Ti-Zr based AB$_2$-type hydrogen storage electrode alloys. *Mater. Sci. Technol.* **2005**, *13*, 166–170. (In Chinese)

180. Yang, H.W.; Wang, Y.Y.; Wan, C.C. Studies of electrochemical properties of Ti$_{0.35}$Zr$_{0.65}$Ni$_x$V$_{2-x-y}$Mn$_y$ alloys with C14 Laves phases for nickel/metal hydride batteries. *J. Electrochem Soc.* **1996**, *143*, 429–435. [CrossRef]

181. Hagström, M.T.; Klyamkin, S.N.; Lund, P.D. Effect of substitution on hysteresis in some high-pressure AB$_2$ and AB$_5$ metal hydrides. *J. Alloys Compd.* **1999**, *293*, 67–73. [CrossRef]

182. Lee, S.F.; Wang, Y.Y.; Wan, C.C. Effect of adding chromium to Ti-Zr-Ni-V-Mn alloy on its cycle life as an Ni/metal-hydride battery material. *J. Power Sources* **1997**, *66*, 165–168. [CrossRef]

183. Lee, H.; Lee, S.; Lee, J. Activation characteristics of multiphase Zr-based hydrogen storage alloys for Ni/MH rechargeable batteries. *J. Electrochem. Soc.* **1999**, *146*, 3666–3671. [CrossRef]

184. Park, H.Y.; Chang, I.; Cho, W.I.; Cho, B.W.; Jang, H.; Lee, S.R.; Yun, K.S. Electrode characteristics of the Cr and La doped AB$_2$-type hydrogen storage alloys. *Int. J. Hydrogen Energy* **2001**, *26*, 949–955. [CrossRef]

185. Liu, B.H.; Li, Z.P.; Kitani, R.; Suda, S. Improvement of electrochemical cyclic durability of Zr-based AB$_2$ alloy electrodes. *J. Alloys Compd.* **2002**, *330*, 825–830. [CrossRef]

186. Young, K.; Ouchi, T.; Fetcenko, M.A. Roles of Ni, Cr, Mn, Sn, Co, and Al in C14 Laves phase alloys for NiMH battery application. *J. Alloys Compd.* **2009**, *476*, 774–781. [CrossRef]

187. Peretti, H.A.; Visintin, A.; Mogni, L.V.; Corso, H.L.; Gamboa, J.A.; Serafini, D.; Triaca, W.E. Hydrogen absorption behavior of multicomponent zirconium based AB$_2$ alloys with different chromium–vanadium ratio. *J. Alloys Compd.* **2003**, *354*, 181–186. [CrossRef]

188. Wang, G.Y.; Xu, Y.H.; Pan, H.G.; Wang, Q.D. Effect of substitution of chromium for manganese on structure discharge characteristics of Ti-Zr-V-Mn-Ni-type multi-phase hydrogen storage electrode alloys. *Int. J. Hydrogen Energy* **2003**, *28*, 499–508. [CrossRef]

189. Pan, H.; Li, R.; Gao, M.; Liu, Y.; Wang, Q. Effects of Cr on the structural and electrochemical properties of TiV-based two-phase hydrogen storage alloys. *J. Alloys Compd.* **2005**, *404*, 669–674. [CrossRef]

190. Yu, J.; Lee, S.; Cho, K.; Lee, J. The cycle life of Ti$_{0.8}$Zr$_{0.2}$V$_{0.5}$Mn$_{0.5-x}$Cr$_x$Ni$_{0.8}$ (x = 0 to 0.5) alloys for metal hydride electrodes of Ni-metal hydride rechargeable battery. *J. Electrochem. Soc.* **2000**, *147*, 2013–2017. [CrossRef]

191. Xu, Y.H.; Chen, C.P.; Wang, X.L.; Lei, Y.Q.; Wang, Q.D. The cycle life and surface properties of Ti-based AB$_2$ metal hydride electrodes. *J. Alloys Compd.* **2002**, *337*, 214–220. [CrossRef]

192. Shinyama, K.; Magari, Y.; Akita, H.; Kumagae, K.; Nakamura, H.; Matsuta, S.; Nohma, T.; Takee, M.; Ishiwa, K. Investigation into the deterioration in storage characteristics of nickel-metal hydride batteries during cycling. *J. Power Sources* **2005**, *143*, 265–269. [CrossRef]

193. Chen, J.; Dou, S.X.; Liu, H.K. Hydrogen desorption and electrode properties of Zr$_{0.8}$Ti$_{0.2}$(V$_{0.3}$Ni$_{0.6}$M$_{0.1}$)$_2$. *J. Alloys Compd.* **1997**, *256*, 40–44. [CrossRef]

194. Chai, Y.J.; Zhao, M.S.; Wang, N. Crystal structural and electrochemical properties of Ti$_{0.17}$Zr$_{0.08}$V$_{0.35}$Cr$_{0.10}$Ni$_{0.30-x}$Mn$_x$ (x = 0–0.12) alloys. *Mater. Sci. Eng. B* **2008**, *147*, 47–51. [CrossRef]

195. Young, K.; Ouchi, T.; Huang, B.; Reichman, B.; Fetcenko, M.A. The structure, hydrogen storage, and electrochemical properties of Fe-doped C14-predominating AB$_2$ metal hydride alloys. *Int. J. Hydrogen Energy* **2011**, *36*, 12296–12304. [CrossRef]

196. Song, M.Y.; Ahn, D.; Kwon, I.; Lee, R.; Rim, H. Development of AB$_2$-type Zr–Ti–Mn–V–Ni–Fe hydride electrodes for Ni–MH secondary batteries. *J. Alloys Compd.* **2000**, *298*, 254–260. [CrossRef]

197. Song, M.Y.; Kown, I.H.; Ahh, D.S.; Sohn, M.S. Improvement in the electrochemical properties of ZrMn$_2$ hydrides by substitution of elements. *Met. Mater. Int.* **2001**, *7*, 257–263. [CrossRef]

198. Kim, S.R.; Lee, J.Y. Electrode characteristics of C14-type Zr-based laves phase alloys. *J. Alloys Compd.* **1994**, *210*, 109–113. [CrossRef]

199. Song, M.Y.; Ahn, D.; Kwon, I.H.; Chough, S.H. Development of AB$_2$-type Zr-Ti-Mn-V-Ni-M hydride electrode for Ni/MH secondary battery. *J. Electrochem. Soc.* **2001**, *148*, A1041–A1044. [CrossRef]

200. Honda, N.; Furukawa, N.; Fujitani, S.; Yonezu, I. Hydrogen Absorbing Modified ZrMn$_2$-Type Alloys. U.S. Patent 4,913,879, 3 April 1990.
201. Ovshinsky, S.R.; Fetcenko, M.A.; Ross, J. A nickel metal hydride battery for electric vehicles. *Science* **1993**, *260*, 176–181. [CrossRef] [PubMed]
202. Kim, S.R.; Lee, K.Y.; Lee, J.Y. Improved low-temperature dischargeability of C14-type Zr–Cr–Ni Laves phase alloy. *J. Alloys Compd.* **1995**, *223*, 22–27. [CrossRef]
203. Kwon, I.; Park, H.; Song, M. Electrochemical properties of ZrMnNi$_{1+x}$ hydrogen-storage alloys. *Int. J. Hydrogen Energy* **2002**, *27*, 171–176. [CrossRef]
204. Lee, H.H.; Lee, K.Y.; Lee, J.Y. The Ti-based metal hydride electrode for NiMH rechargeable batteries. *J. Alloys Compd.* **1996**, *239*, 63–70. [CrossRef]
205. Young, K.; Ouchi, T.; Huang, B.; Reichman, B.; Fetcenko, M.A. Studies of copper as a modifier in C14-predominant AB$_2$ metal hydride alloys. *J. Power Sources* **2012**, *204*, 205–212. [CrossRef]
206. Yamamura, Y.; Seri, H.; Tsuji, Y.; Owada, N.; Iwaki, T. Hydrogen Storage Alloy and Electrode Therefrom. U.S. Patent 5,532,076, 2 July 1996.
207. Young, K.; Ouchi, T.; Lin, X.; Reichman, B. Effects of Zn-addition to C14 metal hydride alloys and comparisons to Si, Fe, Cu, Y, and Mo-additives. *J. Alloys Compd.* **2016**, *655*, 50–59. [CrossRef]
208. Wang, Y.; Zhao, D.; Yuan, H.; Wang, G.; Zhou, Z.; Song, D.; Zhang, Y. Preparation and electrochemical properties of MI(NiCuAlZn)$_5$ hydrogen storage alloys. *Acta Sci. Bat. Univ. Nankaiensis* **2000**, *33*, 120–123. (In Chinese)
209. Young, K.; Ouchi, T.; Huang, B.; Reichman, B.; Fetcenko, M.A. Effect of molybdenum content on structural, gaseous storage, and electrochemical properties of C14-predominant AB$_2$ metal hydride alloys. *J. Power Sources* **2011**, *196*, 8815–8821. [CrossRef]
210. Au, M.; Pourarian, F.; Sankar, S.G.; Wallace, W.E.; Zhang, L. TiMn$_2$-bases alloys as high hydrogen materials. *Mater. Sci. Eng. B* **1995**, *33*, 53–57. [CrossRef]
211. Huang, T.; Wu, Z.; Xia, B.; Xu, N. Influence of stoichiometry and alloying elements on the crystallography and hydrogen sorption properties of TiCr based alloys. *Mater. Sci. Eng. A* **2005**, *397*, 284–287.
212. Erika, T.; Ricardo, F.; Fabricio, R.; Fernando, Z.; Verónica, D. Electrochemical and metallurgical characterization of ZrCr$_{1-x}$NiMo$_x$ AB$_2$ metal hydride alloys. *J. Alloys Compd.* **2015**, *649*, 267–274. [CrossRef]
213. Doi, H.; Yabuki, R. Hydrogen Absorbing Ni, Zr-Based Alloy and Rechargeable Alkaline Battery. U.S. Patent 4,898,794, 6 February 1990.
214. Ruiz, F.C.; Peretti, H.A.; Visintin, A.; Triaca, W.E. A study on ZrCrNiPt$_x$ alloys as negative electrode components for NiMH batteries. *Int. J. Hydrogen Energy* **2011**, *36*, 901–906. [CrossRef]
215. Luan, B.; Cui, N.; Zhao, H.J.; Liu, H.K.; Dou, S.X. Effects of potassium-boron addition on the performance of titanium based hydrogen storage alloy electrodes. *Int. J. Hydrogen Energy* **1996**, *21*, 373–379. [CrossRef]
216. Reddy, A.L.M.; Ramaprabhu, S. Hydrogen diffusion studied in Zr-based Laves phase AB$_2$ alloys. *J. Alloys Compd.* **2008**, *460*, 268–271. [CrossRef]
217. Reddy, A.L.M.; Ramaprabhu, S. Structural and hydrogen absorption kinetics studies of polymer dispersed and boron added Zr-based AB$_2$ alloy. *Int. J. Hydrogen Energy* **2006**, *31*, 867–876.
218. Li, S.; Wen, B.; Li, X. Structure and electrochemical property of ball-milled Ti$_{0.26}$Zr$_{0.07}$Mn$_{0.1}$Ni$_{0.33}$V$_{0.24}$ alloy with 3 mass% B. *J. Alloys Compd.* **2016**, *654*, 580–585. [CrossRef]
219. Young, K.; Fetcenko, M.A.; Koch, J.; Morii, K.; Shimizu, T. Studies of Sn, Co, Al, and Fe additives in C14/C15 Laves alloys for NiMH battery application by orthogonal arrays. *J. Alloys Compd.* **2009**, *486*, 559–569. [CrossRef]
220. Fetcenko, M.A.; Young, K.; Ovshinky, S.R.; Reichman, B.; Koch, J.; Mays, W. Modified Electrochemical Hydrogen Storage Alloy Having Increased Capacity, Rate Capability and Catalytic Activity. U.S. Patent 6,270,719, 7 August 2001.
221. Gamo, T.; Moriwaki, Y.; Iwaki, T. Alloy for Hydrogen Storage Electrodes. U.S. Patent 4,946,646, 7 August 1990.
222. Doi, H.; Yabuki, R. Hydrogen Absorbing Ni-Based Alloy and Rechargeable Alkaline Battery. U.S. Patent 4,983,474, 9 January 1991.
223. Yu, X.B.; Wu, Z.; Huang, T.Z.; Chen, J.Z.; Xia, B.J.; Xu, N.X. Effect of carbon addition on activation and hydrogen sorption characteristics of TiMn$_{1.25}$Cr$_{0.25}$ alloy. *Mater. Chem. Phys.* **2004**, *83*, 273–277. [CrossRef]
224. Young, K.; Ouchi, T.; Huang, B.; Reichman, B.; Blankenship, R. Improvement in −40 °C electrochemical properties of AB$_2$ metal hydride alloy by silicon incorporation. *J. Alloys Compd.* **2013**, *575*, 65–72. [CrossRef]

225. Han, S.; Zhoa, M.; Liu, B. Microstructure and high-temperature electrochemical characteristics of $Zr_{0.9}Ti_{0.1}Ni_{1.0}Mn_{0.7}V_{0.3}Si_x$ ($x = 0.05, 0.10, 0.15, 0.20$) alloy. *Mater. Chem. Phys.* **2005**, *89*, 221–227. [CrossRef]

226. Li, S.; Liu, Y.; Zhu, J. Effect of Si-element on electrochemical properties of Ti-based hydrogen storage alloys. *Titan. Ind. Prog.* **2007**, *24*, 10–22. (In Chinese)

227. Young, K.; Chao, B.; Nei, J. Microstructures of the activated Si-containing AB_2 metal hydride alloy surface by transmission electron microscope. *Batteries* **2016**, *2*, 4. [CrossRef]

228. Drašner, A.; Blažina, Ž. The influence of Si and Ge on the hydrogen sorption properties of the intermetallic compound $ZrCr_2$. *J. Alloys Compd.* **1993**, *199*, 101–104.

229. Xu, L.; Xiao, Y.; Sandwijk, A.; Xu, Q.; Yang, Y. Separation of zirconium and hafnium: A review. In *Energy Materials 2014*; Springer: Heidelberg, Germany, 2014; p. 451.

230. Yamanaka, S.; Higuchi, K.; Miyake, M. Hydrogen solubility in zirconium alloys. *J. Alloys Compd.* **1995**, *231*, 503–507. [CrossRef]

231. Young, K.; Fetcenko, M.A.; Ouchi, T.; Li, F.; Koch, J. Effect of Sn-substitution in C14 Laves phase alloys for NiMH battery application. *J. Alloys Compd.* **2009**, *469*, 406–416. [CrossRef]

232. Morita, Y.; Gamo, T.; Kuranaka, S. Effects of nonmetal addition on hydriding properties for Ti–Mn Laves phase alloys. *J. Alloys Compd.* **1997**, *253*, 29–33. [CrossRef]

233. Giza, K.; Isasieczko, W.; Pavlyuk, V.V.; Bala, H.; Drulis, H.; Adamczyk, L. Hydrogen absorption and corrosion resistance of $LaNi_{4.8}Al_{0.2}$ and $LaNi_{4.8}Al_{0.1}Li_{0.1}$ alloy. *J. Alloys Compd.* **2007**, *429*, 352–356. [CrossRef]

234. Wei, X.; Tang, R.; Liu, Y.; Zhang, P.; Yu, G.; Zhu, J. Effect of small amounts of Li on microstructures and electrochemical properties of non-stoichiometric low-Co AB_5-type alloys. *Int. J. Hydrogen Energy* **2006**, *31*, 1365–1371. [CrossRef]

235. Wang, L.; Yuan, H.; Yang, H.; Zhou, K.; Song, D.; Zhang, Y. Study of the multi-composition AB_2 alloys including Li, made by the diffusion method, and their electrodes. *J. Alloys Compd.* **2000**, *302*, 65–69. [CrossRef]

236. Cracco, D.; Percheron-Guégan, A. Morphology and hydrogen absorption properties of an AB_2 type alloy ball milled with Mg_2Ni. *J. Alloys Compd.* **1998**, *268*, 248–255. [CrossRef]

237. Wang, Y.; Zhao, M. Electrochemical hydrogen storage characteristics of $Ti_{0.10}Zr_{0.15}V_{0.35}Cr_{0.10}Ni_{0.30}-10\%$ $LaNi_3$ composite and its synergetic effect. *Trans. Nonferr. Met. Soc. China* **2012**, *22*, 2000. (In Chinese) [CrossRef]

238. Chu, H.; Zhang, Y.; Sun, L.; Qiu, S.; Xu, F.; Yuan, H.; Wang, Q.; Dong, C. Structure, morphology and hydrogen storage properties of composites prepared by ball milling $Ti_{0.9}Zr_{0.2}Mn_{1.5}Cr_{0.3}V_{0.3}$ with La–Mg-based alloy. *Int. J. Hydrogen Energy* **2007**, *32*, 3363–3369. [CrossRef]

239. Liu, F.J.; Sandrock, G.; Suda, S. Surface and metallographic microstructure of the La-added AB_2 compound $(Ti, Zr)(Mn, Cr, Ni)_2$. *J. Alloys Compd.* **1995**, *231*, 392–396. [CrossRef]

240. Sun, D.; Latroche, M.; Percheron-Guégan, A. Effects of lanthanum or cerium on the equilibrium of $ZrNi_{1.2}Mn_{0.6}V_{0.2}Cr_{0.1}$ and its related hydrogenation properties. *J. Alloys Compd.* **1997**, *248*, 215–219. [CrossRef]

241. Yang, X.G.; Lei, Y.Q.; Shu, K.Y.; Lin, G.F.; Zhang, Q.A.; Zhang, W.K.; Zhang, X.B.; Lu, G.L.; Wang, Q.D. Contribution of rare-earths to activation property of Zr-based hydride electrode. *J. Alloys Compd.* **1999**, *293*, 632–636. [CrossRef]

242. Sun, J.C.; Li, S.; Ji, S.J. The effects of the substitution of Ti and La for Zr in $ZrMn_{0.7}V_{0.2}Co_{0.1}Ni_{1.2}$ hydrogen storage alloys on the phase structure and electrochemical properties. *J. Alloys Compd.* **2007**, *446*, 630–634. [CrossRef]

243. Chen, W.X. Effects of addition of rare-earth element on electrochemical characteristics of $ZrNi_{1.1}Mn_{0.5}V_{0.3}Cr_{0.1}$ hydrogen storage alloy electrodes. *J. Alloys Compd.* **2001**, *319*, 119–123. [CrossRef]

244. Jung, J.H.; Lee, K.Y.; Lee, J.Y. The activation mechanism of Zr-based alloy electrodes. *J. Alloys Compd.* **1995**, *226*, 166–169. [CrossRef]

245. Joubert, J.M.; Sun, D.; Latroche, M.; Percheron-Guegan, A. Electrochemical performances of ZrM_2 (M=V, Cr, Mn, Ni) Laves phases and the relation to microstructures and thermodynamical properties. *J. Alloys Compd.* **1997**, *253*, 564–569. [CrossRef]

246. Han, S.M.; Zhang, Z.; Zhao, M.S.; Zheng, Y.Z. Electrochemical characteristics and microstructure of $Zr_{0.9}Ti_{0.1}Ni_{1.1}Mn_{0.6}V_{0.3}-LaNi_5$ composite hydrogen storage alloys. *Int. J. Hydrogen Energy* **2006**, *31*, 563–567. [CrossRef]

247. Wang, Y.; Zhao, M.; Li, S.; Wang, L. Structure and electrochemical characteristics of melted composite $Ti_{0.10}Zr_{0.15}V_{0.35}Cr_{0.10}Ni_{0.30}$–LaNi$_5$ hydrogen storage alloys. *Electrochim. Acta* **2008**, *53*, 7831–7837. [CrossRef]

248. Lu, Z.; Qin, M.; Jiang, W.; Qing, P.; Liu, S. Effect of AB$_2$-based alloy addition on structure and electrochemical properties of $La_{0.5}Pr_{0.2}Zr_{0.1}Mg_{0.2}Ni_{2.75}Co_{0.45}Fe_{0.1}Al_{0.2}$ hydrogen storage alloy. *J. Rare Earths* **2013**, *31*, 386–394. [CrossRef]

249. HEFA Rare Earth. Available online: http://mineralprices.com (accessed on 8 May 2017).

250. Griessen, R.; Riesterer, T. Heat of Formation Models. In *Hydrogen in Intermetallic Compounds I*; Schlapbach, L., Ed.; Springer: Berlin/Heidelberg, Germany, 1988.

251. Zhu, J.H.; Pike, L.M.; Liu, C.T.; Liaw, P.K. Point defects in Binary NbCr$_2$ Laves-phase alloys. *Scr. Mater.* **1998**, *39*, 833–838. [CrossRef]

252. Kanazawa, S.; Kaneno, Y.; Inoue, H.; Kim, W.Y.; Takasugi, T. Microstructures and defect structures in ZrCr$_2$ Laves phase based intermetallic compounds. *Intermetallics* **2002**, *10*, 783–792. [CrossRef]

253. Wong, D.F.; Young, K.; Ouchi, T.; Ng, K.Y.S. First-principles point defect models for Zr_7Ni_{10} and Zr_2Ni_7 phases. *Batteries* **2016**, *2*, 23. [CrossRef]

254. Massalski, T.B. *Binary Alloy Phase Diagrams*, 2nd ed.; ASM International: Russell, OH, USA, 1990.

255. Notten, P.H.L.; Einerhand, R.E.F.; Daams, J.L.C. How to achieve long-term electrochemical cycling stability with hydride-forming electrode materials. *J. Alloys Compd.* **1995**, *231*, 604–610. [CrossRef]

256. Young, K.; Ouchi, T.; Yang, J.; Fetcenko, M.A. Studies of off-stoichiometric AB$_2$ metal hydride alloy: Part 1. Structural characteristics. *Int. J. Hydrogen Energy* **2011**, *36*, 11137–11145. [CrossRef]

257. Young, K.; Nei, J.; Huang, B.; Fetcenko, M.A. Studies of off-stoichiometric AB$_2$ metal hydride alloy: Part 2. Hydrogen storage and electrochemical properties. *Int. J. Hydrogen Energy* **2011**, *36*, 11146–11154. [CrossRef]

258. Liu, B.H.; Li, Z.P.; Suda, S. Electrochemical cycle life of Zr-based Laves phase alloys influenced by alloy stoichiometry and composition. *J. Electrochem. Soc.* **2002**, *149*, A537–A542. [CrossRef]

259. Zhu, Y.; Pan, H.; Gao, M.; Ma, J.; Lei, Y.; Wang, Q. Electrochemical studies on the Ti–Zr–V–Mn–Cr–Ni hydrogen storage electrode alloys. *Int. J. Hydrogen Energy* **2003**, *28*, 311–316. [CrossRef]

260. Pan, H.; Zhu, Y.; Gao, M.; Liu, Y.; Li, R.; Lei, Y.; Wang, Q. A study on the cycling stability of the Ti–V-based hydrogen storage electrode alloys. *J. Alloys Compd.* **2004**, *364*, 271–279. [CrossRef]

261. Yasuoka, S.; Magari, Y.; Murata, T.; Tanaka, T.; Ishida, J.; Nakamura, H.; Nohma, T.; Kihara, M.; Baba, Y.; Teraoka, H. Development of high-capacity nickel-metal hydride batteries using superlattice hydrogen-absorbing alloys. *J. Power Sources* **2006**, *156*, 662–666. [CrossRef]

262. Young, K.; Fetcenko, M.A. Low Cost, High Power, High Energy, Solid-State, Bipolar Metal Hydride Batteries. U.S. Patent 8,974,948, 10 March 2015.

263. Young, K.; Wong, D.; Nei, J.; Reichman, B.; Chao, B.; Mays, W. Shared Electrode Hybrid Battery-Fuel Cell System. U.S. Patent Appl. 20,150,295,290, 15 October 2015.

264. Zelinsky, M.A.; Koch, J.M. Nickel/metal hydride batteries in stationary applications. *Batteries* **2017**, submitted.

265. Wikipedia. Start-Stop System. Available online: https://en.wikipedia.org/wiki/Start-stop_system (accessed on 31 July 2017).

266. Yan, S.; Meng, T.; Young, K.; Nei, J. Nickel/metal hydride battery in a pouch design. *Batteries* **2017**, in preparation.

267. Young, K.; Nei, J.; Meng, T. Alkaline and Non-Aqueous Proton-Conducting Pouch-Cell Batteries. U.S. Patent Appl. 20,160,233,461, 11 August 2016.

268. Meng, T.; Young, K.; Wong, D.F.; Nei, J. Ionic liquid-based non-aqueous electrolytes for nickel/metal hydride batteries. *Batteries* **2017**, *2*, 4. [CrossRef]

batteries

MDPI

Review

Reviews on Chinese Patents Regarding the Nickel/Metal Hydride Battery

Kwo-Hsiung Young [1,2,*] **, Xiaojuan Cai** [3,4] **and Shiuan Chang** [2,5]

1   Department of Chemical Engineering and Materials Science, Wayne State University, Detroit, MI 48202, USA
2   BASF/Battery Materials-Ovonic, 2983 Waterview Drive, Rochester Hills, MI 48309, USA;
     shiuanc@wayne.edu
3   College of Chemistry and Chemical Engineering, Zhengzhou University of Light Industry,
     No.5 Dongfeng Road, Zhengzhou 450002, Henan, China; amigo.cai@highpowertech.com
4   Shenzhen HighPower Technology, Building 1, 68 Xinxia Road, Pinghu Town, Longgang District,
     Shenzhen 518111, Guangdong, China
5   Department of Mechanical Engineering, Wayne State University, Detroit, MI 48202, USA
*   Correspondence: kwo.young@basf.com; Tel.: +1-248-293-7000

Academic Editor: Catia Arbizzani
Received: 26 May 2017; Accepted: 14 July 2017; Published: 20 August 2017

Abstract: Both the patents issued and applications filed in China regarding nickel/metal hydride (Ni/MH) battery technology are reviewed in the article. Selective works from 39 battery manufactures, 9 metal hydride alloy suppliers, 13 $Ni(OH)_2$ suppliers, 20 hardware suppliers, 19 system integrators, universities, and 12 research institutes are included. China being the country that produces the most Ni/MH batteries is relatively weak in the innovation part of intellectual properties when compared to the US and Japan. However, it produces very many patents in the areas of cell structure optimization and production processes. Designs of high-capacity, high-power, and low-cost cells are compared from different manufacturers.

Keywords: nickel metal hydride battery; Chinese Patent; metal hydride alloy; nickel hydroxide; Chinese battery manufacturer

1. Introduction

The nickel/metal hydride (Ni/MH) battery is an important technology for consumer portable electronics, stationary, and transportation applications [1–3]. Even with the strong competition from the rival Li-ion battery [4–7], Ni/MH still finds its niche market in replacing nickel-cadmium (NiCd) and primary batteries [8–10]. China has been the country that has produced the most (>70%) of the consumer-type Ni/MH batteries for the world since the turn of the century. China has also started to produce Ni/MH batteries for hybrid electrical vehicles produced domestically [11]. While the main mission for the Chinese companies is to make a profit from selling products, many of them have devoted resources to the research and development of Ni/MH batteries—from raw materials, components, the cell, the battery pack, to various applications. Since the intellectual properties belonging to foreign companies have already been reviewed in two prior publications [12,13], only those filed by domestic companies are covered in this article. Three different types of patents are allowed in China: invention patent, utility model, and proactive. In this review, we focus mainly on the invention part unless the content is unique and important. The discussions are classified by the functions of those companies owning the patent right from the battery manufacture, metal hydride (MH) alloy producer, $Ni(OH)_2$ producer, other hardware producer, and system integrator. The relationship between these companies is illustrated in Figure 1. These Chinese issued patents and patent applications can be found online with the English translation of the abstract on the official site of the State Intellectual Property Office of the People's Republic of China (SIPO) [14]. The full patent application in Chinese can be accessed from the website of the patent search engine—SooPAT as well [15]. The patent application number starts with the year of application, for example 200510087099. The issued patent has a shorter number such as 01242474.

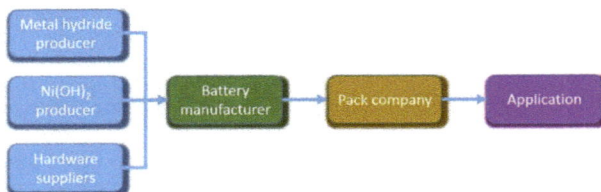

Figure 1. Schematic diagram showing the supply chain for the nickel/metal hydride (Ni/MH) battery industry.

2. Nickel/Metal Hydride (Ni/MH) Battery Manufacturer

In 2012, 74% of the consumer type Ni/MH batteries were produced in China [16]. More than half of them were from the top four manufacturers: Gold Peak (GP, Hong Kong, China), Shenzhen HighPower (SZHP, Shenzhen, Guangdong, China), Corun (Changsha, Hunan, China), and Suppo (Anshan, Liaoning, China) (Figure 2). Brief introductions of the Chinese domestic Ni/MH battery manufacturers are listed in Table 1 and their main inventions in the area of Ni/MH batteries are reviewed in the next few sections. They are mainly located in Guangdong province, especially in the Shenzhen Special District, which just borders Hong Kong (Figure 3). The details of the Ni/MH battery fabrication process were discussed before [12] and a summary flow chart is illustrated in Figure 4.

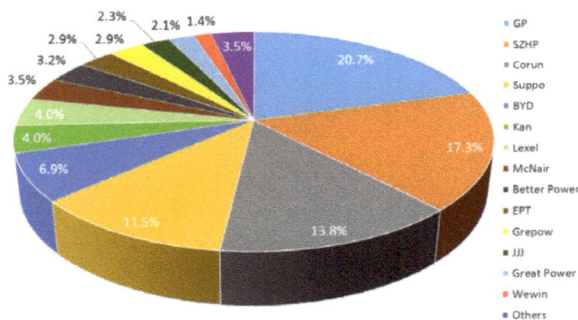

Figure 2. Market share of Chinese domestic Ni/MH battery manufacturer in 2012 [16].

Table 1. Brief introduction of the Chinese Ni/MH battery manufacturers included in this review.

Company Name	Established Date	Headquarter	Main Products	Trademark
Gold Peak 金山/超霸	1964	New Territory, Hong Kong	Ni/MH, Li-ion, primary	GP Batteries
HighPower 豪鹏	2001	Shenzhen, Guangdong	Ni/MH, Li-ion	highpower
Corun 科力远	1998	Changsha, Hunan	Ni/MH	CORUN
Suppo 三普	1996	Anshan, Liaoning	Ni/MH	SUPPO

Table 1. *Cont.*

Company Name	Established Date	Headquarter	Main Products	Trademark
BYD 比亚迪	1995	Shenzhen, Guangdong	Car, Li-ion, Ni/MH	
Kan 凯恩	1993	Suichang, Zhejiang	Ni/MH	
Lexel 力可兴	1997	Shenzhen, Guangdong	Ni/MH	
McNair 迈科	2000	Dongguan, Guangdong	Li-ion, Ni/MH	
BetterPower 倍特力	2002	Shenzhen, Guangdong	Ni/MH, Li-ion	
EPT 量能	2001	Shenzhen, Guangdong	Ni/MH	
Grepow 格瑞普	1998	Shenzhen, Guangdong	Ni/MH, Li-PO	
JJJ 三捷	1989	Jiangmen, Guangdong	Ni/MH, NiCd	
Great Power 鹏辉	2001	Guangzhou, Guangdong	Ni/MH, Li-ion, Li-PO	
Wewin 能一郎	2003	Shenzhen, Guangdong	Li-PO, Ni/MH, Li-ion	
TMK 三俊/大别山	2002	Shenzhen, Guangdong	Ni/MH	
Baosheng 宝生	2006	Chengdu, Sichuan	Ni/MH	
Huangyu 环宇	1982	Xinxiang, Henan	Li-ion	
Fengbiao 沣标/朗泰通	1999	Shenzhen, Guangdong	Ni/MH	
Tianneng 天能	1986	Changxin, Zhejiang	Li-ion, Ni/MH	
Wankaifeng 万凯丰	2010	Mianning, Sichuan	Ni/MH, metal hydride powder	
REO 稀奥科	2001	Baotou, Inner Mongolia	Ni/MH	
BST Power 电科电源	2002	Shenzhen, Guangdong	Ni/MH, NiCd, Li-ion	

Table 1. *Cont.*

Company Name	Established Date	Headquarter	Main Products	Trademark
Zhengda 淄博正大	2009	Zebo, Shandong	Ni/MH	
CENS 赛恩斯	1999	Suzhou, Jiangsu	Ni/MH, NiCd	
Jintion 劲鑫	2002	Quanzhou, Fujian	Ni/MH. NiCd	
Jinhui 金辉	2003	Dongguan, Guangdong	Ni/MH, Li-ion, Li-PO	
Troily 创力	1999	Xinxiang, Henan	NiFe, Ni/MH, NiZn	
Innovation 亿诺	2011	Liuan, Anhui	Ni/MH, Li-PO	
Cel 塞尔	1996	Huaian, Jiangsu	Ni/MH, Li-PO, Li-ion	
Bofuneng 博富能	2003	Shenzhen, Guangdong	Ni/MH, Li-ion, Li-PO	
Ryder 瑞鼎	2005	Shenzhen, Guangdong	Ni/MH, NiCd, charger	
DFEI 德飞	2011	Dali, Shanxi	Ni/MH, Li-ion	
Wintonic 云通	1998	Guangzhou, Guangdong	Ni/MH, Li-ion	
Oceansun 海太阳	2002	Shenzhen, Guangdong	Li-PO, Ni/MH	
Shida 实达	1996	Foshan, Guangzhou	Li-PO, Ni/MH	
Sanik 新力	1995	Foshan, Guangzhou	Ni/MH	
PeaceBay 和平海湾	1996	Tianjin, Tianjin	Ni/MH	
Unitech 联科	2002	Shenzhen, Guangdong	Ni/MH, NiCd, Li-ion	
Taiyi 太一	1996	Zhuhai, Guangdong	Ni/MH	

Figure 3. Locations of domestic Chinese Ni/MH battery manufacturers.

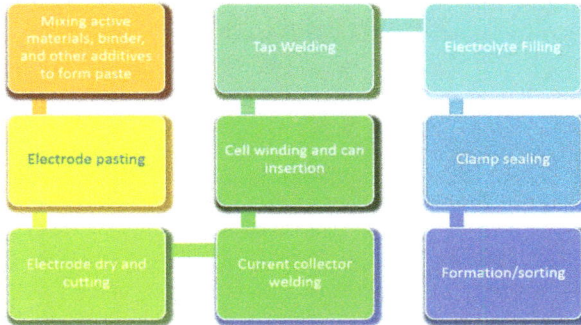

Figure 4. A flow chart showing the Ni/MH battery fabrication processes of a typical manufacturer.

2.1. Gold Peak (GP)

GP obtained two Chinese Patents on the different thicknesses of active material applied on the negative electrode (thicker on the current collector side) to improve the utilization [17] and design of an auto-winding machine [18]. GP also applied patents on a rupture mechanism ensuring a good gas release path when venting occurs [19] (Figure 5), with two separators with different lengths to prevent electrical shortage [20], and a negative electrode with one side scratched to increase the current connectivity to the cell can [21].

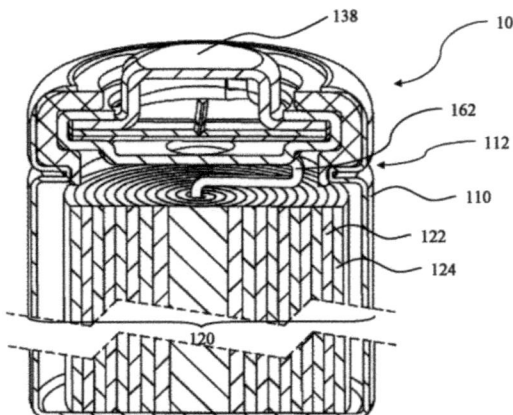

Figure 5. Schematic diagram of a new venting cap design with 10: battery, 138: vent cap, 162: conductive lead, 112: groove, 110: cell can, 122: positive electrode, 124: negative electrode, and 120: electrode assembly [19].

2.2. Shenzhen HighPower

SZHP filed 65 Chinese Patent Applications and was granted 63. In the positive electrode area, a Ni-plated perforated steel (NPPS) as the substrate [22], and a γ-CoOOH coating to extend the cycle life and tolerate overcharge [23] were disclosed. In the negative electrode area, a Ni-coated carbon nanotube as active material [24], a low Pr and Nd AB_5 MH alloy [25], and a hydroxypropyl methylcellulose coating to reduce corrosion [26] were shown. In the electrolyte area, a soluble bromide was added to extend the storage period [27]. In the cell assembly, a multi-cell parallel-connected configuration [28], a multi-venting hole cap design [29], a positive electrode current collector [30], a tap design to self-melt during cell short-circuit [31], a safety short-circuit conduit to prevent explosion [32], a serviceable (electrolyte refilling) cell design [33], an ultra-thin design [34] and a double current-collector design to reduce the cost and production complexity [35] (Figure 6) were introduced. In the fabrication methods, a wet pasting machine [36], a centrifugal electrolyte filling station [37], a capacity sorting machine [38], a tap spot-welding machine [39], and a battery performance testing machine [40] were proposed.

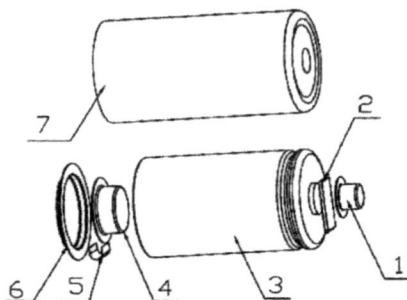

Figure 6. Schematic diagram of a low-cost current collector design with 1: positive terminal, 2: Ni tap, 3: cell assembly, 4: negative electrode bottom plate, 5: Ni tap, 6: negative terminal, and 7: plastic case [35].

2.3. Corun

Corun filed 42 Chinese Patent Application in the area of Ni/MH battery and 18 of them were granted. In the positive electrode area, an extended area covered with a metal wire outside the active material [41], Ca, Y, Co additives [42], and a phosphate [43] and a fluoride [44] coating for high-temperature application were disclosed. In the negative electrode area, a new design of two single-sided electrodes [45] and an electrode composed of single- and double-side coated segments [46] were advocated. In the electrolyte, a tungstate additive for high-temperature charge acceptance [47] was invented. In the cell assembly area, a Ni-plate placed at the bottom of the cell as the negative electrode current collector with a Ni- or Cu-foam in between to increase the high-power capability [48,49], and new type of venting cap designs [50,51] were displayed. A new type of high heat-dissipation battery was proposed with a central conducting rod [52,53] (Figure 7). The production line inventions include an electrode sorting and classification system [54], a battery activation and sorting system [55], an automatic venting cap production machine [56], an automatic electrolyte filling station [57], and a new type of formation process [58].

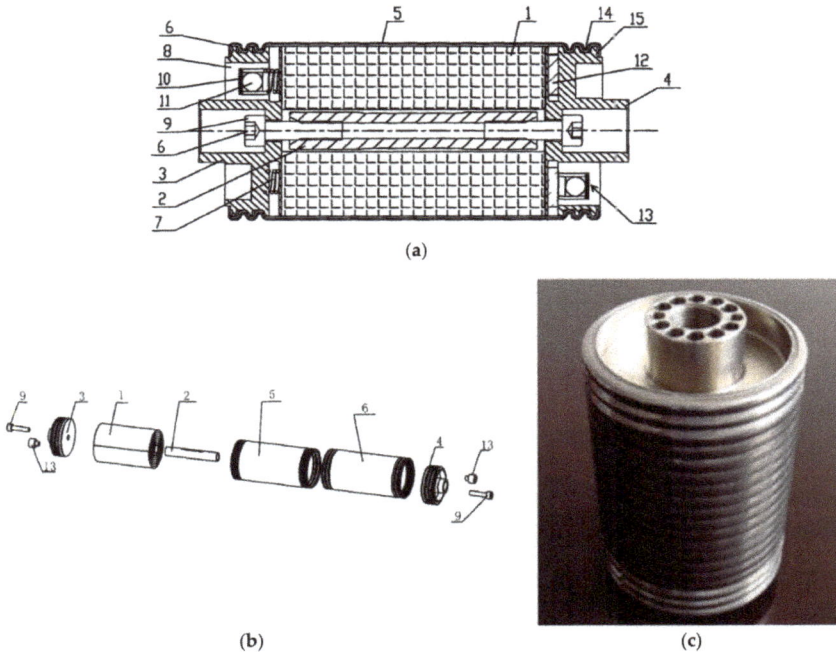

Figure 7. (**a**) A cross-section, (**b**) a side-view of a high heat-dissipation cell design, and (**c**) a real photograph with 1: electrode assembly, 2: central rod, 3: positive electrode terminal, 4: negative electrode terminal, 5: case, 6: insulator plastic cover, 7: spring, 8: safe valve base, 9: screw, 10: vent cap, 11: rubber ball, 12: rubber washer, 13: safety valve, 14: groove, and 15: clamp seal [53].

2.4. Suppo

Suppo filed 10 Chinese Patent Applications and was granted six of them. Suppo patented a liquid phase oxidation method to produce a pre-charged NiOOH positive electrode [59], a perforated Cu-substrate for negative electrode [60], a long-life over-stoichiometric AB_x ($5.05 \leq x \leq 5.5$) MH alloy [61], and a design of a low self-discharge battery [62] (Figure 8).

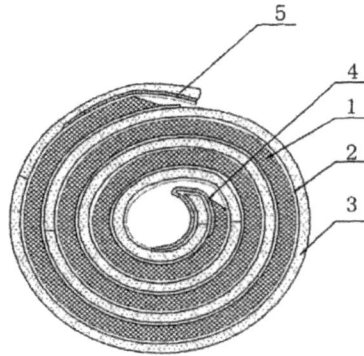

Figure 8. A cross-section view of a low self-discharge cell with an extra spacer at the end of electrode and 1: negative electrode, 2: separator, 3: positive electrode, 4: spacer, and 5: spacer [62].

2.5. BYD

BYD (Shenzhen, Guangdong, China) filed the largest number of Patent Applications in China regarding the Ni/MH battery with 108 applications filed and 84 of them granted. In the positive electrode area, a LiOH solution sprayed Co-coated $Ni(OH)_2$ [63], a Ti-additive to improve the high-temperature charging efficiency [64], a metallic V, Ti, Mo, Mn, Nb, Hf, Zr, Y, or Ta powder additive to improve the capacity and self-discharge performance [65], a poly-oxyalkylene and/or polyhydric alcohol to improve the uniformity in capacity and cycle stability [66], an Al_2O_3, $Al(OH)_3$ or Al-containing salt additives [67], a CoOOH coating [68], a CoO mixture [69] and a depolarizer [70] to increase the charging efficiency and cycle life, and a second coating besides the Co oxides containing the oxide (hydroxide) of Mg, Ba, Zn, Ca, Ti, Al, Ta, Mn, Y, Cr, Cd, Sr, and La-family rare earth element [71] were suggested. In the negative electrode area, surface treatment of $NaBH_4$ [72], NaH_2PO_2 [73], Ni-ion containing solution [74], surface activation by ball milling with alkaline solution [75], Ni or other noble metal [76] coatings, CoO/Co_3O_4 composite additives [77], Y-Ni (or Co) alloy additive to improve the cycle life [78], Cu- [79], Ni- [80], and carbon [81] fine particles coated on the NPPS substrate, addition of Y-containing compounds [82] and tetrapropyl orthosilicate [83], and a metallic thin film on the surface to promote the oxygen recombination during over-charge [84] were proposed. A porous oxide from Ni, Y, Mn, Co, W, Fr, Pb, or Mo was advocated as additives to both positive and negative electrodes to improve the low-temperature performance [85]. In the separator area, a new type of high-temperature (up to 400 °C) material [86] and a new hydrophilic polymer filling [87] were introduced. In the electrolyte area, both an S-containing soluble additive introduced to solve the short-circuit and leakage problems [88] and a gel-type electrolyte for preventing freeze in low-temperature [89] were disclosed. In the cell assembly area, an elongated seal ring [90–92] (Figure 9), a new design of vent cap [93], a venting hole with an oval shape [94], and a multi-layer current collector made of Ni, Cu, stainless steel, Cu and Ni layers [95] were presented. In the alloy design, BYD patterned a NdMgNiAl alloy with a stoichiometry between 3.2 and 3.9 [96], a dual-phase AB_5-based alloy containing a Ti (Zr)-Ni secondary phase [97], a LaNiCuFeCoMnAl alloy containing at least two alkaline earth metals [98], and a RENiCuFeMnSn alloy where RE (rare earth) contains 40–85 wt% La [99] were revealed.

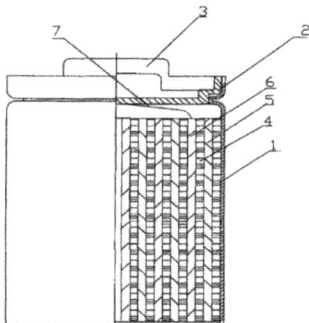

Figure 9. A schematic diagram of a cell with an elongated seal ring and 1: cell wall, 2: sealing ring, 3: vent cap, 4: separator, 5: positive electrode, 6: negative electrode, and 7: tap [91].

2.6. Kan

Kan (Suichang, Zhejiang, China) filed 16 Chinese Patent Applications related to the Ni/MH battery and obtained 13 of them. A MH alloy particle distribution of $D10 \geq 3$ μm, $D50 \leq 80$ μm, and $D90 \leq 130$ μm was claimed to have good high-power capability [100]. An insulator layer was coated on the positive electrode to prevent short-circuit from burr [101]. In the cell assembly, a high-power battery design was repetitively shown in several applications [102–105] (Figure 10). In the fabrication area, an electromagnetic-coupled winder [106], a device for shunting negative electrode [107], an electrode collecting apparatus [108], a connection apparatus to reduce the stress in the automatic spot-welder [109], and an automatic venting cap pressing machine [110] were patented.

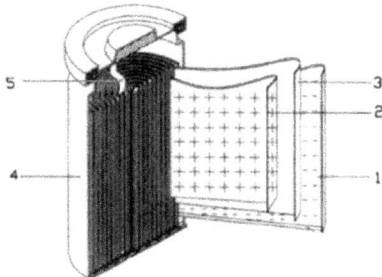

Figure 10. A schematic diagram of a high-power cell design with 1: can, 2: positive electrode, 3: separator, 4: negative electrode, and 5: tap having more than 4 welding spots to the positive electrode [100].

2.7. Lexel

Lexel (Shenzhen, Guangdong, China) filed 13 Chinese Patent applications and received 10 of them as granted. Lexel proposed a composite additive (2% ErO_2, 2% TiO_2, 1% $Ca(OH)_2$) to the positive electrode to increase the energy conversion efficiency during trickle charge [111]. Lexel also advocated the use of oxide or hydroxide of Ti, Y, Er, Tm, Yb, Rf, Ca, or Ba to improve the self-discharge characteristic [112]. In the production facilities, Lexel patented a cell impedance measuring apparatus [113], a cell thermo-resistivity measuring apparatus [114], a cell cleaning machine [115], and a scraping mechanism to remove extra MH powder [116]. In the cell design area, Lexel patented a low-temperature cell [117], a low-cost battery [118], and a high-capacity battery [119] (Figure 11).

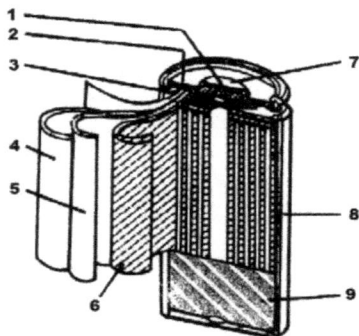

Figure 11. Schematic diagram of a low-cost battery design with 1: rubber, 2: vent cap, 3: plastic gasket, 4: positive electrode, 5: separator, 6: negative electrode, 7: positive terminal, 8: steel can, and 9: space filler [119].

2.8. McNair

McNair (Dongguan, Guangdong, China) disclosed Er, U, and Yb oxides or hydroxides additives in the positive electrode and boron compound in the electrolyte to improve the high-temperature charging efficiency [120], a positive electrode containing 3.7%–4.3% Zn, 1.3%–1.7% Co-co-precipitated $Ni(OH)_2$ with 6% CoO and 1%–10% metallic Ni for a fast charging battery [121], and an additive composed of oxide, hydroxide, fluoride, sulfide, and chloride of Ca, Mn, Er, Y, Yb, and B in the positive electrode with a negative electrode made from superlattice MH for the low self-discharge battery [122].

2.9. BetterPower

There are totally 65 Chinese Patent Applications filed by BetterPower (Shenzhen, Guangdong, China) and 52 got granted. In the positive electrode preparation, an addition of Ca into the paste [123], a vacuum-suction cleaning of the tab area [124], and a Sn-coating Ni-foam [125] were proposed. In the negative electrode preparation, an alkaline pre-activation [126] and an extra current collector layer on the electrode surface [127] were disclosed. In the cell assembly, new materials for the sealing ring [128] and rubber stopper [129], an Al-Cu alloy solder [130], a multi-layer separator [131], a graphite, Ni powder, acetylene black coating inside the wall of the can [132], a varying can wall thickness [133], an open-cell activation method [134], soft oval-shape [135] and cylindrical [136] pouch cells, an automatic powder cleaning station [137], and an automatic folding station [138] were included. A product design of a 9 V battery pack from BetterPower was also granted [139].

2.10. EPT

EPT (Shenzhen, Guangdong, China) filed 52 Chinese Patent Applications on the Ni/MH battery and received 37 of them granted, which cover a fabrication method of a high-capacity cell by adopting electrodes with varying thicknesses [140], a battery with AB_3 MH alloy and pre-oxidized $Ni(OH)_2$ which does not require activation [141], a positive electrode fabrication method using ultrasound atomization [142], a negative electrode containing styrene-butadiene rubber (SBR) or polytetrafluoroethylene (PTFE) as binder [143], a high-power design with additional negative electrode current collector made of Ni-foam [144], and new designs of vent cap [145], sealing ring [146], and current collector [147].

2.11. Grepow

Grepow (Shenzhen, Guangdong, China) filed 21 Chinese Patent Applications and had 10 granted. In the negative electrode area, a Ni-rich (>65%) surface layer prepared by electrodeless plating was introduced to improve the low-temperature/high-rate performance [148]. Fabrication methods for a low-capacity battery [149], a high-power battery [150] (Figure 12), a wide temperature range battery [151], and a low-impedance battery [152] were also disclosed by Grepow. A combination of battery and heating device was proposed to overcome the low-temperature hurdle of the Ni/MH battery [153,154]. Grepow advocated removing the extra active material in the negative electrode area contacting the can [155] and the related fabrication apparatus [156].

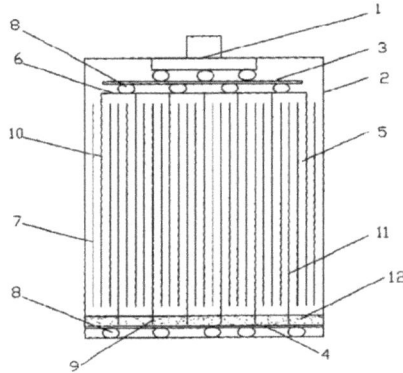

Figure 12. Schematic diagram of a high-power cell with 1: vent cap, 2: cell case, 3: current collector plate, 4: negative electrode plate, 5: electrode assembly, 6: positive electrode plate, 7: separator, 8: spot welding, 9: negative electrode, 10: positive electrode, 11: wound negative electrode, and 12: metallic foam [150].

2.12. JJJ

JJJ (Jiangmen, Guangdong, China) filed nine Chinese Patent Applications and seven of them were accepted. JJJ's inventions include a basic cylindrical cell design [157] (Figure 13), an electrolyte filling apparatus [158], an automatic cell sealing machine [159], a sealing ring coating apparatus [160], a negative electrode waste recycling method [161], a new type of vent cap [162], a low-capacity D-cell design [163], and a current collector design for a high-power battery [164].

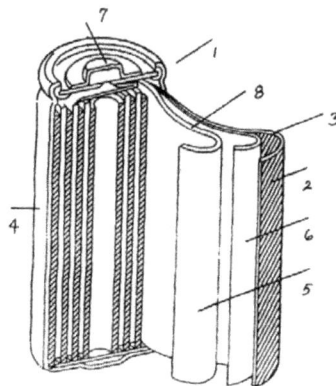

Figure 13. Schematic diagram of a basic structure of a cell with 1: Ni/MH battery, 2: negative electrode, 3: negative substrate, 4: can, 5: positive electrode, 6: separator, 7: vent cap, and 8: positive substrate [147].

2.13. Great Power

Great Power (Guangzhou, Guangdong, China) filed six Chinese Patent Applications and three were allowed. Their inventions include a positive electrode containing CoO, Co, Ni, acetylene black, TiO_2, Y_2O_3, PTFE, and carboxymethyl cellulose [165], a positive electrode substrate using steel web and superfine graphite powder [166], a new type of vent cap [167], and a tri-layer design of negative electrode with a metal plate in between [168].

2.14. Wewin

Only one Chinese Patent Application was filed by Wewin (Xianning, Hubei, China) and is about negative active material using a low-Co AB_5/MgNi composite MH [169].

2.15. Other Battery Manufacturers

Besides the 14 main Chinese Ni/MH manufacturers reported above, there are other smaller companies in China also contributing to the Chinese Patents for the Ni/MH battery and their inventions are summarized in this section. Some of them may be already out of business but their Patent Applications are still reviewed here. Among them, TMK (Shenzhen, China) filed 14 Chinese Patent Applications and received half of them granted. TMK introduced a Ni/MH battery pack with an inspection function [170,171] and a self-balance circuitry [172], a charger preventing overcharge [173], a battery with a good heat dissipation capability [174,175], a control circuitry to prevent over-discharge [176], a low-cost cylindrical cell [177], and a prismatic battery for electrical vehicles (EV) [178]. In the component area, TMK disclosed a new negative electrode current collector [179], a new type vent cap [180], a high-power current collector [181], and a Cu and Ni-coated stainless steel substrate for the negative electrode [182]. In the production facilities, TMK also invented a fixed position electrode holding apparatus [183], a centrifugal electrode dryer [184], and a heated electrolyte filling station [185].

Baosheng (Chengdu, Sichuan, China) applied for 11 Chinese Patents which covered the topics of a high-Fe (3–26 at%) AB_5 MH alloy [186], a Ni/MH battery module for an electrical bus [187], a high-temperature Ni/MH battery with W-containing electrolyte and Ti and Y-added positive electrodes [188], and a high-power Ni/MH battery [189] (Figure 14).

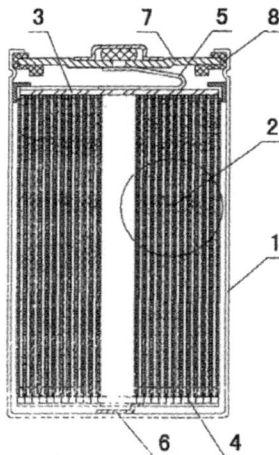

Figure 14. Schematic diagram of a high-power cell with 1: can, 2: electrode assembly, 3: positive electrode, 4: negative electrode, 5: positive current collector, 6: negative current collector, 7: vent cap, and 8: sealing ring [189].

Huanyu (Xinxiang, Henan, China) disclosed an antioxidant made from boric acid [190], a tri-layer separator [191], a Cu-foam [192], a Fe-foam [193], and a Ni-plated expanded-steel [194] (Figure 15) substrates for the negative electrode, and a gas atomization fabrication method of MH powder [195] in their Chinese Patent Applications.

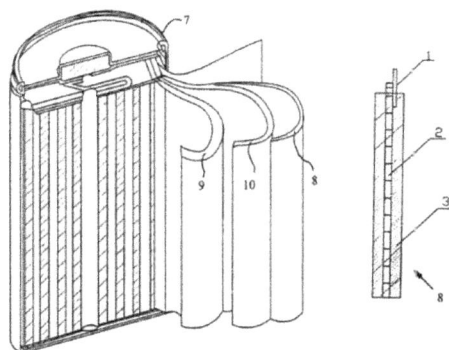

Figure 15. Schematic diagram of a special current collector design with 1: tap, 2: substrate, 3: metal hydride powder, 7: can, 8: negative electrode, 9: positive electrode, and 10: separator [194].

Fengbiao (Shenzhen, Guangdong, China) applied a Chinese Patent mainly in the area of the high-power Ni/MH battery [196–201]. They also proposed a single prismatic cell [202], a multi-prismatic battery pack [203], a low-cost/low-capacity battery [204], a large-capacity battery [205], a wide temperature range battery [206,207] (Figure 16), and a negative electrode recycling apparatus [208] in their additional applications.

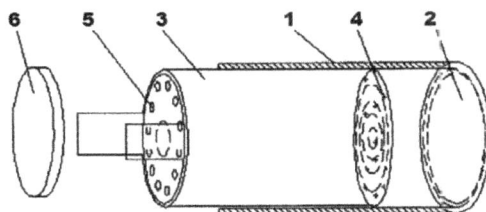

Figure 16. Schematic diagram of a wide temperature range cell with 1: can, 2: Ni-foam mat, 3: electrode assembly, 4: Cu-web with spikes, 5: positive current collector, and 6: vent cap FB [206].

Tianneng (Changxing, Zhejiang, China) invented a bi-polar prismatic Ni/MH battery [209] (Figure 17), a high-power prismatic module [210], a positive electrode fabrication method placing a PTFE layer between the active material and the foam substrate [211], a Y_2O_3-containing low-temperature battery [212], and a Na_3WO_4 containing high-temperature battery [213]. Wangkaifeng (Mianning, Sichuan, China) also owns Chinese Patents about prismatic Ni/MH batteries [214–216] (Figure 18).

Rare-earth ovonic (REO) Ni/MH Power Battery (Baotou, Inner Mongolia, China) filed Chinese Patents on HEV-used battery modules [217] and connectors [218], an EV-used battery module [219], an activation process [220], a short-circuit inspection apparatus [221], a paste mixing apparatus [222], a pasting machine [223,224], an electrode softener apparatus [225], an electrode pressing unit [226], an electrode delivery mechanism [227], a positive current collector ultrasound spot-welder [228], a new type of a sealing ring design [229,230], and an interconnecting mechanism in a battery module [231].

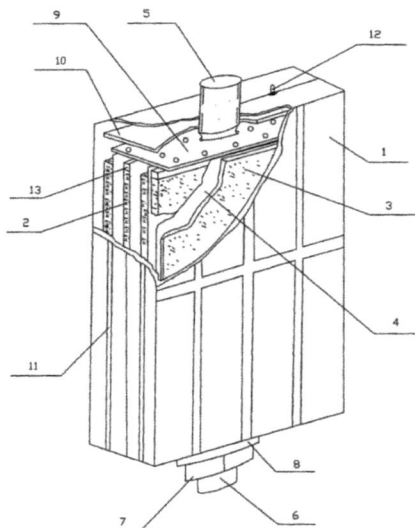

Figure 17. Schematic diagram of a bipolar/prismatic cell with 1: resin case, 2: positive electrode, 3: negative electrode, 4: separator, 5: positive terminal, 6: negative terminal, 7: screw, 8: washer, 9: positive current collector, 10: sealing plate, 11: liquid conduit slot, 12: safety venting valve, and 13: bare substrate [209].

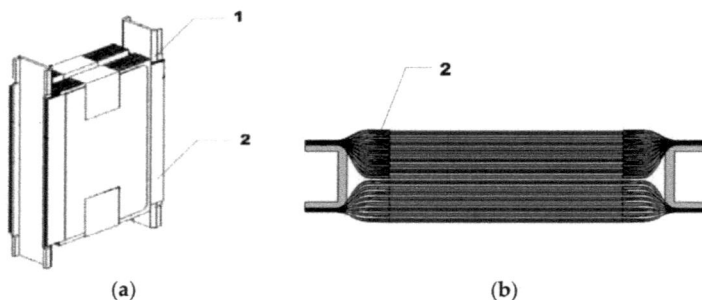

(a)   (b)

Figure 18. (a) A schematic diagram and (b) a cross-section view of a prismatic cell with 1: current collector and 2: tab [215].

BST Power (Shenzhen, Guangzhou, China) filed 16 Chinese Patent Applications and had six granted, which cover the areas of: a high-power battery [232], a wide-temperature range battery [233], a high-temperature overcharge-resistant battery [234], a high-temperature and long life battery [235], a low-capacity battery [236], a flat supporting mechanism for transferring a battery during assembly [237], a method to raise the venting pressure of the safety valve [238], an anti-electrolyte spitting design of the cell [239], a gel-type electrode [240], a new tap design to increase the current collection capability [241], a fast electrolyte-filling station [242], and a powder removal apparatus in the positive electrode softening machine [243].

Zhengda (Zibo, Shandong, China) disclosed a high-power Ni/MH battery with a carbon nanotube coated negative electrode and a Co/Y/Ce hydroxide coated positive electrode [244] (Figure 19), a high-capacity cell design [245] (Figure 20), a constant power charger [246], a prismatic cell design [247–249], and a battery sealing mechanism [250,251] in their Chinese Patent applications.

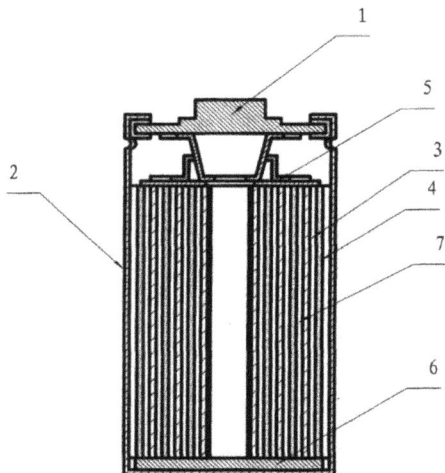

Figure 19. Schematic diagram of a high-power cell with 1: vent cap, 2: can, 3: positive electrode coated with co-precipitated hydroxide of Co, Y, and Ce, 4: negative electrode coated with carbon nanotube, 5: negative current collector, 6: positive current collector, and 7: separator [244].

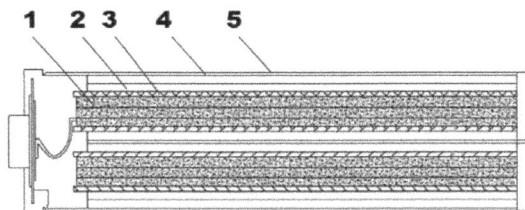

Figure 20. Schematic diagram of a high-capacity cell with 1: positive electrode (three layers), 2: negative electrode (three layers), 3: separator, 4: PTFE anti-corrosion film, and 5: can [245].

Suzhou CENS (Suzhou, Jiangsu, China) filed Chinese Patent Applications on the following subjects: a low-capacity/high-power Ni/MH battery [252], a high-power Ni/MH battery [253], a battery pack connection [254] with a temperature modulation unit [255], a water cooling mechanism [256], and a safe charging device [257].

Jintion (Quanzhou, Fujian, China) owns Chinese Patents in the areas of a high-capacity/high-power [258], high-power [259,260] (Figure 21), low-capacity [261], γ-CoOOH coated high-temperature [262], and β-CoOOH coated long storage [263] Ni/MH batteries, a Y_2O_3 added positive electrode [264], and a new welding design between tap and Ni-foam [265]. It also applied Chinese Patents in these areas as well: a high-power positive electrode with nano-size conductive additives [266], a composite separator for low self-discharge and high-rate applications [267], an electroplating S-containing coating on the negative electrode [268], a high-temperature battery with W-containing additives in the positive electrode [269], a nylon/grid/polypropylene (PP) tri-layer separator [270], and a fast-charge/high-power battery [271].

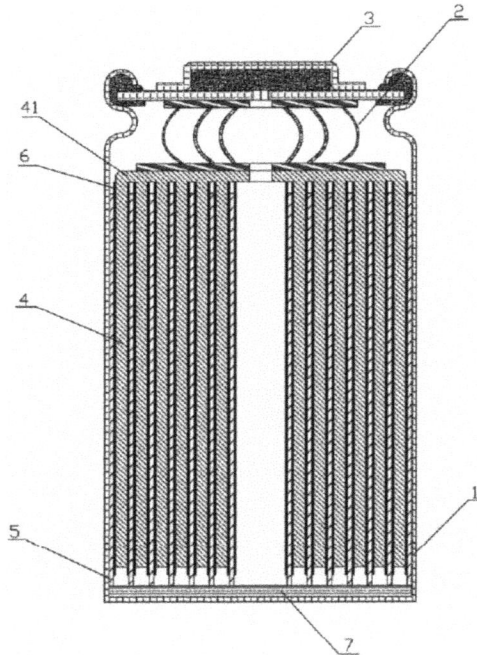

Figure 21. Schematic diagram of a high-capacity/high-power cell with 1: can, 2: current collector, 3: vent cap, 4: positive electrode, 5: negative electrode, 6: separator, 7: Ni-foam, and 41: Ni-foam [260].

Jinhui (Dongguan, Guangdong, China) owns Chinese Patents in the areas of new type button cells [272,273], a low-impedance button cell [274], and a low-cost button cell [275].

Troily New Energy (Xinxiang, Henan, China) has one Chinese Patent granted on a coated positive electrode [276] and one still in examination on a $NaBO_2 \cdot H_2O$ containing electrode for high-temperature application [277].

Innovation New Energy (Liuan, Anhui, China) filed 28 Chinese Patent Applications and had 11 of them granted with the following scope: plastic wrappings for regular [278], low self-discharge [279], high-capacity [280], high-power [281], and high-temperature [282] batteries, sealing glues for regular [283], high-capacity [284], high-temperature [285], low self-discharge [286], and high-power [287] batteries, a positive electrode for a wide temperature-range application [288], a high-temperature/high-power battery [289], a low self-discharge battery [290], a high-capacity non-winding type battery [291] (Figure 22), a high-voltage protection apparatus [292], an energy-efficient battery [293], a high heat-dissipation battery [294], an over-charge protected low-power dissipation battery [295], and an environmentally friendly high-capacity battery [296].

Cel (Huaian, Jiangsu, China) owns two Chinese Patents on a gel coating apparatus for electrode fabrication [297] and a powder scraping apparatus for a dry pasted positive electrode [298].

Bofuneng (Shenzhen, Guangdong, China) owns two Chinese Patents on a prismatic design [299] and an environmental protection cell design [300].

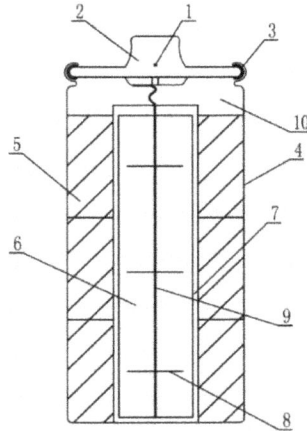

Figure 22. Schematic diagram of a high capacity/non-winding cell with 1: vent hole, 2: cap, 3: sealing ring, 4: can, 5: negative electrode, 6: positive electrode, 7: separator, 8: current collector, 9: tap, and 10: electrolyte [291].

Ryder (Shenzhen, Guangdong, China) owns six Chinese Patents in the subjects of a high-voltage connector design [301], a high-power Ni/MH battery [302,303] (Figure 23), a non-winding battery design [304], a large capacity battery with attached charging unit [305], and a high-voltage module with a high heat dissipation capability [306].

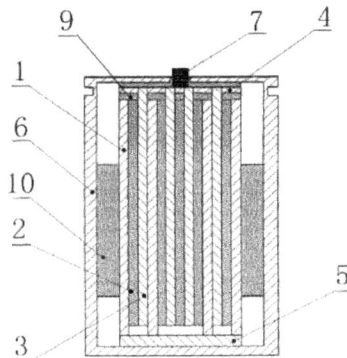

Figure 23. Schematic diagram of a high-power cell with 1: negative electrode, 2: separator, 3: positive electrode, 4: positive current collection, 5: negative current collector, 6: can, 7: cap, 9: insulator, and 10: insulator [302].

DFEI (Dali, Shanxi, China) owns five Chinese Patents in an extra paste removal apparatus in the negative electrode fabrication [307], a dust-removal tape on the positive electrode [308], a high-power design [309] (Figure 24), an automatic transport, vacuum, and de-burr apparatus [310], and a paste stirring apparatus for preparing a positive electrode [311].

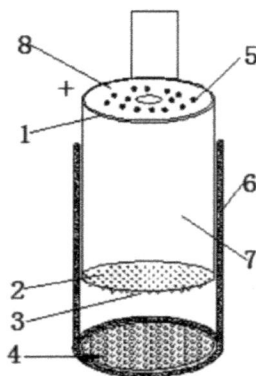

Figure 24. Schematic diagram of a high-power cell with 1: positive current collector plate, 2: Cu mesh, 3: burr, 4: Ni foam, 5: welding spot, 6: insulator, 7: cell, and 8: positive terminal [309].

Wintonic (Guangzhou, Guangdong, China) owns five Chinese Patents on a cylindrical design [312], a nano-carbon containing negative electrode [313], a dual-shell high-temperature can design [314], a high-current positive electrode [315], and a sub-C positive electrode [316].

Oceansun (Shenzhen, Guangdong, China) disclosed a paste positive electrode fabrication method [317], a low-capacity Ni/MH battery [318], and a high-power design [319] in their failed Chinese Patent Applications.

Shida (Foshan, Guangzhou, China) disclosed the following technical ideas in their Chinese Patent Applications: a multi-tab high-power Ni/MH battery [320], a side-folded expanded metal for the electrode substrate [321], and a dry-compaction method for electrode fabrication [322]. Sanik (Foshan, Guangzhou, China) is another Ni/MH battery manufacture in the same city. However, they only filed four patents on the Ni-Fe rechargeable battery and the associated Ni-electrodes.

PeaceBay (Tianjin, China), a company with technology transfer from Toshiba (Tokyo, Japan), filed some early Chinese Patents (all obsolete now) in the areas of fabrications of positive electrodes [323,324], methods of making [325] and winding [326] Ni-foam as positive electrode substrate, a fabrication method for negative electrode [327], a MH/carbon nanotube composite [328] and a MgNi-based alloy [329] as negative active materials, a high-capacity cell design [330] (Figure 25), a reclaiming method of deteriorated MH power [331], and a recycling method for both electrodes [332].

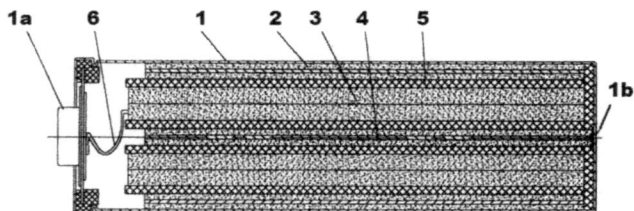

Figure 25. Schematic diagram of a high-capacity cell with 1: can, 1a: positive terminal, 1b: negative terminal, 2: negative electrode, 3: positive electrode, 4: central rod, 5: separator, and 6: current collector [330].

Unitech (Shenzhen, Guangdong, China) filed eight Chinese Patent Applications to cover a protective layer on positive electrode [333], a slot-containing positive electrode [334], a PTFE-coated negative electrode [335], a high-capacity cell design [336], a low-capacity cell design [337], a new cell winding structure [338], and a design of a pasting machine [339].

Taiyi (Zhuhai, Guangzhou, China) has a Chinese Patent proposing a powder container with at least one step inside for the preparing of dry compacted electrode on an expanded metal or a porous metal substrate [340] which was used in some litigation procedures in China. It also applied another application in formation apparatus but is not valid now [341].

3. Metal Hydride Alloy Producer

Chinese MH alloy producers made more than 90% of the MH alloy for Ni/MH battery application (7300 tons in 2016). The top four largest manufactures, Xiamen Tungsten (Xiamen, Fujian, China), Doublewin (Sihui, Guangdong, China), REO Metal Hydride (Baotou, Inner Mongolia, China), and Jiangxi Tungsten (Nanchang, Jiangxi, China) supplied more than 70% of the domestic market in 2016 (Figure 26). Their main product is the rare-earth based AB_5 MH alloy made either by conventional induction melting and casting (Doublewin and REO) or by strip casting (Xiamen Tungsten and Jiangxi Tungsten) (Figure 27). All of them are now working on the high-capacity superlattice MH alloys.

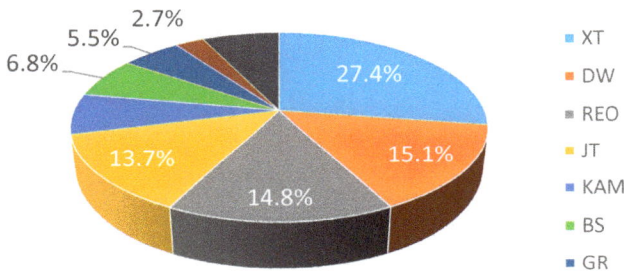

Figure 26. Market shares of MH alloys produced in China in 2016. XT: Xiamen Tungsten, DW: Doublewin, REO: Rare-earth Ovonic, JT: Jiangxi Tungsten, KAM: Kingpowers Advanced Materials, BS: Baotou Santoku, GR: Gansu Rare-earth, and WG: Weishan Gangyan Rare-earth Material.

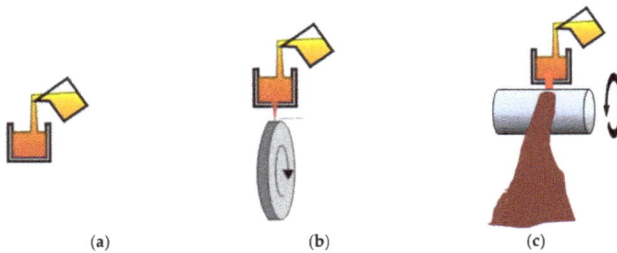

Figure 27. Schematics for (**a**) a conventional melt-and-cast, (**b**) a melt-spin, and (**c**) a strip casting MH alloy fabrication methods.

3.1. Xiamen Tungsten

Xiamen Tungsten, also called Amoi Wuye, has 14 Chinese Patent Applications and had 10 of them granted. In the alloy composition, it disclosed Cu-containing low-cost [342,343], low-Co [344], Co-free and low-Ni [345], Pr-Nd free low self-discharge [346], Mg-containing high-capacity ($\geq$355 mAh g^{-1}) [347], hypo-stoichiometric [348] and Co-free hyper-stoichiometric AB_5 MH alloys [349], and La-Gd [350] and La-Y [351] based A_2B_7 MH alloys. In the alloy production area, it proposed a strip casting of 0.1–0.3 mm thin flake [352–354] (Figure 28) and an electrode fabrication method using Mg-based MH alloy [355].

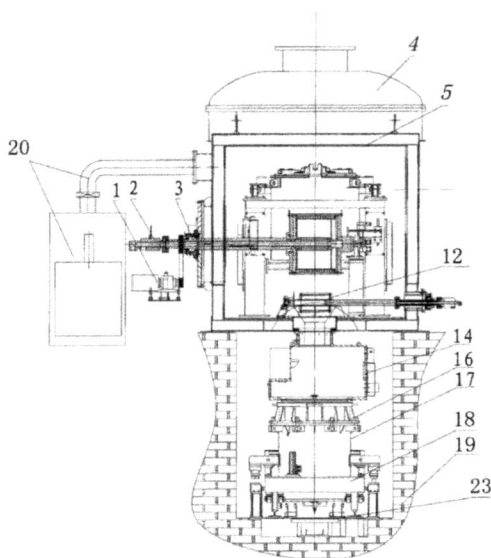

Figure 28. A schematic diagram for a mass-production strip-casting furnace with 1: strip casting motor, 2: rotation coupler, 3: strip casting axis, 4: top cover, 5: furnace, 12: rotating cooling rod, 14: collection box, 16: inside container, 17: outside container, 18: cart, 19: floor foundation, 20: vacuum pump and conduit, and 23: a railing system [354].

3.2. Doublewin

Doublewin filed eight Chinese Patent Applications and all were approved. Their inventions cover mostly the MH alloy production processes, such as a dust collection unit in the powder mixer [356] and in the ingot collection unit [357], an impurity removal unit in the sieving device [358], a cooling mold design [359], a quick electrode fabrication method for alloy quality control [360], and an automatic vacuum casting apparatus [361]. Doublewin also patented MH alloy formula in a (Ag, Sr, Ge)-modified A_2B_7 with improved capacity and low-temperature performance [362] and a (Ag, Sr, Ge, Au) modified AB_5/A_2B_7 mixture [363].

3.3. REO Metal Hydride

REO Metal Hydride filed eight Chinese Patents on MH materials and related fabrication methods which cover a La-rich, Sm-containing superlattice MH alloy [364,365], La-rich, Y, Zr, or Gd-containing over-stoichiometric AB_5 MH alloys [366], an electropolymerized polyaniline coating to protect the alloy surface [367], and a sintered AB_5/AB_3 composite [368].

3.4. Jiangxi Tungsten

Jiangxi Tungsten Haoyun Technology has one issued Chinese Patent on an acid pre-etching process for preparing the negative electrode [369].

3.5. Kingpowers

Kingpowers (Anshan, Liaonining, China) used to be sister company of Suppo and has eight issued Chinese Patents on the alloy formula in a low-Co and (Pr, Nd) free AB_5 MH alloy [370,371] and a Mg-containing A_2B_7 MH alloy [372–374], the alloy production process in annealing [375], and recycling processes from alloy slag [376] and spent batteries [377].

3.6. Others

Guansu Rare-earth (Baiyin, Gansu, China) only has one Chinese Patent Application on a powder grinding, sifting, and mixing process and was rejected [378].

Weishan Gangyan (Jining, Shandong, China) filed 12 Chinese Patent Applications covering areas of a fabrication method of a Ti-containing [379–381] A_2B_7 MH alloy, a nano-graphite/RE_2Mg_{41} type MH composite [382], and a powder packing system [383]. None of them are valid now.

Beijing HarmoFinerY Technology (Beijing, China) owns Chinese Patents of (La, Ce, Dy)-based [384] and (La, Nd, Dy)-based [385] A_2B_7 MH alloys. It also filed a patent about a (La, Y)-based low-temperature A_2B_7 MH alloy.

Vapex (Zhuhai, Guangdong, China) filed 16 Chinese Patent Applications and had 13 granted. In the alloy formula area, Vapex patented a Cu-containing low-Co [386], a Dy-containing [387], a Pt-containing [388], a Fe-containing [389], a Nb-containing [390], a Ga-containing [391], a Sn-Cu containing [386] AB_5 MH alloys, and a Mg-containing A_5B_{19} MH alloy [392]. In the alloy fabrication area, Vapex proposed an induction melting of A_2B_7 alloy in a controlled Mg-vapor [393]. Besides, it also owns patents in the new design of a current collector [394–397], a high-temperature battery design [32], and a Ni/MH battery for solar cell application [398].

4. Ni(OH)$_2$ Producer

The original spherical shape Ni(OH)$_2$ used as the positive electrode active material in the Ni/MH battery was first used in Yuasa (Takatsuki, Osaka, Japan) which was supplied by Tanaka Chemical Co. (Fukui, Japan) [13]. Nowadays in China, almost all spherical shape Ni(OH)$_2$ is produced by the domestic suppliers using a continuous stirring single reactor co-precipitation process invented by Ovonic Battery Co. (Troy, MI, USA) [399]. The top five manufactures, Jinchuan (Lanzhou, Gansu, China), Kingray (Changsha, Hunan, China), Cologne (Xinxiang, Henan, China), Jien (Jilin, Jilin, China), and Aland (Wuwei, Anhui, China) held more than 65% of the market share in 2013 (5.1×10^7 kg , Figure 29). Their contributions to the Ni/MH intellectual properties are summarized in this section.

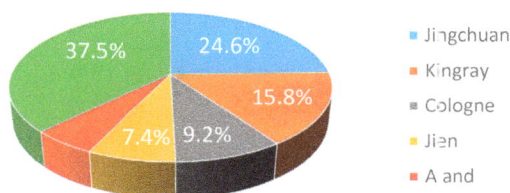

Figure 29. Market shares of Chinese domestic spherical Ni(OH)$_2$ producers in 2013 [13].

4.1. Jinchuan

Jingchuan applied four Chinese Patent Applications on Ni(OH)$_2$ and received one of them. The applications mainly covered the chemical co-precipitation process of making spherical particles [400], a sulfate removal process [401], and a Fe removal process [402].

4.2. Kingray

Kingray New Materials Science and Technology, now Minmetals Capital Co., did not file any Chinese Patent Application on Ni(OH)$_2$.

4.3. Cologne

Cologne filed seven Chinese Patent Applications about Ni(OH)$_2$ and had three of them granted, which cover the areas of a chemical co-precipitation method of making spherical Ni(OH)$_2$ powder [403], a Ni(OH)$_2$ specifically design for high-temperature battery [404], a Co-coated Ni(OH)$_2$ [405], an eletrolytically coated CoOOH on Ni(OH)$_2$ [406,407], and a chemically precipitated CoOOH on Ni(OH)$_2$ [408].

4.4. Jien

Jien owns one Chinese Patent on a mechanical alloying method to coat Co on Ni(OH)$_2$ [409] and one application is still in the examination stage on a Co-coated Ni(OH)$_2$ for high-temperature applications [410].

4.5. Aland

Aland filed 16 Chinese Patent Applications related to Ni(OH)$_2$ and had six of them granted. The scope of these applications covers a gaseous phase [411,412], a liquid phase [413], and a continuous [414] oxidation to prepare a Co-coated Ni(OH)$_2$, a chemical co-precipitation of (Ni, Zn, Mn, Al)(OH)$_2$ [415], a Y, Yb, and Ca containing Ni(OH)$_2$ for high-temperature applications [416], a new method of making Ni(OH)$_2$ with cyclic amine-containing ionic liquid [417], a reactor for making high-power Ni(OH)$_2$ [418], an oxidizer for high-power Co-coated Ni(OH)$_2$ [419], a reactor for making Co(OH)$_2$ coated Ni(OH)$_2$ [420,421], a microwave dryer [422], an electrochemical/chemical pre-charge station for Ni(OH)$_2$, [423], a static impedance measuring apparatus for Co-coated Ni(OH)$_2$ [424], and a scrap recycling process to treat Co-coated Ni(OH)$_2$ [425].

4.6. Others

Zhonghong (Shaoguan, Guangdong, China) owns one Chinese Patent on the chemical co-precipitation of spherical Ni(OH)$_2$ [426] and also has one in the examination stage for the method of making Co-coated Ni(OH)$_2$ [427]. Liyuan (Changsha, Hunan, China), a subsidiary of Corun, owns one Chinese Patent on a chemical co-precipitation method for making spherical Ni(OH)$_2$ powder [428]. Jintian (Xiangtan, Hunan, China) owns one Chinese Patent on the preparation process of the γ-CoOOH coated Ni(OH)$_2$ [429]. Changyu (Jiangmen, Guangdong, China) owns Chinese Patents in the area of a high-capacity and high discharge voltage Ni(OH)$_2$ [430], a high-temperature Ni(OH)$_2$ [431], a Co-coated Ni(OH)$_2$ drying machine [432], and a spherical Ni(OH)$_2$ production method [433,434]. It also filed Chinese Patents on the subjects of a reactor producing Ni(OH)$_2$ [435], a Co-coating reactor for Ni(OH)$_2$ [436], and a γ-CoOOH coating method [437]. Zhongdao Energy Development (Anding, Hainan, China) owns a Chinese Patent on a (Co, Al) hydrotalcite containing positive electrolyte [438]. Zhongjin Metal Powder (Wuxi, Jiangsu, China) has two Chinese Patents on a nano-structured CoOOH [439] and a β-Co(OH)$_2$ [440]. Yixing Xinxing Zirconium (Yixing, Jiangsu, China) owns a patent on a high-temperature battery with CaF$_2$, ZnO, Ba(OH)$_2$, Er$_2$O$_3$, Y$_2$O$_3$, and Zr-compound additives for the positive electrode [441], It also filed Chinese Patents in the areas of a Zr co-precipitated Ni(OH)$_2$ [442] and a rinsing apparatus for spherical Ni(OH)$_2$ [443]. GEM (Shenzhen, Guangdong, China) has one Chinese Patent on a non-stoichiometric Co$_{1.02\sim2}$O additive to increase the electrochemical activity of Ni(OH)$_2$ [444].

5. Other Hardware Suppliers

Besides the active materials from both electrodes (Ni(OH)$_2$ and MH alloy), there is other hardware used in Ni/MH batteries may also protected by Chinese Patents. Instead of grouping by companies, we listed the inventions from various companies under each part category in this section.

5.1. Ni Foam

The Ni-foam used as the substrate mainly for the positive electrode was first patented by INCO (Toronto, ON, Canada, now Vale Canada Limited) with a chemical vapor deposition (CVD) process from nickel carbonyl onto polymer foam which was removed later on by a sintering process [445].

This costly CVD deposition process was replaced by an electrode plating on the polymer form coated with a carbon-containing substance in China. Dr. Zhong, Chairman of Corun, owns a few Chinese Patents on the electroplating container for the continuous fabrication of Ni-foam [446] and a rough surface of Ni-foam [447] which were used in litigation against INCO in China [448]. His company, Liyuan (Changsha, Hunan, China) filed Chinese Patents in the areas of a high-strength Ni-foam [449, 450], a special structured Ni-foam [451], a carbon nanotube coated Ni-foam [452], a Ni-foam design for vehicle use [453], a multi-stage deposited Ni-foam for HEV use [454], a Ni-foam for high-power application [455], a Ni/Co-foam [456], a Fe-foam [457], a crystallization method for deposited Ni [458], a continuous electroplating apparatus [459], and a recycling procedure for making fine Ni powder from Ni-form waste [460].

Tianyu (Heze, Shandong, China) filed Chinese Patent Applications on the subjects of a Ni-alloy foam for high-power application [461,462], an electroplating Ni-foam [463], a super thick Ni- (or Cu-foam) [464], a purification process for Ni-foam [465], a high-uniformity Ni-foam production method [466,467], a low-face density and high-porosity Ni-foam [468], a high efficiency vacuum apparatus [469], and a raw material preparation method for improvement of Ni-foam uniformity [470].

Tiangao (Dalian, Liaoning, China) owns one Chinese Patent on a new type of electroplating basin for Ni-foam [471] and has one in examination about a super high-density Ni-foam [472].

5.2. Separator

The regular separator used in the Ni/MH battery is treated by acrylic acid (white) and a special sulfonated separator (brown) originally patented by Japan Vilene Co. (Tokyo, Japan) [13] has the advantage of low self-discharge since no trace of nitrogen-containing ion is left. Kangjie (Changzhou, Jiangsu, China) filed Chinese Patent Applications on a low-cost and high-yield sulfonation process [473], a multi-layer nylon-based separator [474], and a grafted polyethylene (PE) separator [475]. Lianyou Jinhao (Laizhou, Shandong, China) also has sulfonated separator patents [476–478]. Besides it also filed Chinese Patent Applications on a separator grafting apparatus [479], a separator pore size measurement apparatus [480], and a separator winding machine [481]. Lingqiao Environment Protection Equipment Works (Shanghai, China) has a Chinese Patent on a PP/PTFE/PP composite separator for a Ni/MH battery [482]. Jianxiang Shunxin Technology (Beijing, China) owns a Chinese Patent of a wet impedance measuring apparatus [483]. Meite Environmental Protection Material (Xianyang, Xi'an, China) filed one patent about a low-cost grafting method for making a separator [484]. Kegao (Luoyang, Henan, China) filed one Chinese Patent Application on a high-power separator with high gas permeability and small thermal contraction [485]. Rongsheng (Yizheng, Jiangsu, China) filed one application for its separator manufacturing process [486].

5.3. Stainless Steel Can

Shanghai Jinyang (Shanghai, China) owns Chinese Patents on an incoming parts organizer [487], an electroplating process [488], a dehydration station [489], a water-saving washing station [490], a plating quality control [491] and a thermal energy recycling mechanisms [492] for plating Ni on a stainless steel can. It also applied for Chinese Patents on the automatic plating apparatus [493]. Fujian Jinyang (Sanmin, Fujian, China) field Chinese Patents about double-plating [494] and triple-plating [495] Ni on a stainless steel can. Donggang (Wuxi, Jiangsu, China) filed a Chinese Patent on a can with gradual cylinder wall thickness [496]. Hugang Electronics (Zhangjiagang, Jiangsu, China) claimed an electroplating/chemical plating combination method [497]. Haiyang (Nantong, Jiangsu, China) filed two Chinese Patent Applications on a case with an easy heat dissipation capability [498,499].

5.4. Negative Electrode Substrate

Zhongjin Gaoneng (Shenzhen, Guangdong, China) has three Chinese Patents on the punching die for the fabrication of a perforated non-ferrous metal belt [500], one for the high-power perforated stainless steel substrate [501] and one for the regular NPPS belt [502] for the negative electrode substrate of a Ni/MH battery. Besides, it filed a patent on the fabrication method of NPPS [503]. Jiejing (Dongguan, Guangdong, China) also filed an application for an automatic electroplating apparatus for NPPS [504]. Shuzhen (Shanghai, China) filed an application on the Cu-web used as substrate material for a negative electrode [253]. Kingdom Sifang Metal Product (Qinhuangdao, Hebei, China) owns a Chinese Patent to fabricate NPPS [505]. Shenjian Metallurgical Equipment (Shanghai, China) filed an application on a non-metal fiber/Ni composite material for high-strength/light-weight substrate [506].

6. Applications

Ni/MH batteries were first used in consumer-type applications, such as cell phone, notebook computer, other portable electronic devices (walkman, personal data assistant, etc.), and dry cell replacement. Facing competition from the lighter Li-ion battery, the Ni/MH battery moved to the transportation propulsion type of applications, especially for hybrid electrical vehicles (HEV) and EV. Recently, due to the longevity and wide-temperature tolerance of the Ni/MH battery, it also entered stationary type of applications, for example, uninterrupted power supply (UPS), grid-pairing storage systems, and the EV charging station. The Chinese Patents on the application of Ni/MH area are reviewed here.

6.1. Consumer Type

A few common uses of the Ni/MH battery in consumer application are shown as examples in Figure 30. There are many Chinese inventions on this use of the Ni/MH battery. Some examples are: a moisture-protection battery case [507], a power supply for illumination in a mine [508,509], a power supply with monitoring and communication functions [510], a high-speed power tool [511,512], a new type of flashlight [513], an LED flashlight [514], a pager [515], a portable power modulator [516], a rechargeable radio frequency identification card [517], a portable vacuum cleaner [518], a kinetic energy rechargeable infrared detector [519], a remote humidifier [520], a wall-climbing vacuum cleaner [521], a robot used in a mine [522–525], a trimmer [526], and a wearable battery capacity extender [527].

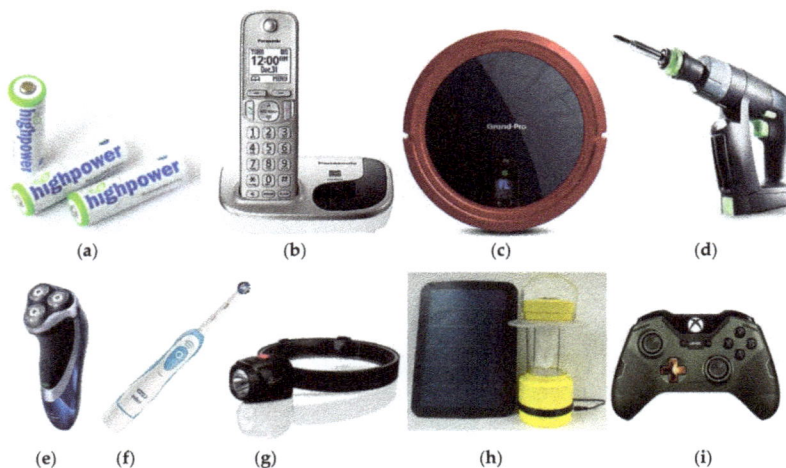

Figure 30. Examples of Ni/MH batteries in consumer applications: (a) retailer, (b) cordless phone, (c) vacuum cleaner, (d) power drill, (e) shaver, (f) toothbrush, (g) flashlight, (h) solar lamp, and (i) remote control for game console.

6.2. Transportation Type

In the transportation area, the Ni/MH battery dominates the HEV market. Three HEVs made in China are shown in Figure 31: a Camry by Guangzhou Auto (Guangzhou, Guangdong, China), a Prius by First Automobile Works (FAW, Changchun, Jilin, China), and an EC7 by Geely Auto (Hangzhou, Zhejiang, China). In the Chinese Patent Applications in this area, an optimized range-extended HEV design [528], a forced-air-cooled HEV battery module [529], a heat dissipation unit [530,531], a temperature-regulated HEV battery system [532], and a HEV-use battery module [533, 534] are available. Besides, Qineng (Yangzhong, Jiangsu, China) filed nine Chinese Patent Applications on the fabrication of an HEV-used Ni/MH battery and module [535–543]. In the EV application, inventions like a high-capacity Ni/MH battery pack [544,545], a high-capacity EV battery [546–548], an EV-used battery module [549–551], an overload protection for an EV battery [552], a pulse-based cycle life testing method for EV application [553], a battery management system for EV with Ni/MH [554], a large-capacity prismatic cell design [555–558] and a cylindrical Ni/MH battery [559] for EV application, an MH alloy designed for EV application [556], a battery module for EV [550,551], a Ni/MH battery pack [560], a special high-power cylindrical cell design [561–564], and an EV charging method [565] were demonstrated. In other transportation areas, a lawn mower [566], a fork lift truck [567], a solar-cell powered car, boat, or plane [568], a wind-powered car [569], an underwater propeller [570], a battery assembly for an E-bicycle [571–575], and a fast-charge moving cart [576] can be found.

(a)   (b)   (c)

Figure 31. Examples of Ni/MH batteries in transportation applications: (**a**) a Camry HEV (Guangzhou Auto), (**b**) a Prius HEV (FAW), and (**c**) an EC7 HEV (Geely Auto).

6.3. Stationary Type

In the un-interrupted power supply (UPS) applications, a video recording system with self-supporting power supply [577], a firefighting emergent light [578], a high-rise building emergency power system [579], a charging circuitry [580], and a mobile communication tower [581] are listed. In the energy storage associated with solar and wind energy, a solar cell/battery combination [582–585], and a Ni/MH rechargeable by solar power [398], a solar charger [586], a light emitting diode (LED) solar lamp [587], a solar rechargeable flashlight [588], and a solar-charged EV battery [589] are included. An example of a solar station equipped with Ni/MH energy storage is shown in Figure 32. In grid-companion energy storage, an energy distribution and EV used charging station [590], a monitoring system for energy storage [591], a double-source power unit [592], a monitoring system [593], a micro-grid Ni/MH system [594], a mega-watt class energy storage system [595], and an energy balancing system for Ni/MH storage [596] were introduced.

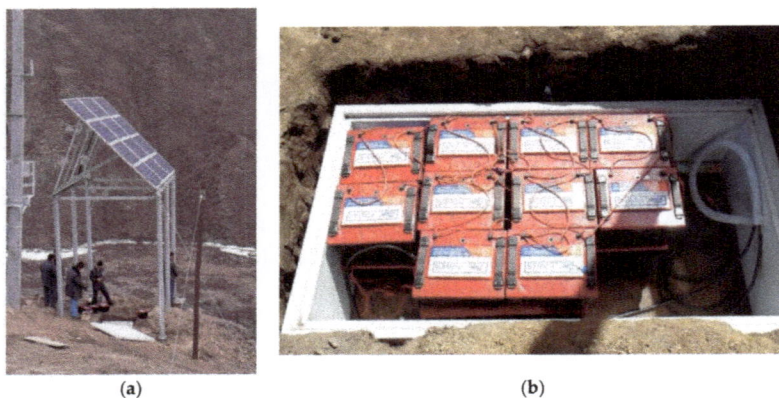

(a)   (b)

Figure 32. Example of Ni/MH batteries in alternative energy storage applications: (a) a solar cell system and (b) the accompanying Ni/MH battery pack installed and tested in Qinghai Plateau. (Courtesy of TMK).

7. Universities and Research Institutes

Many Chinese universities and state-owned research institutes have worked in the Ni/MH battery area, especially those involved in the 863 special project [597]. We reviewed the status of the Patent Applications of a few representative examples and present them here.

7.1. Beijing Institute of Technology

Beijing Institute of Technology (Beijing, China) filed four Chinese Patent Applications related to the Ni/MH battery and had one granted. Their invention covers the areas of a NiO-containing positive electrode [598], an inner-pressure reduction additive [599,600], and a battery non-destructive regeneration method [601].

7.2. Beijing University of Chemical Technology

Beijing University of Chemical Technology (Beijing, China) filed six Chinese Patent Applications about $Ni(OH)_2$ and had two of them granted. The inventions are composed of an activated graphene/needle-shaped $Ni(OH)_2$ composite [602], nano-thin films of α-$Ni(OH)_2$ [603] and β-$Ni(OH)_2$ [604], a nano-structured/doped α-$Ni(OH)_2$ [605], an electrochemical/chemical process of making doped nano-$Ni(OH)_2$ [606], a low-cost fabrication method of nano-$Ni(OH)_2$ using oxygen in air [607], and a nano-C/$Ni(OH)_2$ composite [608].

7.3. Tsinghua University

Tsinghua University (Beijing, China) owns one Chinese Patent on an estimation of state-of-charge based on a standard battery model [609]. It also applied for the following inventions: a real-time estimation of the temperature difference between inside and outside of a cell [610], a magnetic treatment of MH alloy to improve its high-rate performance [611], and an air-cooled apparatus for an HEV-used Ni/MH battery module [612].

7.4. Tianjin University

Tianjin University (Tianjin, China) filed four Chinese Patents Application on these subjects: a negative electrode surface treatment to improve electrochemical performance [613], an apparatus to measure inner pressure of a Ni/MH cell [614], a tri-cellulose acetate/metal oxide composite separator [615], and α-$Ni(OH)_2$ prepared by a ball-milling process [616].

7.5. Nankai University

Nankai University (Tianjin, China) filed patents jointly with Peacebay. It also filed patents on their own in the areas of a C_{60} or boron nitride added negative electrode [617], electrode recycling methods [332], a large-capacity EV-used Ni/MH battery [555], and a negative electrode pasting line [618].

7.6. Tianjin Polytechnic University

Tianjin Polytechnic University (Tianjin, China) has two Chinese Patents on a sulfonated separator [619,620].

7.7. Shanghai Jao Tong University

Shanghai Jao Tong University (Shanghai, China) has one Chinese Patent on a primary (non-rechargeable) Ni/MH battery and withdrew one on a Ni/MH battery with a self-heat-dissipation mechanism [621].

7.8. Donghua University

Donghua University (Shanghai, China) has two Chinese Patents on a composite separator [622] and a nano-coating separator [623].

7.9. Jilin University

Jilin University (Changchun, Jilin, China) filed five Chinese Patent Applications on MH alloy/graphene [624], MH alloy/Co_3O_4 [625], and MH alloy/nanoporous Ni [626] composite negative electrodes, a fast cycle life estimation method for MH alloy with a super long life [627], and a MoS_2 surface treatment for MH alloy [628].

7.10. Yanshan University

Yanshan University (Qinhuangdao, Hebei, China) mainly worked with REO MH Alloy Company in their inventions. In addition, it also filed Chinese Patent Applications on the subjects of a single phase Nd-Mg-Ni-based $PuNi_3$ MH alloy [629] and a surface polypyrrole treatment to improve the electrochemical performance of MH alloy [630] by itself.

7.11. Sichuan University

Sichuan University (Chengdu, Sichuan, China) owns two Chinese Patents on a Nd-free/low-temperature AB_5 MH alloy [631] and a low-Co LaPrCe-Ni based AB_5 MH alloy [632]. It also filed one application on an Nd-free A_2B_7 MH alloy [633].

7.12. University of Electronic Science and Technology of China (UESTC)

UESTC (Chengdu, Sichuan, China) owns three Chinese Patents in the areas of TiO_2 [634], Al_2O_3 ceramic [635], and NiO_2 [636] coatings on a positive electrode.

7.13. Zhejiang University

Most of the Chinese Patents from Zhejiang University are associated with KAN Battery. Zhejiang University (Hangzhou, Zhejiang, China) filed 45 Chinese Patent Applications themselves on an amorphous Mg-Ni [637] and a Ti-Cu-Ni [638] MH alloy as negative electrode active materials and ionic nitridation [639] and B-diffusion [640] methods to improve the anti-corrosion properties of the Mg-based alloy surface.

7.14. Zhejiang Normal University

Zhejiang Normal University (Hangzhou, Zhejiang, China) filed Chinese Patent Applications in the following subjects: a new binder [641], a new electrolyte additive [642], tartaric [643] and mucus [644] salts containing positive electrodes, and a saccharic salt containing a positive electrode [645].

7.15. South East University

South East University (Wuxi, Jiangsu, China) owns two Chinese Patents on a weak acid surface treatment for MH alloy [646] and a high-capacity Mg-Co based MH alloy [647]. It also filed two applications in the areas of an Mg-Pd-Co MH alloy [648] and a recycling method of a spent battery [649].

7.16. Henan Polytechnic University

Henan Polytechnic University (Jiaozuo, Henan, China) filed two Chinese Patent Applications on a fabrication method of RE-Mg-B MH alloy [650] and a multi-phase Mg-based MH alloy [651].

7.17. Guangdong University of Technology

Guangdong University of Technology (Guangzhou, Guangdong, China) owns three Chinese Patents on a graphene-based [652] and a nano-carbon composite [653] high-capacity negative electrodes for high-power battery, and a heat measuring apparatus for a high-power cylindrical cell [654]. It also applied patents in the areas of α-Ni(OH)$_2$ [655], nano-Ni(OH)$_2$ [656], and a multi-element doped nano α-Ni(OH)$_2$ [657,658].

7.18. South China University of Technology

South China University of Technology (Guangzhou, Guangdong, China) owns one Chinese Patent on a Sm-containing A$_2$B$_7$ MH alloy [659]. It also filed applications based on a multilayer thinfilm electrode [660], a thin-film electrode [661], and a Sm-containing Pr, Nd-free MH alloy [662].

7.19. Lanzhou University

Lanzhou University (Lanzhou, Gansu, China) filed one Chinese Patent Application on a three-dimensional cellular Ni(OH)$_2$ [663].

7.20. Institute of Metal Research (IMR) in the Chinese Academy of Sciences

IMR (Shenyang, Liaoning, China) filed four Chinese Patents in the areas of a Ni-foam using vacuum evaporation [664], a thick electrodeposition Ni-foam [665], and a nano-carbon decorated current collector for the negative electrode [666].

7.21. Changchun Institute of Applied Chemistry (CIAC) in the Chinese Academy of Sciences

CIAC (Changchun, Jilin, China) filed five Chinese Patents in the areas of a La-based AB$_3$ MH alloy [667,668], a high-power negative electrode with active carbon or graphite additives [669], a safe design of a button cell [670], and a battery/supercapacitor composite negative electrode [671].

7.22. Shanghai Institute of Microsystem and Information Technology (SIMIT) in the Chinese Academy of Sciences

SIMIT (Shanghai, China) filed Chinese Patents in the areas of a method for improving the long-term storage performance of a Ni/MH battery [672], a high-capacity Ti-V-based MH alloy [673], and a bipolar battery design [674].

7.23. Dalian Institute of Chemistry and Physics (DICP) in the Chinese Academy of Sciences

DICP (Dalian, Liaoning, China) filed a Chinese Patent on a conductive polymer/AB_3 MH alloy composite [675].

7.24. General Research Institute for Nonferrous Metals (GRINM)

GRINM (Beijing, China) filed 11 Chinese Patents to cover a high-capacity, long-life, and high-power A_2B_7/A_5B_{19} based MH alloy [676], a low self-discharge Ce_2Ni_7/Cd_2Co_7 based MH alloy [677], a long life Pr (Nd) free A_2B_7 (>85%) MH alloy [678], a Nd-free low-temperature AB_5 MH alloy [679], a conducting composite for negative electrode [680], a zirconia based separator [681], high-temperature/high-power positive electrode active materials [682,683], a disordered composite positive electrode active material for power application [684], a Ni, Al $(OH)_2$-based positive electrode [685], RE_2O_3-based additives to positive electrodes [686], and a $Ni(OH)_2$ for high-temperature/high-power application [687].

7.25. Guangzhou Research Institute of Non-Ferrous Metals (GZRINM)

GZRINM (Guangzhou, Guangdong, China) filed four Chinese Patents on a low self-discharge Nd-based A_2B_7 MH alloy [688], a La, Sm-based A_2B_7 MH alloy [659], a Co-free AB_5 MH alloy [689], and a low-temperature MH alloy containing a secondary $CeCo_4B$ phase [690].

7.26. Baotou Rare Earth Research Institute (BRERI)

BRERI (Baotou, Inner Mongolia, China) owns a Chinese Patent on a Y-containing A_2B_7 MH alloy [691]. Besides, it also applied for the following inventions: Y containing A_5B_{19} [692] and AB_3 [693] MH alloys, Zr, Ti-modified AB_3 [694], A_2B_7 [695], and A_5B_{19} MH alloys [696], a Y-Ni based MH alloy [697], an amorphous Mg-Ni alloy [698], and a method to improve the capacity stability of Mg-based MH alloy [699].

7.27. National Engineering Research Center for Nanotechnology (NERCN)

NERCN (Shanghai, China) applied for two Chinese Patents on a superlattice MH alloy/BCC alloy composite [700] and an addition of acetylene back to the negative electrode [701].

7.28. Guilin Geology and Mining Academe (GGMA)

GGMA (Guilin, Guangxi, China) has one Chinese Patent on a fabrication method of a paste used in the negative electrode [702].

7.29. Changsha Research Institute of Mining and Metallurgy (CRIMM)

CRIMM (Changsha, Hunan, China) owns a Chinese Patent in the Ce, Co-coated $Ni(OH)_2$ composite material [703].

7.30. China Electronics Technology Group Corporation (CETC) 18th Research Institute

CETC (Tianjin, China) owns one Chinese Patent on a chemical fabrication method of $Ni(OH)_2$ [704] and applied one on the Ni/MH battery with a temperature control apparatus [705].

7.31. Central Iron & Steel Research Institute (CISRI)

CISRI (Beijing, China) applied Chinese Patents in the areas of W-containing high-capacity and V-high-temperature AB_5 MH alloys [706,707], a single-role quench method to prepare MH alloy for high-temperature application [708], a fabrication of a Ni/MH battery [709], an MH alloy pulverization by hydrogenation [710], and an electrolysis method of making misch metal for the raw material of MH alloy [711].

8. Comparisons

So far, we have reviewed Ni/MH battery-related patents (applications) filed in the USA (371) [12], Japan (275) [13], and China (692). The areas of these patents (applications) are categorized into negative electrode (metal hydride alloy), positive electrode (Ni(OH)$_2$), battery structure, manufacturing facilities, applications, and others (other parts, charging algorithm, recycling, etc.) and are plotted in Figure 33. While the US Patents are distributed evenly in each category, Japanese and Chinese Patents are heavy on the negative electrode and battery structure, respectively (Figure 33). As for the assignees, almost all US Patents are from foreign companies and domestic research institutes and those for Japan and China Patents are from several sources: Battery companies, MH alloy suppliers, spherical nickel hydroxide suppliers, vendors of other parts, universities, and research institutes (Figure 34).

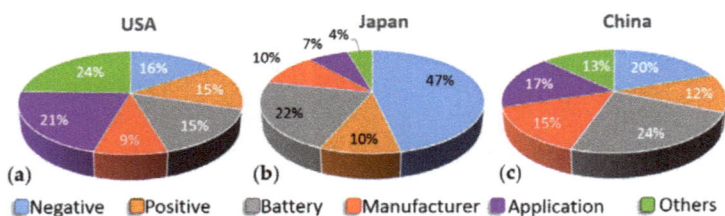

Figure 33. Areas of Patents (Applications) in (**a**) USA, (**b**) Japan, and (**c**) Chinese from negative electrode components (MH alloy), positive electrode component (nickel hydroxide), battery structure, manufacturing facilities, applications, and other areas.

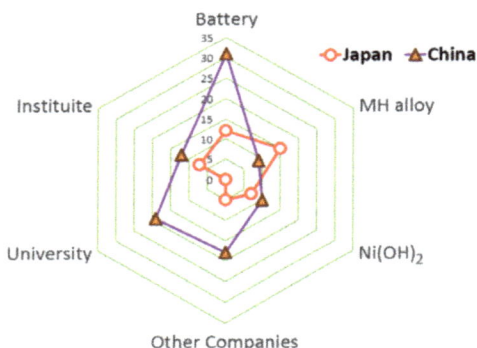

Figure 34. Comparison of numbers of company/institute contributing to the Patents (Applications) in Japan and China.

9. Conclusions

692 Chinese Patent Applications in the nickel/metal hydride battery area are summarized in this review articles categorized by their assignees: battery manufacturers, metal hydride alloy manufacturers, nickel hydroxide manufactures, vendors of other marts, universities, and state-owned research institutes. Some representative schematic diagrams of cell structure designs are included to stimulate new ideas in this area.

Acknowledgments: The authors would like to thank the following individuals for contribution to the manuscript: David Pawlick from BASF, Xiangjun Liu from TMK, Chathan Ho from Coran, and Aishen Wu from Shida.

Author Contributions: Xiaojuan Cai collected the information. Kwo-Hsiung Young and Shiuan Chang prepared the manuscript.

Conflicts of Interest: The authors declare no conflict of interest.

Abbreviations

Ni/MH	Nickel/Metal Hydride
NiCd	Nickel-Cadmium
HEV	Hybrid Electrical Vehicle
MH	Metal Hydride
SIPO	State Intellectual Property Office of the People's Republic of China
GP	Gold Peak
SZHP	Shenzhen HighPower
Li-PO	Lithium-Polymer
NPPS	Ni-Plated Perforated Steel
RE	Rare Earth
SBR	Styrene-Butadiene Rub
PTFE	Polytetrafluoroethylene
EV	Electrical Vehicle
REO	Rare-Earth OVONIC
PP	Polypropylene
CVD	Chemical Vapor Deposition
PE	Polyethylene
UPS	Un-Interrupted Power Supply
LED	Light Emitting Diode
UESTC	University of Electronic Science and Technology of China
CIAC	Changchun Institute of Applied Chemistry
SIMIT	Shanghai Institute of Microsystem and Information Technology
DICP	Dalian Institute of Chemistry and Physics
GNINM	General Research Institute for Nonferrous Metals
GZRINM	Guangzhou Research Institute of Non-Ferrous Metals
BRERI	Baotou Rare Earth Research Institute
NERCN	National Engineering Research Center for Nanotechnology
GGMA	Guilin Geology and Mining Academe
CRIMM	Changsha Research Institute of Mining and Metallurgy
CETC	China Electronics Technology Group Corporation
CISRI	Central Iron & Steel Research Institute

References

1. Zelinsky, M.; Koch, J. Batteries and Heat—A Recipe for Success? Available online: www.battcon.com/PapersFinal2013/16-Mike%20Zelinsky%20-%20Batteries%20and%20Heat.pdf (accessed on 5 July 2017).
2. Young, K.; Ng, K.Y.S.; Bendersky, L. A Technical Report of the Robust Affordable Next Generation Energy Storage System-BASF Program. *Batteries* **2016**, *2*, 2. [CrossRef]
3. Trapanese, M.; Franzitta, V.; Viola, A. Description of hysteresis of nickel metal hydride battery. In Proceedings of the 38th Annual Conference on IEEE Industrial Electronics Society (IECON) 2012, Montreal, QC, Canada, 25–28 October 2012; pp. 967–970.
4. Köhler, U.; Kümpers, J.; Ullrich, M. High performance nickel-metal hydride and lithium-ion batteries. *J. Power Sources* **2002**, *105*, 139–144. [CrossRef]
5. NiMH vs. Li-ion: A Battery Comparison. Available online: https://turbofuture.com/misc/Which-is-better-Nickel-Metal-Hydride-NiMH-or-Lithium-Ion-Li-ion-batteries (accessed on 5 July 2017).
6. Trapanese, M.; Franzitta, V.; Viola, A. Description of hysteresis in lithium battery by classical Preisach model. *Adv. Mater. Res.* **2013**, *622*, 1099–1103.
7. Trapanese, M.; Franzitta, V.; Viola, A. The Jiles Atherton model for description on hysteresis in lithium battery. In Proceedings of the 2013 Twenty-Eighth Annual IEEE Applied Power Electronics Conference and Exposition (APEC), Long Beach, CA, USA, 17–21 March 2013; pp. 2772–2775.
8. Teraoka, H. Development of Ni-MH ESS with Lifetime and Performance Estimation Technology. In Proceedings of the 34th International Battery Seminar & Exhibit, Fort Lauderdale, FL, USA, 20–23 March 2017.
9. Teraoka, H. Ni-MH Stationary Energy Storage: Extreme Temperature & Long Life Developments. In Proceedings of the 33rd International Battery Seminar & Exhibit, Fort Lauderdale, FL, USA, 21–24 March 2016.
10. Teraoka, H. Development of Highly Durable and Long Life Ni-MH Batteries for Energy Storage Systems. In Proceedings of the 32th International Battery Seminar & Exhibit, Fort Lauderdale, FL, USA, 9–12 March 2015.
11. Ogawa, K. Toyota to start local production of NiMH battery cells in China by end of 2016. *Nikkei Technology Online*. 9 March 2016. Available online: http://techon.nikkeibp.co.jp/atclen/news_en/15mk/030900433/ (accessed on 2 November 2016).

12. Chang, S.; Young, K.; Nei, J.; Fierro, C. Reviews on the U.S. Patents regarding nickel/metal hydride batteries. *Batteries* **2016**, *2*, 10. [CrossRef]

13. Ouchi, T.; Young, K.; Moghe, D. Reviews on the Japanese Patent Applications regarding nickel/metal hydride batteries. *Batteries* **2016**, *2*, 21. [CrossRef]

14. Patent Search and Analysis of SIPO. Available online: http://english.sipo.gov.cn/ (accessed on 2 November 2016).

15. SooPAT Website. Available online: http://www2.soopat.com/Home/IIndex (accessed on 2 November 2016).

16. Bai, W. The Current Status and Future Trends of Domestic and Foreign NiMH Battery Market. Available online: http://cbe.com/u/cms/201406/06163842rc01.pdf (accessed on 4 April 2017).

17. You, P.; Cai, J.; Yu, H. Rechargeable Battery. Chinese Patent Application No. 201120521690, 13 December 2011.

18. Wu, C.; Ling, P.; Chen, R. Method for Preparing Battery Roll Cores by Automatic Film Winding and Automatic Film Winder. Chinese Patent Application No. 201310582763, 18 November 2013.

19. You, P.; Liu, B.; Su, Y. Battery. Chinese Patent Application No. 201210019480, 20 January 2012.

20. Wu, C.; Chen, R.; Yu, H.; Lei, G. Battery Anode-Cathode Diaphragm, Preparation Method and Applications Thereof. Chinese Patent Application No. 201310482443, 15 October 2013.

21. Wu, C.; Ling, P.; Chen, R.; Lei, G. Novel Battery Negative Electrode as Well as Preparation and Application Thereof. Chinese Patent Application No. 201310482441, 15 October 2013.

22. Li, W.; Kong, L.; Wen, H.; Yang, W.; Li, X.; Xiong, S.; Han, X. Method for Making Battery Positive Plate and Positive Electrode Slurry Thereof. Chinese Patent Application No. 200710305063, 26 December 2007.

23. Li, W.; Kong, L.; Wen, H.; Tian, X.; Han, X. Nickel-Hydrogen Battery and Preparation Method for Anode Material. Chinese Patent Application No. 200810068362, 8 July 2008.

24. Zhou, Z.; Kong, L.; Li, W. Secondary battery anode piece and preparation method thereof. Chinese Patent Application No. 201110048145, 28 February 2011.

25. Zou, J.; Yang, R.; Zhang, L.; Cai, X.; Liao, X.; Li, W.; Jiang, Y.; Yang, J.; Qian, W.; Chen, Y.; et al. Hydrogen Storage Alloy Material for Nickel-Hydrogen Battery and Preparation Method. Chinese Patent Application No. 201210117357, 19 April 2012.

26. Liang, D.; Zou, J.; Zhang, L.; Liao, X.; Cai, X.; Kong, L. Battery Plate and Manufacturing Method Thereof. Chinese Patent Application No. 201110202077, 18 July 2011.

27. Li, W. Long Time Stored Nickel-Hydrogen Battery and Manufacturing Method Thereof. Chinese Patent Application No. 200510034847, 31 May 2005.

28. Li, W.; Guo, Y.; Han, X.; Huang, W.; Liao, X. Refitting Cylindrical Nickel-Hydrogen Battery. Chinese Patent Application No. 200920129531, 16 January 2009.

29. Wen, H.; Li, X.; Kong, L.; Yin, S.; Li, W. Battery Positive Pole Cap. Chinese Patent Application No. 200920136056, 24 March 2009.

30. Yang, W.; Liao, X.; Kong, L.; Li, W. Battery. Chinese Patent Application No. 201010505937, 12 October 2010.

31. Yang, W.; He, R.; Cai, X.; Liao, X.; Yu, L.; Zhou, T.; Li, W. Tab and Battery with Using the Same. Chinese Patent Application No. 201320217146, 25 April 2013.

32. Zou, J.; Liang, D.; Zhang, L.; Cai, X.; Liao, X.; Kong, L. Nickel-Metal Hydride Battery. Chinese Patent Application No. 201110194572, 12 July 2011.

33. Wu, W.; Yang, W.; Cai, X.; Liao, X. Maintainable Long-Service-Life Secondary Battery. Chinese Patent Application No. 201520120349, 28 February 2015.

34. Mo, X.; Yu, L.; Cheng, J.; Li, W. Novel Ultrathin Battery and Preparation Method Thereof. Chinese Patent Application No. 201510201345, 24 April 2015.

35. Li, W.; Guo, Y.; Han, X.; Huang, W.; Liao, X. Refitted Cylindrical Hydrogen Nickel Battery. Chinese Patent Application No. 200920129530, 16 January 2009.

36. Wang, X.; Su, Y.; Zhou, D.; Wu, Q.; Wen, H.; Li, W. Method for Manufacturing Battery Negative Pole Piece. Chinese Patent Application No. 200910107202, 30 April 2009.

37. Li, W.; Ma, W. Battery Centrifugal Liquid Filling Method Using Clamp. Chinese Patent Application No. 200510034697, 25 May 2005.

38. Tang, Z. Capacitance Dividing Cabinet Accuracy Testing Device. Chinese Patent Application No. 201420019634, 13 January 2014.

39. Cao, D.; Yang, K.; Sun, S.; Liao, X. Spot Welding Tool of Piece is Drawn Forth to Battery. Chinese Patent Application No. 201520645622, 25 August 2015.

40. Mo, X.; Yu, L.; Cheng, J.; Li, W. A Battery Test Apparatus. Chinese Patent Application No. 201520463569, 30 June 2015.
41. Zhong, F.; He, C.; Chen, F. Nickel-Metal Hydride Battery and Positive Plate Thereof. Chinese Patent Application No. 201020111306, 9 February 2010.
42. Xiong, X.; Yang, S.; Zhou, J.; Zhong, F. A Making Technology for Nickel Hydrogen High-Temperature Charging Battery. Chinese Patent Application No. 200610136707, 17 November 2006.
43. Zhang, J.; He, C.; He, Y.; Dai, G.; Chen, X.; Jiang, S.; Pi, Y. Spherical Shape Nickel Hydroxide for High-Temperature Nickel Electric Cell. Chinese Patent Application No. 200710035470, 30 July 2007.
44. Zhang, J.; He, C.; He, Y.; Dai, G.; Chen, X.; Jiang, S.; Pi, Y. Spherical Shape Nickel Hydroxide for High-Temperature Nickel Electric Cell. Chinese Patent Application No. 200710035471, 30 July 2007.
45. He, C.; Chen, F.; Yang, Y.; Zhang, Y. Manufacturing Methods of Secondary Battery Negative Electrode Slice, Negative Electrode and Secondary Battery, and Secondary Battery Manufactured by Same. Chinese Patent Application No. 201010585009, 13 December 2010.
46. Zhong, F.; He, C.; Chen, F. Preparation Method of Negative Plate of Chemical Battery. Chinese Patent Application No. 201010107805, 9 February 2010.
47. Gao, J.; Yang, S.; Zhou, J.; Yao, S.; Wang, Y. Special Electrolytic Solution for Alkaline Nickel-Hydrogen Battery. Chinese Patent Application No. 200810143053, 7 October 2008.
48. Yang, S.; Wang, X.; Zhou, J.; Zhong, F. High-Power Charging Battery. Chinese Patent Application No. 200620052560, 20 October 2006.
49. Yan, S.; Wang, X.; Zhou, J.; Zhou, F. High Power Charging Battery Manufacture Process. Chinese Patent Application No. 200610032432, 20 October 2006.
50. Zhou, C.; Luo, Y.; Yang, S.; Zhou, J. Cover-Cap for Cylindrical Battery and Cylindrical Battery. Chinese Patent Application No. 201410485861, 22 September 2014.
51. Zhou, C.; Lu, D.; Zhou, J.; Liu, L. Cap for Circular Battery. Chinese Patent Application No. 201420207528, 25 April 2014.
52. Zhong, F.; Ho, C.; Chen, F. Cylindrical battery. Chinese Patent Application No. 200980000348, 25 September 2009.
53. Chen, F.; He, C.; Zhong, F. Cylindrical battery. Chinese Patent Application No. 200920066023, 25 September 2009.
54. He, Y.; Lin, Y.; Tang, Z.; Zhang, C. Method and System for Classifying Battery Pole Pieces. Chinese Patent Application No. 201010202222, 18 June 2010.
55. Chen, F.; He, C.; Zhong, F. Forming and Capacity Dividing Method. Chinese Patent Application No. 201010107736, 9 February 2010.
56. Xiao, J.; Zhong, F.; Quan, G.; Yang, C.; Liu, B. Automatic Cap Welding All-In-One Machine. Chinese Patent Application No. 201510553196, 1 September 2015.
57. Xiao, J.; Zhong, F.; Quan, G.; Yang, C.; Liu, B. Automatic Integrated Liquid Filling Machine. Chinese Patent Application No. 201510551892, 1 September 2015.
58. Yang, S.; Liu, T.; Li, L.; Zhou, C. Forming Method of Nickel-Metal Hydride Battery. Chinese Patent Application No. 201410490021, 23 September 2014.
59. Pang, L.; Chen, X.; Ye, D. Method for Preparing Cobalt Clad Beta-NiOOH by Liquid Phase Oxidation Method. Chinese Patent Application No. 201010565217, 30 November 2010.
60. Huang, N.; Pang, L.; Fu, L. Conductive Skeleton for Negative Pole of Nickel-Hydrogen Battery. Chinese Patent Application No. 201020536041, 20 September 2010.
61. Guo, J.; Guo, S.; Xia, C. High-Capacity Hydrogen-Bearing Alloy Powder with Long Service Life for High-Performance Secondary MH-Ni Battery. Chinese Patent Application No. 00123177, 31 October 2000.
62. Huang, N.; Pang, L. Cylindrical Nickel-Hydrogen High Charge Retention Battery. Chinese Patent Application No. 200920014000, 27 May 2009.
63. Li, W.; Zhou, W.; Wang, C. Method for Processing Cobalt Coated Nickel Hydroxide. Chinese Patent Application No. 02151988, 19 November 2002.
64. Lu, H. High-Temperature Ni/H$_2$ Battery and Its Manufacture. Chinese Patent Application No. 99116113, 5 April 1999.
65. Zhou, W. Alkaline Secondary Battery Positive Electrode Material and Alkaline Secondary Battery. Chinese Patent Application No. 200510126001, 28 November 2005.

66. Gu, M.; Zhang, X.; Zhou, W. Nickel Anode Accumulator Anode Slurry and Method for Preparing Anode Accumulator Using the Same. Chinese Patent Application No. 200710122892, 9 July 2007.
67. Li, W.; Wang, C. Alkaline Secondary Cell. Chinese Patent Application No. 03140045, 2 August 2003.
68. Zhou, W.; Chen, S.; Yu, F. Method for Preparing Alkaline Secondary Cell Anode Active Matter. Chinese Patent Application No. 200610109529, 4 August 2006.
69. Cao, C. Composite Nickel Powder, Preparation Thereof, Nickel Anode and Alkaline Accumulator. Chinese Patent Application No. 200710198727, 10 December 2007.
70. Zhu, Z. Positive Electrode Material, Positive Electrode and Battery Containing the Material and Preparing Process Thereof. Chinese Patent Application No. 200510087099, 26 July 2005.
71. Zhou, W. Anode Active Substance and Its Preparing Method and Anode and Battery. Chinese Patent Application No. 200610141071, 29 September 2006.
72. Deng, X.; Cheng, X.; Zhang, F.; Gong, Q. Hydrogen Storage Alloy Type Hydrogenation Catalyst and Preparation Method Thereof. Chinese Patent Application No. 200710077544, 3 December 2007.
73. Geng, W. Hydrogen Storage Alloy Powder Surface Processing Method. Chinese Patent Application No. 200610111583, 28 August 2006.
74. Feng, H. Surface Processing Method for Hydrogen Storage Alloy Powder. Chinese Patent Application No. 200810097956, 16 May 2008.
75. Feng, H.; Geng, W. Method for Preparing Hydrogen Storage Alloy. Chinese Patent Application No. 200910106438, 27 March 2009.
76. Yan, H. High Power Nickel-Hydrogen Accumulator Negative Electrode Active Substance and Its Preparation Method and Nickel-Hydrogen Accumulator. Chinese Patent Application No. 200510035315, 13 June 2005.
77. Geng, W. Composite Particle, Nickel-Hydrogen Secondary Cell Negative-Pole and Cell and Method for Making Same. Chinese Patent Application No. 200510126115, 30 November 2005.
78. Geng, W. Active Material for Nickel-Hydrogen Battery Cathode and Preparation Method Thereof. Chinese Patent Application No. 200710147646, 31 August 2007.
79. Liu, T.; Tan, S. Steel Belt, Cell Negative Pole Using Said Steel Belt and Cell and Their Preparing Method. Chinese Patent Application No. 200510068305, 30 April 2005.
80. Liu, T.; Tan, S. Steel Strip, Cell Cathode Using Same and Cell and Their Manufacture Methods. Chinese Patent Application No. 200510117681, 8 November 2005.
81. Feng, H. Cathode Active Substance and Preparation Method Thereof, Cathode and Battery. Chinese Patent Application No. 200810146427, 28 August 2008.
82. Geng, W. Nickel-Hydrogen Secondary Battery Negative Electrode and Battery and Producing Method. Chinese Patent Application No. 200510126002, 28 November 2005.
83. Deng, X.; Cheng, X.; Zhang, F.; Gong, Q. Preparation Method of Hydrogen Storage Material. Chinese Patent Application No. 200810094115, 4 May 2008.
84. Li, W.; Yan, H. Nickel-Hydrogen Battery Negative Pole and Battery Using Same and Preparing Method. Chinese Patent Application No. 200410101265, 17 December 2004.
85. Xie, H.; Huang, Z. Electrode Material of Nickel-Hydrogen Battery and Preparation Method Thereof and Nickel-Hydrogen Battery. Chinese Patent Application No. 201010294722, 28 September 2010.
86. Li, C.; Jiang, L.; Li, Q.; Gong, Q. Battery Isolating Film and Method for Producing the Same. Chinese Patent Application No. 200610170396, 29 December 2006.
87. Xie, H.; Huang, Z. Nickel-Hydrogen Battery Diaphragm, and Preparation Method and Nickel-Hydrogen Battery Thereof. Chinese Patent Application No. 201010219090, 30 June 2010.
88. Xie, H. Electrolyte of Nickel Hydrogen Battery and Nickel Hydrogen Battery. Chinese Patent Application No. 201010260261, 19 August 2010.
89. Xie, H. Nickel Hydrogen Secondary Battery. Chinese Patent Application No. 200910188446, 27 November 2009.
90. Zhang, J.; Wang, H.; Wang, C. Cylindrical Alkaline Secondary Battery and its Seal Ring Module. Chinese Patent Application No. 03238936, 10 March 2003.
91. Lun, Y.; He, Y.; Deng, S. Secondary Battery. Chinese Patent Application No. 200820213846, 21 November 2008.
92. Lun, Y. Secondary Battery. Chinese Patent Application No. 200920129200, 12 January 2009.
93. Zhu, Z.; Sun, L. Battery Assembling Cap. Chinese Patent Application No. 200620017913, 23 August 2006.
94. Lun, Y. Electrode Terminal, Battery Cover Cap Containing the Electrode Terminal and Battery Thereof. Chinese Patent Application No. 200820094941, 21 June 2008.

95. He, Z.; Liu, W. Battery Metal Flow Collection Band. Chinese Patent Application No. 200420102334, 3 December 2004.

96. Huang, Z.; Chen, Q. Hydrogen Storage Alloy and Hydrogen Storage Alloy Cathode Including the Same. Chinese Patent Application No. 200610127740, 1 September 2006.

97. Zhang, F. Hydrogen Storage Alloy and Its Preparation Method. Chinese Patent Application No. 200610167384, 31 December 2006.

98. Pan, Y.; Geng, W. Hydrogen Storage Alloy, Preparation Thereof, and Cathode and Battery Using the Hydrogen Storage Alloy. Chinese Patent Application No. 200710147642, 31 August 2007.

99. Yang, Y. Hydrogen Storage Alloy, Preparation Method Thereof and Negative Pole and Battery Containing Hydrogen Storage Alloy. Chinese Patent Application No. 200810126837, 26 June 2008.

100. Wang, B.; Gao, M.; Liu, Y.; Chen, J.; Feng, R. High-Power Nickel-Metal Hydride Battery and Manufacturing Method Thereof. Chinese Patent Application No. 201210405985, 23 October 2012.

101. Jin, Z.; Wang, Z.; He, R.; Wu, Y.; Wu, D.; Liu, W.; Liu, J.; Fan, C. Nickel-Hydrogen Battery Positive Plate. Chinese Patent Application No. 201520589228, 7 August 2015.

102. Wang, B.; Pan, H.; Gao, M. Secondary Battery with High-Rate Discharge Characteristic. Chinese Patent Application No. 201220543651, 23 October 2012.

103. Wang, B.; Pan, H.; Gao, M. High-Power Nickel-Metal Hydride Battery. Chinese Patent Application No. 201220545556, 23 October 2012.

104. Wang, B.; Pan, H.; Guo, M. Secondary Battery Capable of Improving High-Rate Discharge Property of Secondary Battery. Chinese Patent Application No. 201220543416, 23 October 2012.

105. Wang, B.; Gao, M.; Liu, Y.; Chen, J.; Feng, R. Secondary Battery with Improved High-Ratio Discharge Performance and Production Method of Secondary Battery. Chinese Patent Application No. 201210406152, 23 October 2012.

106. Pan, Y.; Che, J.; Zhu, W.; Pan, W. Electromagnetic Clutch Driving Winder. Chinese Patent Application No. 201120327340, 2 September 2011.

107. Pan, Y.; Chen, J.; Zhu, W.; Pan, W. Chute Device for Shunting of Negative Plates. Chinese Patent Application No. 201120327287, 2 September 2011.

108. Pan, Y.; Chen, J.; Zhu, W.; Pan, W. Polar Plate Receiving Device. Chinese Patent Application No. 201120327407, 2 September 2011.

109. Pan, Y.; Chen, J.; Zhu, W.; Pan, W. Connecting Device for Eliminating Stress of Automatic Spot Welding Machine. Chinese Patent Application No. 201120327257, 2 September 2011.

110. Jin, Z.; Liu, J.; Wu, Y.; Pan, G.; Zhang, W. Automatic Battery Cap Press. Chinese Patent Application No. 201520168069, 24 March 2015.

111. Gao, X.; Xie, X.; Zhang, P.; Liao, S.; Yang, Y. High Temperature Composite Additive Agent for Nickel Base Charging Battery. Chinese Patent Application No. 200610061895, 27 July 2006.

112. Gao, X.; Liao, S.; Xie, X.; Yang, Y. Nickel Hydrogen Charging Battery. Chinese Patent Application No. 200710073554, 16 March 2007.

113. Long, Q.; Xie, J.; Xie, X. Voltage Internal Resistance General-Purpose Tester. Chinese Patent Application No. 201220068181, 28 February 2012.

114. Long, Q.; Xie, J.; Yuan, H.; Wang, J. Nickel-Hydrogen Battery Thermistor of Group Tester. Chinese Patent Application No. 201520856256, 30 October 2015.

115. Peng, H.; Zeng, S.; Xie, H.; Xie, J. Battery Cleaning Machine. Chinese Patent Application No. 201220056525, 21 February 2012.

116. Peng, H.; Yan, Y.; Wang, J. Whitewashed Device is Scraped to Nickel-Hydrogen Battery Negative Pole. Chinese Patent Application No. 201520845461, 28 October 2015.

117. Luo, Q.; Xie, X. Nickel-Hydrogen Battery Capable of Being Used in Low-Temperature Environment and Preparation Method Thereof. Chinese Patent Application No. 201210043538, 24 February 2012.

118. Yu, Y.; Peng, H. Rechargeable Nickel-Hydrogen Battery. Chinese Patent Application No. 201220056523, 21 February 2012.

119. Gao, X.; Che, Y. Sealed Ni/MH Battery. Chinese Patent Application No. 99243344, 23 September 1999.

120. Li, Z.; Li, Q.; Yang, S. A Nickel-Hydrogen High-Temperature Battery and Its Making Method. Chinese Patent Application No. 200810027050, 27 March 2008.

121. Li, Z.; Li, Q.; Chen, H. Nickel Hydrogen Quick Charging Battery and its Making Method. Chinese Patent Application No. 200810027053, 27 March 2008.
122. Li, Z.; Li, Q.; Tian, X.; Liu, Y. Nickel-Hydrogen Low Self-Discharge Battery. Chinese Patent Application No. 200810027054, 27 March 2008.
123. Long, X.; Ye, K.; Guan, T.; Zhang, Y.; Bin, J. Method of Manufacturing Nickel-Metal Hydride Battery. Chinese Patent Application No. 201110191333, 8 July 2011.
124. Fan, Z.; Guan, D.; Long, X.; Ye, K.; Zhang, Y. Method for Preparing Nickel Battery. Chinese Patent Application No. 200910188936, 15 December 2009.
125. Long, X.; Ye, K.; Fan, Z. Positive Base Material of Battery. Chinese Patent Application No. 200920144753, 27 February 2009.
126. Long, X.; Ye, K.; Fan, Z. Nickel-Metal Hydride Battery Cathode Alkali Treatment Method. Chinese Patent Application No. 200910136112, 30 April 2009.
127. Long, X.; Ye, K.; Fan, Z. Nickel-Hydrogen Power Battery Cathode. Chinese Patent Application No. 200920151587, 30 April 2009.
128. Fan, Z.; Long, X.; Ye, K. Battery Anti-Explosion System. Chinese Patent Application No. 200910036407, 4 January 2009.
129. Fan, Z.; Long, X.; Ye, K. High Temperature Failure Rubber. Chinese Patent Application No. 200910036691, 15 January 2009.
130. Long, X.; Ye, K.; Fan, Z.; Guan, D.; Zhang, Y. Welding of Battery Nickel Sheets. Chinese Patent Application No. 200910041975, 19 August 2009.
131. Long, X.; Ye, K. Nickel Battery with Multi-Layer Membrane. Chinese Patent Application No. 200820214287, 5 December 2008.
132. Long, X.; Ye, K. Cell Casing with Active Layer and Cell Using Same. Chinese Patent Application No. 200820235079, 15 December 2008.
133. Long, X.; Ye, K.; Fan, Z. Steel Shell of Cylindrical Cell. Chinese Patent Application No. 200920050216, 15 January 2009.
134. Fan, Z.; Guan, D.; Long, X.; Ye, K.; Zhang, Y. Manufacturing Method of Cylindrical Battery. Chinese Patent Application No. 200910189079, 18 December 2009.
135. Long, X.; Xu, X.; Zhou, W.; Cheng, Z. Flexibly-Packaged Elliptic Battery. Chinese Patent Application No. 201120441364, 9 November 2011.
136. Long, X.; Xu, X.; Zhou, W.; Cheng, Z. Soft Packaging Cylindrical Battery. Chinese Patent Application No. 201120441377, 9 November 2011.
137. Long, X.; Ran, G.; Long, Z. Automatic Dust Removal Machine. Chinese Patent Application No. 201510782747, 16 November 2015.
138. Long, X.; Ran, G.; Long, Z. Automatic Edge-Folding Machine. Chinese Patent Application No. 201510786877, 16 November 2015.
139. Long, X.; Li, J.; Guan, D.; Zhao, S.; Li, X. Combined Type Cell (9V). Chinese Patent Application No. 201430441530, 11 November 2014.
140. Guan, S.; Wang, L.; Zhao, S.; Shi, S.; Xiong, K.; Guo, J.; Yuan, X.; Li, J. Electrolyte and Electrochemical Component Using Same. Chinese Patent Application No. 201310216070, 3 June 2013.
141. Liang, D.; Meng, Z.; Wang, S.; Tang, Y. Formation-Free Nickel-Metal Hydride Battery and Manufacturing Method Thereof. Chinese Patent Application No. 201310293737, 12 July 2013.
142. Yang, Z.; Peng, D.; Wang, S.; Liu, X.; Wang, W.; Ji, J. Manufacturing Process for Positive Plate of Nickel-Hydrogen Rechargeable Battery. Chinese Patent Application No. 200910190780, 10 October 2009.
143. Liang, S.; Ciao, S.; Wang, W.; Ciao, H. A Negative Electrode Sheet for a Nickel-Hydrogen Battery and a Method for Producing the Same and a Nickel-Hydrogen Battery. Chinese Patent Application No. 201510452395, 28 July 2015.
144. Zhou, J.; Wang, S. Manufacturing Technology of High Power Charging Battery. Chinese Patent Application No. 200310112211, 20 November 2003.
145. Zheng, L. Cap Pressing Device for Batteries. Chinese Patent Application No. 201110103881, 25 April 2011.
146. Liang, D.; Wang, W.; Tang, Y. Cylindrical Alkaline Battery and Sealing Ring Thereof. Chinese Patent Application No. 201420673268, 12 November 2014.

147. Wei, N.; Peng, D.; Wang, S.; Yu, D.; Mo, Y. Charging Battery and Assembling Structure of Charging Battery Cover and Current Collector. Chinese Patent Application No. 201020690810, 30 December 2010.

148. Hu, Y.; Jia, H.; Liu, M. Nickel-Hydrogen Battery Discharging at Low Temperature and High Multiplying Power and Preparation Method and Negative Plate Thereof. Chinese Patent Application No. 200910119762, 26 March 2010.

149. Liu, M.; Jia, H.; Hu, Y. Low-Capacity Nickel-Metal Hydride Secondary Battery and Assembly Method Thereof. Chinese Patent Application No. 200910261710, 28 December 2009.

150. Xu, Z.; Fan, Z. Novel Nickel-Hydrogen Power Battery. Chinese Patent Application No. 200520053766, 11 January 2005.

151. Liu, M.; Liu, J. Wide-Temperature Range and High-Power Nickel-Hydrogen Battery and Preparation Technology Thereof. Chinese Patent Application No. 201511015536, 30 December 2015.

152. Liu, M.; Liu, J. Low Internal Resistance Nickel-Hydrogen Battery. Chinese Patent Application No. 201521125143, 30 December 2015.

153. Zhu, M.; Tao, G.; Zhu, Y.; Deng, X. Battery System. Chinese Patent Application No. 201620013831, 6 January 2016.

154. Zhu, M.; Gao, D.; Guo, Q.; Huang, C. Battery System. Chinese Patent Application No. 201610009992, 6 January 2016.

155. Liu, M.; Liu, J. A Cylindrical Nickel Metal Hydride Battery. Chinese Patent Application No. 201521118986, 12 December 2015.

156. Liu, M.; Liu, J. Cylindrical Nickel-Metal Hydride Battery and Negative Pole Wet-Method Powder Scraping Process. Chinese Patent Application No. 201511010831, 30 December 2015.

157. Li, M.; Lu, Y.; Ou, X.; Lian, W.; Xie, H.; Lin, Z.; Yang, F.; Tan, B. Nickel-Hydrogen Battery and Manufacturing Technique Thereof. Chinese Patent Application No. 92101326, 9 March 1992.

158. Xu, J.; Zhang, H.; Ouyang, J. Injection Method for Battery Electrolyte. Chinese Patent Application No. 201110237409, 18 August 2011.

159. Lin, Y.; Tan, W.; Tan, Y.; Li, W.; Chen, X.; Tan, G. Battery Sealing Machine. Chinese Patent Application No. 201110237406, 18 August 2011.

160. Lin, Y.; Tan, W.; Tan, Y.; Li, W.; Chen, X.; Tan, G. Spreading Machine for Battery. Chinese Patent Application No. 201110237390, 18 August 2011.

161. Chen, J.; Huang, Z.; Zhang, H.; Ouyang, J. Recovery Method of Ni-MH Battery Negative Pole Waste. Chinese Patent Application No. 201110237411, 18 August 2011.

162. Huang, Z.; Huang, J.; Chen, J.; Huang, B.; Zhang, H.; Ouyang, J. An anti-Explosion Battery Cap. Chinese Patent Application No. 201120301061, 18 August 2011.

163. Huang, J.; Huang, Z.; Chen, J.; Huang, B.; Zhang, H.; Ouyang, J. D-Type Low Capacity Battery Structure. Chinese Patent Application No. 201120301104, 18 August 2011.

164. Zhang, H.; Huang, J.; Chen, J.; Huang, B.; Ouyang, J. Power Battery Current-Collecting Disc. Chinese Patent Application No. 201120301102, 18 August 2011.

165. Nan, J.; Hou, X.; Xue, J. Nickel-Hydrogen Alkaline Battery and Preparation Method Thereof. Chinese Patent Application No. 200410077333, 9 December 2004.

166. Zeng, X.; Xia, X.; Xue, J. Battery Plus Plate Current-Collector and Producing Technology Thereof. Chinese Patent Application No. 200610122367, 25 September 2006.

167. Xia, X.; Nan, J.; Meng, X. Alkaline Battery Cover. Chinese Patent Application No. 200320117076, 13 October 2003.

168. Liu, J.; Xia, X.; Xue, J. Negative Electrode of NiMH Battery and Its Making Process. Chinese Patent Application No. 02134370, 17 July 2002.

169. Huang, T.; Wang, J. Composite Hydrogen Storage Alloy for NiMH Battery Cathode. Chinese Patent Application No. 200810197606, 12 November 2008.

170. Wu, H.; Liu, X.; Huang, J. Nickel-Hydrogen Battery Pack Provided with Detection Device. Chinese Patent Application No. 201010229034, 16 July 2010.

171. Liu, X.; Wu, H.; Huang, J. Nickel-Metal Hydride Battery Set with Detecting Device. Chinese Patent Application No. 201020261323, 17 July 2010.

172. Wu, H.; Sun, J.; Yang, R.; Wang, T. Equalization Charging Device of Nickel-Metal Hydride Battery. Chinese Patent Application No. 201220070633, 29 February 2012.

173. Wang, T.; Tu, J.; Yang, R. Charger with Nickel-Hydrogen Battery Overcharging Prevention Function. Chinese Patent Application No. 201220241474, 25 May 2012.

174. Sun, J.; Jiang, Z. Nickel-Metal Hydride Battery Facilitating Heat Dissipation. Chinese Patent Application No. 201220241313, 25 May 2012.

175. Sun, J.; Wu, H.; Yang, R. Nickel-Metal Hydride Battery with Anti-Overheating Function and Stable Structure. Chinese Patent Application No. 201220180479, 25 April 2012.

176. Wu, H.; Sun, J.; Wang, T. Control Device for Preventing Nickel-Hydrogen Battery Overdischarge. Chinese Patent Application No. 201220180491, 25 April 2012.

177. Sun, J.; Wu, H. Low-Cost Nickel-Metal Hydride Battery Provided with Electrodes of High Winding Performance. Chinese Patent Application No. 201220180477, 25 April 2012.

178. Wu, H.; Sun, J.; Yang, R.; Wang, T. Square Nickel-Metal Hydride Battery, Battery Pack and Electric Vehicle Using Battery Pack. Chinese Patent Application No. 201220070631, 29 February 2012.

179. Huang, J.; Huang, J.; Wu, H. Cathode Current Collecting Structure of Nickel-Hydrogen Battery. Chinese Patent Application No. 200920131330, 20 April 2009.

180. Wu, H.; Sun, J.; Huang, J. Rechargeable Battery and Cap Thereof. Chinese Patent Application No. 201220292409, 21 June 2012.

181. Sun, J.; Wu, H.; Jiang, M. High-Power Rechargeable Battery Current Collector and Rechargeable Battery Thereof. Chinese Patent Application No. 201220290378, 20 June 2012.

182. Huang, J.; Huang, J.; Wu, H. Battery Negative Plate. Chinese Patent Application No. 200920131329, 20 April 2009.

183. Sun, J.; Jiang, Z.; Huang, J. Positioning Clamping Device of Battery Pole Piece. Chinese Patent Application No. 201220290379, 20 June 2012.

184. Sun, J.; Wu, H.; Jiang, M. Centrifugal Dryer Used for Battery Pole Piece. Chinese Patent Application No. 201220290372, 20 June 2012.

185. Sun, J.; Wu, H.; Huang, J. Electrolyte Injector with Heating Function. Chinese Patent Application No. 201220290380, 20 June 2012.

186. Chen, Y.; Yang, H.; Wu, C.; Chao, D. High Iron Hydrogen Storage Electrode Alloy, Preparation Method Thereof and Nickel-Hydrogen Battery Cathode Material. Chinese Patent Application No. 201110144878, 31 May 2011.

187. Meng, Z.; He, Y.; Sun, W. Nickel-Metal Hydride Battery Pack for Pure Electric Bus. Chinese Patent Application No. 201410159453, 21 April 2014.

188. Zhang, X.; Sun, W.; Lin, X.; Wu, Y. High-Temperature Nickel-Metal Hydride Battery. Chinese Patent Application No. 201010287314, 20 September 2010.

189. Zhang, X.; Wu, Y.; Li, H.; Li, X.; Sun, W. High-Power Nickel-Hydrogen Battery. Chinese Patent Application No. 201120023136, 25 January 2011.

190. Tang, C.; Cheng, S.; Hu, C.; Zhang, M.; Guo, H.; Wang, F.; Zhou, Z.; Li, P. Antioxidant for Alkaline Cell and Alkaline Cell Using Antioxidant. Chinese Patent Application No. 201310272023, 2 July 2013.

191. Wang, L.; Sun, Y. Alkaline Secondary Cell Diaphragm and Its Producing Method. Chinese Patent Application No. 200410056887, 27 August 2004.

192. Cao, Y.; Yang, T.; Guo, J. Manufacturing Method of Alkaline Battery Electrode and Battery Produced Using Said Method. Chinese Patent Application No. 01144538, 19 December 2001.

193. Cao, Y.; Yang, T.; Guo, J. Manufacturing method of alkaline battery electrode and battery produced using said method. Chinese Patent Application No. 01144539, 19 December 2001.

194. Cao, Y.; Xie, F. Secondary NiMH Battery. Chinese Patent Application No. 200620115719, 31 May 2006.

195. Li, W. Manufacturing Method of Nickel Hydrogen Battery Negative Electrode Material. Chinese Patent Application No. 03100898, 27 January 2003.

196. Wu, Y.; Deng, Z.; Zeng, Y.; He, L.; Deng, Z.; Zhu, D. Nickel-Hydrogen Power Battery and Preparation Method Therefor. Chinese Patent Application No. 201510434308, 22 July 2015.

197. Wu, Y.; Zhu, D.; Lv, D.; Chang, J.; Wang, C.; Wen, X.; He, Y.; Xia, Z. Cylindrical Nickel-Hydrogen Power Cell. Chinese Patent Application No. 201220096555, 15 March 2012.

198. Wu, Y.; Zhu, D.; Lv, D.; Chang, J.; Wang, C.; Wen, X.; He, Y.; Xia, Z. Nickel-Hydrogen Battery. Chinese Patent Application No. 201220096530, 15 March 2012.

199. Wu, Y.; Zhu, D.; Lv, D.; Jing, J.; Wang, C.; Wen, X.; He, Y.; Xia, Z. Dual-Purpose Power Battery. Chinese Patent Application No. 201220096554, 15 March 2012.
200. Wu, Y.; Zhu, D.; Lv, D.; Chang, J.; Wang, C.; Wen, X.; He, Y.; Xia, Z. Novel Nickel-Hydrogen Power Battery. Chinese Patent Application No. 201220096553, 15 March 2012.
201. Wu, Y. High-Efficiency Nickel Metal Hydrogen Power Battery. Chinese Patent Application No. 201420456274, 14 August 2014.
202. Wu, Y.; Zhu, D.; Lv, D.; Chang, J.; Wang, C.; Wen, X.; He, Y.; Xia, Z. Novel Square Ni-MH Power Battery. Chinese Patent Application No. 201220096552, 15 March 2012.
203. Wu, Y.; Zhu, D.; Lv, D.; Chang, J.; Wang, C.; Wen, X.; He, Y.; Xia, Z. Composite Novel Nickel-Metal Hydride Battery. Chinese Patent Application No. 201220096558, 15 March 2012.
204. Wu, Y.; Zhu, D.; Lv, D.; Chang, J.; Wang, C.; Wen, X.; He, Y.; Xia, Z. Novel Nickel-Metal Hydride Battery. Chinese Patent Application No. 201220096556, 15 March 2012.
205. Deng, Z.; Zeng, Y.; He, L.; Wu, Y.; Wang, L.; Zhu, D. Large Capacity Ni-MH Power Battery. Chinese Patent Application No. 201520322087, 18 May 2015.
206. Wu, Y. High-Temperature and Low-Temperature Nickel-Hydrogen Power Cell with Super-High Multiplying Power. Chinese Patent Application No. 201420456286, 14 August 2014.
207. Deng, Z.; Zeng, Y.; He, L.; Wu, Y.; Wang, L.; Zhu, D. A High-Power Ni-MH Battery Suitable for a Wide Temperature Range. Chinese Patent Application No. 201520322089, 18 May 2015.
208. Li, X. Negative Pole Polar Plate Recycle Device. Chinese Patent Application No. 201620139343, 23 February 2016.
209. Zhang, T.; Chen, Y.; Zhang, P.; Gu, Y. Double Electric Pole Columns Power Type Square Body Nickel-Hydrogen Battery. Chinese Patent Application No. 200710156932, 20 November 2007.
210. Zhang, T.; Chen, Y.; Zhang, P.; Gu, Y. Power Type Cube Nickel-Hydrogen Batteries of Module Electric Core. Chinese Patent Application No. 200710156935, 20 November 2007.
211. Gu, Y.; Zhang, J.; Bai, Y.; Zhu, Y.; Mo, Z.; Qian, L. Nickel-Hydrogen Battery Positive Electrode Sheet Manufacturing Method and Manufacturing Apparatus Thereof. Chinese Patent Application No. 201210431784, 1 November 2012.
212. Gu, Y.; Zhang, J.; Zhu, Y.; Mo, Z.; He, Y.; Qian, L. Low-Temperature Nickel-Hydrogen Battery and Preparation Method Thereof. Chinese Patent Application No. 201210431080, 1 November 2012.
213. Gu, Y.; Zhang, J.; Chen, A.; Mo, Z.; Qian, L. High Temperature Nickel-Hydrogen Battery and Preparation Method Thereof. Chinese Patent Application No. 201210430545, 1 November 2012.
214. Tian, X.; Zhou, S. Square Battery Shell. Chinese Patent Application No. 201320083480, 25 February 2013.
215. Tian, X.; Zhou, S. Rectangular Battery and Manufacturing Method Thereof. Chinese Patent Application No. 201310058048, 25 February 2013.
216. Tian, X.; Zhou, S. Square Battery. Chinese Patent Application No. 201320083860, 25 February 2013.
217. Cao, S.; Jiang, Z.; Yin, L.; Huangfu, Y.; Wang, Y.; Jia, C.; Wu, F.; Zhao, X.; Zhang, L.; Zhang, W.; et al. Hybrid Cars Battery Module and its Battery Pack. Chinese Patent Application No. 201610875601, 30 September 2016.
218. Cao, S.; Huangfu, Y.; Yin, L.; Wang, Y.; Jia, C.; Zhao, X.; Zhang, L.; Ji, Y.; Zhang, W.; Zheng, Z.; et al. Cell Connector. Chinese Patent Application No. 201610875603, 30 September 2016.
219. Li, J.; Cao, S.; Jia, C.; Wu, F.; Yan, Q.; Zhang, W.; Hou, S. Nickel-Metal Hydride Battery Pack for Electric Bicycle. Chinese Patent Application No. 201020158887, 19 March 2010.
220. Xu, S.; Liang, W.; Zhang, Z.; Yang, Y.; Cao, S.; Zhang, W.; Ma, Y.; Zhang, X.; Xu, G.; Xing, Z.; et al. Precharging Technique for Forming Nickel Hydrogen Battery. Chinese Patent Application No. 200610151962, 3 September 2006.
221. Li, J.; Cao, S.; Jia, C.; Zhang, L.; Huangfu, Y.; Wang, Y.; Zhang, W.; Yang, T. Battery Micro-Short Low-Temperature Detection Method. Chinese Patent Application No. 201310119797, 8 April 2013.
222. Li, J.; Cao, S.; Jia, C.; Chang, Z.; Huangfu, Y.; Wang, Y.; Yin, L.; Ma, Y. Electrode Slurry Mixing Method. Chinese Patent Application No. 201310118746, 8 April 2013.
223. Li, J.; Cao, S.; Jia, C.; Huangfu, Y.; Zhang, W.; Chen, S.; Miao, R.; Wu, H. Cathode Coating Device. Chinese Patent Application No. 201320127460, 20 March 2013.
224. Zhang, Z.; Zhang, X.; Xu, G. Multi-Way Paste-Applicator with Welding Headspace. Chinese Patent Application No. 200420065793, 1 July 2004.

225. Li, J.; Cao, S.; Jia, C.; Wu, F.; Huangfu, Y.; Wang, Y.; Chen, S.; Du, J. Method for Eliminating Battery Plate Stress. Chinese Patent Application No. 201310118460, 8 April 2013.

226. Cao, S.; Chang, Z.; Huang, F.; Li, J.; Li, J.; Shi, Q.; Xu, S.; Yang, Y. Nickel-Metal Hydride Battery Electrode Burr Punching Machine. Chinese Patent Application No. 201020158886, 19 March 2010.

227. Cao, S.; Jiang, Z.; Chang, Z.; Huangfu, Y.; Yin, L.; Wang, Y.; Jian, C.; Zhao, X.; Li, J.; Ji, Y. Pole Piece Conveying Positioner. Chinese Patent Application No. 201520224145, 15 April 2015.

228. Zhang, Z.; Xu, S.; Yang, Y.; Chang, Z.; Ma, Y.; Zhao, X.; Feng, H.; Liu, J.; Liu, Z. Ultrasonic Welding Machine of Current Collecting Belt of Positive Plate. Chinese Patent Application No. 201020154887, 19 March 2010.

229. Cao, S.; Jiang, Z.; Chang, Z.; Huangfu, Y.; Yin, L.; Wu, F.; Wang, Y.; Jia, C.; Hu, D.; Zheng, Z. Battery Insulated Ring Assembly Device. Chinese Patent Application No. 201520224142, 15 April 2015.

230. Li, J.; Cao, S.; Jia, C.; Huangfu, Y.; Wang, Y.; Yin, L.; Zhang, W. Cylindrical Battery Insulation Space Ring. Chinese Patent Application No. 201320128321, 20 March 2013.

231. Li, J.; Cao, S.; Xu, G.; Ma, Y.; Jia, C.; Wang, Y.; Shi, Q.; Chang, Z. Connecting Component for Nickel-Hydrogen Battery Groups. Chinese Patent Application No. 201020154903, 19 March 2010.

232. Zhao, P.; Chen, P.; Ma, G. High Power Nickel-Metal Hydride Battery. Chinese Patent Application No. 201120029381, 28 January 2011.

233. Ma, G.; Cao, Y. Super-Wide-Temperature-Range Nickel-Hydrogen Battery and Manufacturing Method Therefor. Chinese Patent Application No. 201510746318, 15 November 2015.

234. Chen, P.; Hu, Z.; Ma, G. Method for Manufacturing Ultra-High Temperature Overcharging-Resistance Long Service Life Nickel-Hydrogen Battery. Chinese Patent Application No. 201310173019, 11 May 2013.

235. Hu, Z.; Chen, P.; Ma, G. Method for Manufacturing Ultra-High Temperature Long-Service Life Nickel-Hydrogen Batteries. Chinese Patent Application No. 201410816955, 24 December 2014.

236. Cao, Y.; Ma, G. Novel Low Capacity Nickel-Hydrogen Battery of Structure. Chinese Patent Application No. 201520803268, 15 October 2015.

237. Ma, G. A Flat Supporting Flame to Produce Cylindrical Ni-MH Battery. Chinese Patent Application No. 201710018936, 11 January 2017.

238. Gao, P.; Huang, D.; Fan, H.; Wang, F.; Lou, L. Method for Improving Air Impermeability of Cylindrical Battery and Cylindrical Battery. Chinese Patent Application No. 201410791776, 19 December 2014.

239. Hu, Z. Novel Structure of Safety Type Nickel-Metal Hydride Battery. Chinese Patent Application No. 201320565787, 12 September 2013.

240. Hu, Z.; Chen, P.; Ma, G. Nickel-Metal Hydride Battery. Chinese Patent Application No. 201120029358, 28 January 2011.

241. Ma, G.; Chen, P. Nickel Hydrogen-Nickel Cadmium Battery. Chinese Patent Application No. 201120029357, 28 January 2011.

242. Peng, L. Rapid Liquid Feeding Device for Nickel-Metal Hydride Battery. Chinese Patent Application No. 201320565780, 12 September 2013.

243. Tang, B.; Ma, G. A Gumming Device for on Battery Positive Plate Softening Machine. Chinese Patent Application No. 201520827309, 22 October 2015.

244. Zhang, Y.; Zhu, Y.; Li, L. Super Power Nickel-Hydrogen Battery Overturning Traditional Technology. Chinese Patent Application No. 201310083759, 4 March 2013.

245. Zhang, Y.; Ge, K.; Si, W. Novel Nickel-Metallic Oxide Cell Structure. Chinese Patent Application No. 201220358743, 23 July 2012.

246. Zhang, Y. Nickel-Metal Hydride Battery Charger. Chinese Patent Application No. 201220358739, 23 July 2012.

247. Zhang, Y.; Ge, K. A Prismatic Sealed Battery Module. Chinese Patent Application No. 201220358692, 23 July 2012.

248. Zhang, Y. Sealing Structure for Square Metal-Casing Storage Battery Post. Chinese Patent Application No. 201220357573, 23 July 2012.

249. Zhang, Y. Square Plastic-Casing Sealed Storage Battery. Chinese Patent Application No. 201220358740, 23 July 2012.

250. Zhang, Y.; Ge, K. Steel Shell Battery Seal Structure. Chinese Patent Application No. 201220357555, 23 July 2012.

251. Zhang, Y. Compound Sealing Device of Battery Electrode Column. Chinese Patent Application No. 201020609670, 16 November 2010.

252. Xu, Y.; Qiu, J.; Ge, C.; He, X.; Liu, Q. Low-Capacity High-Power Nickel-Hydrogen Battery. Chinese Patent Application No. 200520073477, 6 July 2005.

253. Xu, Y.; Ge, C.; Jiang, X.; Qiu, J.; He, X.; Xia, B.; Liu, Q. High Power Ni/MH Battery. Chinese Patent Application No. 200620069267, 8 February 2006.

254. Li, G.; Xu, Y.; Dong, M.; Li, X.; Zhang, L. Battery Connecting Apparatus. Chinese Patent Application No. 200810168497, 30 September 2008.

255. Li, G.; Xu, Y.; Yu, Z. Battery Group with Temperature Regulating Device. Chinese Patent Application No. 201010283138, 16 September 2010.

256. Xu, Y.; Li, G.; Dong, M.; Li, X.; Lu, Y.; Yue, J. Liquid Cooling Device of Battery Set. Chinese Patent Application No. 200910115238, 20 April 2009.

257. Dong, M.; Li, G.; Li, X.; Xu, Y.; Zhong, Z. Method for Charging Battery Pack. Chinese Patent Application No. 200910186454, 3 November 2009.

258. Chen, D. Dynamic High-Capacity Nickel-Hydrogen Battery and Production Process Thereof. Chinese Patent Application No. 200910225295, 13 November 2009.

259. Chen, D.; Chen, W.; Li, P.; Zhang, G.; Yin, C. Ni/MH Power Battery and Production Technology Thereof. Chinese Patent Application No. 201110201987, 19 July 2011.

260. Chen, D.; Chen, W.; Li, P.; Zhang, G.; Yin, C. Nickel-Hydrogen Power Battery. Chinese Patent Application No. 201120254890, 19 July 2011.

261. Chen, D.; Chen, W.; Li, P.; Zhang, L. Ultralow-Capacity Battery and Powder Pulling Die for Manufacturing Battery. Chinese Patent Application No. 201310268984, 1 July 2013.

262. Chen, D.; Chen, W.; Li, P.; Zhang, G.; Yin, C. Method for Producing High Temperature Nickel-Metal Hydride Battery. Chinese Patent Application No. 201110455378, 31 December 2011.

263. Chen, D.; Chen, W.; Li, P.; Zhang, G.; Yin, C. Manufacturing Method of Nickel-Metal Hydride Battery Capable of Being Stored for Long Time. Chinese Patent Application No. 201110455379, 31 December 2011.

264. Chen, W.; Chen, D.; Li, P.; Zhang, G. Production Method for Positive Pole of High-Power Nickel-Metal Hydride Battery. Chinese Patent Application No. 201210289894, 15 August 2012.

265. Chen, D.; Chen, W.; Li, P. Novel Welding Method for Anode Tab and Foamed Nickel of Battery. Chinese Patent Application No. 201310367366, 22 August 2013.

266. Chen, D.; Chen, W.; Li, P.; Zhang, G.; Yin, C. Method for Preparing Anode of High-Power Nickel-Hydrogen Battery. Chinese Patent Application No. 201110174707, 27 June 2011.

267. Chen, D.; Chen, W.; Li, P.; Zhang, G.; Yin, C. Battery Diaphragm with Improved Structure and Production Technology Thereof. Chinese Patent Application No. 201110211199, 27 July 2011.

268. Chen, D.; Chen, W.; Li, P.; Zhang, G.; Yin, C. Nickel-Metal Hydride Battery Cathode Preparing Method. Chinese Patent Application No. 201110350094, 8 November 2011.

269. Chen, D.; Chen, W.; Li, P.; Zhang, G.; Yin, C. High-Temperature Nickel-Metal Hydride Battery and Manufacturing Method Thereof. Chinese Patent Application No. 201110381832, 28 November 2011.

270. Chen, W.; Chen, D.; Chen, X.; Lyu, G.; Liu, Y. Novel Battery Separation Membrane. Chinese Patent Application No. 201510249811, 16 May 2015.

271. Chen, D.; Chen, W.; Chen, C.; Zhang, G. Rapid Charging Power Battery. Chinese Patent Application No. 201610571747, 20 July 2016.

272. Zou, B.; Hu, X. Novel Knot Formula Nickel-Hydrogen Battery. Chinese Patent Application No. 201520557708, 29 July 2015.

273. Zou, B.; Hu, X. Knot Formula Nickel-Hydrogen Battery. Chinese Patent Application No. 201520557757, 29 July 2015.

274. Zou, B.; Hu, X. Formula Nickel-Hydrogen Battery is Detained in Low Internal Resistance. Chinese Patent Application No. 201520557650, 29 July 2015.

275. Zou, B.; Hu, X. Low-Cost Knot Formla Nickel-Hydrogen Battery. Chinese Patent Application No. 201520557651, 29 July 2015.

276. Yang, Y.; Li, Q.; Li, X. Cladding Nickel Electrode and Preparation Method Thereof. Chinese Patent Application No. 201410272553, 18 June 2014.

277. Yang, Y.; Li, X.; Lei, Y.; Wang, X. NiMH Battery Electrolyte Additive, and an Electrolyte Containing a Nickel Hydrogen Battery Additives. Chinese Patent Application No. 201610925948, 31 October 2016.

278. Di, E.; Qiu, C.; Chen, T.; Guan, S.; Li, J.; Zheng, X.; Yin, D. Ordinary Battery External Wrapping Film and Preparation Method Thereof. Chinese Patent Application No. 201610248734, 19 April 2016.

279. Zhai, E.; Qiu, C.; Chen, T.; Guan, S.; Li, J.; Zheng, X.; Yin, D. Low-Self-Discharging Battery Outer Wrapping Thin Film and Preparing Method Thereof. Chinese Patent Application No. 201610248753, 19 April 2016.

280. Zhai, E.; Qiu, C.; Chen, T.; Guan, S.; Li, J.; Zheng, X.; Yin, D. Film for Wrapping Exteriors of High Capacity Batteries and Preparation Process Thereof. Chinese Patent Application No. 201610248752, 19 April 2016.

281. Zhai, E.; Qiu, C.; Chen, T.; Guan, S.; Li, J.; Zheng, X.; Yin, D. External Coating Thin Film of Power Battery and Preparation Method Thereof. Chinese Patent Application No. 201610248748, 19 April 2016.

282. Zhai, E.; Qiu, C.; Chen, T.; Guan, S.; Li, J.; Zheng, X.; Yin, D. Film for Wrapping Exteriors of High-Temperature Batteries and Preparation Method Thereof. Chinese Patent Application No. 201610248738, 19 April 2016.

283. Zhai, E.; Qiu, C.; Chen, T.; Guan, S.; Li, J.; Zheng, X.; Yin, D. Special Sealant for Ordinary Batteries and Preparation Process Thereof. Chinese Patent Application No. 201610248732, 19 April 2016.

284. Zhai, E.; Qiu, C.; Chen, T.; Guan, S.; Li, J.; Zheng, X.; Yin, D. Special Sealant for High Capacity Batteries and Preparation Process Thereof. Chinese Patent Application No. 201610248740, 19 April 2016.

285. Zhai, E.; Qiu, C.; Chen, T.; Guan, S.; Li, J.; Zheng, X.; Yin, D. Special Sealant for High-Temperature Batteries and Preparation Process Thereof. Chinese Patent Application No. 201610248747, 19 April 2016.

286. Zhai, E.; Qiu, C.; Chen, T.; Guan, S.; Li, J.; Zheng, X.; Yin, D. Seal Gum Specially Used for Low-Self-Discharge Battery and Preparation Process of Same. Chinese Patent Application No. 201610248757, 19 April 2016.

287. Zhai, E.; Qiu, C.; Chen, T.; Guan, S.; Li, J.; Zheng, X.; Yin, D. Sealant Special for Power Batteries and Preparation Process of Sealant. Chinese Patent Application No. 201610248751, 19 April 2016.

288. Zhai, E. Anode Material of Nickel-Metal Hydride Battery. Chinese Patent Application No. 201410270953, 17 June 2014.

289. Li, J. Preparation Method of High-Temperature Ni-MH Power Battery. Chinese Patent Application No. 201210538083, 13 December 2012.

290. Huang, P.; Li, H.; Zhai, E. Preparation Method of Low Self-Discharge Nickel-Metal Hydride Battery. Chinese Patent Application No. 201410541530, 14 October 2014.

291. Li, J. Pole-Piece-Free Non-Sinding High-Power-Capacity Nickel-Hydrogen Battery and Processing Technology Thereof. Chinese Patent Application No. 201210355458, 18 September 2012.

292. Zhai, E.; Chen, T.; Li, J. High Voltage Nickel-Metal Hydride Battery Safety Protective Device. Chinese Patent Application No. 201510352510, 24 June 2015.

293. Zhai, E.; Chen, T.; Li, J. Energy-Saving Type High-Power Nickel Metal Hydride Battery. Chinese Patent Application No. 201510352144, 24 June 2015.

294. Zhai, E.; Chen, T.; Li, J. Heat Dissipation Type Nickel-Metal Hydride Battery. Chinese Patent Application No. 201510352233, 24 June 2015.

295. Zhai, E.; Chen, T.; Li, J. Low-Power-Consumption Overvoltage Protective Device for Nickel-Metal Hydride Battery. Chinese Patent Application No. 201510352123, 24 June 2015.

296. Zhai, E.; Chen, T.; Li, J. Environment-Friendly High-Capacity Nickel-Hydrogen Battery. Chinese Patent Application No. 201510352143, 24 June 2015.

297. Zhai, D.; Guo, D.; Han, N.; Zhu, R.; Cui, W.; Wang, R. Glue Solution for Surface Treatment of Nickel-Metal Hydride (NiMH) Battery Plate and Gum Dipping Device. Chinese Patent Application No. 201010189698, 28 May 2010.

298. Wang, R.; Wang, X.; Guo, D.; Fan, J. An Assembled Powder Scraping Apparatus for the Dary Method of Preparing Positive Electrode of Ni-MH Batteries. Chinese Patent Application No. 201120065653, 14 March 2011.

299. Guan, Y.; Su, W.; Yuan, T. Square Nickel-Metal Hydride Battery. Chinese Patent Application No. 201120319631, 29 August 2011.

300. Guan, Y.; Su, W.; Yuan, T. Environment-Friendly Nickel-Metal Hydride Battery. Chinese Patent Application No. 201120319658, 29 August 2011.

301. Chen, P. High-Voltage Nickel-Metal Hydride Battery Structure. Chinese Patent Application No. 201320338290, 7 June 2013.

302. Chen, P. High-Power Nickel-Hydrogen Battery. Chinese Patent Application No. 201120116046, 19 April 2011.

303. Chen, P. A High-Power Ni-MH Battery. Chinese Patent Application No. 201420355227, 30 June 2014.

304. Chen, P. A Non-Winding High-Capacity Ni-MH Battery. Chinese Patent Application No. 201420350094, 29 June 2014.

305. Chen, P. High-Capacity Nickel-Hydrogen Battery Pack with Self-Provided Charger. Chinese Patent Application No. 201120168375, 24 May 2011.

306. Chen, P. Multifunctional Charger with Various Input and Output Modes. Chinese Patent Application No. 201120124612, 25 April 2011.

307. Liu, Q.; Gao, G.; He, J.; Liu, J.; Zhu, Z.; Jia, F. Nickel-Hydrogen Battery Negative Pole Slurry Scraping Device. Chinese Patent Application No. 201620012233, 7 January 2016.

308. Gao, G.; He, J.; Liu, J.; Liu, Q.; Zhu, Z.; Jia, F. Nickel-Hydrogen Battery Positive Plate Dust Removal Rubberizing Device. Chinese Patent Application No. 201620076470, 26 January 2016.

309. Wu, Y.; Liu, Q.; Gao, G.; Liu, J.; Zhu, Z.; Jia, F. Power Nickel-Hydrogen Battery. Chinese Patent Application No. 201620013206, 7 January 2016.

310. Zhu, Z.; He, J.; Gao, G.; Liu, J.; Liu, Q.; Jia, F. Automatic Transportation of Nickel-Hydrogen Battery, Dust Absorption, Flash Removed. Chinese Patent Application No. 201620077158, 26 January 2016.

311. Wu, Y.; He, J.; Liu, J.; Liu, Q.; Zhu, Z.; Jia, F. A Agitating Unit that is Used for Nickel-Hydrogen Battery Positive Pole to Join in Marriage Whitewashed Process. Chinese Patent Application No. 201620011848, 7 January 2016.

312. Hu, H.; Hua, Y.; Li, J.; Lin, D.; Ye, H.; Zhan, B.; Zhan, J.; Zhou, H. Cylinder-Type Nickel-Metal Hydride Battery. Chinese Patent Application No. 201010114307, 11 February 2010.

313. Zhan, B.; Luo, S.; Zhan, J.; Tang, B.; Wang, J. Preparation Method of Negative Electrode Plate of Nickel-Metal Hydride Battery. Chinese Patent Application No. 201210272493, 1 August 2012.

314. Zhang, B.; Zhan, J. Temperature-Resisting Cylindrical Nickel-Metal Hydride Battery. Chinese Patent Application No. 201010114308, 11 February 2010.

315. Shen, M.; Xia, P.; Yan, Q.; Zheng, L.; He, Z.; Cheng, F.; Zhou, H.; Zhan, B.; Zhan, J.; Luo, S. Nickel Hydroxide Anode Pole Piece of Large Current Nickel-Metal Hydride Battery. Chinese Patent Application No. 201120255986, 19 July 2011.

316. He, Z.; Zheng, L.; Zhou, H.; Yan, L.; Zhan, B.; Zhan, J.; Luo, S. Special Nickel Hydroxide Positive Plate for SC Type Nickel-Metal Hydride Batteries. Chinese Patent Application No. 201020289229, 6 August 2010.

317. Pan, C.; Zhou, H.; Chen, X. Method for Fabricating Nickel Electrode in Nickel Hydrogen Battery. Chinese Patent Application No. 200510036556, 18 August 2005.

318. Xiang, K.; Chen, X.; Li, Y. Low-Capacity Nickel-Metal Hydride Battery. Chinese Patent Application No. 201020577092, 22 October 2010.

319. Zhou, H.; Pan, C.; Xiang, K.; Cheng, X. Nickel-Hydrogen Dynamic Cell. Chinese Patent Application No. 200520121356, 30 December 2005.

320. Wu, A.; Kuang, D. A High-Power Nickel-Hydrogen Cell. Chinese Patent Application No. 200320117714, 5 November 2003.

321. Wu, A.; Kuang, D. Flanged Metal Expanding Guipure for Manufacturing Electrode. Chinese Patent Application No. 200320117712, 5 November 2003.

322. Wu, A.; Kuang, D. Large Content High Power Nickel-Cadmium Charging Cell. Chinese Patent Application No. 200320117685, 4 November 2003.

323. Yan, D.; Wang, J. Manufacture Technology of Positive Plate for Nickel-Hydrogen Cell. Chinese Patent Application No. 97120114, 4 November 1997.

324. Yan, D.; Cui, W.; Chen, Q. Method for Manufacturing Ni(OH)$_2$ Electrodes of Ni-H Battery. Chinese Patent Application No. 98101896, 26 May 1998.

325. Ma, K.; Yan, D. Metal Strap Covered with Foam Nickel Material and Making Method Thereof. Chinese Patent Application No. 00109696, 22 June 2000.

326. Wang, J.; Yang, X.; Cheng, Q. Winding Process of Foamed Nickel Electrode. Chinese Patent Application No. 97120113, 4 November 1997.

327. Yan, D.; Cui, W.; Cheng, X. Method for Manufacturing Metal Hydride Electrode of Nickel-Hydrogen Battery. Chinese Patent Application No. 97121781, 22 December 1997.

328. Gao, X.; Qin, X.; Lan, Y. Composite Hydrogen-Storing Electrode Material of Hydrogen-Storing Alloy/Nano Carbon Material and its Preparing Process. Chinese Patent Application No. 00107426, 12 May 2000.

329. Yuan, H.; Wang, Y.; Wang, L. Magnesium-Base Hydrogen Storing Alloy Material. Chinese Patent Application No. 00109349, 31 May 2000.

330. Gong, W.; Gong, Z.; Gen, M. High Energy Nickel-Metal Hydride Cell. Chinese Patent Application No. 03258177, 31 July 2003.

331. Yan, J.; Wang, R.; Yan, D. Regeneration Method of Deactivated Negative Alloy Powder of Secondary Nickel-Hydrogen Battery. Chinese Patent Application No. 00132165, 19 December 2000.

332. Yan, J.; Wang, R.; Yan, D. Recovering Method of Defective Positive and Negative Pole Material of Secondary Nickel-Hydrogen Battery. Chinese Patent Application No. 00132164, 19 December 2000.

333. Chu, C.; Du, J.; Luo, Y. Anode Piece of Nickel-Metal Hydride Battery. Chinese Patent Application No. 200920168290, 4 September 2009.

334. Yu, Y. Energy-Saving Monitoring System of Power Line Carrier Street Lamp. Chinese Patent Application No. 200920169294, 21 August 2009.

335. Du, J.; Luo, Y.; Tu, Y. Nickel-Metal Hydride Battery Pole Piece and Battery Using the Same. Chinese Patent Application No. 200920168292, 4 September 2009.

336. Du, J.; Luo, Y.; Tu, Y. Nickel-Metal Hydride High-Capacity Cylindrical Battery. Chinese Patent Application No. 200920168291, 4 September 2009.

337. Du, J.; Luo, Y.; Ma, C. Low-Capacity Battery Bottom Mat. Chinese Patent Application No. 200920168289, 4 September 2009.

338. Lin, J.; Yan, Y.; Luo, Y. Winding Structure for Nickel-Hydrogen Battery. Chinese Patent Application No. 201220272255, 11 June 2016.

339. Tu, Y.; Yan, Y.; Luo, Y. Mold for Wet Slurry-Pulling of Nickel-Metal Hydride Battery Cathodes. Chinese Patent Application No. 201220272309, 11 June 2012.

340. Wang, Y.; Li, W.; Sun, L. Electrode of Charging Battery and Method and Equipment for Making Electrode. Chinese Patent Application No. 97122056, 19 December 1997.

341. Wang, J.; Wang, Y. Integrated Battery Device. Chinese Patent Application No. 98248410, 18 November 1998.

342. Lin, J.; Yang, J.; Qian, W.; Zhang, Y.; Jiang, Y.; Zhang, P.; Chen, Y. Cu-Contained Rare Earth System AB$_5$-Type Hydrogen Storage Alloy and Preparation Method Thereof. Chinese Patent Application No. 201210178222, 31 May 2012.

343. Zhang, P.; Yang, J.; Qian, W.; Zhang, Y.; Jiang, Y.; Chen, Y.; Chen, Y.; Lin, J. Low Cost Hydrogen Storage Alloy, Preparation Method and Application Thereof. Chinese Patent Application No. 200910112283, 21 July 2009.

344. Zhou, Z.; Huang, C.; Qian, W.; Yang, J.; Song, Y.; Cui, S.; Lin, C.; Zhang, Y.; Zhang, P. Low-Cost High-Performance Rare-Earth-Based AB$_5$-Type Hydrogen Storage Alloy and Preparation Method Thereof. Chinese Patent Application No. 200910112323, 22 July 2009.

345. Jiang, Y.; Chen, Y.; Chen, Y.; Zhang, Y.; Wu, Y. Non-Cobalt Low-Nickel Hydrogen Storage Alloy. Chinese Patent Application No. 201110032195, 28 January 2011.

346. Chen, Y.; Yang, J.; Qian, W.; Zhang, Y.; Jiang, Y.; Chen, Y.; Zhang, P.; Lin, J. Low-Self-Discharge Hydrogen Storage Alloy and Preparation Method Thereof. Chinese Patent Application No. 201110032749, 28 January 2011.

347. Zhang, P.; Yang, J.; Qian, W.; Jiang, L.; Lin, Z. Low-Cost AB$_5$ Type Hydrogen Storage Alloy Having Ultrahigh Capacity Characteristic, and Preparation Method and Application Thereof. Chinese Patent Application No. 201210412250, 25 October 2012.

348. Lin, J.; Yang, J.; Qian, W.; Zhang, Y.; Jiang, Y.; Zhang, P.; Chen, Y. Low Temperature Power Type Hydrogen Storage Alloy for Nickel-Metal Hydride Battery. Chinese Patent Application No. 201210143770, 10 May 2012.

349. Jiang, Y.; Yang, J.; Zhang, Y.; Ding, X.; Chen, Y.; Chen, Y.; Lin, Y.; Zhou, Y. Over-Stoichiometric Low-Cost Hydrogen Storage Alloy, and Preparation Method and Application Thereof. Chinese Patent Application No. 200810210684, 8 August 2008.

350. Zhang, P.; Yang, J.; Qian, W.; Zhang, Y.; Jiang, Y. Rare Earth Magnesium-Based Hydrogen Storage Alloy with Low Cost and Long Life and Applications Thereof. Chinese Patent Application No. 201110319482, 29 October 2011.

351. Kakeya, T.; Kodama, M.; Kanemoto, M.; Okuda, D.; Zhang, P.; Lin, Z.; Yang, J.; Qian, W.; Jiang, L. Hydrogen Storage Alloy and Manufacturing Method Thereof. Chinese Patent Application No. 201410409393, 19 August 2014.

352. Huang, Z.; Qian, W.; Chen, H.; Zhang, Y.; Zhu, G. Method for Preparing Hydrogen Storing Alloy Powder. Chinese Patent Application No. 200710008544, 2 February 2007.

353. Huang, Z.; Qian, W.; Chen, H.; Zhang, Y.; Zhu, G. Vacuum Induction Smelting Furnace. Chinese Patent Application No. 200820145393, 29 August 2008.

354. Huang, C.; Qian, W.; Chen, H.; Zhang, Y.; Zhu, G. Vacuum Induction Smelting Furnance. Chinese Patent Application No. 200810071672, 29 August 2008.

355. Chen, Y.; Zhang, Y.; Jiang, Y. Method for Preparing Magnesium-Based Cathode of Nickel-Hydrogen Battery. Chinese Patent Application No. 200910111626, 27 April 2009.

356. Min, D.; Chen, Y.; Yang, L.; He, C. Discharging Dust-Collecting Device for Hydrogen Storage Alloy Powder Mixer. Chinese Patent Application No. 201120206564, 17 June 2011.

357. Min, D.; Chen, Y.; Yang, L.; Tao, J. Dust Collection Control System Beneficial to Production. Chinese Patent Application No. 201120206503, 17 June 2011.

358. Min, D.; Chen, Y.; Luo, T.; Yang, L.; He, C. Cleaning Device Capable of Effectively Disposing Impurities on Vibrating Screen. Chinese Patent Application No. 201120206511, 17 June 2011.

359. Min, D.; Chen, Y.; Luo, T.; Yang, L.; Tao, J. Cooling Mould Set Beneficial for Crystallization of Alloys. Chinese Patent Application No. 201120206500, 17 June 2011.

360. Min, D.; Chen, Y.; Luo, T.; He, C. Sheet Producing Device for Fast Testing Electrochemistry Performance of Hydrogen Storage Alloy Powder. Chinese Patent Application No. 201120206636, 17 June 2011.

361. Min, D.; Chen, Y.; Luo, T.; Tao, J. Automated Pouring System for Vacuum Melting Furnace. Chinese Patent Application No. 201120206514, 17 June 2011.

362. Chen, Y.; Liu, L.; Min, D. A_2B_7 Hydrogen Storage Alloy for Nickel-Hydride Battery and Preparation Method Thereof. Chinese Patent Application No. 201310228766, 8 June 2013.

363. Min, D.; Liu, L.; Chen, Y. Rare Earth-Magnesium-Nickel System Heterogeneous Hydrogen Storage Alloys Used for Nickel-Hydrogen Batteries and Preparing Method Thereof. Chinese Patent Application No. 201410222437, 23 May 2014.

364. Zhu, X.; Zhao, X.; Ji, L.; Wang, Y.; Li, Q. High-Capacity Hydrogen Storage Alloy Electrode Material and Production Method Thereof. Chinese Patent Application No. 201310268222, 28 June 2013.

365. Ji, L.; Zhao, X.; Zhu, X.; Han, S.; Wang, Y.; Li, Q.; Liu, Y.; Jing, Y.; Xu, J. Hydrogen Storage Alloy for Nickel-Metal Hydride Battery. Chinese Patent Application No. 201310728257, 26 December 2013.

366. Zhu, X.; Xu, J.; Zhao, X. Hydrogen Storage Alloy for Nickel-Metal Hydride Battery and Manufacturing Method of Hydrogen Storage Alloy. Chinese Patent Application No. 201510879635, 3 December 2015.

367. Han, S.; Shen, W.; Zhu, X.; Zhou, Y. Method for Improving Electrochemical Performance of Hydrogen Storage Alloy Powder by Utilizing Electropolymerization Polyaniline. Chinese Patent Application No. 201110158364, 2 June 2011.

368. Han, S.; Zhang, Z.; Zhu, X.; Li, J.; Xu, S.; Yang, Y.; Jing, T.; Wang, X. Preparation Method of High Content Rare Earth-Magnesium Base Composite Hydrogen Storage Alloy for NiMH Battery. Chinese Patent Application No. 200410092078, 2 November 2004.

369. Zhang, P. Method for Preparing AB_5 Type Mixed Rare-Earth Hydrogen-Bearing Alloy Powder. Chinese Patent Application No. 200710123188, 2 July 2007.

370. Wang, C.; Jiang, B.; Chi, X.; Shen, X.; Liu, L.; Zhang, Y. Low-Cobalt and Praseodymium-Neodymium-Free AB_5 Type Hydrogen Storage Alloy with Low Cost and Preparation Method Thereof. Chinese Patent Application No. 201210215478, 27 June 2012.

371. Wang, C.; Jiang, B.; Chi, X.; Shen, X.; Liu, L.; Zhang, Y. Praseodymium-Neodymium-Free Low-Cost Superlong Life Type Hydrogen Storage Alloy and Preparation Method Thereof. Chinese Patent Application No. 201210362335, 26 September 2012.

372. Guo, J.; Jiang, B.; Sun, C.; Wang, Z. Rare Earth Hydrogen Storage Alloy Containing Mg and Preparing Method Thereof. Chinese Patent Application No. 200710157452, 15 October 2007.

373. Wang, C.; Guo, J.; Jiang, B.; Shen, X. Magnesium-Containing Superlattice Hydrogen Storage Alloy and Preparation Method Thereof. Chinese Patent Application No. 201110152699, 8 June 2011.

374. Wang, C.; Jiang, B.; Chi, X.; Shen, X.; Liu, L.; Zhang, Y. Praseodymium-Free, Neodymium-Free and Cobalt-Free High Capacity Superlattice Hydrogen Storage Alloy Containing Magnesium. Chinese Patent Application No. 201310362901, 16 August 2013.

375. Guo, J.; Jiang, B.; Sun, C.; Wang, C. Preparation Method of AB$_5$ Type Hydrogen-Storage Alloy Used on Ni/MH Battery. Chinese Patent Application No. 200810246900, 30 December 2008.
376. Wang, C.; Guo, J.; Jiang, B.; Shen, X. Method for Treating Hydrogen Storage Alloy Waste Residues. Chinese Patent Application No. 201110151658, 8 June 2011.
377. Wang, C.; Guo, J.; Jiang, B.; Shen, X. Method for Recycling Metals from Waste Nickel-Hydrogen Batteries. Chinese Patent Application No. 201110173754, 25 June 2011.
378. Miao, G.; Ren, H.; Zhou, X.; Sun, X. Preparation Method of High Performance Hydrogen Storage Alloy Powder. Chinese Patent Application No. 200810150430, 23 July 2008.
379. Gao, J.; Zhang, Y.; Shang, H.; Yang, T.; Zhai, T.; Li, Y. High-Capacity RE-Mg-Ti-Co-Al System AB$_2$ Type Hydrogen Storage Electrode Alloy Applied to NiMH Batteries of Hybrid Electric Vehicles and Preparation Method of Alloy. Chinese Patent Application No. 201510008846, 8 January 2015.
380. Gao, J.; Yong, H.; Zhang, Y.; Shang, H.; Li, Y.; Yuan, Z. RE-Mg-Ni-Ti-Cu-Al-B Series AB Type Electrode Alloy Used for Ni-MH Battery and Preparation Method. Chinese Patent Application No. 201610454477, 22 June 2016.
381. Gao, J.; Shang, H.; Li, J.; Xu, C. Low-Magnesium RE-Mg-Ti-Al-B Series Hydrogen-Storage Alloy for Ni-MH Secondary Battery and Preparation Method. Chinese Patent Application No. 201410085723, 11 March 2014.
382. Gao, J.; Yuan, Z.; Yang, T.; Liao, Y.; Ma, X. Nano-Graphite Compoujnded High-Capacity RE-Mg-Ni-Based Hydrogen Storage Material and Preparation Method Thereof. Chinese Patent Application No. 201510554498, 2 September 2015.
383. Gao, J. Automatic Split Charging Device of Hydrogen-Storage Alloy Powder. Chinese Patent Application No. 201320585007, 23 September 2013.
384. Huang, Y.; Wu, J.; Liu, H.; Li, R.; Xie, Y.; Yang, K. La-Mg-Ni Negative Electrode Hydrogen Storage Material Used in Power-Type Nickel-Metal Hydride Battery. Chinese Patent Application No. 201110339760, 1 November 2011.
385. Liu, H.; Li, R.; Wu, J. La-Mg-Ni Negative Hydrogen Storage Material for Nickel-Hydrogen Batteries. Chinese Patent Application No. 200910163893, 14 August 2009.
386. Liu, H.; Wu, J. AB$_5$ Type Negative Pole Hydrogen Storing Material. Chinese Patent Application No. 200810175124, 30 October 2008.
387. Liu, H.; Wu, J. AB$_5$ Type Negative Pole Hydrogen Storing Material and Preparation Method Thereof. Chinese Patent Application No. 200510083535, 8 July 2005.
388. Liu, H.; Wu, J.; Liu, J. AB$_5$ Negative Pole Hydrogen Storing Material. Chinese Patent Application No. 200510083537, 8 July 2005.
389. Liu, H.; Wu, J.; Wang, M. AB$_5$ Type Negative Pole Hydrogen-Storage Material. Chinese Patent Application No. 200510099899, 9 September 2005.
390. Liu, H.; Wu, J. AB$_5$ Type Negative Pole Hydrogen-Storage Material. Chinese Patent Application No. 200510099901, 9 September 2005.
391. Liu, H.; Wu, J. AB$_5$ Type Negative Pole Hydrogen-Storage Material. Chinese Patent Application No. 200510099902, 9 September 2005.
392. Liu, H.; Wu, J. Pr$_5$Co$_{19}$ Type Cathode Hydrogen Storage Material and Application Thereof. Chinese Patent Application No. 200810181720, 4 December 2008.
393. Liu, H.; Tian, B. AB$_{3.5}$ Type Hydrogen-Storing Negative Pole Material and its Preparation Process and Use. Chinese Patent Application No. 200610148600, 22 November 2006.
394. Liu, H.; Liu, J. Novel Collector of Hydrogen Battery. Chinese Patent Application No. 201020676945, 23 December 2010.
395. Liu, H.; Liu, J. Current-Collecting Sheet with Novel Structure for Hydrogen Battery. Chinese Patent Application No. 201020676956, 23 December 2010.
396. Liu, H.; Liu, J. Easy-To-Weld Current-Collecting Sheet of Hydrogen Cell. Chinese Patent Application No. 201020676958, 23 December 2010.
397. Liu, H.; Liu, J. Easily-Welded Current-Collecting Sheet for Hydrogen Battery. Chinese Patent Application No. 201020676935, 23 December 2010.
398. Yang, S.; Wang, X.; Zhou, J.; Zhou, F. Solar Energy Nickel-Hydrogen Charging Battery Manufacture Process. Chinese Patent Application No. 200610032433, 20 October 2006.

399. Fierro, C.; Fetcenko, M.A.; Young, K.; Ovshinsky, S.R.; Sommers, B.; Harrison, C. Nickel Hydroxide Positive Electrode Material Exhibiting Improved Conductivity and Engineered Activation Energy. U.S. Patent 6,228,535, 8 May 2001.

400. Gao, W.; Wang, Q.; Gong, J.; Meng, X.; Wang, T.; Zhang, X.; Yang, Z.; Fan, G.; Wang, W.; Chen, J. Method for Preparing Pure-Phase Spherical Nickel Hydroxide. Chinese Patent Application No. 201410551376, 17 October 2014.

401. Gao, W.; Wang, Q.; Fan, G.; Wang, W.; Chen, J.; Ma, M.; Zhao, S.; Tao, S. Method for Removing Sulfate Radicals in Nickel Hydroxide. Chinese Patent Application No. 201510803603, 20 November 2015.

402. Zhou, T.; Geng, W.; Xin, H.; Wu, Y.; Zhao, M.; Zhang, J.; Lai, J.; Wang, R.; Wang, Y.; Cui, W. Nickel Hydroxide Raw Material Iron Removing Process and Pneumatic Stirring and Precipitating Impurity Removing Tank Adopting Nickel Hydroxide Raw Material Iron Removing Process. Chinese Patent Application No. 201510812812, 20 November 2015.

403. Yin, Z.; Xu, Y.; Zhang, H.; Han, H. Method for Producing Spherical Nickel Hydroxide. Chinese Patent Application No. 200510017529, 20 April 2005.

404. Xu, Y.; Yin, Z.; Kang, B. Method for Producing Nickel Hydroxide Used for High Temperature Battery. Chinese Patent Application No. 200510017528, 20 April 2005.

405. Xu, Y.; Kang, B.; Yin, Z.; Wei, L.; Wang, M.; Han, H.; Zhang, H.; Yuan, J. Spherical Nickel Hydroxide with Composite Cobalt Layer Being Coated and Preparation Method. Chinese Patent Application No. 200510017530, 20 April 2005.

406. Chang, Z.; Shangguan, E.; Wu, F.; Tang, H.; Cheng, D.; Xu, Y.; Yin, Z. Method for Electrolyzing Nickel Hydroxide Covering Hydroxyl Cobalt Oxide Layer. Chinese Patent Application No. 200710054761, 28 June 2007.

407. Chang, Z.; Shangguan, N.; Wu, F.; Tang, H.; Cheng, D.; Xu, Y.; Yin, Z. Method for Electrolytic Preparing Hydroxy Cobalt Nickel Oxide. Chinese Patent Application No. 200710054762, 28 June 2007.

408. Yin, Z.; Xu, Y. Method of Cladding Hydroxy Cobalt Oxide on Spherical Nickel Hydroxide Surface. Chinese Patent Application No. 200510048557, 11 November 2005.

409. Zhang, B.; Zhao, Q.; Wang, Y.; Dong, Y.; Zhang, H. Cobalt Coating Method for Spherical Nickel Hydroxide by Mechanical Fusion. Chinese Patent Application No. 201210405104, 22 October 2012.

410. Wang, Y.; Zhao, Q.; Zhang, H.; Yang, X.; Zhang, B.; Dong, Y. Spherical Cobalt-Coated Nickel Hydroxide Applied to Power Battery and Preparation Method of Spherical Cobalt-Coated Nickel Hydroxide. Chinese Patent Application No. 201510024014, 19 January 2015.

411. Wang, D.; Sun, W.; Wang, X.; Li, S.; Wang, Y.; Sun, C.; Lu, H.; Zhang, Z.; Zhang, C. Dry Gaseous Oxidation Preparation of Spherical Cobalt-Coated Nickel Hydroxide. Chinese Patent Application No. 201110344412, 4 November 2011.

412. Wang, D.; Li, S.; Wang, Y.; Lu, H.; Sun, W. Special Vertical Oxidizer for Gas-Phase Oxidation Cobalt-Coated Spherical Nickel Hydroxide. Chinese Patent Application No. 201120310715, 24 August 2011.

413. Sun, W.; Wang, X.; Li, S.; Wang, D.; Wang, Y.; Lu, H.; Zhang, Z.; Zhang, C. Liquid Phase Oxidation Method for Preparing Co-Coated Spherical Nickel Hydroxide. Chinese Patent Application No. 201110344391, 4 November 2011.

414. Wang, D.; Wang, X.; Lu, H.; Li, S.; Wang, Y.; Sun, C.; Sun, W. Technology for Producing Covering Cobalt Spherical Nickelous Hydroxide through Continuous Method. Chinese Patent Application No. 201210344973, 15 September 2012.

415. Wang, D.; Sun, W.; Wang, X.; Li, S.; Lu, H.; Wang, Y.; Zhang, C.; Zhang, Z.; Sun, C. Method for Preparing Spherical Nickel Hydroxide by Doping Zinc, Manganese and Aluminum. Chinese Patent Application No. 201110382222, 25 November 2011.

416. Sun, W.; Li, S.; Qu, Z.; Wang, X.; Sun, C.; Hu, X.; Lu, H.; Wang, D. Spherical Nickel Hydroxide for High-Temperature Battery Anode and Preparation Method Thereof. Chinese Patent Application No. 201010557786, 24 November 2010.

417. Li, X.; Liu, P.; Luo, M.; Xiao, Z.; Sun, W.; Li, S. Preparation Method of Spherical Nickel Hydroxide Used as Battery Cathode Material. Chinese Patent Application No. 201110180469, 30 June 2011.

418. Sun, W.; Li, S.; Wu, X.; Lu, H.; Wang, D.; Sun, C.; Jiang, H. Special Reactor for Producing Dynamic Spherical Nickel Hydroxide. Chinese Patent Application No. 201120199983, 14 June 2011.

419. Wang, D.; Sun, W.; Sun, C.; Wang, X.; Li, S.; Lu, H.; Wang, Y. Power-Type Dedicated Oxidizer for Cobalt-Coated Spherical Nickel Hydroxide Production. Chinese Patent Application No. 201120310679, 24 August 2011.
420. Lu, H.; Sun, W.; Li, S.; Wang, X.; Wang, D.; Wang, Y. Reaction Kettle Used for the Production of Coating Cobalt Hydroxide on Spherical Nickel Hydroxide. Chinese Patent Application No. 201120310734, 24 August 2011.
421. Wang, D.; Sun, W.; Sun, C.; Li, S.; Wang, Y.; Lu, H. Reaction Kettle System for Producing Cobalt-Covered Spherical Nickel Hydroxide through Continuous Method. Chinese Patent Application No. 201220510191, 8 October 2012.
422. Ding, H.; Wu, X.; Jun, W.; Sun, W. Spherical Nickel Hydroxide Microwave Drying System. Chinese Patent Application No. 201020555163, 30 September 2010.
423. Li, S.; Sun, W.; Wang, X.; Lu, H.; Wang, Y.; Wang, D.; Sun, C.; Zhang, C.; Zhang, Z. Electrochemistry-Chemistry Mixed Oxidation Tank of Preoxidized Spherical Nickel Hydroxide. Chinese Patent Application No. 201120347053, 16 September 2011.
424. Li, S.; Sun, W.; Wang, X.; Jiang, H.; Lu, H.; Wang, D.; Sun, C.; Li, X. Static Impedance Tester for Positive Powder Materials of Cobalt-Overlapped Type Spherical Nickel Hydroxide. Chinese Patent Application No. 201220048827, 13 February 2012.
425. Li, S.; Sun, W.; Ding, H.; Wang, X.; Sun, C.; Lu, H.; Wang, D. Method of Recovering and Treating Edge Scraps of Cobalt-Coated Spherical Nickel Hydroxide. Chinese Patent Application No. 201110159518, 14 June 2011.
426. Du, Y.; Lu, K.; Yang, L.; Xie, Q. Production Process for Spherical Nickel Hydroxide. Chinese Patent Application No. 201510864847, 1 December 2016.
427. Du, Y.; Lu, K.; Yang, L.; Xie, Q. Production Technology of Spherical Cobalt-Coated Nickel Hydroxide. Chinese Patent Application No. 201510917705, 14 December 2015.
428. Zhang, J.; Tao, W.; Tang, Z.; Dai, G.; Jiang, S.; He, Y.; Chen, X.; Deng, M.; Peng, W.; Zhou, X. Preparation Method of Spherical Nickel Hydroxid. Chinese Patent Application No. 200610031587, 28 April 2006.
429. Zhou, Q.; Tan, S.; Zhang, H.; Hu, Z.; Yuan, Q. Making Method of Nickel Hydroxide with Coated Gamma Hydroxy Cobalt Oxide. Chinese Patent Application No. 200710035313, 9 July 2007.
430. Tian, J.; Liu, H.; Guo, S.; Shu, Z. High Specific Capacity Discharge Platform Spherical Nickel Hydroxide. Chinese Patent Application No. 200910303136, 10 June 2009.
431. Tian, J.; Guo, S.; Liu, H.; Shu, C. Spherical Nickel Hydroxide for High-Temperature Nickel Batteries and Preparation Method Thereof. Chinese Patent Application No. 200910303135, 10 June 2009.
432. Liu, H.; Liu, L.; Liang, Z.; Tan, Y.; Shu, C.; Tian, J.; Wang, Q. Cobalt-Coated Ball Nickel Drying Unit. Chinese Patent Application No. 201120001640, 5 January 2011.
433. Tan, Y.; Li, C.; Yuan, S.; Liu, L.; Tian, J. Device for Preparing Spherical Nickel Hydroxide. Chinese Patent Application No. 201220297326, 25 June 2012.
434. Shu, C.; Li, C.; Yuan, S.; Liu, L.; Tian, J. Device for Preparing Spherical Nickel Hydroxide. Chinese Patent Application No. 201220297328, 25 June 2012.
435. Shu, C.; Li, C.; Yuan, S.; Liu, L.; Tian, J. Device for Preparing Spherical Nickel Hydroxide. Chinese Patent Application No. 201210209117, 25 June 2012.
436. Li, C.; Liu, L.; Yuan, S.; Tan, Y.; Tian, J. Reaction Kettle Device for Preparing and Coating Spherical Nickel Hydroxide. Chinese Patent Application No. 201210209131, 25 June 2012.
437. Tian, J.; Liu, L.; Liu, H.; Shu, Z.; Tang, Z. Preparation Method of Spherical Nickel Hydroxide Anode Material Coated with Gamma-Hydroxy Cobalt Oxide. Chinese Patent Application No. 201110031535, 29 January 2011.
438. Fu, C.; Yang, Z. Method for Preparing Nickel Positive Electrode of Nickel-Metal Hydride Secondary Battery by Using Cobalt-Aluminum Hydrotalcite and Application Thereof. Chinese Patent Application No. 201410726996, 3 December 2014.
439. Gao, F.; Wu, L. Synthesis Technology of Nano Cobalt Oxyhydroxide. Chinese Patent Application No. 201310375503, 26 August 2013.
440. Gao, F. Synthesis Process of Beta-Type Nanoscale Cobalt Hydroxide. Chinese Patent Application No. 201210294360, 17 August 2012.
441. Yang, X.; Chen, M. High Temperature Nickel-Hydrogen Battery. Chinese Patent Application No. 200510040591, 17 June 2005.

442. Yang, X.; Wang, H.; Yang, Y.; Xu, Y. Zirconium Added Nickel Hydroxide and its Preparing Method. Chinese Patent Application No. 200610041099, 22 July 2006.
443. Endo, T.; Satoh, K.; Suzuki, M.; Uchiyama, S. Imformation Processing Method and Apparatus for Finding Position and Orientation of Targeted Object. Chinese Patent Application No. 200510069367, 13 May 2005.
444. Xu, K.; Guo, X.; Nie, Z.; Lai, D. Additive of Electrode Material for Nickel-Hydrogen Cell and Preparing Method Thereof. Chinese Patent Application No. 02156516, 16 December 2002.
445. Babjak, J.; Ettel, V.A.; Paserin, V. Method of Forming Nickel Foam. U.S. Patent 4,957,543, 18 September 1990.
446. Tian, J.; Liu, H.; Guo, S.; Shu, Z. High Specific Capacity Discharge Platform Spheical Nickel Hydroxide. Chinese Patent Application No. 98230333, 10 June 2009.
447. Tian, J.; Guo, S.; Liu, H.; Shu, C. Spherical Nickel Hydroxide for High-Temperature Nickel Batteries and Preparation Method Thereof. Chinese Patent Application No. 99115542, 10 June 2009.
448. Hunan Corun Sues Vale Inco for Patent Infringement. Available online: http://www.chinaknowledge.com/Newswires/News_Detail.aspx?type=1&NewsID=19435 (accessed on 12 April 2017).
449. Xiao, J.; Hu, P.; Li, G. Nickel Foam Material with High Tensile Strength. Chinese Patent Application No. 201410589622, 29 October 2014.
450. Xiao, J.; Li, L. Electroplating Technology for Preparing High-Tensile-Strength Foamed Nickel. Chinese Patent Application No. 201410576054, 26 October 2014.
451. Zhong, F.; Tao, W. Special Foamed Nickel Material. Chinese Patent Application No. 200410022873, 6 February 2004.
452. Xiao, J.; Hu, P.; Li, L. Preparation Method of Carbon Nanotube-Coated Foam Nickel. Chinese Patent Application No. 201410566347, 23 October 2014.
453. Chen, H.; Xiao, J.; Hu, P.; Zhong, J.; Yu, X.; Zhu, J.; Yu, K. Method for Preparing Cathode Substrate Material, Foamed Nickel of Vehicle-Mounted Power Batteries. Chinese Patent Application No. 201410484828, 22 September 2014.
454. Hu, P.; Zhu, J.; Zhou, H. Multi-Stage Electroplating Device for Foamed Nickel for HEV (Hybrid Electrical Vehicle). Chinese Patent Application No. 201420072988, 20 February 2014.
455. Chen, H.; Xiao, J.; Hu, P.; Zhong, J.; Yu, X.; Zhu, J. Positive Plate of Nickel-Hydrogen Power Battery, Preparation Method Thereof and the Nickel-Hydrogen Power Battery. Chinese Patent Application No. 201410421575, 26 August 2014.
456. Xiao, J.; Li, G. Porous Foam Nickel Cobalt Material Preparation Method. Chinese Patent Application No. 201410576524, 26 October 2014.
457. Zhong, Z.; Tao, W.; Dong, X. Porous Foam Iron Alloy and Porous Foam Fe-Based Materials and their Fabrication Method. Chinese Patent Application No. 01114535, 14 June 2001.
458. Tang, Y.; He, Y.; Tao, W.; Hu, Z.; Xiao, T.; Duan, Z.; Zhu, J.; Zhou, G.; Long, W.; Zhou, X. Technology of Preparing Foamed Nickel by Earlier Stage Crystallization Treatment. Chinese Patent Application No. 03124796, 5 September 2003.
459. Cheng, H.; Zhu, J.; Zhong, J.; Xiao, J.; Hu, P.; Li, J. Combined Type Continuous Electroplating Device. Chinese Patent Application No. 201420073016, 20 February 2014.
460. Xiao, J.; Hu, P.; Li, G. Method for Preparing Microfine Nickel Powder by Using Waste Foam Nickel Material. Chinese Patent Application No. 201410566205, 23 October 2014.
461. Wang, N.; Bi, Y.; Song, X.; Huang, X.; Su, I.; Li, C.; Jiang, I.; Luo, I. Preparation of Alloy Foaming Nickel. Chinese Patent Application No. 200410035761, 21 September 2004.
462. Song, X.; Wang, N.; Bi, Y.; Jiang, I.; Luo, I.; Meng, G.; Li, C.; Su, I.; Zhang, Q. Foam Nickel for Powder Nickel-Hydrogen Cell and Preparing Method Thereof. Chinese Patent Application No. 200810237966, 4 December 2008.
463. Li, G.; Fu, L. High Efficiency Producing Method for Purified Porcine Leukocyte Interferon by Using Antibody Affinity Chromatography Technology. Chinese Patent Application No. 201310008039, 10 January 2013.
464. Wang, N.; Bi, Y.; Song, X.; Jiang, I.; Li, C. Producton of Superthick Foaming Nickel or Copper. Chinese Patent Application No. 200610044768, 14 June 2006.
465. Wang, N.; Song, X.; Li, C.; Bi, Y.; Jiang, Y.; Liu, J.; Li, W.; Su, Y.; Meng, G.; Jiang, S. Production Process for Increasing Purity of Foamed Nickel and Special Forming Press. Chinese Patent Application No. 201310742388, 30 December 2013.

466. Song, X.; Bi, Y.; Wang, N.; Jiang, Y.; Liu, J.; Li, W.; Su, Y.; Li, C.; Meng, C.; Jiang, S. High-Consistency Foam Nickel Preparation Equipment and Prepartion Method. Chinese Patent Application No. 201310746234, 30 December 2013.

467. Bi, Y.; Liu, J.; Song, X.; Wang, N.; Jiang, Y.; Li, W.; Li, C.; Tian, L. Preparation Equipment of High-Consistency Nickel Foam. Chinese Patent Application No. 201320884245, 30 December 2013.

468. Song, X.; Wang, N.; Bi, Y.; Liu, H.; Jiang, Y.; Luo, Y.; Li, C.; Meng, G. Foam Nickel with Ultra-Low Surface Density and High Aperture Ratio and Manufacturing Method Thereof. Chinese Patent Application No. 201110434825, 22 December 2011.

469. Bi, Y.; Wang, N.; Liu, J.; Song, X.; Li, W.; Jiang, Y.; Meng, G.; Tian, L. Efficeient Energy-Saving Vacuum Acquisition Equipment for Nickel Foam Production. Chinese Patent Application No. 201320861669, 25 December 2013.

470. Song, X.; Bi, Y.; Li, W.; Wang, N.; Jiang, Y.; Liu, J.; Su, Y.; Li, C.; Meng, G.; Jiang, S. Production Process Capable of Improving Consistency of Foamed Nickel and Used Based Material Pretreatment Device. Chinese Patent Application No. 201310742319, 30 December 2013.

471. He, F.; Ren, C.; Feng, X.; Liu, Y. Novel Foam Nickel Electroplating Groove. Chinese Patent Application No. 201520560084, 30 July 2015.

472. Liu, Y.; He, F.; Jia, L. An Ultra-High Surface Density of Nickel Foam Processing. Chinese Patent Application No. 201610885713, 11 October 2016.

473. Kong, D.; Kong, J.; Du, Y. Non-Woven Fabric for Manufacturing Ni-MH Battery Sulfonated Membrane, Ni-MH Battery Sulfonated Membrane and Manufacturing Method Thereof. Chinese Patent Application No. 201010124563, 3 March 2010.

474. Kong, D. Compound Nylon Diaphragm, and Manufacturing Method. Chinese Patent Application No. 200610086130, 1 September 2006.

475. Kong, D. Grafted Polypropylene Diaphragm and Manufacturing Method. Chinese Patent Application No. 200610088413, 23 August 2006.

476. Fang, K.; Cai, N.; Xu, J. Battery Diaphragm Sulphonation Equipment and Sulphonation Treatment Technique. Chinese Patent Application No. 200810249733, 27 May 2009.

477. Fang, K.; Xu, J. Sulfonated Battery Diaphragm. Chinese Patent Application No. 200910255811, 30 December 2009.

478. Gang, K.; Xu, J. Novel Nickel-Hydrogen Battery Diaphragm. Chinese Patent Application No. 201010589184, 15 December 2010.

479. Xu, J.; Fang, K.; Xu, Y.; Zhong, Y. Grafting Treatment Apparatus and Grafting Treatment Method for Battery Diaphragm Material. Chinese Patent Application No. 201310295941, 16 July 2013.

480. Fang, K.; Xu, J. Testing Device Used for Maximum Apertures of Diaphragms of Nickel-Metal Hydride Batteries. Chinese Patent Application No. 201020660909, 15 December 2010.

481. Xu, J.; Fang, K.; Xu, Y.; Zhou, Y. Multifunctional Battery Diaphragm Rewinding Equipment. Chinese Patent Application No. 201320419479, 16 July 2013.

482. Huang, B.; Huang, L.; Zhong, Z.; Shen, Y.; Chen-guan, F.; Li, Z. Composite Diaphragm of Nickel-Metal Hydride Battery. Chinese Patent Application No. 201320053446, 31 January 2013.

483. Liu, Z. Method for Detecting Nickel-Hydrogen Battery Separator Wet Electric Resistance and Device Thereof. Chinese Patent Application No. 200910147639, 10 June 2009.

484. Wang, P. Preparation Method of Nickel-Hydrogen Battery Separator. Chinese Patent Application No. 201610949179, 26 October 2016.

485. Cui, G.; Cui, P.; Zhao, H. High-Power Nickel-Metal Hydride Battery Diaphragm. Chinese Patent Application No. 201610859718, 29 September 2016.

486. Gao, J. Nickel-Metal Hydride Battery Diaphragm Base Material Production Method. Chinese Patent Application No. 201310416317, 13 September 2013.

487. Wang, W. Feeding Device of Electroplating Material Filling Roller. Chinese Patent Application No. 201420333161, 20 June 2014.

488. Gu, Y.; Liu, Y. Plating Solution for Tumble-Plating and Tumble-Plating Method of Steel Battery Shell. Chinese Patent Application No. 201010185248, 26 May 2010.

489. Wang, W.; Li, D. Intermittent Type Control System of Electroplating Production Line. Chinese Patent Application No. 201320401753, 8 July 2013.

490. Sun, X.; Yang, W. Combined Type Water-Saving Cleaning Tank for Automatic Nickel Barrel Plating Produciton Line. Chinese Patent Application No. 201420715505, 25 November 2014.
491. Wang, W. Electroplated Battery Negative Electrode Bottom Cover Quality Detecting Plate. Chinese Patent Application No. 201420329710, 19 June 2014.
492. Sun, X. Heat Energy Recovery Device Of Nickel Plating Production Line. Chinese Patent Application No. 201520006502, 6 January 2015.
493. Liu, Y. Battery Shell Discharge Disc Barrel Plating Device. Chinese Patent Application No. 201511001904, 29 December 2015.
494. Liu, Y.; Pu, J.; Deng, X.; Le, G. Duplex Nickel Plating Technology for Steel Battery Shells and Steel Battery Shells Prepared Through Same. Chinese Patent Application No. 201510965886, 22 December 2015.
495. Liu, Y.; Pu, J.; Deng, X.; Le, G. Three-Layer Nickel Plating Process for Battery Steel Shell and Battery Steel Shell Manufactured by Process. Chinese Patent Application No. 201510965899, 27 April 2015.
496. Zhang, C. Battery Steel Shell Structure. Chinese Patent Application No 201210438295, 6 November 2012.
497. Cai, Y. Chemical Plating Method Used for Battery Steel Shell. Chinese Patent Application No. 201610689958, 19 August 2016.
498. Li, C.; Wang, Z. Nickel-Hydrogen Battery Outer Shell Benefitting for Heat Radiation. Chinese Patent Application No. 200710021199, 3 April 2007.
499. Li, C.; Wang, Z. Heat Radiation Favourable Nickel-Hydrogen Battery Case. Chinese Patent Application No. 200720036503, 19 March 2008.
500. Hu, J.; Wang, M.; Li, Y.; Lu, K.; Jiang, X. Processing Method of Perforated Non-Ferrous Metal Band for Battery Plate. Chinese Patent Application No. 200610033972, 22 February 2006.
501. Xu, K.; Xue, J. A Punched Steel Belt for High-Power Battery Electrode. Chinese Patent Application No. 00134512, 8 December 2000.
502. Deng, G.; Li, Q.; Wang, D.; Nan, C.; Liu, D.; Guo, J. Nickel Plating Steel Strip. Chinese Patent Application No. 201320416124, 12 July 2013.
503. Deng, F.; Li, Q.; Wang, D.; Nan, C.; Liu, D.; Guo, J. Nickel-Plated Steel Strip and Preparation Method Thereof. Chinese Patent Application No. 201310294031, 12 July 2013.
504. Lei, S. Continuous Material Feeding and Electroplating Equipment for the Connector Case and its Continuous Electroplating Technique. Chinese Patent Application No. 201610909773, 19 October 2016.
505. Wang, Q.; Zhao, X. Rolling Equipment for Manufacturing Substrate Strip for Cathode of Nickel-Metal Hydride Battery. Chinese Patent Application No. 201010297387, 30 September 2010.
506. Chen, Y.; Shen, J.; Jiang, X.; Cai, Y.; Zhao, X. High-Strength Light-Weight Conducting Base Electrode of Ni-H Battery. Chinese Patent Application No. 99121593, 9 October 1999.
507. Chen, W.; Chen, D.; Che, W.; Li, P. Damp-Proof Nickel-Metal Hydride Battery Protective Decive for Underway Beacon Lamps. Chinese Patent Application No. 201610571260, 20 July 2016.
508. Chen, D.; Chen, C.; Chen, W.; Liu, Y. Nickel-Metal Hydride Battery Power Source of Mine Lamp. Chinese Patent Application No. 201610571749, 20 July 2016.
509. Chen, D.; Chen, C.; Chen, W.; Liu, I. Mine Lamp Nickel-Hydrogen Battery Power. Chinese Patent Application No. 201620763499, 20 July 2016.
510. Fan, H.; Zhang, Z.; Dong, D.; Xin, E.; Che, B. NiMH Explosion-Sroof Power Box with Remote Monitoring Function. Chinese Patent Application No. 201420617872, 24 October 2014.
511. Chen, W.; Chen, D.; Chen, C.; Zhang, G. High-Speed Electric Tool Storage Power Supply Battery. Chinese Patent Application No. 201610571368, 20 July 2016.
512. Chen, W.; Chen, D.; Chen, C.; Zhang, G. High-Speed Electric Tool Deposit Supply Battery. Chinese Patent Application No. 201620763441, 20 July 2016.
513. Sun, L. Rechargeable Flashlight. Chinese Patent Application No. 200920200487, 23 November 2009.
514. Huang, Y. Nickel-Hydrogen Battey LED Light for Keyholes. Chinese Patent Application No. 201610536477, 8 July 2016.
515. Chen, L. Pager Without Batteries. Chinese Patent Application No. 96201763, 26 January 1996.
516. Chen, T.; Li, D.; Teng, D.; Bu, X.; Tao, J.; Sheng, J.; Luan, F; Tan, Y.; Jiao, C.; Shan, H. Electric Power Management Circuit for Hand-Held Equipment. Chinese Patent Application No. 200520037199, 16 December 2005.

517. Liu, M. Rechargeable Active Radio Frequency Identification Card. Chinese Patent Application No. 200920277920, 10 December 2009.
518. Wu, Y. Portable Dust Collector. Chinese Patent Application No. 201310609820, 27 November 2013.
519. Wen, X. Essential Safety Type Infrared Temperature Measuring Aparatus. Chinese Patent Application No. 01265963, 7 December 2001.
520. Tan, G.; Huang, B. Wireless Humidifier. Chinese Patent Application No. 201510372005, 29 June 2015.
521. Wang, J.; Zhang, X. Wall-Climbing Robot. Chinese Patent Application No. 201010147738, 15 April 2010.
522. Cai, H.; Yu, H. Be Applied to Power Battery of Colliery Robot. Chinese Patent Application No. 201521085907, 8 June 2016.
523. Cai, H.; Yu, L. Intelligence Power Battery Group of Mining Robot. Chinese Patent Application No. 201521085823, 22 December 2015.
524. Cai, H.; Yu, L. Intelligence High Performance Power Battery Group of Mining Robot. Chinese Patent Application No. 201521085824, 22 December 2015.
525. Cai, H.; Yu, L. A Power Battery Group that Robot that is Arranged in Colliery to Carry Out Safety Inspection and Recover. Chinese Patent Application No. 201521077403, 22 December 2015.
526. Wei, J. Electric Lopping Device. Chinese Patent Application No. 200820003370, 1 February 2008.
527. Chen, M.; Yu, J.; Zheng, Y.; Ma, Z. Expandable Battery Assembly of Wearable Equipment. Chinese Patent Application No. 201310288077, 10 July 2013.
528. Zhang, B.; Xu, G.; Hou, Q. Cost Based Method for Optimizing External PHEV (Plug-in Hybrid Electric Vehicle) Power Assembly and Application Thereof. Chinese Patent Application No. 201110095690, 15 April 2011.
529. Yan, Y.; Wang, W.; Lou, Y.; Xia, B. Accumulator Battery for Mixed Motor Vehicle of Wind Cooling Forced Radiating Structure. Chinese Patent Application No. 200620047078, 25 October 2006.
530. Yang, Y.; Zhang, X.; Qin, D.; Ren, Y.; Zhou, A.; Hu, M.; Su, L.; Zhao, C. Heat Radiation System of Nickel-Hydrogen Battery Set for Hybrid Power Vehicle. Chinese Patent Application No. 200810069859, 20 June 2008.
531. Li, J.; Zhao, C.; Ren, Y.; Qiao, X. Nickel-Metal Hydride Battery Pack Heat Dissipation Device for Hybrid Electric Vehicle. Chinese Patent Application No. 200910191514, 20 November 2009.
532. Chen, X.; Huang, Q.; Wang, Q. Hybrid Electric Vehicle Battery System with Temperature Adjustment Function. Chinese Patent Application No. 200920053771, 31 March 2009.
533. Duan, Q.; He, J. Nickel Hydrogen Battery Component for Hybrid Electric Vehicle (240QNYD6). Chinese Patent Application No. 200730312745, 22 September 2007.
534. Duan, Q.; He, J. Nickel Hydrogen Battery Component for Hybrid Electric Vehicle (A21). Chinese Patent Application No. 200730315661, 25 October 2007.
535. Fei, Y. Method for Processing Negative Plate of Nickel-Hydrogen Battery. Chinese Patent Application No. 200710194192, 4 December 2007.
536. Fei, Y. Inline Method for Single Battery of Nickel-Hydrogen Battery Module. Chinese Patent Application No. 200910138016, 4 November 2009.
537. Fei, Y. Novel Nickel-Hydrogen Battery Safety Device. Chinese Patent Application No. 200920233796, 24 July 2009.
538. Fei, Y. Novel Nickel-Hydrogen Battery Air-Pressure Safety Valve. Chinese Patent Application No. 200920233795, 2 June 2010.
539. Fei, Y. Nickel-Hydrogen Battery Pressure Safety Valve. Chinese Patent Application No. 200820215927, 25 November 2008.
540. Fei, Y. Making Method of Alkalescent Accumulator Anode and Cathode Board. Chinese Patent Application No. 200710024527, 21 June 2007.
541. Feng, Y. An Inter-Linked Method for Monocase Battery of Dynamic Battery Unit. Chinese Patent Application No. 200710194193, 4 December 2007.
542. Fei, Y. Method for Modifying Spherical Nickel Hydroxide by Nano Carbon Technique. Chinese Patent Application No. 200710024530, 21 June 2007.
543. Fei, Y. Battery Module Device Capable of Balancing Air Pressure of Monomer. Chinese Patent Application No. 201020022872, 8 January 2010.

544. Chen, W.; Chen, W.; Chen, D.; Li, P. High-Volume Nickel Metal Hydride Assembly Battery of Electric Automobile. Chinese Patent Application No. 201610571145, 20 July 2016.

545. Chen, W.; Chen, W.; Chen, D.; Li, P. An Electric Car High-Capacity Nickel-Metal Hydride Battery Assembly. Chinese Patent Application No. 201620764331, 20 July 2016.

546. Zhou, X.; Chen, R.; Huang, W. High-Capacity Nickel-Hydrogen Battery. Chinese Patent Application No. 201110310096, 14 October 2011.

547. Kong, L.; Qiao, Z.; Ni, J.; Yang, T.; Feng, X. Large-Capacity Nickel Hydrogen Battery Pack Structure. Chinese Patent Application No. 201210106541, 12 April 2012.

548. Kong, L.; Qiao, Z.; Ni, J.; Yang, T.; Feng, X. Large-Capacity Nickel-Metal Hydride Battery Pack Structure. Chinese Patent Application No. 201220153590, 12 April 2012.

549. Shen, J.; Huang, J.; Xing, Z.; Ding, G.; Li, X.; Rong, R.; Cao, G.; Song, X. Nickel-Hydrogen Battery Module Structure. Chinese Patent Application No. 02286503, 4 December 2002.

550. Huang, J.; Shen, J.; Ge, J.; Rong, R.; Zhao, J. Electric Vehicle Nickel-Hydrogen Batteries Module. Chinese Patent Application No. 02286509, 4 December 2002.

551. Huang, J.; Shi, C.; Ding, G.; Ge, J.; Rong, R.; Zhao, J. Nickel-Hydrogen Batteries Assembly Structure for Electric Driven Vehicle. Chinese Patent Application No. 02286507, 26 November 2003.

552. Chen, D.; Chen, W.; Chen, W.; Liu, Y. Electric Vehicle High-Capacity Nickel Hydrogen Power Battery Overload Protector. Chinese Patent Application No. 201610571171, 20 July 2016.

553. Nie, J.; Dai, R.; Du, X.; Chen, K. A Battery Cycle Life Testing Method Using Pulse Charging for the Ni-MH Battery Used in the EV Application. Chinese Patent Application No. 201610975942, 7 November 2016.

554. Ran, J. Nickel-Hydrogen Battery Management System for Electric Vehicle. Chinese Patent Application No. 200710120418, 17 August 2007.

555. Zhang, Y.; Chen, Y.; Song, D.; Chen, J.; Cao, X.; Wong, G.; Zhou, Z. Nickel-Hydride Accumulator for Large Capacity Electric Vehicle. Chinese Patent Application No. 92112168, 19 October 1992.

556. Chen, Y. Nickel-Hydride Secondary Battery Alloy Material Storing Hydrogen for Electro Car. Chinese Patent Application No. 95100252, 27 January 1995.

557. Chen, Y.; Cai, Y.; Yu, D. Large Capacity Rectangular Nickel Hydrogen Cell for Electric Vehicle. Chinese Patent Application No. 95227246, 30 December 1995.

558. Chen, Y.; Shen, J.; Jiang, X. Electrode of Laminated Bipolar Power-Type NiMH Battery and its Manufacture. Chinese Patent Application No. 99121592, 9 October 1999.

559. Chen, Y.; You, Y.; Yu, D.; Zhao, X. Large-Volume Column-Shaped Ni-MH Battery for Electric Vehicle. Chinese Patent Application No. 200520130130, 25 October 2005.

560. Shi, H. Vehicle Nickel-Hydrogen Battery Pack. Chinese Patent Application No. 201510367973, 29 June 2015.

561. Bian, L.; Zhang, H. Circular Nickel-Metal Hydride Battery and Manufacturing Method Thereof. Chinese Patent Application No. 201410170872, 25 April 2014.

562. Bian, L.; Zhang, H. Circular Nickel-Hydrogen Battery. Chinese Patent Application No. 201420207643, 25 April 2014.

563. Peng, J.; Su, G.; Guo, W.; Hua, M. Nickel-Metal Hydride Battery Plate Structure. Chinese Patent Application No. 201320579219, 18 September 2013.

564. Peng, J.; Guo, W.; Su, G.; Lin, Q. Ni-MH (Nickel-Metal Hydride) Battery with Heat Dissipation Function. Chinese Patent Application No. 201320579328, 18 September 2013.

565. Liu, Y.; Zhou, L.; Ni, R. Nickel-Hydrogen Battery Charging Method. Chinese Patent Application No. 201410176573, 29 April 2014.

566. Zhang, M. Combined Universal Automatic Green Trimmer. Chinese Patent Application No. 02158332, 21 December 2002.

567. Liang, J.; Ye, M.; Li, S.; Wu, X.; Cao, B. Electric Forklift Running Driving Control System Having Energy Recovery. Chinese Patent Application No. 200610042604, 31 March 2006.

568. Xu, K. Endurance Type Photovoltaic-Ni-MH-Air Force Energy Power Supply Equipment (System). Chinese Patent Application No. 201010206071, 9 June 2010.

569. Xu, K. Air Power-Photovoltaic Automobile. Chinese Patent Application No. 201110083444, 25 March 2011.

570. Niu, A. Shallow-Water Miniature-Underwater Robot System. Chinese Patent Application No. 201310356205, 15 August 2013.

571. Zhao, W.; Zhang, R.; Tian, X. A Novel Nickel-Hydrogen Battery Assembly. Chinese Patent Application No. 200310107600, 18 December 2003.

572. Che, Y. Ni-MH Secondary Battery for Vehicle with Electric Power as Auxiliary Power. Chinese Patent Application No. 96105382, 7 June 1996.

573. Sheng, H. Long-Mileage Electro Mobile with Batteries of Relatively Small Size and Low Weight. Chinese Patent Application No. 201210002351, 5 January 2012.

574. Liao, B. Middle Axle and Motor Two-In-One Electric Bicycle. Chinese Patent Application No. 200620013609, 18 April 2006.

575. Chen, Y.; Cai, Y.; Yu, D. High-Performance Light-Weight Electric Bicycle. Chinese Patent Application No. 99111629, 25 August 1999.

576. Wei, Q.; Che, X.; Xie, J.; Guo, W. Fill Electronic Commodity Circulation Car of Formula Soon. Chinese Patent Application No. 201520263789, 28 April 2015.

577. Yan, Y.; Zheng, L. Self-Maintained Power Supply Type Camera Device. Chinese Patent Application No. 200620026709, 17 June 2006.

578. Cheng, L.; Chen, C.; Mu, M.; Zheng, Q.; Zheng, X. Light Sparing Device for Fire-Fighting Emergency Indicating Lamp. Chinese Patent Application No. 201210548318, 12 December 2012.

579. Liu, H.; Zhou, S.; Chen, J.; Hu, S.; Zhang, C.; Xia, M.; Wang, H. Building Energy Storage Emergence Energy-Saving System. Chinese Patent Application No. 201310387634, 31 August 2013.

580. Dong, Y.; Guan, E.; Zhao, S.; Lu, Y.; Wang, J.; Guo, X.; Wang, D.; Ge, P.; Hao, X.; Meng, X. Nickel-Metal Hydride Battery Charge-Discharge Circuit in Uninterruptible Power Supply. Chinese Patent Application No. 201310324538, 30 July 2013.

581. Zheng, F.; Wan, C. Nickel-Hydrogen Battery Charging Method for Mobile Communication Apparatus. Chinese Patent Application No. 200810013905, 11 January 2008.

582. Shang, D.; Liu, D.; Sun, S.; Chen, Y.; Liu, L.; Han, J. Novel Solar Energy Street Lighting. Chinese Patent Application No. 201520936839, 16 November 2015.

583. Yang, C. Combination Type Battery and Solar Streetlight Using the Same. Chinese Patent Application No. 200810058442, 26 May 2008.

584. Yang, W.; Wu, Z.; Sun, J.; Wei, Z.; Yuan, F.; Feng, Z.; Xu, S.; Zhou, X.; Zhou, M. Photovoltaic Secondary Battery. Chinese Patent Application No. 200810117527, 1 August 2008.

585. Yang, W.; Wu, B. Photovoltaic Secondary Battery. Chinese Patent Application No. 201110330376, 26 October 2011.

586. Sun, L.; Ning, Z. Solar Charger for Nickel-Hydrogen Rechargeable Battery or Lithium Battery. Chinese Patent Application No. 201010581978, 10 December 2010.

587. Niu, L. Solar Light Bulb. Chinese Patent Application No. 201110233216, 11 August 2011.

588. Ni, J. Solar Flashlight. Chinese Patent Application No. 201010611393, 29 December 2010.

589. Huang, C. Automotive Nickel-Metal Hydride Battery Charging Device. Chinese Patent Application No. 201010292812, 27 September 2012.

590. Yao, Y.; Zhou, Q.; Yu, Z. Distributed Electric Energy Storage and Power Supply Method for Storing Energy and Supplying Power by Utilizing Batteries of Electric Automobile. Chinese Patent Application No. 200910242313, 11 December 2009.

591. Zhang, Y.; Shen, S.; Qi, Z.; Bao, H. Ni-MH Battery Energy Storage Monitoring System. Chinese Patent Application No. 201010550710, 19 November 2010.

592. Zhou, S.; Liu, H.; Ling, J.; Dai, Q.; Wu, R. Double-Power Device and Power Supplying Method Thereof. Chinese Patent Application No. 201110235549, 17 August 2011.

593. Zhang, Y.; Shen, S.; Qi, Z.; Bao, H. Nickel-Metal Hydride Battery Energy Storage Monitor System. Chinese Patent Application No. 201020614908, 19 Novmber 2010.

594. Liu, H.; Zhou, H.; Huang, L.; Zhou, S.; Xia, M. Microgrid Nickel-Metal Hydride Battery Energy Storage System. Chinese Patent Application No. 201310635057, 2 December 2013.

595. Guo, J.; Kuang, D.; Song, X.; Wang, B.; Yang, T. MW-Class Ni-MH Battery Energy Storage System. Chinese Patent Application No. 200910264102, 30 December 2009.

596. Liu, H.; Che, X.; Hu, S.; Zhou, H.; Huang, L.; Xu, J.; Zhong, F. Battery Pack Equalization Method for Energy Storage System Adopting Nickel-Series Storage Batteries. Chinese Patent Application No. 201510351344, 24 June 2015.

597. Wikipedia. 863 Program. Available online: https://en.wikipedia.org/wiki/863_Program (accessed on 12 April 2017).

598. Wu, F.; Wu, B.; Chen, S.; Wang, J.; Yang, K.; Mu, D.; An, W.; Gan, F. Nickelous Compound Anode Material Used in Nickel-Hydrogen Battery and Preparation Technique. Chinese Patent Application No. 200910085937, 8 June 2009.

599. Wu, F.; Wang, F.; Chen, S.; Zhang, C.; Wang, G. Additive for Reducing Nickel-Hydrogen Cell Internal Pressure. Chinese Patent Application No. 200410080263, 29 September 2004.

600. Wu, F.; Wang, F.; Chen, S.; Zhang, C.; Wang, G. Additive for Reducing Internal Pressure of NiMH Battery. Chinese Patent Application No. 200610140775, 27 September 2004.

601. Wu, F.; Li, L.; Chen, S.; Shan, Z.; Yang, K.; Wong, J.; Su, Y.; Wang, G. A Novel Method Used for Non-Destructive Battery Regeneration. Chinese Patent Application No. 200310117260, 10 December 2003.

602. Zhao, D.; Lin, H.; Yao, R.; Li, C.; Qiang, Z. Activated Graphene/Needle-Shaped Nickel Hydroxide Nanocomposite Material and Preparation Method Thereof. Chinese Patent Application No. 201610299450, 6 May 2016.

603. Xiang, X.; Tian, Z.; Li, F. Alpha-Nickel Hydroxide Nano/Micro Structure Material and Preparation Method Thereof. Chinese Patent Application No. 201010108802, 5 February 2010.

604. Sun, X.; Lu, Z.; Chang, Z.; Zhu, W.; Duan, X. Nickel Hydroxide Nanosheet Thin-Film Material as Well as Preparation Method and Application Thereof. Chinese Patent Application No. 201110205586, 21 July 2011.

605. Pan, J.; Ka, S.; Sun, Y.; Wan, P.; Wang, Z. Preparation Method of Multivariate Doping Spherical Alpha-Ni(OH)$_2$ with Nanometer Secondary Structure. Chinese Patent Application No. 200910237128, 11 November 2009.

606. Pan, J.; Zhang, J.; Sun, Y.; Wang, Z. Multi-Component Doped Spherical Nano Nickel Hydroxide Synthesized by Chemical-Electrochemical Combined Method. Chinese Patent Application No. 201110105234, 26 April 2011.

607. Pan, J.; Yue, X.; Sun, Y.; Wang, Z. Method for Synthesizing Spherical Nano-Structure Ni(OH)$_2$ through Atom Economical Method. Chinese Patent Application No. 201210201770, 19 June 2012.

608. Pan, J.; Yang, H.; Nei, Y.; Sun, Y. Nanometer Ni(OH)$_2$ Composite Material and Its Preparation Method. Chinese Patent Application No. 201610909487, 18 October 2016.

609. Lin, C.; Tian, G.; Chou, B.; Chen, Q.; Han, X. Method for Estimating Nickel-Hydrogen Power Battery Charged State Based on Standard Battery Model. Chinese Patent Application No. 200710064294, 9 March 2007.

610. Lin, C.; Tian, G.; Qiu, B.; Chen, Q.; Han, X. Method for Real-Time Evaluating Internal-External Temperature Difference of Nickel-Hydrogen Electro-Kinetic Cell. Chinese Patent Application No. 200710063061, 26 January 2007.

611. Pan, C.; Yu, R. Method and Apparatus for Improving High Rate Performance of Rare-Earth Series Nickel-Hydrogen Cell Minus Pole. Chinese Patent Application No. 200810224675, 23 October 2008.

612. Jiao, H.; Song, J.; Cai, S. Air Cooling Device of Nickel-Hydrogen Battery for Mixed Power Vehicle. Chinese Patent Application No. 01270981, 16 November 2001.

613. Shan, Z.; Tian, G.; Bi, X. Nickel-Hydrogen Cell Cathode Surface Treating Method. Chinese Patent Application No. 200410072718, 15 November 2004.

614. Chen, Y.; Han, Z.; Zheng, Y. Inner Pressure Measurer for Nickel-Hydrogen Battery. Chinese Patent Application No. 97228116, 9 October 1997.

615. Xu, X.; Liu, Q. Alkaline Battery Cellulose Triacetate-Metal Oxide Composite Membrane and Making Method Thereof. Chinese Patent Application No. 200510013199, 21 February 2005.

616. Tang, Z.; Liu, Y. Ball Milling Preparation Method for Alpha-Ni(OH)$_2$. Chinese Patent Application No. 200710057114, 10 April 2007.

617. Chen, J.; Li, S.; Tao, Z. Nickel-Hydrogen Power Battery. Chinese Patent Application No. 03130136, 18 June 2003.

618. Li, S.; Wu, F.; Yuan, H.; Zhou, W.; Han, L.; Qu, J.; Niu, G.; Gao, F.; Song, D.; Gao, Y. Apparatus for Making Negative Electrode of Nickel-Hydrogen Cell. Chinese Patent Application No. 98204215, 5 May 1998.

619. Jiao, X.; Cheng, B.; Yang, W.; Liu, Y.; Wang, X. Method for Producing Septum Substrate Cloth of Nickel-Hydrogen Battery with Sulphonation Method and Septum Substrate Cloth of Nickel-Hydrogen Battery. Chinese Patent Application No. 200610015843, 25 September 2006.

620. Chen, B.; Jiao, X.; Yang, W.; Ren, Y.; Liu, Y.; Li, J.; Geng, M. Production of nickel-hydrogen diaphragm and products thereof. Chinese Patent Application No. 200510014633, 26 July 2005.
621. Cui, W.; Wang, W. Effectively Cooling Nickel-Hydrogen Battery Structure. Chinese Patent Application No. 200810200650, 27 September 2008.
622. Wang, H.; Jin, X.; Wu, H.; Sha, C. Nickel-Metal Hydride Battery Diaphragm Material and Forming Method Thereof. Chinese Patent Application No. 201210258618, 24 July 2012.
623. Wang, H.; Jin, X.; Wu, H.; Zhang, Y. Nano-Coating Diaphragm Material and Forming Method Thereof. Chinese Patent Application No. 201210485694, 23 November 2012.
624. Yang, C.; Li, M.; Cui, R.; Chen, L.; Wen, Z.; Zhu, Y.; Zhao, M.; Li, J.; Jiang, Q. Preparation Method for Hydrogen Storage Alloy and Graphene Composite Material and Application of Composite Material. Chinese Patent Application No. 201510996700, 28 December 2015.
625. Yang, C.; Li, M.; Zhang, D.; Jing, W.; Jin, B.; Lang, X.; Jiang, Q. Method and Application for Improving Discharge Capacity and High Rate Discharge Performance of Hydrogen Storage Alloy through Co_3O_4 in-site Compounding. Chinese Patent Application No. 201610031157, 18 January 2016.
626. Yang, C.; Li, M.; Zhang, D.; Wang, C.; Wen, Z.; Zhu, Y.; Zhao, M.; Li, J.; Jiang, Q. Preparation Method of Hydrogen Storage Alloy and Nano-Porous Nickel Composite (HSAs/NPNi) and Application Thereof. Chinese Patent Application No. 201510996773, 28 December 2015.
627. Wang, C.; Yang, C.; Li, M.; Chen, L.; Wen, Z.; Zhu, Y.; Zhao, M.; Li, J.; Jiang, Q. Quick Identifying Method for Hydrogen Storage Alloy with Super Long Service Life. Chinese Patent Application No. 201610487597, 29 June 2016.
628. Yang, C.; Chen, L.; Zai, S.; Zhu, Y.; Wen, Z.; Zhao, M.; Li, J.; Jiang, Q. Method for Surface Modification on Hydrogen Storage Alloy by Molybdenum Disulfide, and Application Thereof. Chinese Patent Application No. 201610422629, 13 June 2016.
629. Han, S.; Du, W.; Liu, B. Preparation Method of $PuNi_3$ Type Single Phase Neodymium-Magnesium-Nickel Alloy Electrode Material. Chinese Patent Application No. 201610055742, 27 January 2016.
630. Han, S.; Shen, W.; Wang, Y.; Pei, Y. Method for Improving Electrochemical Performance of Hydrogen Storing Alloy by Using Polypyrrole. Chinese Patent Application No. 201210264478, 30 July 2012.
631. Chen, Y.; Tao, M.; Tu, M.; Wu, C. Non-Neohymium Rare-Earth System Electrode Material with Low-Temperature Hydrogen-Storage Alloy. Chinese Patent Application No. 200410040649, 9 September 2004.
632. Tu, M.; Chen, Y.; Tang, D.; Li, Q.; Li, N.; Yan, K. Hydrogen-Storage Alloy Electrode Material of Low-Co Lanthanum-Praseodymium-Cerium-Nickel Series. Chinese Patent Application No. 99115177, 28 September 1999.
633. Chen, Y.; Shen, X.; Tao, M.; Wu, C. Free Neodymium Rare-Earth Magnesium Series Low Temperature Store Hydrogen Electrode Alloy. Chinese Patent Application No. 200810045057, 26 March 2008.
634. Wang, C.; Tang, H.; Xu, B.; Chen, L. Method for Preparing Titanium Oxide Coating on Surface of Nickel Electrode of Nickel-Metal Hydride Battery. Chinese Patent Application No. 201310690581, 16 December 2013.
635. Wang, C.; Tang, H.; Xu, B.; Chen, L. Method for Preparing Aluminum Trioxide Coating on Surface of Nickel Electrode. Chinese Patent Application No. 201310687991, 16 December 2013.
636. Wang, C.; Tang, H.; Xu, B.; Chen, L. Method for Preparing Aluminum Trioxide-Nickel Oxide Ceramic Coating on Nickel Electrode Surface. Chinese Patent Application No. 201310689993, 16 December 2013.
637. Chen, C.; Chen, L.; Chen, Y.; Wang, X.; Wang, Q. Non Crystal State Hydrogen Storage Composite Material and Its Producing Method. Chinese Patent Application No. 200310122827, 21 December 2003.
638. Chen, L.; Yu, K.; Xiao, X.; Jiang, W.; Jia, Y.; Lei, Y.; Che, C.; Ying, T.; Ge, H. Amorphous State Titanium-Cuprum-Nickel-Base Hydrogen-Storing Composite Material and Preparation Method thereof. Chinese Patent Application No. 200810163647, 22 December 2008.
639. Liu, B.; Li, Z. Method for Improving Corrosion Resistance Performance of Magnesium-Based Hydrogen Storage Alloy by Using Ion Nitriding Method. Chinese Patent Application No. 201010137917, 24 August 2010.
640. Li, Z.; Liu, B. Method for Improving Corrosion Resistance of Hydrogen Storage Alloy by Surface Boronising Method. Chinese Patent Application No. 201010137919, 24 August 2010.
641. Feng, Y.; Lv, Y. Novel Nickel-Hydrogen Secondary Battery Adhesive. Chinese Patent Application No. 200810122172, 10 November 2008.
642. Lv, Y.; Chen, J.; Wang, X.; Su, H.; Feng, Y. Electrolyte Additive of Novel Nickel-Metal Hydride Secondary Battery. Chinese Patent Application No. 200910099669, 15 June 2009.

643. Lv, Y.; Feng, Y. Tartaric Acid Metal Complex-Added Anode of Nickel Hydrogen Secondary Battery. Chinese Patent Application No. 200810122171, 10 November 2008.

644. Lv, Y.; Chen, J.; Wang, X.; Su, H.; Feng, Y. Anode Provided with Mucus Acid Metal Compound for Nickel-Hydrogen Secondary Battery. Chinese Patent Application No. 200910099666, 22 December 2010.

645. Lu, Y.; Feng, Y. Nickel-Hydrogen Battery Anode Material Prepared from Glucaric Acid Metal Complex Doped Beta-Ni(OH)$_2$ and Method Thereof. Chinese Patent Application No. 200810062361, 19 May 2008.

646. Lei, L.; Kuang, G.; Wang, Q. Surface Treatment Method for Hydrogen Storage Alloy. Chinese Patent Application No. 201410040817, 28 January 2014.

647. Zhang, Y.; Zhuang, X.; Zhan, L.; Chen, J. High-Capacity Magnesium-Cobalt-Series Hydrogen-Storage Electrode Material and Preparation Method Thereof. Chinese Patent Application No. 201410690517, 25 November 2014.

648. Zhang, Y.; Zhan, L.; Zhuang, X.; Chen, J. Magnesium-Palladium-Cobalt Ternary Nickel-Hydrogen Battery Negative Material and Preparation Method Thereof. Chinese Patent Application No. 201410688254, 25 November 2014.

649. Lei, L.; Yu, X. Resource Separation and Recycling Production Method for Waste Nickel Hydrogen Battery Content. Chinese Patent Application No. 200810020326, 29 February 2008.

650. Liu, B.; Fan, Y.; Wang, K.; Zhang, B.; Wang, Y. Equipment and Method for Preparing Nickel-Base Hydrogen Storage Alloy Containing Rare-Earth, Magnesium and Boron. Chinese Patent Application No. 201610519616, 1 July 2016.

651. Liu, B.; Fan, Y.; Chen, Q.; Shang, X.; Zhang, B. Multi-Phase Magnesium Rare Earth Nickel Based Hydrogen Storage Alloy and Application Thereof. Chinese Patent Application No. 201610702384, 22 August 2016.

652. Feng, Z.; Wu, Y.; Peng, Z. Preparation Method of Negative Pole of Graphene-Based High-Capacity Nickel-Hydrogen Power Battery. Chinese Patent Application No. 201410452157, 5 September 2014.

653. Zhang, H.; Chen, Y.; Chen, Y.; Zhang, G. Carbon Nanometer Composite Nickel-Hydrogen Power Battery Cathode Sheet Preparation Method and Its Uses. Chinese Patent Application No. 200710030263, 17 September 2007.

654. Zhang, G.; Huang, X.; Rao, Z. Measuring Device for Heat Yield of Cylinder-Type Power Battery Material. Chinese Patent Application No. 201010101760, 25 January 2010.

655. Jian, X.; Chu, Y.; Li, W.; Feng, Z.; Zhao, T.; Zhang, W.; Luo, L. Preparation Method of Alpha Nickel Hydroxide and Application Thereof. Chinese Patent Application No. 201610552931, 12 July 2016.

656. Zhu, Y.; Huang, L.; Qiu, H.; Zhao, W.; Li, Q. Preparation Method and Application of Nanometer Nickel Hydroxide and Composite Electrode Therof. Chinese Patent Application No. 200910038501, 8 April 2009.

657. Zhang, Z.; Zhu, Y.; Zhou, C.; Bao, J.; Ye, X.; Wu, S.; Zheng, H.; Lin, X. Multi-Element Doped Nano Alpha-Ni(OH)$_2$ Material and Preparation Method Thereof. Chinese Patent Application No. 201010281338, 10 September 2010.

658. Zhou, Z.; Zhu, Y.; Zhang, Z.; Lin, X.; Ye, X.; Wu, S.; Zheng, H.; Bao, J. Nano Multiphase Nickel Hydroxide Containing Rare-Earth Elements and Synthetic Method of Nano Multiphase Nickel Hydroxide. Chinese Patent Application No. 201110152223, 8 June 2011.

659. Ouyang, L.; Li, L.; Zhu, M.; Min, D.; Wang, H.; Liu, J.; Chen, Y.; Xiao, F.; Sun, T.; Tang, R. A$_2$B$_7$ Nickel Hydrogen Battery Cathode Material and Preparation Method Thereof. Chinese Patent Application No. 201210444590, 8 November 2012.

660. Ouyang, L.; Wang, H.; Peng, C.; Zhu, M. Multi-Layer Film Electrode for Nickel-Hydrogen Cell and Preparation Method Thereof. Chinese Patent Application No. 200410077527, 22 December 2004.

661. Zhu, H.; Zhong, Z.; Ouyang, L.; Zeng, M.; Wang, H. Film Electrode for Nickel-Hydrogen Battery and Its Preparation Method Thereof. Chinese Patent Application No. 02115117, 22 April 2002.

662. Sun, T.; Xiao, F.; Tang, R.; Zhu, M.; Ouyang, L.; Wang, H.; Wang, Y.; Xiao, Z.; Liu, J.; Huang, L. Samarium-Containing Praseodymium and Neodymium-Free Hydrogen Storage Alloy for Nickel-Hydrogen Power Battery. Chinese Patent Application No. 201310035757, 30 January 2013.

663. Wang, Y.; Huo, H.; Xu, C. Three-Dimensional Honeycomb-Shaped Ni(OH)$_2$ Battery Material as Well as Preparation Method and Application Thereof. Chinese Patent Application No. 201310248778, 21 June 2013.

664. Ren, Y.; Tong, M.; Xiao, K.; Yang, K.; He, Y. Preparation Method of Nickel Foam or Nickel Foam Substrate Alloy. Chinese Patent Application No. 201410181448, 30 April 2014.

665. Qu, W.; Zhang, G.; Lou, L.; Yang, K. Electrodeposition Thick Dimension Foam Nickel and Preparation Thereof. Chinese Patent Application No. 200710158158, 15 November 2007.

666. Guo, Y.; Ye, H.; Yin, Y. Current Collector for Metal Secondary Battery Negative Electrode and Preparation Method and Application for Current Collector. Chinese Patent Application No. 201610034943, 19 January 2016.

667. Zhao, M.; Sun, C.; Wu, Y. Process for Preparing Alloy Material as Negative Electrode of Ni-H Battery. Chinese Patent Application No. 00118976, 6 September 2000.

668. Zhao, M.; Zhang, X. Process for Preparing Alloy Material as High-Power Cathode of Ni-Metal Hydride Battery. Chinese Patent Application No. 02117955, 27 May 2002.

669. Wang, D.; Wu, Y.; Wang, L.; Li, X.; Cheng, Y.; Tang, C.; Shi, Z.; Yang, Y.; Zhang, X.; Gao, S. Recarburization Negative Electrode of Metal Hydride-Nickel Battery Method for Improving Power. Chinese Patent Application No. 201210048023, 28 February 2012.

670. Jin, Y.; Wu, Y.; Zhao, D.; Wang, L.; Xiao, S.; Cheng, Y.; Sun, L.; Lin, J.; Yin, D. Polymer Application Method of Improving Safety and Capacity of Button Cell. Chinese Patent Application No. 201210570162, 25 December 2012.

671. Wang, D.; Wu, Y.; Wang, L.; Li, X.; Cheng, Y.; Tang, C.; Yang, Y.; Shi, Z.; Gao, S.; Dong, X. Method for Arranging Bar Code Type Capacitive Electrode Plate in Electrode of Alkaline Cell as Well as Mixed Negative Electrode. Chinese Patent Application No. 201210049944, 29 February 2012.

672. Xia, B.; Lou, Y.; Zhang, J.; Xu, N. Method for Improving Shelf Characteristic of Nickel-Hydrogen Battery. Chinese Patent Application No. 200710046704, 29 September 2007.

673. Feng, S.; Wu, Z.; Yu, X. High Volume Titanium-Vanadium Base Hydrogen Storage Electrode Alloy Material. Chinese Patent Application No. 200410052933, 16 July 2004.

674. Xia, B.; Xiang, Y.; Ma, L. Improved Type Double Polar Nickel-Hydrogen Battery Group. Chinese Patent Application No. 02218026, 7 June 2002.

675. Xu, F.; Qi, Y.; Sun, L.; Yang, L.; Zhang, J.; Qiu, S.; Zhang, Y. Composite Material of Conducting High Polymers/Alloy for Nickel-Hydrogen Battery and Preparation Thereof. Chinese Patent Application No. 200710011049, 20 April 2007.

676. Wei, H.; Jian, X.; Zhu, L.; Weng, Z.; Jiang, L.; Liu, X. Rare Earth-Magnesium-Nickel-Aluminum Base Hydrogen Storage Alloy for Nickel-Hydrogen Battery and Manufactured Nickel-Hydrogen Battery. Chinese Patent Application No. 201010597158, 10 December 2010.

677. Wei, H.; Jian, X.; Bai, Z.; Zhu, L.; Wang, Z.; Jiang, L. Nickel-Metal Hydride Battery and Low-Self-Discharge Rare Earth-Magnesium-Nickel-Aluminum Series Hydrogen Storage Alloy Thereof. Chinese Patent Application No. 201010286096, 17 September 2010.

678. Yuan, S.; Xin, G.; Jiang, L.; Liu, X.; Wang, M. A Pr, Nd-Free Hydrogen Storage Alloy for Long-Life Ni-MH Batteries. Chinese Patent Application No. 201510573817, 10 September 2015.

679. Zhu, L.; Wu, B.; Dong, G.; Chen, H.; Liu, M.; Jian, X.; Du, J.; Jiang, L.; Mao, C. Cathode Hydrogen-Stored Material Used for Low-Temperature Ni-H Battery and Battery Thereof. Chinese Patent Application No. 200510123747, 22 November 2005.

680. Jian, X.; Jiang, L.; Liu, X.; Wang, Z.; Wei, H.; Zhong, X.; Zhu, L. Compound Conductive Agent for Negative Pole of NiMH Power Battery. Chinese Patent Application No. 200810227091, 21 November 2008.

681. Li, H.; Lu, F.; Jiang, L.; Wang, S.; Liu, X.; Li, G. Method for Preparing Zirconium Oxide Diaphragm Material for Nickel-Hydrogen Battery. Chinese Patent Application No. 200710065078, 2 April 2007.

682. Wu, B.; Zhu, L.; Cheng, H.; Jian, X.; Liu, M.; Du, J.; Jiang, L.; Mao, C. Positive-Pole Base Material Capable of Being Ased for High-Temp. Nickel-Hydrogen Power Cell and Cell Thereof. Chinese Patent Application No. 200510126186, 1 December 2005.

683. Cheng, Y.; Xie, S.; Jian, X.; Zhu, L.; Wang, Z.; Wei, H.; Jiang, L. High Temperature/High-Capacity Nickel-Hydrogen Battery Anode Active Material and Production Method and Use Thereof. Chinese Patent Application No. 200710177037, 8 November 2007.

684. Jiang, W.; Yu, C.; Fu, Z. Active Positive Electrode Material of Composite NiMH Battery and NiMH Power Accumulator. Chinese Patent Application No. 01143468, 28 December 2001.

685. Fu, Z.; Jiang, W.; Yu, L. Nickel-Hydrogen Cell Positive-Pole Material-NiAl Double Hydrogen Oxide and Preparation Method. Chinese Patent Application No. 200410029888, 31 March 2004.

686. Fang, Q.; Jian, X.; Zhu, L.; Wei, H.; Wang, Z.; Jiang, L.; Liu, X. Additive for Positive Electrode of High-Temperature Nickel-Hydrogen Power Battery and Preparation Method Thereof as Well as Positive Electrode Substance of Battery. Chinese Patent Application No. 200910090879, 11 August 2009.

687. Wu, B.; Fang, X.; Zhu, L.; Jian, X.; Cheng, Y.; Chen, H.; Jiang, L. Ni(OH)$_2$ Anode Active Materials for High Temperature NiMH Electrokinetic Cell and Preparation Method Thereof. Chinese Patent Application No. 200610165433, 20 December 2006.

688. Wang, Y.; Xiao, F.; Peng, N.; Lu, Q.; Tang, R. A Hydride Storage Alloy for Low Self Discharge Nikel-Metal Hydride Battery. Chinese Patent Application No. 200710031184, 31 October 2007.

689. Wang, Y.; Xiao, F.; Tang, R.; Peng, N.; Lu, Q. Non-Cobalt AB$_5$ Hydrogen Storage Alloy. Chinese Patent Application No. 200810028733, 12 June 2008.

690. Teng, R.; Wang, Y.; Xiao, F.; Lu, Q.; Peng, N. Hydrogen Storage Alloy for Low Temperature NiMH Power Cell. Chinese Patent Application No. 200810027969, 8 May 2008.

691. Yan, H.; Xiong, W.; Wang, L.; Li, B.; Li, J. Yttrium-Nickel Rare Earth-Based Hydrogen Storage Alloy. Chinese Patent Application No. 201410429202, 28 August 2014.

692. Yan, H.; Li, B.; Xiong, W.; Wang, L.; Li, J. Yttrium-Nickel Rare Earth-Based Hydrogen Storage Alloy. Chinese Patent Application No. 201410429187, 28 August 2014.

693. Yan, H.; Wang, L.; Xiong, W.; Li, B.; Li, J. Rare Earth-Yttrium-Nickel Family Hydrogen Storage Alloy, and Secondary Battery Containing Hydrogen Storage Alloy. Chinese Patent Application No. 201410427259, 28 August 2014.

694. Wang, L.; Yan, H.; Xiong, W.; Li, B.; Li, J. Zirconium and Titanium-Doped AB$_3$ Type Earth-Yttrium-Nickel Family Hydrogen Storage. Chinese Patent Application No. 201410427179, 28 August 2014.

695. Xiong, W.; Yan, H.; Li, B.; Wang, L.; Li, J. Zirconium and Titanium-Doped A$_2$B$_7$ Type Rare Earth-Yttrium-Nickel Family Hydrogen Storage Alloy. Chinese Patent Application No. 201410427220, 28 August 2014.

696. Li, B.; Yan, H.; Wang, L.; Xiong, W.; Li, J. A$_5$B$_{19}$ Type Rare Earth-Yttrium-Nickel System Hydrogen Storage Alloy Added with Zirconium and Titanium Elements. Chinese Patent Application No. 201410427199, 28 August 2014.

697. Wang, L.; Yan, H.; Chen, G.; Yu, B.; Xiong, W.; Li, B.; Li, J.; Zhao, X. A Fabrication Method for Y, Ni-Containing Hydrogen Storage Alloy. Chinese Patent Application No. 201611134225, 10 December 2016.

698. Zeng, H. Method for Recognizing Different-Phase Cross-Line Ground Fault of Double-Circuit Lines Based on Actual Measurement of Voltage of Different-Phase Cross-Line Grounding Point. Chinese Patent Application No. 201510096986, 4 March 2015.

699. Kong, F.; Yan, H.; Xiong, W.; Li, B. Method for Improving Magnesium-Based Hydrogen-Storage Electrode Capacity Attenuation for Nickel-Hydrogen Battery. Chinese Patent Application No. 200510096984, 30 August 2005.

700. Wang, D.; Wang, B.; Jin, C.; He, D. Superlattice Hydrogen Storage Alloy Material for Nickel Hydrogen Battery and Preparation Method Thereof. Chinese Patent Application No. 201010220084, 6 July 2010.

701. Li, Z.; Huang, T.; Cheng, L.; Sheng, X.; Liu, R.; Wang, R.; He, D. Method for Preparing Nickel-Hydrogen Battery Cathode. Chinese Patent Application No. 200810038898, 12 June 2008.

702. Qin, H.; Lv, Z.; Liu, W.; Lin, F.; Zhang, Z.; Zhang, J.; Lei, X.; Lu, A.; Meng, G.; Su, Y. Cathode Paste for Manufacturing Cathode of Nickel-Metal Hydride Battery, Cathode of Nickel-Metal Hydride Battery as Well as Nickel-Metal Hydride Battery. Chinese Patent Application No. 201310606368, 26 November 2013.

703. Zhang, H.; Zhou, Q.; Gan, S.; Xu, H.; Hu, Z.; Sun, Z.; Tan, S.; Zhang, Z.; Peng, H. Cerium-Cobalt-Coated Nickel Hydroxide Composite Material and Preparation Method and Application Thereof. Chinese Patent Application No. 201410351399, 23 July 2014.

704. Zhang, J.; Wang, L.; Zhang, X.; Fan, L.; Li, J. Nickelous Hydroxide in Positive Pole Material of Alkaline Secondary Cell and Producing Method Thereof. Chinese Patent Application No. 200710059995, 24 October 2007.

705. Zhang, J. Nickel-Hydrogen Cell with Temperature Control Device. Chinese Patent Application No. 200520123630, 30 November 2005.

706. Li, R.; Wu, J.; Zhou, S. Cathode Material in Use for Nickel-Hydrogen Battery in High Capacity. Chinese Patent Application No. 200510073051, 31 May 2005.

707. Li, R.; Wu, J.; Zhou, S. Negative Pole Material for High-Temp Nickel-Hydrogen Battery with Long-Service Life. Chinese Patent Application No. 200410004750, 3 March 2004.

708. Li, R.; Wu, J.; Zhou, S. Negative Pole Material for High-Temperature NiH and its Preparation Process. Chinese Patent Application No. 03153957, 25 August 2003.

709. Cheng, J.; Xu, D.; Wang, X. Manufacture of Alkaline Secondary Nickel-Hydrogen Cell and its Positive and Negative Pole. Chinese Patent Application No. 98125274, 15 December 1998.

710. Wu, J.; Li, R. Apparatus for Preparing Negative-Electrode Hydrogen-Storage Alloy Powder for Nickel-Hydrogen Cell. Chinese Patent Application No. 98201981, 16 March 1998.

711. Zhang, H. Process for Preparing Mixed Rare-Earth Metals. Chinese Patent Application No. 99121650, 12 October 1999.

batteries

MDPI

Review

Reviews of European Patents on Nickel/Metal Hydride Batteries

Shiuan Chang [1,2], Kwo-Hsiung Young [2,3,*] and Yu-Ling Lien [4]

[1]   Department of Mechanical Engineering, Wayne State University, Detroit, MI 48202, USA;
      shiuanc@wayne.edu
[2]   BASF/Battery Materials—Ovonic, 2983 Waterview Drive, Rochester Hills, MI 48309, USA
[3]   Department of Chemical Engineering and Materials Science, Wayne State University, Detroit, MI 48202, USA
[4]   Department of Chemistry, Michigan State University, East Lansing, MI 48824, USA; yulinglien@gmail.com
*   Correspondence: kwo.young@basf.com; Tel.: +1-248-293-7000

Academic Editor: Andreas Jossen
Received: 22 June 2017; Accepted: 27 July 2017; Published: 26 August 2017

Abstract: Patent applications in the field of nickel/metal hydride (Ni/MH) batteries are reviewed to provide a solid technology background and directions for future developments. As the fourth review article in the series of investigations into intellectual properties in this area, this article focuses on 126 patent applications filed by European companies at the European Patent Office, while the earlier articles dealt with those from USA, Japan, and China. The history and current status of the key companies in the Ni/MH battery business are briefly discussed. These companies are categorized by their main roles in the industry, i.e., battery manufacturer, metal hydride alloy supplier, separator supplier, and others. While some European companies are pioneers in bringing the Ni/MH product to customers, others have made significant contributions to the development of the technology, especially in the button cell, bipolar cell, and separator areas.

Keywords: nickel metal hydride battery; European patent; metal hydride alloy; nickel hydroxide; European battery manufacturer

1. Introduction

Nickel/metal hydride (Ni/MH) batteries are widely used in consumer electronics, stationary energy storage, and hybrid electric vehicles (HEV). Their advantages are in energy density (compared to the replaced NiCd), cycle stability at deep discharge (>3000), robust construction, ability to be quickly charged, environmental friendliness, and higher abuse tolerance (compared to the Li product) [1]. Two types of Ni/MH batteries are available: the small consumer type (usually cylindrical with a capacity <4 Ah) and the larger transportation type (usually prismatic with a capacity >6 Ah). Nowadays, most consumer type and HEV Ni/MH batteries are produced in China and Japan. Nevertheless, European companies still have a small portion of the market share (for example, about 10% in 2013 [2]) (Figure 1a). On the other hand, European countries are a large market for consumer type Ni/MH batteries. For instance, in 2013, Germany, the Netherlands, Belgium, and the UK accounted for a combined 8.3% of the Ni/MH batteries exported from China [2] (Figure 1b), and in February 2014 Europe received 32.7% of the Ni/MH batteries from Japanese manufacturers (mainly FDK, Tokyo, Japan) [2] (Figure 1c). European companies made great contributions to the debut of Ni/MH batteries in their early days. For example, Klaus-Dieter Beccu from Battelle Memorial Institute (Geneva, Switzerland) contributed to the earliest patent using MH alloy as the active material in the negative electrode to construct a Ni/MH battery in 1970 [3]; Annick Percheron-Guégen and her coworkers from the Agence Nationale de Valorisation de la Recherche (Neuilly sur Seine, France) filed the first patent of using $LaNi_5$ modified with Ti, Ca, Ba, Cr, and/or Cu with improved capacity as negative electrode material in 1978 [4]; and Johannes Willems, Johann van Beek, and Buschow Kurt from Philips (Eindhoven, Netherlands) filed a USA patent adding Co and/or Cu in LaNi$_5$ to reduce the lattice expansion during hydride process and consequently extend the cycle life in 1984 [5]. In the past, we have reviewed the Ni/MH-related patents (or patents applications) filed in the USA [6], Japan [7], and China [8]. In this review, we focus on patent applications to the European Patent Office submitted by European companies; these applicants are grouped as battery manufacturers, metal hydride (MH) alloy suppliers, separator suppliers, and others. The patent applications can be found online through the website of World Intellectual Property Organization (WIPO) [9]. In the case of a non-English description, a machine translation option supplied by WIPO is available free of charge. A brief introduction of European companies contributing to the European Patent Application (EPA) is included in Table 1.

Table 1. European companies related to Ni/MH battery manufacturing included in this review, in order of their presence.

Company Name	Headquarter	Products Related to Ni/MH	Trademark	Website
Saft	Bagnolet, France	C, D, F-size cells		http://saftbatteries.com/
Arts Energy	Nersac, France	C, D, F-size cells		http://www.arts-energy.com
Varta	Ellwangen, Germany	AAA to FA, button cell		http://www.varta-microbattery.be/
Nilar	Täby, Sweden	Prismatic, bipolar cell		http://www.nilar.com/
Alcatel	Paris, France	Cell phone battery		http://www5.alcatel-lucent.com
Hoppecke	Brilon-Hoppecke, Germany	Prismatic cell		https://www.hoppecke.com/
GfE	Nuremberg, Germany	MH alloy powder		http://www.gfe.com
Treibacher	Althofen, Austria	MH alloy powder		https://www.treibacher.com
SciMat	Swindon, UK	Separator		Not available now
Freudenberg	Weinheim, Germany	Separator		http://www.freudenberg.com

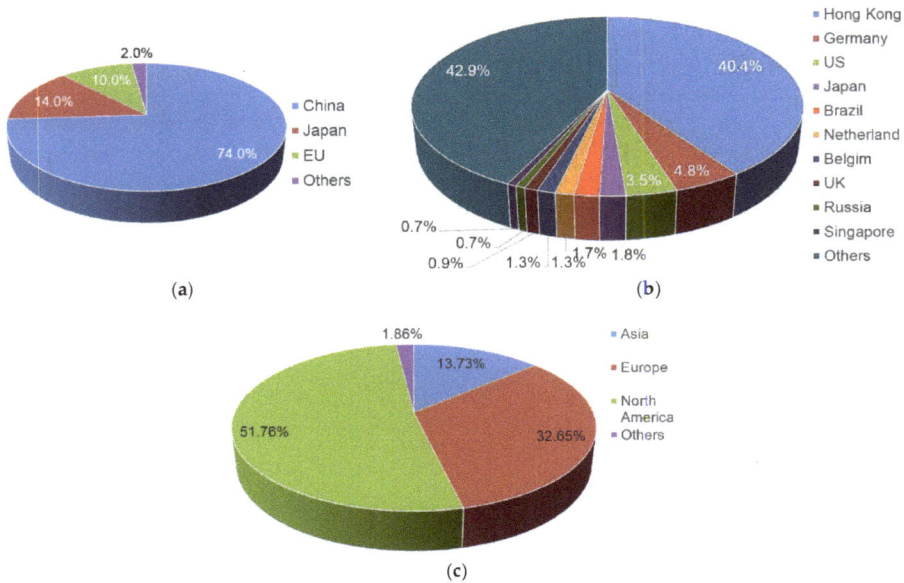

Figure 1. Consumer Ni/MH battery market shares for (**a**) Sources in 2013, (**b**) Exports from China in 2013, and (**c**) Exports from Japan in February 2014. Data originate from [2].

2. Ni/MH Battery Manufacturer

Patent applications from five European battery/chemical companies related to the fabrication of Ni/MH batteries are reviewed in this section. There are two large European companies, Varta (Ellwangen, Germany) and SAFT (Bagnolet, France), that manufactured consumer-type Ni/MH batteries and submitted patent applications in related fields over the past two decades. In 2002, Varta microbattery spun off from Varta and focused on the button cell business. In 2013, SAFT sold its Small Nickel Battery division to Arts Energy (Nersac, France). Alcatel (later merged into Lucent Technology, Paris, France) made Ni/MH batteries for cell phones and was active in filing patent applications. Recently, Nilar (Täby, Sweden) began to offer bi-polar prismatic Ni/MH batteries for transportation applications. Hoppecke (Brilon-Hoppecke, Germany) made prismatic flooded-type Ni/MH batteries for transportation applications.

2.1. SAFT and Arts Energy

The *Société des Accumulateurs Fixes et al Traction* (SAFT) was founded in 1918 and manufactured batteries for the luggage carts and lighting in locomotives. In 2006, Saft and Johnson Controls (Milwaukee, WI, USA) formed a joint-venture pursuing batteries for electric vehicles (EV) and HEV. It was acquired by Total (Courbevoie, France) and delisted from the stock exchange in 2016. SAFT filed 24 and 19 European patent applications (EPAs) from 1990 to 2013 in the name of Accumulateurs Fixes and SAFT SA, respectively. In the positive electrode area, SAFT filed EPAs on the following subjects: a fabrication method of co-precipitated metal hydroxide [10], a description of characteristic of β-Ni(OH)$_2$ [11], a fabrication method of pasted electrode [12,13], and uses of several additives including Celestine [14], CoO with a spinal structure [15], active carbon [16], Li(OH)$_2$ [17], Ba(OH)$_2$ and Sr(OH)$_2$ [18], Nb$_2$O$_5$ [19], non-conductive fiber [20], and plastic binders [21–23]. Out of these inventions, those with additives and binders are particularly helpful in improving the performance of positive electrodes. In the negative electrode area, SAFT filed EPAs on the following subjects: active materials made from a single phase TiNi-based alloy [24], a Zr-based C14 AB_2 alloy [25], a mixed-metal based AB_5 alloy [26], superlattice alloys containing Y [27], Mg [28], or Zr (Ti) [29] and A_2B_7 [30] and A_5B_{19} [31] phases, a catalytic substrate [32], a hydrophilic additive [33], and a paste process [34]. Among these EPAs, the superlattice alloy-related patents lead to a new direction of material research in MH alloy field. SAFT filed multiple EPAs on the subject of a foam substrate and associated current collector design [35–41] (Figure 2), which built a strong foundation for further research in the area. In the separator area, SAFT filed EPAs on microporous grafted membranes [42,43], a tri-layer composite [44], a double-layer with different orientations [45], and a N-retaining substrate [46]. SAFT also filed three EPAs on a charging algorithm [47–49] and one on a leak-tight box for holding battery modules [50].

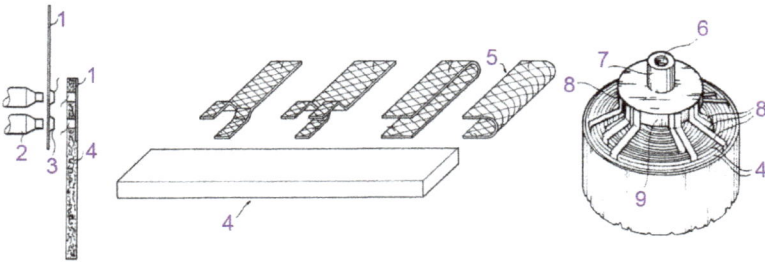

Figure 2. Schematics of three difference designs of current collectors by SAFT. The parts are 1: metal plate, 2: plug, 3: indent, 4: electrode [37]; 5: expanded metal or perorated metal strip [38]; 6: threaded holes, 7: terminal assembly, 8: conducting strip, and 9: shoulder [40].

Arts Energy filed two EPAs in the early days; one on the re-sealable safety valve [51] and the other one on a battery cooling module design [52]. Arts Energy continues to perform research and produce Ni/MH batteries after it was purchased from SAFT in 2013, but has not filed an EPA since then. Its products are mainly cylindrical NiCd and Ni/MH batteries (Figure 3).

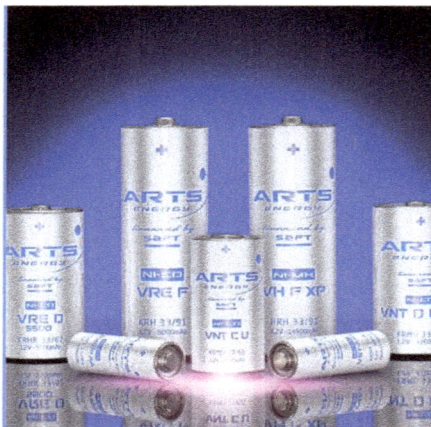

Figure 3. Cylindrical cells from Arts Energy [53].

2.2. Varta and Varta Microbattery

Varta (originally *Vertrieb, Aufladung, Reparatur Transportabler Akkumulatoren*) was founded in 1887 and produced the batteries that powered the first EV in Germany. For the production of Ni/MH batteries, Varta formed a joint-venture, 3C Alliance, with Toshiba (Tokyo, Japan) and Duracell (Bethel, CT, USA) in 1996. Varta filed 10 EPAs in 1994 and 1995, which cover an oxidation-resistant graphite additive in the positive electrode [54,55], a use of foam or felt electrode in the button cell [56], a multi-segment cylindrical cell design [57] (Figure 4a), a button cell design [58,59] (Figure 4b), Fe- [60], Ti(Zr)- [61] containing AB_5 MH alloys, and metal recovery from spent batteries [62,63]. Out of these EPAs, those related to the new cell structure designs are important to extend the applicability of Ni/MH into various fields.

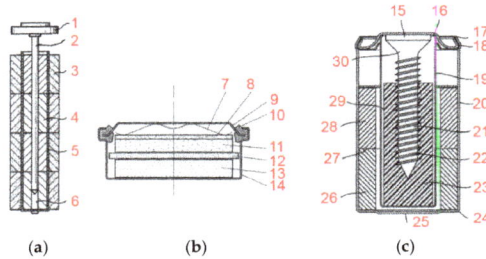

Figure 4. Schematics of (**a**) A multi-segment cylindrical cell, (**b**) A button cell, and (**c**) A hollow-cathode cylindrical cell from Varta. The parts are 1: terminal, 2: conducting pin, 3: positive electrode, 4: negative electrode, 5: separator, 6: central hole [57]; 7: pressure spring, 8: current collector plate, 9: cover, 10: gasket, 11: negative electrode, 12: electrolyte-saturated separator, 13: positive electrode, 14: case [58]; 15: head, 16: disc shape lid with radial profile, 17: rim (bent radially inward), 18: seal, 19: separator, 20: cylindrical wall, 21: shank, 22:thread, 23: negative electrode, 24: can, 25: flat circular bottom, 26: one positive electrode, 27: positive electrode assembly, 28: another segment of positive electrode, 29: cavity, 30: screw shape current collector [64].

In 2002, Johnson Controls acquired the automotive battery division of Varta. Immediately before the sale, Varta Automotive GMGB filed an EPA on an electrolyte containing soluble Al-compound added in the positive electrode to improve high-temperature charge acceptance [65].

Varta Microbattery was a spin-off company from Varta in 2002 and is now owned by Montana Tech Components (Menziken, Switzerland). Their main products are button cells (Figure 5). Between 1998 and 2015, five EPAs were filed with Varta Microbattery as the assignee. They covered the areas of a charging device for Ni/MH battery [66,67], a hollow-cathode cylindrical cell design [64] (Figure 4c), and a battery/supercapacitor composite design [68,69]. The combination of rechargeable battery and supercapacitor increases the high-rate charging capability of Ni/MH battery and opens the door for some power-demanding applications.

Figure 5. Button cells from Varta Microbattery [70].

2.3. Alcatel

Alcatel (originally short fir the *Société Alsacienne de Constructions Atomiques, de Télécommunications et d'Électronique*) was merged into CGE in 1966, which was later merged again with Lucent Technologies in 2006 and formed Alcatel-Lucent. It was finally merged into the Nokia Group in 2016. Alcatel made Ni/MH batteries for cell phones and cordless phone (Figure 6). Between 1994 and 2003, Alcatel filed 8 EPAs under the names of Alsthom CGE Alcatel and CIT Alcatel. They disclosed two designs for metal current collectors [71,72] (Figure 7), a minus ΔV cut-off charge control method [73], a C14 AB_2 MH alloy [74], a coated, perforated metal substrate for a negative electrode [75], a double-layered separator [76], a cell design with a negative/positive ratio >1.15 [77], and a $Ni(OH)_2$ coating on a MH alloy to improve low-temperature and cycle performance [78]. The last EPA is very important for the use of Ni/MH battery in areas with very harsh winter, Siberia, for example.

Figure 6. Alcatel One Touch 535 with Ni/MH battery [79].

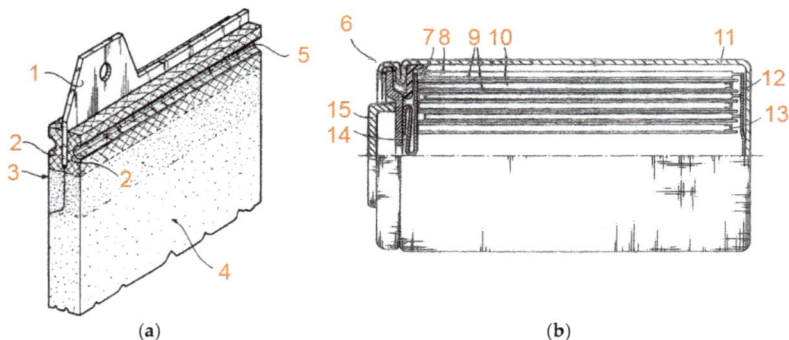

(a)   (b)

Figure 7. Schematics of (**a**) A metal-foam connection and (**b**) A cylindrical cell design from Alcatel. The parts are 1: metal strip, 2: foam without $Ni(OH)_2$, 3: auxiliary foam, 4: main foam, 5: slot formed by compression [71]; 6: folded part of can, 7: metal tab, 8: negative electrode, 9: separator, 10: positive electrode, 11: can, 12: a piece of nickel plated steel disc, 13: adhesive tape, 14: lug, and 15: cover [72].

2.4. Nilar

Nilar was founded in Sweden in 2011. Before moving to Sweden, Nilar used to operate its research facility in Aurora, CO, USA. Nilar's main product is a prismatic, bi-polar Ni/MH battery used in transportation applications (Figure 8). Nilar owns 14 USA Patents covering components, casing, assembly, module construction of a bi-polar design Ni/MH battery. Nilar filed 13 EPAs between 2002 and 2012, which included assembly of bipolar plates design [80] and manufacture [81], design [82] and manufacture [83–86] of a current collector, a case [87], a stacking arrangement [88], a pressure sensor [89], a powder reduction unit [90], a gasket design [91], and an energy managing system [92] for bipolar Ni/MH batteries. These bi-polar Ni/MH batteries are very important for large-scale transportation applications, which require both large capacity and high-rate drain capability. Recently, Prof. Dag Noréus filed an international Patent Cooperation Treaty (PCT) with Nilar regarding the addition of H_2, O_2, and/or H_2O_2 in sealed Ni/MH batteries to extend cycle life [93].

(a)   (b)

Figure 8. (**a**) Nilar's bipolar prismatic Ni/MH batteries and (**b**) A ferry powered by them [94].

2.5. Hoppecke

Hoppecke was founded in 1927 to produce industrial batteries. It supplied lead-acid, Ni/MH (Figure 9), and Li-ion batteries to railway applications. Hoppecke filed two EPAs in 2011 disclosing a positive electrode manufacturing process [95] and a flooded design for a Ni/MH battery [96]. The latter is unusual and opens the way for stationary storage associated with alternative energies, such as solar and wind power.

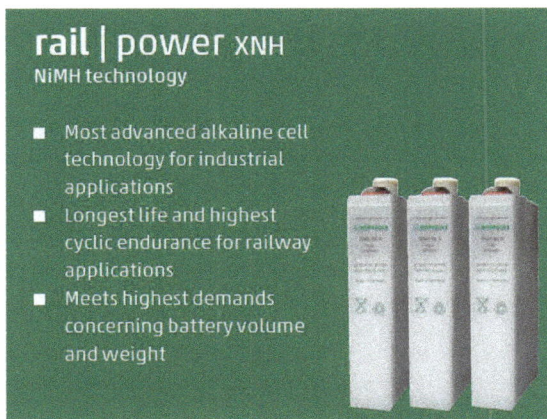

Figure 9. Prismatic flooded Ni/MH batteries from Hoppecke [97].

3. MH Alloy Suppliers

GfE (Nuremberg, Germany) and Treibacher (Althofen, Austria) were previously the only two European companies producing MH materials used as active materials in the negative electrode of Ni/MH battery. Nowadays, all the Ni/MH battery manufacturers have switched to vendors from Japan and China.

3.1. GfE

GfE was founded in 1911 to produce metal products with electric furnaces. Today, it still offers an AB_2-based MH alloy, mainly for the use of solid hydrogen storage [98]. GfE filed one EPS in 1999 regarding addition of a metal oxide (Cr_2O_3, for example) to hydrogen storage material (Mg, for example) as a catalyst [99].

3.2. Treibacher

Treibacher Industries AG was founded in 1898 to produce ferroalloy for the steel industry. Today, it still offers AB_2 and AB_5 MH alloys for hydrogen storage and battery applications, respectively [100]. Treibacher filed an EPA in 2001 to cover a Cr-containing AB_5 MH alloy [101]. It also filed an international PCT on a two-chamber hydrogen storage canister [102].

4. Separator Suppliers

Scimat (Swindon, UK) and Freudenberg (Weinheim, Germany) were once two major separator companies making non-woven grafted polyethylene/polypropylene separators for Ni/MH battery producers worldwide. In 2006, Scimat was acquired by Freudenberg and, together, they still supply separators for high-performance Ni/MH batteries, especially in the HEV area. In 2015, Freudenberg also became the largest (75%) stockholder for Nippon Vilene—the top nonwoven producer in Japan.

4.1. Scimat

Scimat was founded in 1987, focusing on the production and development of automated controlled separator production process. It was acquired by its main competitor, Freudenberg, in 2006. Between 1985 and 1998, Scimat filed 7 EPSs on a grafting method preparing microporous particles [103], a halopolymer film [104,105], a double-layer (one porous, the other one partially blocking the pores) membrane [106], a grafted sheet composed of copolymer and polyolefin [107], an ultraviolet radiation process to increase the surface hydrophilicity of the separator [108], and a laminate composed of fiber and vinyl monomer on the surface [109]. The Scimat separator is very important in the HEV application.

4.2. Freudenberg

Freudenberg was founded in 1849 to make leather products. Combining the intellectual property and production capacity of past rivals Scimat and Nippon Vilene, Freudenberg became the only separator producer outside of China. Freudenberg filed 8 EPAs from 1993 to 2010, covering the subjects of a non-woven hydrophilic separator material [110], a grafting process making separator material from polymers [111], an addition of alkaline hydroxide or carbonate in the separator before use [112], a separator material containing a titrimetrically determined ammonia binding property [113], a chemically active separator material with an alkaline electrolyte [114], a bi-layer separator with different fiber diameters [115], a core-shell duel-component fiber as a separator [116], and a separator material with coating to enhance its puncturing resistance [117]. Together with Eveready Battery Company (St. Louis, MO, USA), Freudenberg filed an EPA on a cell design using a separator less than 0.15 mm thick in the dry state [118] (Figure 10).

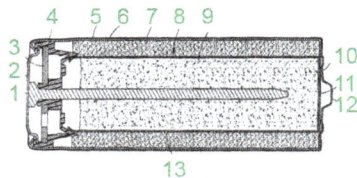

Figure 10. Schematic of a cylindrical cell design from Eveready/Freudenberg. The parts are 1: negative current collector, 2: negative terminal, 3: metal plate, 4: polymer seal, 5: metalized plastic film, 6: steel can, 7: tubular positive electrode, 8: separator, 9: negative electrode, 10: bottom of negative electrode, 11: bottom of folding cylinder, 12: positive terminal, and 13: folding can [118].

5. Other Companies

In addition to the companies mentioned above, which are (or were) involved directly with the manufacturing of Ni/MH batteries, there are other companies and universities in Europe that occasionally applied for EPAs in the Ni/MH battery field. In battery design, Deutsche Automobilgesellschaft (now a subsidiary of Daimler AG, Stuttgart, Germany) filed 3 EPAs on a prismatic cell [119] with the negative electrode covered by polytetrafluoroethylene (PTFE) threads [120], and a bi-polar design of electrode stacking [121].

In the $Ni(OH)_2$ area, the University of Southampton (Southampton, UK) filed one EPA [122] and two PCTs [123,124] on the subject of a porous positive electrode active material with regularly spaced pores. In the MH alloy area, Höganäs AB (Höganäs, Sweden) filed two EPAs on an organically modified silicic acid coating for MH alloys [125], and MH alloys prepared by gas atomization with a unique dendritic microstructure [126]. Fraunhofer-Gesellschaft (Munich, Germany) also filed an EPA on an organically modified silicic acid polycondensated coating for MH alloys [127].

In the charging algorithm and devices, Peugeot (Paris, France) presented a method relating to inner pressure to the state-of-charge (SOC) [128]; Friwo Geräetebau (Ostbevern, Germany) patented a temperature sensor equipment charging device [129]; SGS Thomson Microelectronics (Geneva, Switzerland) disclosed a $-\Delta V$ cut-off method [130,131]; and Solarc Innovative Solarprodukte (Berlin, Germany) disclosed a solar-cell compatible battery charging unit [132].

EU countries are very concerned about battery recycling activities [133]. A special program, COLABATS, was initiated in Europe in 2013 to conduct the development of recycling technology for Ni/MH and Li-ion batteries [134]. In terms of battery recycling, Umicore (Brussels, Belgium) patented a smelting process to reclaim Ni and Co [135]; Montanuniversität Leoben (Leoben, Austria) disclosed a method of reclaiming rare earth metals by using a non-oxidizing acid [136]; Akkuser Ltd. (Nivala, Finland) has a Patent issued on the separation of spent batteries by chopping or crushing followed by a refining or smelting process [137]; CT Umwelttechnik AG (Winterthur, Switzerland) patented a recycling process for the mixture of various kind of batteries [138]; Titalyse S.A. (Croix-de-Rozon, Switzerland) patented an apparatus to sort spent batteries with different size and shape by an optical camera with imaging software [139]; Enviro EC AG (Zug, Switzerland) published a method of reclaiming MH from spent batteries by magnetic separation [140]; SNAM (Saint-Quentin-Fallavier, France) also reclaimed MH by spraying surfactant and later sedimentation [141]; and Eco Recycling SRL (Rome, Italy) reported a physical/chemical combinational method to reclaim Cu and plastic materials [142]. Additionally, French Alternative Energies and Atomic Energy Commission (Gif-sur-Yvette, France) filed a PCT on a process to reclaim rare earth elements, specifically Ni, Co, Mn, and Fe, by selective salt precipitation from a mixture of acid solution for spent Ni/MH battery [143]. Finally, only one EPA was found for the application of Ni/MH batteries, a low self-discharge battery used in soap dispensers filed by Vivian Blick [144].

6. Comparisons to Other Countries

The development of MH alloy and associated Ni/MH battery started from European companies, such as Battelle, Philips, Benz (Stuttgart, Germany) and followed by the successful commercialization by American and Japanese companies [145]. Now the Ni/MH batteries are mainly fabricated in China and Japan. Throughout the years, many important patents were filed in these four areas/countries, namely EU, USA, Japan, and China. The nature of these patent applications is compared in Table 2. Compared to other area/countries, European companies filed the fewest patents (application) in the Ni/MH field. However, their patents in areas such as button cells (owned by Varta Micronbattery), bi-polar design (owned by Nilar), and separators (owned by SciMat and now Freudenberg) are all very important. USA has the largest consumer market in the world and most of its patents related to Ni/MH batteries were filed by Japanese and European companies.

Table 2. Comparison of numbers of patents (applications) filed in EU, USA, Japan, and China and number of associated companies (institutes). * The number of companies used in USA only counts for American companies (institutes).

Country	EU	USA	Japan	China
Number of patents (application) related to Ni/MH	126	371	275	692
Number of patents in negative electrode	22	58	128	136
Number of patents in positive electrode	18	54	27	83
Number of patents in manufacturing technology	10	33	28	104
Number of companies (institutes) filed patents	29	18 *	46	99

7. Conclusions

A total of 126 patent applications related to Ni/MH batteries and filed by European companies to the European Patent Office were reviewed. The scope covers metal hydride formulations and processing, nickel hydroxide preparation, electrode fabrication, cell design, charging method, and recycling. Together with reviews of patents (applications) in the Ni/MH field filed in the USA, Japan, and China, this work completes our studies on the intellectual properties in this field. Compared to companies (institutes) in other countries, European companies started the Ni/MH battery, but were not as active in the development of the production technology. They still made significant contributions in the separator and prismatic cell areas.

Author Contributions: Shiuan Chang collected the information. Kwo-Hsiung Young and Yu-Ling Lien prepared the manuscript.

Conflicts of Interest: The authors declare no conflict of interest.

Abbreviations

The following abbreviations are used in this manuscript:

Ni/MH	Nickel/metal hydride
HEV	Hybrid electric vehicle
MH	Metal hydride
WIPO	World Intellectual Property Organization
EPA	European Patent Application
EV	Electric vehicle
PCT	Patent Cooperation Treaty
PTFE	Polytetrafluoroethylene
SOC	State-of-charge

References

1. Battery and Energy Technologies. Available online: http://www.mpoweruk.com/nimh.htm (accessed on 25 May 2017).
2. Bai, W. The Current Status and Future Trends of Domestic and Foreign NiMH Battery Market. Available online: http://cbea.com/u/cms/www/201406/06163842rc0l.pdf (accessed on 25 May 2017).
3. Beccu, K.-D. Electrical Accumulator with a Metal Hydride Serving as the Cathodic Reactive Material Arranged in Suspension in the Electrolyte. U.S. Patent 3,520,728, 14 July 1970.
4. Percheron-Guégen, A.; Achard, J.C.; Loriers, J.; Bonnemay, M.; Bronoël, G.; Sarradin, J.; Schlapbach, L. Electrode Materials Based on Lanthanum and Nickel, and Electrochemical Uses of Such Materials. U.S. Patent 4,107,405, 15 August 1978.
5. Willems, J.J.G.S.A.; van Beek, J.R.G.C.M.; Buschow, K.H.J. Electrochemical cell comprising stable hydride-forming material. U.S. Patent 4,487,817, 11 December 1984.
6. Chang, S.; Young, K.; Nei, J.; Fierro, C. Reviews on the U.S. Patents regarding nickel/metal hydride batteries. *Batteries* **2016**, *2*, 10. [CrossRef]
7. Ouchi, T.; Young, K.; Moghe, D. Reviews on the Japanese Patent Applications regarding nickel/metal hydride batteries. *Batteries* **2016**, *2*, 21. [CrossRef]
8. Young, K.; Cai, X.; Chang, S. Reviews on the Chinese Patent regarding Nickel/metal hydride battery. *Batteries* **2017**, *3*, 24. [CrossRef]
9. World Intellectual Property Organization. Available online: https://patentscope.wipo.int/search/en/search.jsf (accessed on 23 May 2017).
10. Blanchard, P.; Klein, J.P. Process for the Preparation of a Metal Hydride Powder and the Powder Obtained. Eur. Pat. Appl. 91401640, 18 June 1991.
11. Bernard, P.; Audry, C.; Lecerf, A.; Senyarich, S. Nickel Electrode for Alcaline Accumulator. Eur. Pat. Appl. 96400423, 28 February 1996.
12. Bernard, P.; Dennig, C.; Cocciantelli, J.-M.; Coco, I.; Alcorta, J. Non-Sintered Nickel Electrode. Eur. Pat. Appl. 99400049, 11 January 1999.
13. Bernard, P.; Simonneau, O.; Bertrand, F. Pasted Nickel Electrode. Eur. Pat. Appl. 97401011, 5 May 1997.
14. Bernard, P.; Goubault, L.; Guiader, O. Positive Electrode for an Electrochemical Generator with an Alkaline Electrolyte. Eur. Pat. Appl. 08290869, 16 September 2008.
15. Bernard, P.; Audry, C. Conductive Material for Electrode of Secondary Battery with Alkali Electrolyte. Eur. Pat. Appl. 01401672, 25 June 2001.
16. Chevalier, S.; Debdary, M.; Desprez, P. Negative Electrode for an Asymmetric Supercapacitor with Nickel Hydroxide Positive Electrode and Alkali Electrolyte and Method for Manufacturing Same. Eur. Pat. Appl. 12167763, 11 May 2012.
17. Bernard, P.; Baudry, M.; Jan, O.; Audry, C. Non-Sintered Nickel Electrode for Secondary Battery with Alkali Electrolyte. Eur. Pat. Appl. 99403260, 23 December 1999.
18. Bernard, P.; Goubault, L.; Hezeque, T. Electrochemical Alkaline Generator with Improved Service Life. Eur. Pat. Appl. 05291300, 17 June 2005.
19. Bernard, P.; Goubault, L.; Gillot, S. Electrode for Alkaline Storage Battery. Eur. Pat. Appl. 12175342, 6 July 2012.
20. Bernard, P. Positive Electrode for Alkaline Battery. Eur. Pat. Appl. 06291402, 5 September 2006.
21. Coco, I.; Cocciantelli, J.-M.; Villenave, J.-J. Electrode of non Sintered Type for Accumulator with Alkaline Electrolyte. Eur. Pat. Appl. 97401723, 17 July 1997.
22. Bernard, P.; Goubault, L. Plastified Electrode for Alkaline Battery. Eur. Pat. Appl. 07848215, 11 September 2007.
23. Bernard, P.; Goubault, L.; Gillot, S.; Feugnet, T. Plastic-Coated Electrode for Alkaline Storage Battery. Eur. Pat. Appl. 10162920, 17 May 2010.
24. Bouet, J.; Knosp, B.; Percheron-Guegan, A.; Jordy, C. Hydridable Material for a Nickel-Hydride Battery Negative Electrode and Process for Production. Eur. Pat. Appl. 93401399, 2 June 1993.
25. Bouet, J.; Knosp, B.; Percheron-Guegan, A.; Canet, O. Hydridable Material for a Nickel-Hydride Battery Negative Electrode. Eur. Pat. Appl. 93402198, 9 September 1993.

26. Bouet, J.; Knosp, B.; Percheron-Guegan, A.; Cocciantelli, J.-M. Hydridable Material for Nickel-Hydride Accumulator Negative Electrode. Eur. Pat. Appl. 0608646, 3 August 1994.

27. Bernard, P.; Knosp, B.; Baudry, M. Composition for Negative Electrode of Accumulator with Alkaline Electrolyte. Eur. Pat. Appl. 07291035, 23 August 2007.

28. Lavaud, C.; Dillay, V.; Bernard, P. Negative Active Material for Hydride metal Nickel Accumulator. Eur. Pat. Appl. 08012162, 4 July 2008.

29. Bernard, P.; Knosp, B.; Latroche, M.; Zhang, J.; Serin, V.; Hytch, M. Active Material for a Negative Electrode of a Nickel-Metal Hydride Alkaline Accumulator. Eur. Pat. Appl. 11188182, 8 November 2011.

30. Bernard, P.; Knosp, B.; Baudry, M. Active Material Composition and Alkaline Electrolyte Accumulator. Eur. Pat. Appl. 06290808, 18 May 2006.

31. Bernard, P.; Knosp, B.; Latroche, M.; Ferey, A. Hydridable Alloy for Alkaline Storage Battery. Eur. Pat. Appl. 07290197, 16 February 2007.

32. Senyarich, S.; Cocciantelli, J.-M. Hydrophylic Electrode for Alkaline Accumulator and Method of Preparation. Eur. Pat. Appl. 97402510, 23 October 1997.

33. Sjövall, R.; Greis, M. Low Maintenance Alkaline Electrochemical Cell. Eur. Pat. Appl. 12190882, 31 October 2012.

34. Coco, I.; Cocciantelli, J.-M.; Villenave, J.-J. Negative Electrode for Ni-Metal Hydride Accumulator. Eur. Pat. Appl. 96400850, 22 April 1996.

35. Grange-Cossou, M.; Torregrossa, W. Process of Manufacturing an Electrode with Spongiform Support for Electrochemical Generator and Electrode Obtained Thereby. Eur. Pat. Appl. 90117834, 17 September 1990.

36. Caillon, G.; Lebarbier, C. Method for Covering an Electrode with a Sponge like Support for Electrochemical Generator and Electrode Obtained by This Process. Eur. Pat. Appl. 90121037, 2 November 1990.

37. Guerinault, J.-M.; Brunarie, J. Method for Joining a Metallic Current Collector to an Electrode Comprising a Spongeous Supporting Structure for Electrochemical Generator and Electrode Obtained Thereby. Eur. Pat. Appl. 91121081, 9 December 1991.

38. Grange-Cossou, M.; Guerinault, J.-M.; Carteau, B.; Brunarie, J. Method for Joining a Metallic Collector to an Electrode Comprising a Spongeous Supporting Structure for Electrochemical Generator and Electrode Obtained Thereby. Eur. Pat. Appl. 92401446, 26 May 1992.

39. Bouet, J.; Pichon, B. Process for Making a Foam-Type Support for an Electrode of a Secondary Battery. Eur. Pat. Appl. 92401861, 30 June 1992.

40. Fradin, J. Cylindrical Electrochemical Generator with Spirally Wounded Electrodes. Eur. Pat. Appl. 97401780, 24 July 1997.

41. Goubault, L.; Bernard, P.; Gillot, S. Plastic-Coated Electrode for Alkaline Storage Battery. Eur. Pat. Appl. 14170069, 27 May 2014.

42. Gineste, J.-L.; Pourcelly, G.; Brunea, J.; Perton, F.; Broussely, M. Grafted Microporous Separator for Electrochemical Generator and Its Manufacturing Process. Eur. Pat. Appl. 93402037, 10 August 1993.

43. Caillon, G.; Crochepierre, B. Alkaline Open Battery Comprising a Microporous Membrane. Eur. Pat. Appl. 06290869, 30 May 2006.

44. Green, A.; Champalle, P.; Liska, J.-L. Separator for Alkaline Accumulator. Eur. Pat. Appl. 91402721, 11 October 1991.

45. Senyarich, S.; Leblanc, P. Separator for Accumulator with Spirally Wounded Electrodes and Alkaline Electrolyte. Eur. Pat. Appl. 97401667, 10 July 1997.

46. Senyarich, S.; Viaud, P. Alkaline Electrolyte Secondary Battery. Eur. Pat. Appl. 97402264, 29 September 1997.

47. Berlureau, T.; Liska, J.-L. Method for Charging Maintenance-Free Nickel Metal Hydride Batteries. Eur. Pat. Appl. 98400455, 26 February 1998.

48. Berlureau, T.; Bariand, M.; Liska, J.-L. Method for Controlling the Rapid Charging of an Alkaline Electrolyte Industrial Accumulator. Eur. Pat. Appl. 99401903, 26 July 1999.

49. Gardes, F.; Juan, A.; Maloizel, S. Method for Managing the Charge of a Battery. Eur. Pat. Appl. 08291208, 18 December 2008.

50. Vasti, P.; Tridon, X. Leaktight Box for Alkaline Batteries. Eur. Pat. Appl. 13773814, 9 October 2013.

51. Payen, S.; Raymond, A. A Resealable Valve and Electrochemical Generator Comprising Said Valve. Eur. Pat. Appl. 01402994, 22 November 2001.

52. Vigier, N.; Besse, S. Electrochemical Generator with Surface of Revolution. Eur. Pat. Appl. 03290969, 18 April 2003.
53. Arts Energy. Available online: http://www.arts-energy.com/ (accessed on 25 May 2017).
54. Lichtenberg, F.; Kleinsorgen, K.; Hofmann, G. Nickel/Metallic Hydride Secondary Cell. Eur. Pat. Appl. 94116439, 19 October 1994.
55. Lichtenberg, F.; Kleinsorgen, K.; Hofmann, G. Electrical Nickel-Metal Hydride Accumulator with Electrode of Nickelhydroxide Comprising Graphite. Eur. Pat. Appl. 94115951, 10 October 1994.
56. Klaus, C. Gastight Alkaline Button-Type Accumulator. Eur. Pat. Appl. 94116622, 21 October 1994.
57. Klaus, C.; Simon, G.; Koehler, U.; Kannler, H. Gastight Nickel/Hydride-Accumulation of Round Cell Type. Eur. Pat. Appl. 95194924, 18 March 1995.
58. Koehler, U.; Chen, G.; Lindner, J. Gastight Metaloxide-Metalhydride Accumulator. Eur. Pat. Appl. 95106832, 5 May 1995.
59. Koehler, U.; Klaus, C.; Hofmann, G.; Lichtenberg, F. Alkaline Gastight Accumulator in the Shape of a Button Cell. Eur. Pat. Appl. 95106833, 5 May 1995.
60. Lichtenberg, F.; Koehler, U.; Kleinsorgen, K.; Foelzer, A.; Bouvier, A. Alkaline Metal Oxide, Metal Hydride Battery. Eur. Pat. Appl. 96102419, 17 February 1996.
61. Lichtenberg, F. Alloys for Use as Active Material for the Negative Electrode of an Alkaline, Rechargeable, Nickel Metal-Hybride Battery and Its Method of Preparation. Eur. Pat. Appl. 96110281, 26 June 1996.
62. Kleinsorgen, K.; Koehler, U.; Bouvier, A.; Foelzer, A. Process for Recovery of Metals from Used Nickel-Metal Hydride Accumulators. Eur. Pat. Appl. 95940265, 1 December 1995.
63. Kleinsorgen, K.; Koehler, U.; Bouvier, A.; Foelzer, A. Process for Recovery of Metals from Used Nickel-Metal Hydride Accumulators. Eur. Pat. Appl. 95941642, 1 December 1995.
64. Brenner, R. Electrochemical Cell with a Positive Electrode in the Form of a Hollow Cylinder and a Negative Electrode Assembled Therein. Eur. Pat. Appl. 13178468, 30 July 2013.
65. Baeuerlein, P. Nickel-Metal Hydride Secondary Battery. Eur. Pat. Appl. 01116524, 7 July 2001.
66. Knop, I. Charging Device Adapted for Charging NiMeH Batteries. Eur. Pat. Appl. 98102036, 6 February 1998.
67. Schein, H.; Kron, N.; Ilic, D. Mobile Charger for Charging Secondary Batteries from Secondary Batteries. Eur. Pat. Appl. 06776545, 2 August 2006.
68. Schula, C.; Scholz, S.; Pytlik, E.; Ensling, D. Secondary Electrochemical Cell and Charging Method. Eur. Pat. Appl. 15735963, 9 July 2015.
69. Schula, C.; Scholz, S.; Pytlik, E.; Ensling, D. Secondary Electrochemical Cell and Charging Method. Eur. Pat. Appl. 15735964, 9 July 2015.
70. VARTA. Available online: http://www.varta-microbattery.com (accessed on 25 May 2017).
71. Loustau, M.-T.; Verhoog, R.; Precigout, C. Method for Joining a Metallic Current Collector to an Electrode Comprising a Fibrous or a Spongeous Supporting Structure for Electrochemical Generator and Electrode Obtained Thereby. Eur. Pat. Appl. 94401155, 25 May 1994.
72. Amiel, O.; Belkhir, I.; Freluche, J.-P.; Pineau, N.; Dupuy, C.; Babin, S. Non-Sintered Electrode with Tridimensional Support for Secondary Battery with Alkaline Electrolyte. Eur. Pat. Appl. 00403112, 9 November 2000.
73. Cuesta, R.; Rouverand, C. Charge Control Method of a Sealed Nickel Storage Cell and Charger Therefor. Eur. Pat. Appl. 94401127, 20 May 1994.
74. Knosp, B.; Mimoun, M.; Bouet, J.; Gicquel, D.; Jordy, C. Hydridable Material for Nickel-Hydride Accumulator Negative Electrode. Eur. Pat. Appl. 95400803, 10 April 1995.
75. Cypel, L. Metal Hydride Negative Electrode Comprising a Coated Perforated Sheet. Eur. Pat. Appl. 98401150, 14 May 1998.
76. Senyarich, S.; Viaud, P. Alkaline Electrolyte Secondary Battery, Particularly of the Type nickel-Cadmium or Nickel Metal Hydride. Eur. Pat. Appl. 98402291, 17 September 1998.
77. Berlureau, T.; Liska, J.-L. Sealed Nickel-Metal Hydride Storage Battery. Eur. Pat. Appl. 00402019, 13 July 2000.
78. Bernard, P.; Goubault, L.; Le, G.L. Electrochemically Active Material for Negative Electrode of an Alkaline Electrolyte Secondary Battery. Eur. Pat. Appl. 03290119, 17 January 2003.
79. Alcatel-Lucent. Available online: https://en.wikipedia.org/wiki/Alcatel-Lucent (accessed on 29 July 2017).
80. Fredriksson, L.; Puester, N. A Method for Manufacturing a Biplate Assembly, a Biplate Assembly and a Bipolar Battery. Eur. Pat. Appl. 02746283, 8 July 2002.

81. Hug, L.; Puester, N.H.; Fredriksson, L. An Apparatus for Manufacturing an Electrode. Eur. Pat. Appl. 03770206, 7 November 2003.

82. Fredriksson, L.; Puester, N. A Bipolar Battery and a Biplate assembly. Eur. Pat. Appl. 02798876, 13 September 2002.

83. Fredriksson, L.; Puester, N. A Bipolar Battery, a Method for Manufacturing a Bipolar Battery and a Biplate Assembly. Eur. Pat. Appl. 02798875, 13 September 2002.

84. Fredriksson, L.; Puester, N.H. An Electrode, a Method for Manufacturing an Electrode and a Bipolar Battery. Eur. Pat. Appl. 03810729, 7 November 2003.

85. Fredriksson, L.; Puester, N.H. A Bipolar Battery and a Method for Manufacturing a Bipolar Battery. Eur. Pat. Appl. 03812398, 7 November 2003.

86. Hock, D.; Puester, N.; Fredriksson, L. A Method for Manufacturing a Bipolar Battery with a Gasket. Eur. Pat. Appl. 04800257, 3 November 2004.

87. Jensen, K.; Puester, N.; Hock, D. A Casing for a Sealed Battery. Eur. Pat. Appl. 06717032, 20 March 2006.

88. Fredriksson, L.; Puester, N.H.; Howlett, R. A Battery Stack Arrangement. Eur. Pat. Appl. 07703860, 15 January 2007.

89. Fredriksson, L.; Howlett, R. A Bipolar Battery Including a Pressure Sensor. Eur. Pat. Appl. 07712225, 15 February 2007.

90. Fredriksson, L.; Puester, N.H. A System for Reducing the Power in a Battery Stack Arrangement. Eur. Pat. Appl. 07852114, 17 December 2007.

91. Hock, D.; Fredriksson, L.; Puester, N.H. A Gasket, a Bipolar Battery and a Method for Manufacturing a Gasket. Eur. Pat. Appl. 08712719, 18 February 2008.

92. Howlett, R. An Energy Management System. Eur. Pat. Appl. 12738289, 28 June 2012.

93. Noréus, D. A Metal Hydride Battery with Added Hydrogen Gas, Oxygen Gas or Hydrogen Peroxide. International Application No. PCT/SE2016/051020, 27 April 2017.

94. Nilar. Available online: http://www.nilar.com (accessed on 25 May 2017).

95. Schaffrath, U.; Ohms, D.; Benczur-Uermoessy, G.; Markolf, R.; Schmelter, K. Method for Producing a Positive Nickel-Hydroxide Electrode for a Nickel-Metal Hydrid or Nickel-Cadmium Battery. Eur. Pat. Appl. 11007464, 14 September 2011.

96. Ohms, D.; Schaedlich, G.R.N.; Kleinschnittger, B.D.-I.; Markolf, R.; Schmelter, K. Nickel-Metal Hydrid Battery. Eur. Pat. Appl. 11007465, 14 September 2011.

97. Hoppecke Power from Innovation. Available online: https://www.hoppecke.com/ (accessed on 25 May 2017).

98. Master Alloys. Available online: http://www.gfe.com/fileadmin/user_upload/pdfs/Produkt spezifikationen_Alloys/HYDRALLOY_C_2004732_2005169_2019929_V4.pdf (accessed on 25 May 2017).

99. Klassen, T.; Bormann, R.; Oelerich, W.; Guether, V.; Otto, A. Metalliferous Storage Material for Hydrogen and Method for Producing Same. Eur. Pat. Appl. 99955727, 17 September 1999.

100. Non-Ferrous Metal Alloys. Available online: https://www.treibacher.com/en/products/non-ferrous-metal-alloys.html (accessed on 25 May 2017).

101. Knosp, B.; Arnaud, O.; Hezeque, T.; Barbic, P.; Bouvier, A. Hydridable Alloy. Eur. Pat. Appl. 01402059, 30 July 2001.

102. Hebenstreit, G.; Barbic, P. Metal Hydride Reservoir. International Application No. PCT/AT2006/000241, 28 December 2006.

103. Dubrow, R.S.; Froix, M.F. Method of Making a Microporous Article. Eur. Pat. Appl. 85902861, 20 May 1985.

104. Park, G.B.; Cook, J.A. Microporous Films. Eur. Pat. Appl. 90112666, 18 December 1985.

105. Dorling, M.G.L.; Barker, D.J.; Mcloughlin, R.H. Microporous Films. Eur. Pat. Appl. 89906800, 13 June 1989.

106. Singleton, R.W.; Cook, J.A.; Gargan, K. Polymer Membrane. Eur. Pat. Appl. 91916836, 19 September 1991.

107. Gargan, K.; Singleton, R.W.; Cook, J.A. Polymeric Sheet. Eur. Pat. Appl. 92914965, 9 July 1992.

108. Mcloughlin, R.H.; Gentilcore, G.; Cook, J.A. Non-Woven Fabric Treatment. Eur. Pat. Appl. 98925861, 8 June 1998.

109. Gentilcore, G.; Lancaster, I.M. Non-Woven Fabric Laminate. Eur. Pat. Appl. 98925860, 8 June 1998.

110. Hoffmann, H.; Schwoebel, R. Nonwoven Hydrophilic Material for Electrochemical Battery Separator and Method of Manufacturing. Eur. Pat. Appl. 93109577, 16 June 1993.

111. Wenneis, W.; Hoffmann, H.; Severich, B.; Keul, H.R.N. Alkaline Cell or Battery. Eur. Pat. Appl. 02009348, 3 May 2002.
112. Kritzer, P.; Trautmann, C. Hydrophylized Separator Material. Eur. Pat. Appl. 02023842, 24 October 2002.
113. Kritzer, P.; Schwoebel, R.-P. Method for the Production of a Separator Material for Alkaline Batteries. Eur. Pat. Appl. 03003011, 12 February 2003.
114. Kritzer, P.; Feistner, H.-J.; Schilling, H.; Kalbe, M. Nonwoven Fabric, Fiber and Electrochemical Cell. Eur. Pat. Appl. 05023520, 27 October 2005.
115. Kritzer, P. Separator for Housing in Batteries and Battery. Eur. Pat. Appl. 07006211, 27 March 2007.
116. Kritzer, P.; Schilling, H. Layer with Insulated Fibres and Electrochemical Cell. Eur. Pat. Appl. 07020102, 15 October 2007.
117. Roth, M.; Weber, C.; Berg, M.; Geiger, S.; Hirn, K.; Waschinski, C.; Falusi, S.; Kasai, M. Separator with Increased Puncture Resistance. Eur. Pat. Appl. 10749599, 11 August 2010.
118. Audebert, J.-F.; Feistner, H.-J.; Frey, G.; Farer, R.; Thrasher, G.L.; Weiss, M. Nonwoven Separator for Electrochemical Cell. Eur. Pat. Appl. 02789528, 8 November 2002.
119. Benczur-Uermoessy, G. Gastight Prismatic Nickel-Metal Hydride cell. Eur. Pat. Appl. 00942051, 10 June 2000.
120. Benczur-Uermoessy, G.; Ohms, D.; Waidelich, D. Electrode Capable of Storing Hydrogen and a Method for the Production of the Same. Eur. Pat. Appl. 00942050, 10 June 2000.
121. Benczur-Uermoessy, G.; Gensierich, M.; Ohms, D.; Wiesener, K. Battery in Bipolar Stacked Configuration and Method for the Production Thereof. Eur. Pat. Appl. 00927157, 6 May 2000.
122. Bartlett, P.N.; Owen, J.R.; Nelson, P.A. Electrochemical Cell. Eur. Pat. Appl. 03767996, 12 December 2003.
123. Bartlett, P.N.; Owen, J.R.; Nelson, P.A. Electrochemical Cell. International Application No. PCT/GB2003/005441, 28 October 2004.
124. Bartlett, P.N.; Owen, J.R.; Nelson, P.A. Electrochemical Cell Suitable for Use in Electronic Device. International Application No. PCT/GB2003/005442, 24 July 2004.
125. Bruchmann, B.; Klok, H.; Scholl, M.T. Preparation and Use of Modified Polylysines. Eur. Pat. Appl. 10152919, 15 November 2006.
126. Arvidsson, J.; Carlström, R.; Hallen, H.; Löfgren, S. Powder Composition and Process for the Preparation Thereof. Eur. Pat. Appl. 98928768, 5 June 1998.
127. Popall, M.; Olsowski, B.; Cochet, S. Particles Coated with an Organically Modified (Hetero) Silicic Acid Polycondensate and Containing a Metal Core Suited for Storing Hydrogen, Batteries Produced Therewith and Method for the Production Thereof Using the Particles. Eur. Pat. Appl. 10726978, 29 June 2010.
128. Maugy, C.; Porcellato, D.; Tarascon, J.-M.; Sainton, P. Method and System for Charging a Battery Wherein the Internal Pressure Varies Depending on State of Charge and Temperature. Eur. Pat. Appl. 04292653, 9 November 2004.
129. Bothe, M.; Schröder, G.B.R.; Breuch, G. Battery Charger with State of Charge Determination on the Primary Side. Eur. Pat. Appl. 05012135, 6 June 2005.
130. Nicolai, J. Quick Charge Method for Battery and Integrated Circuit for Performing the Method. Eur. Pat. Appl. 94400575, 16 March 1994.
131. Yuen, T.K. Battery Charger. Eur. Pat. Appl. 94300655, 28 January 1994.
132. Lang, O. Method for Regulating Charging of Nickel Cadmium and Nickel Metal Hydride Batteries, and Power Supply Unit. Eur. Pat. Appl. 06762920, 26 July 2006.
133. Umicore. Battery Recycling in the EU. Presented in the NaatBatt Battery Recycling Workshop, Ann Arbor, MI, USA. 30 November 2016. Available online: http://naatbatt.org/wp-content/uploads/2016/12/What%E2%80%99s-the-Rest-of-the-World-Doing_Caffarey_A.pdf (accessed on 24 July 2017).
134. About the Colabats Project. Available online: http://www.colabats.eu/about-the-colabats-project (accessed on 27 July 2017).
135. Cheret, D.; Santén, S. Battery Recycling. Eur. Pat. Appl. 05075736, 30 March 2005.
136. Luidold, S.; Antrekowitsch, H. Recovery of Rare Earth Metals from Waste Material by Leaching in Non-Oxidizing Acid and by Precipitation Using Sulphates. Eur. Pat. Appl. 10188264, 20 October 2010.
137. Pudas, J.; Erkkila, A.; Viljamaa, J. Battery Recycling Method. Eur. Pat. Appl. 11708474, 16 March 2011.
138. Hanulik, J. Battery Recycling Process, in Particular for Dry Batteries. Eur. Pat. Appl. 95920722, 16 June 1995.
139. Wiaux, J.; Lazouni, A.; Indaco, A. Used Battery and Cell Sorting Method and Apparatus. Eur. Pat. Appl. 94906358, 24 February 1994.

140. Alavi, K.; Salami, B. Method of Treating Nickel-Cadmium of Nickel-Hydride Cells. Eur. Pat. Appl. 93113054, 14 August 1993.
141. Traverse, J.-P.; Bressolles, J.-C.; Archier, P. Method for Processing Scrap Containing One or More Alloys that React to Form Hydrides, to Enable Recycling Thereof. Eur. Pat. Appl. 96926454, 24 July 1996.
142. Toro, L.; Veglio, F.; Beolchini, F.; Pagnanelli, F.; Furlani, G.; Granata, G.; Moscardini, E. Plant and Process for the Treatment of Exhausted Accumulators and Batteries. Eur. Pat. Appl. 11185532, 18 October 2011.
143. Laucournet, R.; Billy, E. Method for Recovering Metals Contained in an Ni-MH Battery. International Application No. PCT/IB2014/066008, 21 May 2015.
144. Black, V. Soap Dispenser Conversion Kit. Eur. Pat. Appl. 15156912, 27 February 2015.
145. Young, K. Metal Hydride. In *Reference Module in Chemistry, Molecular Science and Chemical Engineering*; Reedijk, J., Ed.; Elsevier: Waltham, MA, USA, 2013. [CrossRef]

batteries

MDPI

Article

Increase in the Surface Catalytic Ability by Addition of Palladium in C14 Metal Hydride Alloy

Kwo-Hsiung Young [1,2,*], Taihei Ouchi [2], Jean Nei [2] and Shiuan Chang [3]

1   Department of Chemical Engineering and Materials Science, Wayne State University, Detroit, MI 48202, USA
2   BASF/Battery Materials—Ovonic, 2983 Waterview Drive, Rochester Hills, MI 48309, USA;
    taihei.ouchi@basf.com (T.O.); jean.nei@basf.com (J.N.)
3   Department of Mechanical Engineering, Wayne State University, Detroit, MI 48202, USA;
    ShiuanC@wayne.edu
*   Correspondence: kwo.young@basf.com; Tel.: +1-248-293-700

Received: 6 June 2017; Accepted: 9 August 2017; Published: 9 September 2017

Abstract: A combination of analytic tools and electrochemical testing was employed to study the contributions of Palladium (Pd) in a Zr-based AB_2 metal hydride alloy ($Ti_{12}Zr_{22.8}V_{10}$ $Cr_{7.5}Mn_{8.1}Co_7Ni_{32.2}Al_{0.4}$). Pd enters the A-site of both the C14 and C15 Laves phases and shrinks the unit cell volumes, which results in a decrease of both gaseous phase and electrochemical hydrogen storage capacities. On the other hand, the addition of Pd benefits both the bulk transport of hydrogen and the surface electrochemical reaction. Improvements in high-rate dischargeability and low-temperature performances are solely due to an increase in surface catalytic ability. Addition of Pd also decreases the surface reactive area, but such properties can be mediated through incorporation of additional modifications with rare earth elements. A review of Pd-addition to other hydrogen storage materials is also included.

Keywords: metal hydride; nickel metal hydride battery; Laves phase alloy; palladium; electrochemistry; pressure concentration isotherm

1. Introduction

Zr-based AB_2 metal hydride (MH) alloy is an important research subject since it provides a possible improvement to the relatively low gravimetric energy density of nickel/metal hydride batteries [1,2]. Work regarding substitution of C14 Laves phase MH alloys started at the first row of transition metals [3–6] and proceeded to several non-transition metals (for example, Mg [7], La [8], Ce [9], and Nd [10]). Palladium (Pd), one of the two elements (the other is Vanadium (V)) with hydrogen-storage (H-storage) capabilities at room temperature (the heats of hydride formation for Pd and V are −20 [11] and −33.5 $kJ \cdot mol^{-1}$ [12], respectively), is very special among all possible substitution candidates. Pd's ability to absorb a large volume of hydrogen was first reported more than 150 years ago by Thomas Graham in 1866 [13], which built the foundation for modern MH research work [14–16]. In addition to use as a pure material, Pd also participates in H-storage research in many ways, such as a main ingredient in Pd-based alloys [17–20], an additive in the form of a nanotube [21], nanoparticle [22–25], or polycrystalline powder [26,27], a component in Pd [28–42] and Pd-containing thin films [43–46], and an alloying ingredient [41–101]. The major results accomplished by incorporating Pd in MH alloys are summarized in Table 1, and consist mainly of improvements in gaseous hydrogen absorption and desorption kinetics, electrochemical discharge capacity, high-rate dischargeability (HRD), activation, and cycle life performance in several MH alloy systems, including Mg, C, A_2B, AB, AB_2, AB_5, and body-centered-cubic solid solutions. In the two papers dealing with Pd alloyed in AB_2 MH alloys, one only discussed the HRD performance [56,79] and the other one is focused on the C15-dominated MH alloy [54]. Therefore, it is important to further investigate the influences of Pd-addition to the structural, gaseous phase, electrochemical properties, and their correlations in the C14-based AB$_2$ MH alloys.

Table 1. Summary of the Pd-substitution research based on different preparation methods, including arc melting (AM), replacement diffusion (RD), mechanical alloying by ball milling (MA), thermal annealing (TA), induction melting (IM), melt spinning (MS), levitation melting (LM), and wet impregnation (WI), in chronological order. GP and EC denote gaseous phase and electrochemical applications, respectively. HRD and I_o represent high-rate dischargeability and surface reaction current, respectively.

Host	Preparation	Application	Amount	Major Effect(s) of Pd	Reference
LaNi$_5$	AM	GP	16 at %	Increased plateau pressure	[47]
Mg$_2$Ni	RD	GP	8.3 at %	Increased absorption kinetics	[48]
Mg$_2$Ni	MA	GP	<1 wt. %	Increased absorption kinetics	[49]
TiFe	MA	GP	<1 wt. %	Increased activation	[50]
V$_3$TiNi$_{0.56}$	AM	EC	1 & 5 at %	Increased capacity	[51]
TiFe	AM + TA	GP	2.5 to 15 at %	Increased activation	[52]
Ti$_2$Ni	AM	EC	9.6 at %	Increased HRD and cycle life	[53]
AB$_2$	AM	EC	3.3 at %	Increased cycle life	[54]
Mg	MA	GP	14 wt. %	Increased desorption kinetics	[55]
AB$_2$	IM	EC	1 to 4 at %	Increased HRD	[56]
Mg$_2$Ni	MS	EC	5 to 20 at %	Increased capacity and cycle life	[57,58]
MgNi$_x$	MS	EC	10 at %	Easy amorphization	[59]
MgNi	MA	EC	1 to 10 at %	Increased cycle life	[60]
MgNi	MA	EC	10 at %	Increased cycle life	[61]
TiFe	AM	EC	5 to 10%	Increased EC activity	[62]
Mg$_{0.9}$Ti$_{0.1}$Ni	MA	EC	0 to 7.5 at %	Increased cycle life and I_o	[63–66]
Li$_3$BN$_2$	MA	GP	5 to 10 wt. %	Increased desorption kinetics	[67]
Mg	MA	GP	5 wt. %	Decreased absorption kinetics	[68]
Mg	MA	GP	10 wt. %	Increased desorption kinetics	[69]
MgTi$_x$	MA	EC	5 at %	Increased activation	[70]
Mg$_6$Pd$_7$Si$_3$	TA	GP	44 at %	Increased cycle life	[71]
LaMg$_2$Pd	TA	GP	25 at %	Novel MH alloy	[72]
TiVCr	AM	GP	0 to 0.5 at %	Increased capacity and activation	[73]
TiVCr	LM	EC	0 to 3 at %	Increased capacity, cycle life, and activation	[74]
TiZrNi	AM	EC	0 to 7 at %	Increased capacity, HRD, and cycle life	[75]
MgNi	MA	EC	0 to 5 at %	Increased HRD and cycle life	[76]
C	WI	GP	0 to 6 at %	Increased capacity	[77]
MgTi	MA	EC	3.3 at %	Increased capacity and I_o	[78]
AB$_2$	AM + TA	EC	5 to 10 wt. %	Increased HRD	[79]
Mg$_2$Ni	MA	EC	10 wt. %	Increased capacity	[80]
TiNi	IM	GP	0 to 2.5 at %	Decreased Capacity	[81]
MgNi	MA	EC	3.5 at %	Increased capacity, HRD, and cycle life	[82]
LaNi$_5$	AM + TA	GP	4 to 25 at %	Increase in plateau pressure	[83]
Mg$_2$Ni	MA	EC	3.3 at %	Increased capacity and cycle life	[84]
C	WI	GP	5 wt. %	Decreased absorption kinetics	[85]
MgNi	MA	EC	5 at %	Increased HRD and cycle life	[86]
MgNi	MA	EC	0 to 5 at %	Increased capacity and cycle life	[87]
Mg	MA	GP	0.1 to 5 wt. %	Increased absorption and desorption	[88]
MgNi	MA	EC	0 to 4 at %	Increased capacity	[89]
LaMg$_2$Ni	IM	GP	5 at %	Increased absorption and desorption	[90]
TiNi	MA	EC	5 wt. %	Increased capacity and cycle life	[91]
Mg$_2$Co	MA	EC	5 at %	Increased capacity, I_o, and cycle life	[92,93]
WMCNT	WI	GP	5 wt. %	Increased capacity	[94]
Na$_2$SiO$_3$	TA	GP	2.5 to 5 wt. %	Increased capacity	[95]
Graphene	WI	GP	5 to 10 wt. %	Increased capacity	[96]
TiNi, Ti$_2$Ni	MA + TA	EC	5 wt. %	Increased capacity and cycle life	[97]
C	WI	GP	0 to 13 wt. %	Increased capacity	[98]
Mg$_6$Pd	TA	GP	14 at %	Novel MH alloy	[99]
TiVCr	AM	GP	0.05 to 0.1 at %	Increased capacity	[100]
MgCo	MA	EC	5 at %	Increased HRD	[101]
PdCu, PdCuAg	MA	GP	15 to 100 at %	Increased capacity	[102]

In order to improve the electrochemical performance of C14-based MH alloy, especially at an ultra-low temperature (-40 °C), effects of Pd-incorporation were investigated. We fabricated the alloys, analyzed their microstructures with X-ray diffractometer (XRD) and scanning electron microscope (SEM) studied the gaseous phase reaction with hydrogen by pressure-concentration-temperature (PCT) isotherms, measured the electrochemical and magnetic properties, and correlated the results.

2. Experimental Setup

Arc melting under a 0.08 MPa Ar protective atmosphere was employed to prepare the sample ingots. To improve the homogeneity of the composition, the samples were flipped five times during the melting/cooling procedure. After cooling, each sample went through a hydriding/dehydriding process to created fissures and cracks to facilitate the later grinding process. The final product was a −200 mesh powder ready for the electrochemical testing. A Varian *Liberty* 100 inductively coupled plasma optical emission spectrometer (ICP-OES, Agilent Technologies, Santa Clara, CA, USA) was used to examine the chemical composition of each sample. For the structural analysis, a Philips *X'Pert Pro* XRD (Philips, Amsterdam, The Netherlands) and a JEOL-*JSM6320F* SEM (JEOL, Tokyo, Japan) with energy dispersive spectroscopy (EDS) were used. Since EDS analysis is only semi-qualitative in nature, results were used only for comparison purpose. For the gaseous phase H-storage study, a multi-channel PCT (Suzuki Shokan, Tokyo, Japan) was used. PCT measurements were performed at 30, 60, and 90 °C after a 2-h thermal cycle between room temperature and 300 °C under 2.5 MPa H_2 pressure. Electrode and cell preparations, as well as the electrochemical measurement methods, used for the experiments in the current study were the same as the ones used in our previous studies on the AB_2 MH alloys [103,104]. Electrochemical testing was performed in an open-to-air flooded cell configuration against a partially pre-charged sintered $Ni(OH)_2$ counter electrode at room temperature. A test electrode was made by dry compacting the MH powder directly onto an expanded Ni substrate (1 cm × 1 cm) without the use of any binder or conductive powder, and the average weight of active material per electrode was approximately 50 mg. The electrolyte used for testing was 30 wt. % KOH solution. Each electrode was charged with a current of 50 mA·g^{-1} for 10 h and then discharged at the same rate until a cut-off voltage of 0.9 V was reached. Two more pulls at 12 and 4 mA·g^{-1} then followed. For the surface reaction exchange current measurement (I_o), linear polarization was performed by first fully charging the system, then discharging to 50% of depth-of-discharge, and followed by scanning the current in the potential range of −20 to +20 mV of the open circuit voltage at a rate of 0.1 mV s^{-1}. For the bulk hydrogen diffusion coefficient (D) measurement, the system in a fully charged state was polarized at 0.6 V for 7200 s. All electrochemical measurements were performed on an Arbin Instruments BT-2143 Battery Test Equipment (Arbin Instruments, College Station, TX, USA). A Solartron 1250 Frequency Response Analyzer (Solartron Analytical, Leicester, UK) with a sine wave amplitude of 10 mV and a frequency range of 0.5 mHz to 10 kHz was used to conduct the AC impedance measurements. A Digital Measurement Systems Model 880 vibrating sample magnetometer (MicroSense, Lowell, MA, USA) was used to measure the magnetic susceptibility (M.S.) of the activated alloy surface (activation was performed by immersing the sample powder in 30 wt. % KOH solution at 100 °C for 4 h).

3. Results and Discussions

3.1. Properties of Pd

Several key physical properties of Pd are compared with those of transition metal elements commonly used in AB_2 MH alloys in Table 2. Pd is the heaviest among the reported elements (i.e., the highest atomic number) and thus does not have a significant weight advantage in H-storage applications. Moreover, Pd is in the same column and has the same number of outer-shell electrons as Ni (10), but it is located a row below in the periodic table (4d instead of 3d for Ni). Table 2 also shows the scarcity of Pd, which makes it very expensive, with a cost more than 2000 times higher than Ni (US$20,580 kg^{-1} for Pd [105] vs. US$10.4 kg^{-1} for Ni [106]). Furthermore, the atomic radius of Pd in the Laves phase is between those of the conventional A-site (Zr and Ti) and B-site elements (other elements in Table 2). The preferred ratio of average atomic radius of the A-site atoms to that of the B-site atoms in the Laves phase is approximately $\sqrt{3/2} \approx 1.225$ [107]. A Laves phase alloy with Pd in the A-site must incorporate a B-site element with an atomic radius of approximately 1.242 Å (1.521/1.225), which is too small for the commonly used B-site elements (Table 2). Besides Pd has a very high electronegativity value, which indicates that Pd attracts electrons, and is expected to occupy the B-site in intermetallic compounds like other commonly used modifying elements. Therefore, a Laves phase with Pd in the A-site is unlikely to happen. The only known Pd-containing Laves phase binary alloys are $CaPd_2$, $SrPd_2$, and $BaPd_2$ (all C15 structures) when alloyed with large alkaline earth elements [108,109]. It is also known that Pd, together with Cr, Mn, and Co, form a solid solution with Ni, indicating that a high solubility of Pd in Ni-based phases (TiNi and AB_2 for battery application) can be expected. The heat of hydride formation (ΔH_h), an indication of the metal-to-hydrogen bond strength, for Pd is slightly higher than that of V, meaning the hydride of Pd is more stable than that of V and causing the H-storage capacity of Pd to be lower than that of V ($PdH_{0.75}$ [110] vs. VH). Finally, due to its superior H_2 dissociative properties, Pd serves as a common catalyst in facilitating hydrogen absorption and desorption for MH alloys [111].

Table 2. Properties of **Pd** and other constituent elements in the alloys of this study. The radius quoted here is the atomic radius found in the Laves phase. Hcp, fcc, and bcc stand for hexagonal, face-centered-cubic, and body-centered cubic structures, respectively. ΔH_h is the heat of hydride formation. IMC denotes intermetallic compound. Ni forms a solid solution with a continuous composition range and has no IMC with Pd, Cr, Mn, or Co.

Property	Zr	Ti	Pd	V	Cr	Mn	Co	Ni	Al
Atomic Number	40	22	46	23	24	25	27	28	13
Number of Outer-layer e^-	4	4	10	5	6	7	9	10	3
Earth Crust Abundance (%)	0.013	0.66	6×10^{-7}	0.019	0.014	0.11	0.003	0.009	8.1
Radius (Å) [112]	1.771	1.614	1.521	1.491	1.423	1.428	1.385	1.377	1.582
Electronegativity	1.33	1.54	2.20	1.63	1.66	1.55	1.88	1.91	1.61
Crystal Structure [113]	hcp	hcp	fcc	bcc	bcc	bcc	hcp	fcc	fcc
Melting Point (°C) [114]	1855	1668	1555	1910	1907	1246	1495	1455	660
ΔH_h (kJ·mol^{-1}) [11]	−94	−67	−20	−35	−8	−8	15	−3	3
Number of IMC with Ni [115]	8	3	0	3	0	0	0	-	4

3.2. Chemical Composition

Six alloys (Pd0, Pd1, Pd2, Pd3, Pd4, and Pd5) with compositions of $Ti_{12}Zr_{22.8-x}V_{10}Cr_{7.5}$ $Mn_{8.1}Co_7Ni_{32.2}Al_{0.4}Pd_x$ (x = 0, 1, 2, 3, 4, and 5) were prepared by arc melting within a water-cooled Cu crucible. The Pd-free Pd0 alloy was also the base alloy used previously in studies of La- [8], Ce- [9], and Nd-substituted [10] AB_2 MH alloys, and was selected due to its balanced electrochemical performances with regard to capacity, rate, and cycle stability. In the composition design, Pd was assumed to occupy the A-site, due to its relatively large size (Table 2), and therefore the Zr-content was reduced to maintain the slightly hypo-stoichiometry (B/A = 1.87). ICP results are compared with the design compositions in Table 3. Only small deviations in the Mn-content were found, due to the Mn overcompensation in the case of evaporation loss. The average electron density (e/a), an important factor determining the ratio of C14 to C15 phase abundances [116–120], is calculated from the constituent elements' numbers of outer-shell electrons. Since Pd has more outer-shell electrons (10), compared to the replaced Zr (4), e/a increases with increasing Pd. While the observed e/a is very close to the designed e/a, the B/A ratios determined by the ICP results of the Pd-containing alloys are slightly higher than those determined by the design compositions, due to the slight loss of Pd and correspondingly increased in the Mn-content.

Table 3. Design compositions (in **bold**) and ICP results in at %. The average electron density is shown as e/a, and B/A is the ratio of B-atoms to A-atoms (Ti, Zr, and Pd).

Alloy	Source	Ti	Zr	V	Cr	Mn	Co	Ni	Pd	Al	e/a	B/A
Pd0	**Design**	**12.0**	**22.8**	**10.0**	**7.5**	**8.1**	**7.0**	**32.2**	**0.0**	**0.4**	**6.771**	**1.87**
	ICP	11.9	22.9	10.0	7.5	8.0	7.1	32.2	0.0	0.4	6.773	1.87
Pd1	**Design**	**12.0**	**21.8**	**10.0**	**7.5**	**8.1**	**7.0**	**32.2**	**1.0**	**0.4**	**6.831**	**1.87**
	ICP	12.0	21.3	10.3	7.5	8.5	7.0	31.9	1.1	0.4	6.834	1.91
Pd2	**Design**	**12.0**	**20.8**	**10.0**	**7.5**	**8.1**	**7.0**	**32.2**	**2.0**	**0.4**	**6.891**	**1.87**
	ICP	12.0	20.5	9.9	7.5	8.6	6.9	32.2	2.0	0.4	6.900	1.90
Pd3	**Design**	**12.0**	**19.8**	**10.0**	**7.5**	**8.1**	**7.0**	**32.2**	**3.0**	**0.4**	**6.951**	**1.87**
	ICP	12.0	19.5	10.1	7.5	8.6	7.0	32.1	2.8	0.4	6.949	1.92
Pd4	**Design**	**12.0**	**18.8**	**10.0**	**7.5**	**8.1**	**7.0**	**32.2**	**4.0**	**0.4**	**7.011**	**1.87**
	ICP	11.9	18.7	10.1	7.6	8.7	7.0	31.9	3.7	0.4	6.996	1.92
Pd5	**Design**	**12.0**	**17.8**	**10.0**	**7.5**	**8.1**	**7.0**	**32.2**	**5.0**	**0.4**	**7.071**	**1.87**
	ICP	11.9	17.8	10.2	7.4	8.6	7.1	32.0	4.6	0.4	7.055	1.92

3.3. XRD Analysis

Alloy structures were studied using XRD, and the resulting patterns are shown in Figure 1. Besides the C14 (MgZn$_2$-type, hexagonal, $hP12$ with a space group of $P6_3/mmc$) and overlapped C15 (MgCu$_2$-type, cubic, cF24, with a space group of $Fd\overline{3}m$) peaks, a small peak at around 41.5° was identified and assigned as a TiNi-based cubic phase (with a B2 structure, cubic, cI2, a space group of $Pm3m$). With the increased Pd-content, the Laves phase peaks shift to higher angles (indicating a decrease in the lattice constants), and the TiNi peak moves in the opposite direction. Through a full-pattern analysis using the Jade 9.0 software (MDI, Livermore, CA, USA), the lattice constants and abundances of the C14, C15, and TiNi phases were calculated, and the results are listed in Table 4. In the C14 phase, both the lattice constants a and c decrease, and the a/c ratio increases with increasing Pd-content. Since the size of Pd is between those of the A-atoms (Ti, Zr) and those of the B-atoms (except for Al), the lattice constants increase if Pd occupies the B-site and decrease if Pd sits in the A-site. Thus, the evolution of the C14 lattice constants clearly indicates that Pd occupies the A-site, despite its relatively high electronegativity (Table 2). V, with a slightly smaller size than Pd, was shown to occupy the B-site in the C14 structure [121]. However, size is apparently not the only determining factor in site selection because Al, which is larger than Pd, was found to occupy the B-site in the C14 structure [122] and Sn, with a much larger size compared to Pd, first occupies the A-site when its concentration is less than or equal to 0.1 at %, but then moves to the B-site at higher concentrations [123]. The lattice constant of the C15 phase also decreases with increasing Pd-content, suggesting that Pd is also in the A-site in C15. We will continue this discussion with the phase compositions revealed by EDS in the next section. Different from the observations made in the Laves phases, the lattice constant of the cubic TiNi phase increases with increasing Pd-content (as indicated by the shift of peak at around 41.5° to lower angles), which shows that Pd is in the B-site (Ni-site) in the TiNi (B2) structure. TiNi and TiPd, which share the same B2 structure, form a continuous solid solution, as demonstrated in the Ni-Pd-Ti ternary phase diagram [124,125]. Therefore, it is not surprising to observe that Pd partially replaces Ni in the TiNi phase. The partial replacement of Fe by Pd in TiFe (with the B2 structure) also leads to an expansion in the unit cell [52]. Evolution of the lattice constants from the C14, C15, and TiNi phases are plotted in Figure 2 and illustrate the linear dependencies on Pd-content in the design. The phase abundances obtained from the XRD analysis are plotted in Figure 3. Since the major peaks of C15 overlap with several peaks of C14, the C14 and C15 phase abundances were calculated from the integration of diffraction peaks using a calibration with previous samples performed by the Rietveld method. With increasing Pd-content, the C14 phase abundance experiences an initial drop, followed by a flat plateau, and finally another drop; the C15 phase abundance increases slightly in the beginning, then decreases, and finally increases at the highest Pd-content; the TiNi phase abundance continues to increase. Evolution of the C14 and C15 phase abundances are not monotonic as the evolution of *e/a* (Table 3) since Pd has a much higher chemical potential, which increases the *e/a* value at the C14/C15 threshold (C14:C15 = 1:1) [120]. Therefore, a higher *e/a* value does not necessarily correlate to a higher C15 phase abundance in Pd-containing alloys. Furthermore, the impact of adding Pd at different concentrations to the C14 crystallite size is insignificant, suggesting that all the alloys have very similar liquid-solid compositions at elevated temperatures, due to the high affinity between Pd and Ni.

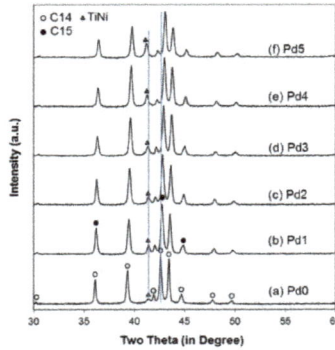

Figure 1. X-ray diffraction (XRD) patterns using Cu-K$_\alpha$ as the radiation source for alloys (**a**) Pd0, (**b**) Pd1, (**c**) Pd2, (**d**) Pd3, (**e**) Pd4, and (**f**) Pd5. In addition to the two Laves phases, another cubic phase is also identified. Vertical lines indicate the shifts of the TiNi and main C14/C15 peaks into lower and higher angles, respectively, with increasing Pd-content in design.

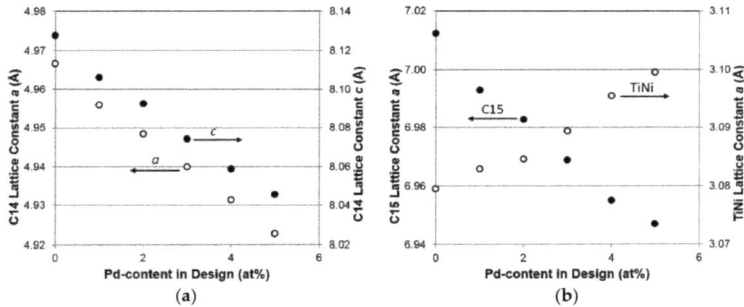

Figure 2. Evolution of the lattice constants of the (**a**) C14 and (**b**) C15 and TiNi phases with increasing Pd-content in design.

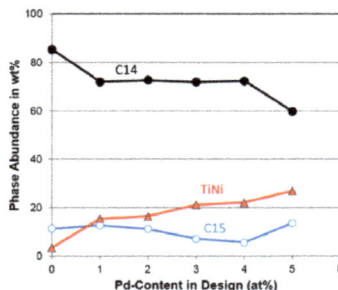

Figure 3. Evolution of the C14, C15, and TiNi phase abundances with increasing Pd-content in design.

Table 4. Lattice constants *a* and *c*, *a*/*c* ratio, and unit cell volume (V_{C14}) for the C14 phase, lattice constant *a* for the C15 and TiNi phases, Full width at half maximum (FWHM), and phase abundances in wt. % calculated from the XRD analysis.

Structural Property	Pd0	Pd1	Pd2	Pd3	Pd4	Pd5
a, C14 (Å)	4.9739	4.9631	4.9561	4.9471	4.9394	4.9328
c, C14 (Å)	8.1134	8.0915	8.0767	8.0598	8.0427	8.0254
a/*c*, C14 (Å)	0.61305	0.61337	0.61363	0.61380	0.61415	0.61465
V_{C14} (Å^3)	173.83	172.61	171.81	170.83	169.93	169.12
a, C15 (Å)	7.0121	6.9929	6.9827	6.9689	6.9550	6.9468
a, TiNi (Å)	3.0795	3.0829	3.0846	3.0893	3.0955	3.0995
FWHM, C14 (103)	0.237	0.25	0.249	0.241	0.234	0.243
C14 Crystallite Size (Å)	482	446	448	469	491	465
C14 Abundance (%)	85.4	72.1	72.7	72.0	72.4	59.8
C15 Abundance (%)	11.2	12.6	11.0	7.0	5.5	13.4
TiNi Abundance (%)	3.4	15.3	16.3	21.0	22.1	26.8

3.4. SEM/EDS Analysis

The distribution and composition of the constituent phases in all the alloys were studied by SEM/EDS. Representative SEM backscattering electron images (BEI) of alloys Pd1 to Pd5 are shown in Figure 4, while that of the base alloy Pd0 was previously published (Figure 3a in [8]). The composition of the numbered spots in each micrograph was further analyzed by EDS, and the results are listed in Table 5. Areas with the brightest contrast have a B/A ratios between 0.9 and 1.1 and are identified as the cubic TiNi phase. It should be noted that for the B/A ratio calculation of TiNi, Pd is treated as a B-site element since TiNi and TiPd share the same structure and form a continuous solid solution in the Ni-Pd-Ti ternary phase diagram [124]. Among the constituent phases, Pd has the highest solubility in TiNi, which explains the increase in TiNi phase abundance with increasing Pd-content (Figure 3). Concentrations of the major elements (Ti, Zr, Ni, and Pd) in the TiNi phase are plotted in Figure 5a as functions of Pd-content in design. The observed replacement of the smaller Ni with the larger Pd enlarges the TiNi unit cell, as shown by XRD analysis. Moreover, the main matrix with a B/A ratio between 2.1 and 2.3 and a relatively low *e*/*a* value (6.7 to 6.9) was assigned to a slightly hyper-stochiometric C14 phase with Pd residing in the A-site. Pd resides in the A-site for the C14 phase, or the B/A ratio would be even higher and beyond the practical range [126]. The dilemma of site selection for Pd is the same as for the case of V, which resides in the A- and B- sites in AB and AB$_2$ phases, respectively [127,128]. Concentrations of the major elements (Ti, Zr, Ni, and Pd) in the C14 phase are plotted in Figure 5b as functions of Pd-content in design. The major changes observed with increasing Pd-content in the design include a decrease in Zr and an increase in Pd. Pd is smaller than Zr and consequently causes a shrinkage in the C14 unit cell, as indicated by XRD analysis (Figure 2a). Although Pd and V have similar atomic radii (Table 2), they act differently in the multi-phase MH alloy; while Pd occupies the A-site in C14 and the B-site in TiNi, V does the opposite [10,121]. The large differences in numbers of outer-shell electron and electronegativities of Pd and V must play an important role in their site-selecting outcomes. One additional thing worth mentioning is the increase in lattice constant ratio *a*/*c* with increasing Pd-content (Table 4). This has been reported previously that the occupancy of B-site atoms (2*a* and/or 6*h*—Wykoff notation—in Figure 6) has an impact on the *a*/*c* ratio [128–130]. However, the correlation between where the A-site occupancy and the *a*/*c* ratio has not been reported, since only one possible site is available for the A-atoms (4*f* in Figure 6). When the *a*/*c* ratios of alloys in the current study and those of alloys in a previous Ti/Zr study [103] are plotted against the Zr-contents in C14 in Figure 7, we found that the *a*/*c* ratio increases with increasing Zr-content, except for when the Zr-content is greater than 15.5% in the Ti/Zr study. Therefore, the A-site arrangement on the A$_2$B plane must affect the *a*/*c* ratio, which warrants further computational studies.

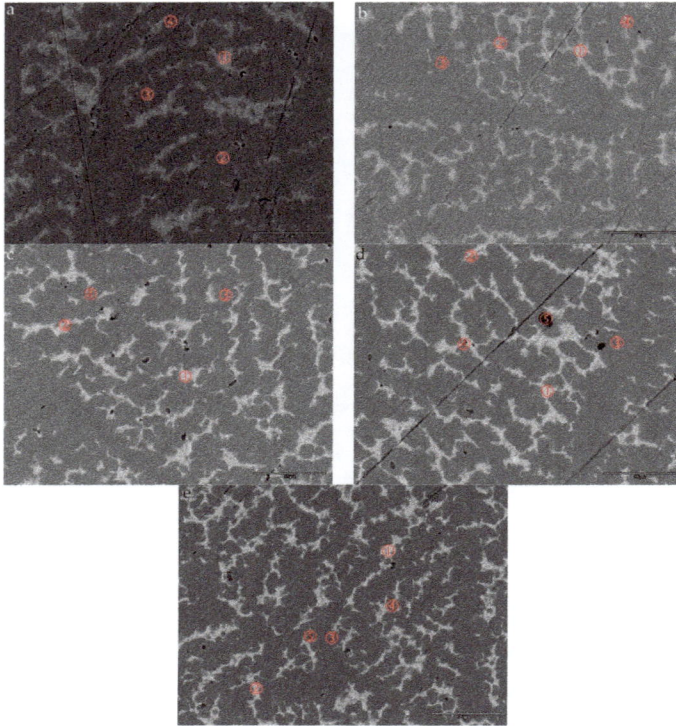

Figure 4. Scanning Electron Microscope Backscattering Electron Images (SEM BEI) micrographs from alloys (**a**) Pd1, (**b**) Pd2, (**c**) Pd3, (**4**) Pd4, and (**e**) Pd5. The composition of the numbered areas was analyzed by EDS, and the results are shown in Table 5. The bar at the lower right corner in each micrograph represents 25 μm.

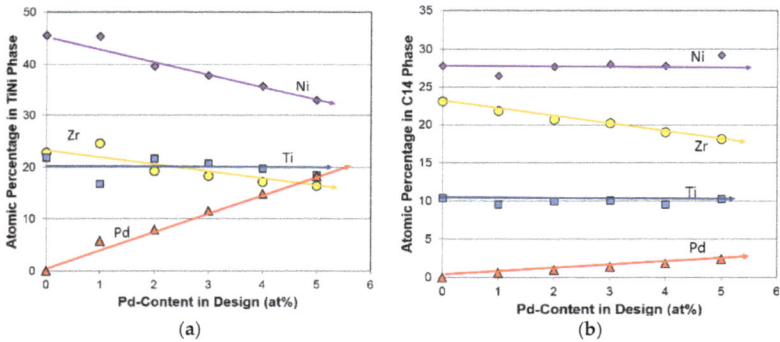

Figure 5. Evolution of the contents of the major constituting elements in the (**a**) TiNi and (**b**) C14 phases with increasing Pd-content in design.

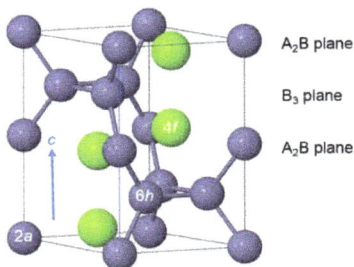

Figure 6. Schematic of the C14 structure. Green and blue spheres represent the A- (Zr, Ti, and Pd) and B- (V, Cr, Mn, Co, Ni, and Al) atoms, respectively. While two sites (*2a* on the A_2B plane and *6h* on the B_3 plane) are available for the B-atoms, only one site (*4f*) on the A_2B plane is available for the A-atoms.

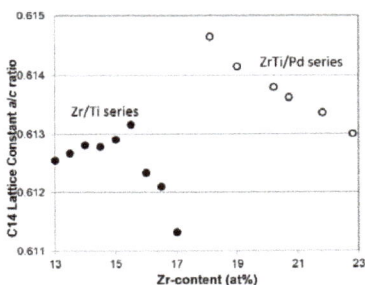

Figure 7. Evolution of the C14 lattice constant a/c ratio with increasing Zr-content in the phase. Data for Zr/Ti and ZrTi/Pd series are from a prior work [103] and the current study, respectively.

Table 5. Summary of the EDS results from several selective spots in the SEM-BEI micrographs shown in Figure 4. All compositions are in at %. The main C14 phase is identified in **bold**.

Alloy	Location	Ti	Zr	V	Cr	Mn	Co	Ni	Pd	Al	*e/a*	B/A	Phase
	Pd1-1	16.8	24.5	1.4	0.5	2.3	2.7	45.4	5.8	0.5	7.3	1.1	TiNi
	Pd1-2	13.3	20.8	9.2	5.1	8.2	6.7	35.2	1.0	0.5	6.9	1.8	C15
Pd1	Pd1-3	9.6	21.8	12.4	10.6	10.2	7.8	26.5	0.6	0.5	6.7	2.1	**C14**
	Pd1-4	7.8	58.9	5.0	2.6	4.2	3.0	17.4	0.9	0.3	-	-	ZrO_2
	Pd2-1	21.6	19.2	2.1	0.9	3.6	4.3	39.6	8.0	0.7	7.2	1.0	TiNi
	Pd2-2	11.6	20.8	7.2	4.0	7.1	5.1	40.9	2.8	0.5	7.2	1.8	C15
Pd2	Pd2-3	10.0	20.7	12.4	9.6	10.1	8.0	27.7	1.0	0.5	6.7	2.2	**C14**
	Pd2-4	5.7	69.1	3.3	2.3	3.0	2.3	12.5	1.6	0.3	-	-	ZrO_2
	Pd3-1	20.7	18.3	2.0	0.8	4.3	3.8	37.8	11.6	0.7	7.3	1.0	TiNi
	Pd3-2	12.7	19.7	6.0	3.0	6.4	4.5	41.9	5.3	0.5	7.4	1.7	C15
Pd3	Pd3-3	10.1	20.2	11.9	10.1	9.7	8.1	28.0	1.4	0.4	6.8	2.2	**C14**
	Pd3-4	19.5	51.8	3.3	1.5	3.2	2.2	15.3	3.0	0.2	-	-	ZrO_2
	Pd4-1	19.7	17.2	2.0	0.8	5.4	3.4	35.7	14.9	0.8	7.4	0.9	TiNi
	Pd4-2	10.7	18.9	7.5	4.2	6.9	4.8	40.8	5.7	0.6	7.4	1.8	C15
Pd4	Pd4-3	9.6	19.0	12.1	10.9	10.2	8.1	27.8	1.8	0.5	6.8	2.3	**C14**
	Pd4-4	3.6	80.1	1.8	1.0	1.3	1.3	7.6	3.2	0.1	-	-	ZrO_2
	Pd4-5	1.7	0.2	42.8	41.1	8.8	2.4	2.8	0.1	0.1	-	-	BCC
	Pd5-1	18.5	16.4	2.3	1.0	6.6	3.1	33.0	18.3	0.8	7.5	0.9	TiNi
	Pd5-2	12.4	18.0	7.7	4.1	6.7	4.7	38.7	7.2	0.5	7.3	1.7	C15
Pd5	Pd5-3	12.2	16.6	10.7	6.8	8.8	7.4	33.6	3.5	0.4	7.1	2.1	**C14**
	Pd5-4	10.3	18.1	11.5	10.4	9.4	8.3	29.2	2.4	0.4	6.9	2.2	**C14**
	Pd5-5	7.3	51.4	7.3	5.0	5.2	4.2	17.4	2.0	0.2	-	-	ZrO_2

The region between the main C14 matrix and TiNi secondary phase shows a contrast between C14 and TiNi and has been assigned as the C15 phase, due to its relatively high e/a (6.9–7.4) [118,120]. Transmission electron microscopy [131,132] and electron backscattering diffraction [133] confirmed that the C15 phase solidifies between the formations of the C14 and B2 phases in the multi-phase MH alloys. Unlike the C14 phase, the C15 phase is hypo-stoichiometric with the B/A ratio between 1.7 and 1.8. Solubility of the off-stoichiometric phase is caused by either the anti-site defect or vacancy [134]. Figure 8 provides a comparison of solubilities for the C14 and C15 alloys with Ti, Zr, or Hf as the A-site element. While the C14 alloy leans slightly toward being hyper-stoichiometric, the C15 alloy has an approximately equal opportunity to become either hyper- or hypo-stoichiometric. Therefore, we do not have a clear explanation for the stoichiometry preferences of the Laves phases in the current study. Furthermore, a shift in the C14/C15 threshold with increasing Pd-content is observed in Figure 9 and is thought to be due to the high chemical potential of Pd in the A-site, as predicted previously [120]. Compared to the C14 phase, the C15 phase has a higher solubility of Pd and Ni (Table 5), which have the highest number of outer-shell electrons (10) and consequently contribute to a higher e/a value. Lastly, areas with the darkest contrast consist of ZrO_2, which is the product of oxygen scavenging commonly seen in the Zr-based AB_2 MH alloys [130,135,136].

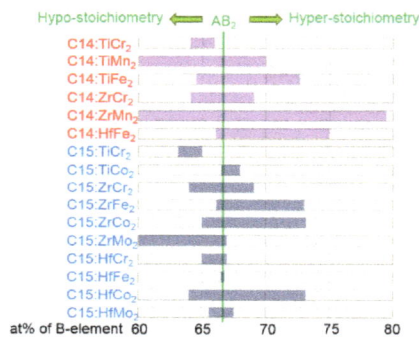

Figure 8. Comparison of solubilities (flexibility in stoichiometry) for the C14 and C15 binary alloys with Ti, Zr, or Hf as the A-site element (date from [115]).

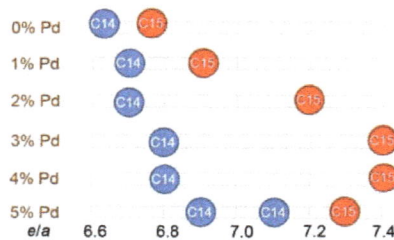

Figure 9. Plots of the e/a values for the C14 and C15 phases. A shift in the C14/C15 threshold to a higher e/a value with increasing Pd-content in design is observed.

3.5. PCT Analysis

PCT isotherms were used to study the interaction between the alloys and hydrogen gas. Both the 30 and 60 °C isotherms for each alloy are plotted in Figure 10. These PCT curves lack an appreciable amount of plateau and are similar to those of the multi-phase MH alloys due to the synergetic effects between the main working phase and catalytic secondary phase(s) [137]. In general, plateau pressure increases and the storage capacity and absorption/desorption hysteresis decrease as Pd-content increases. The gaseous phase H-storage properties obtained from the PCT isotherms are summarized in Table 6. As the Pd-content increases, both the maximum and reversible capacities first increase slightly with 1 at % Pd but then decrease. The desorption pressure at 0.75 wt. % (plateau pressure) H-content shows a monotonically increasing trend. This reduction in hydride stability by adding Pd was also observed in the AB$_5$ alloy previously [83]. Moreover, both the maximum capacity and log (desorption pressure at 0.75 wt. %) show linear dependencies on the C14 unit cell volume for all Pd-containing alloys, as demonstrated in Figure 11. Therefore, we believe that the gaseous H-storage characteristics are mainly determined by the main C14 phase. One point that does not follow the trend seen in Figure 11 is from alloy Pd1. Although alloy Pd1 has a smaller C14 unit cell and a lower C14 abundance compared to the Pd-free base alloy Pd0, its capacity increases slightly due to a large increase in the TiNi phase abundance. However, when prepared as an alloy, $Ti_{1.04}Ni_{0.86}Pd_{0.1}$ exhibits a mixed B19'/R/B2/Ti$_2$Ni structure and yields a discharge capacity of only 148 mAh·g^{-1} at C/5 rate [82]. Therefore, the direct influence of the TiNi phase on H-storage capacity should be minimal. The contribution from the TiNi phase most likely occurs through the synergetic effects that arise from TiNi and other phases, as observed previously [104,137,138]. The remaining capacities during desorption at 0.002 MPa of each alloy are listed in the third row in Table 6 and decrease with increasing Pd-content. Raising the plateau pressure would not necessary decrease the remaining capacity, as seen from a study on a series of pure $Zr_{1-x}Ti_xMnFe$ C14 MH alloys [139]. Therefore, we believe the decrease in remaining capacity during desorption (that correlates to a more complete desorption) results from the presence of the catalytic TiNi phase (either through an increase in abundance or an increase in the Pd-content in TiNi). Similar phenomenon has also been found in the study of the Mg-incorporated C14-predominated alloys [7]. Slope factor is defined as the ratio of desorption capacity between 0.01 and 0.5 MPa to total desorption capacity, and a higher slope factor corresponds to a flatter PCT isotherm. From the data listed in Table 6, slope factor in this series of alloys decreases with increasing Pd-content in design, which means the isotherm becomes more slanted—an indication of increased synergetic effects between the main storage phase and catalytic secondary phase(s) [10]. Due to the lack of an identifiable plateau in the PCT isotherm, hysteresis is defined as log (ratio of absorption to desorption pressures at 0.75 wt. % H-storage) and listed in Table 6. PCT hysteresis is commonly accepted as correlating to the energy needed to overcome the reversible lattice expansion in the metal (the αphase)/MH (the βphase) phase boundary during hydrogen absorption [119]. The catalytic TiNi phase facilitates the hydrogen absorption in the storage phase by pre-expanding the lattice near the interface and thus decreasing the energy needed to propagate hydrogen through the bulk [133]. Finally, the thermodynamic properties specifically changes in hydride enthalpy (ΔH_h) and entropy (ΔS_h), were calculated using the equilibrium pressures at 0.75 wt. % H-storage and the Van't Hoff equation,

$$\Delta G = \Delta H_h - T\Delta S_h = R\,T\,ln\,P, \tag{1}$$

where T and R are the absolute temperature and ideal gas constant, respectively. The calculated values for alloys Pd0 to Pd4 are listed in the last two rows of Table 6. Those for alloy Pd5 are not available since its high hydrogen equilibrium pressure is beyond the limit of our PCT apparatus (>2 MPa) and therefore cannot be measured. With increasing Pd-content in design, both ΔH_h and ΔS_h increase. While the increase in ΔH_h is due to shrinkage of the C14 unit cell, the increase in ΔS_h is caused by an increase in disorder in the hydride, correlating well with the observed decrease in slope factor.

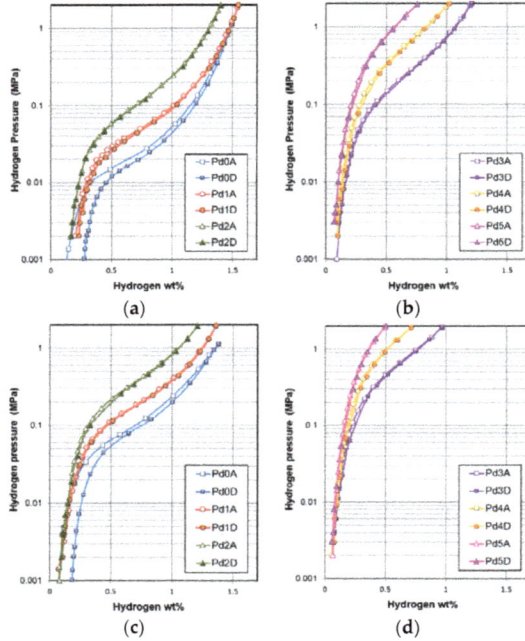

Figure 10. 30 °C PCT isotherms of alloys (**a**) Pd0, Pd1, and Pd2 and (**b**) Pd3, Pd4, and Pd5 and 60 °C PCT isotherms of alloys (**c**) Pd0, Pd1, and Pd2 and (**d**) Pd3, Pd4, and Pd5. Open and solid symbols are for absorption and desorption curves, respectively.

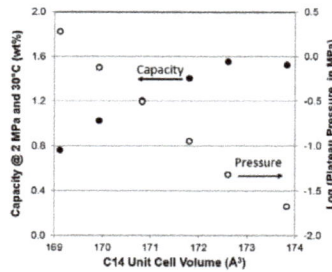

Figure 11. Plot of the H-storage capacity and log (plateau pressure) vs. the C14 unit cell volume.

Table 6. Summary of gaseous phase properties: maximum and reversible capacities, plateau pressure, slope factor, hysteresis, and changes in enthalpy and entropy.

Gaseous Phase Property	Unit	Pd0	Pd1	Pd2	Pd3	Pd4	Pd5
Maximum capacity @ 2 MPa and 30 °C	wt. %	1.52	1.55	1.40	1.20	1.02	0.76
Reversible Capacity @ 30 °C	wt. %	1.25	1.32	1.23	1.11	0.92	0.69
Capacity @ 0.002 MPa and 30 °C	wt. %	0.30	0.23	0.17	0.10	0.09	0.06
Desorption Pressure @ 0.75 wt. % and 30 °C	MPa	0.021	0.048	0.112	0.306	0.745	1.882
Slope Factor @ 30 °C	%	78	81	78	66	53	44
Hysteresis @ 30 °C		0.21	0.05	0.03	0.05	0.03	0.00
$-\Delta H_h$	kJ·mol $H_2{}^{-1}$	41.6	40.5	35.6	30.9	28.0	-
$-\Delta S_h$	J·mol $H_2{}^{-1}$·K^{-1}	127	125	118	111	109	-

3.6. Electrochemical Analysis

Electrochemical testing was performed in an open-to-air flooded cell configuration against a partially pre-charged sintered $Ni(OH)_2$ counter electrode at room temperature. Each electrode was charged with a current of 50 mA·g^{-1} for 10 h and then discharged at the same rate until a cut-off voltage of 0.9 V was reached. The capacity obtained at this rate is assigned as the high-rate discharge capacity. Two more pulls at 12 and 4 mA·g^{-1} then followed. The capacities at the three different rates were summed, and the sum is designated as the full capacity. The ratio of the high-rate to full capacities is reported as HRD. The activation behaviors in the electrochemical environment of alloys in the current study are compared in Figure 12. Judging from the full capacities and HRD in the first 13 cycles, the addition of Pd improves the activation performances of both properties. While the degradation in full capacity was negligible for all alloys, degradations in HRD are obvious and become more severe with increasing Pd-content. The deterioration in HRD with cycling is due to the formation of a passive layer on the surface of TiNi, whose abundance also increases as Pd-content increases. The Pd-addition in many MH alloys results in improvement in cycle stability (Table 1), a positive contribution from the dense nature of TiO_2 [140] and stability of Pd/PdO in alkaline solution [141].

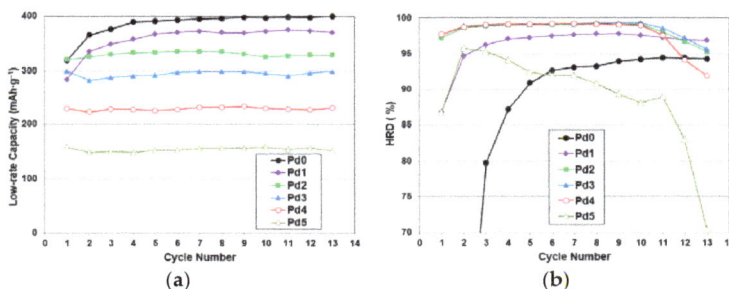

Figure 12. Activation behaviors observed from (**a**) full capacity and (**b**) HRD for the first 13 electrochemical cycles measured at room temperature.

All the electrochemical properties obtained from the alloys are summarized in Table 7. With increasing Pd-content, the following trends are observed: the high-rate capacity first increases with the addition of catalytic Pd, but then decreases due to the reduction in unit cell volume of C14; the full capacity decreases monotonically; HRD increases; and the activation performance is ultimately improved. The increase in capacity for the gaseous phase in alloy Pd1 was not observed in the electrochemical capacity. Although the TiNi phase is considered highly catalytic in the gaseous phase, it is also prone to surface passivation and, consequently, may not be as effective in the electrochemical environment. Electrochemical discharge capacity is plotted against the gaseous phase maximum H-storage capacity, shown in Figure 13. Gaseous phase maximum H-storage capacity is composed of reversible and irreversible capacities and considered to be the upper bound for the electrochemical discharge capacity. Therefore, although a close correlation between electrochemical discharge capacity and gaseous phase maximum H-storage capacity can be observed, it falls below the conversion of 1 wt. % = 268 mAh·g^{-1} due to some capacity irreversibility. Moreover, the linear relationship of electrochemical discharge capacity vs. gaseous phase maximum H-storage capacity indicates that the origin for the decrease in electrochemical discharge capacity with increasing Pd-content is the same as that in the gaseous phase, specifically a decrease in the C14 unit cell volume. For all the alloys, the discharge capacity is smaller than the gaseous phase H-storage since the open-to-air configuration and high plateau pressure cause an incomplete charging in the electrochemical environment.

Table 7. Summary of electrochemical half-cell measurements: capacities at the 3rd cycle, HRD at the 3rd cycle, cycles needed to achieve 92% HRD, bulk diffusion coefficient, surface exchange current, and results from AC impedance and magnetic susceptibility measurements. AC impedance measurement was performed at −40 °C while all other properties were measured at room temperature.

Electrochemical and Magnetics Properties	Unit	Pd0	Pd1	Pd2	Pd3	Pd4	Pd5
3rd Cycle High-rate Discharge Capacity	mAh·g^{-1}	300	335	327	285	226	143
3rd Cycle Full Discharge Capacity	mAh·g^{-1}	376	349	330	288	228	150
3rd Cycle HRD	%	80	96	99	99	99	98
Activation Cycle # to Achieve 92% HRD		6	1	1	1	1	1
Diffusion Coefficient, D	10^{-10} cm^2·s^{-1}	2.1	4.4	6.2	2.0	4.1	4.5
Surface Reaction Current, I_o	mA·g^{-1}	12.8	24.7	28.8	25.2	22.1	17.1
Charge-transfer Resistance @ −40 °C	Ω·g	158.6	29.1	28.3	22.9	15.7	39.9
Double-layer Capacitance @ −40 °C	F·g^{-1}	0.18	0.16	0.15	0.16	0.13	0.10
RC Product @ −40 °C	s	28.4	4.8	4.2	3.6	2.0	4.0
Total Saturated Magnetic Susceptibility, M_S	emu·g^{-1}	0.035	0.015	0.008	0.018	0.011	0.013
Applied Field @ M.S. = $\frac{1}{2}$ M_S, $H_{1/2}$	kOe	0.50	0.47	0.34	0.61	0.77	0.36

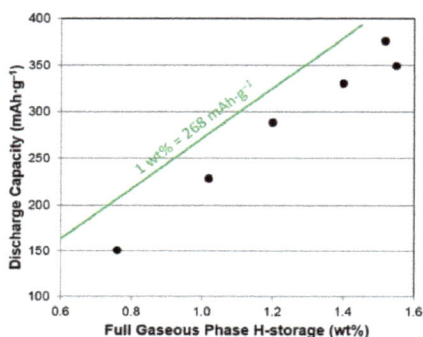

Figure 13. Comparison of the electrochemical discharge capacity vs. gaseous phase H-storage capacity. The green line shows the conversion between two properties, which sets the upper bound for electrochemical discharge capacity.

To trace the source of Pd's contribution to HRD, both D (bulk-related) and I_o (surface-related) were measured, and the results are listed in Table 7. Details on these two measurements can be found in our previous publication [8]. The reported values were averaged from the values measured from three samples prepared in parallel. Except for alloy Pd3, the D values from the Pd-containing alloys are at least double of that from the Pd-free Pd0 alloy. We repeated the same experiments three times for alloy Pd3, and the results are very close to the first measurement. At the present time, we cannot explain the relatively low D value for alloy Pd3 and speculate that it may be related to the distribution and orientation alignment of the C14 and C15 grains. The I_o value increases in the first two Pd-containing alloys (Pd1 and Pd2) but decreases with further increase in the Pd-content. In general, both D and I_o are improved by the addition of Pd, so we can conclude that the origin of enhanced HRD in Pd-containing C14-based MH alloys is a combination of transportation of hydrogen in the alloy bulk and facilitation of the surface electrochemical reaction.

Low-temperature performance of alloys in the current study was evaluated by AC impedance analysis. Both the charge-transfer resistance (R) and double-layer capacitance (C) were obtained from the semi-circle in the Cole-Cole plot (plot of the negative imaginary part vs. the real part of impedance with varying frequency). While R is closely related to the speed of electrochemical reaction, C is proportional to the reactive surface area, and their product (RC) can be interpreted as the surface catalytic ability without any contribution from the surface area [7,9,142]. These calculated values are listed in Table 7 and plotted with the amounts of Pd, Ce [9], and Nd [10] present in the C14 MH alloys in Figure 14. The *R* values are reduced dramatically with all the additives, but Ce and Nd are demonstrate the most dramatic decrease in *R*, compared to Pd. Figure 14b shows that the surface area increases by a large amount with Ce-addition, also does not increases as much with Nd-addition, and decreases slightly with Pd-addition. While adding Ce and Nd results in the formation of a soluble AB phase and a consequent increase the surface area in alkaline solution [9,10], the TiNi phase is more protective and lowers the amount of reactive surface area in the Pd-containing alloys. Figure 14c demonstrates that although all three additives increase the surface catalytic ability by lowering the *RC* product (corresponding to a faster reaction), the Nd- and Pd-containing alloys (especially alloys Pd3 and Pd4) are more catalytic than the Ce-containing alloys. In conclusion, Pd, Ce, and Nd increase the surface electrochemical reaction rate by improving the catalytic ability, reactive surface area, and both, respectively. Future substitution work should combine the highly catalytic Pd and effective surface area promoter Ce.

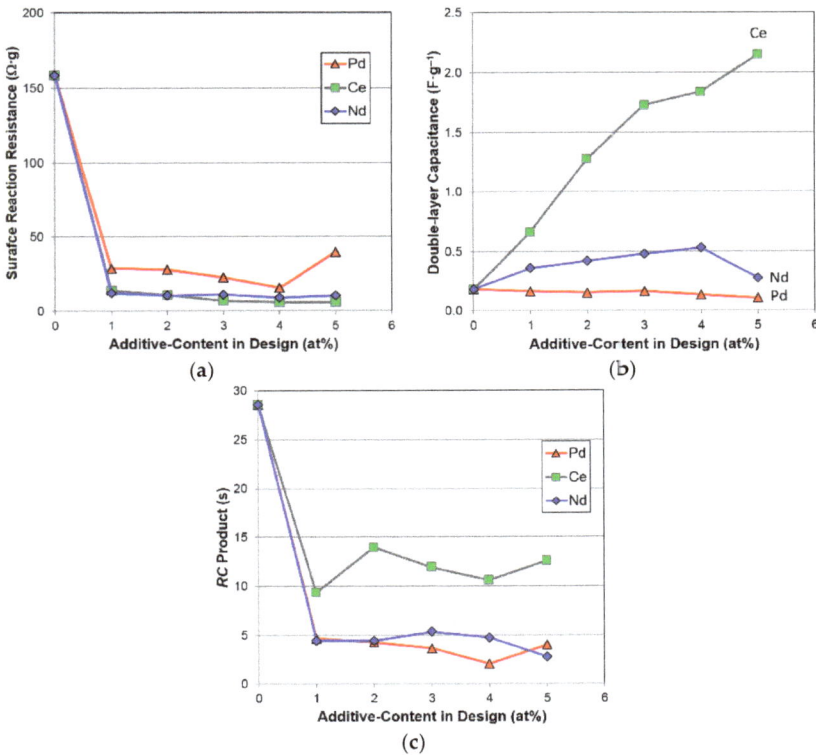

Figure 14. Plots of (**a**) surface reaction resistance (*R*), (**b**) double-layer capacitance (*C*), and (**c**) their product (*RC*) as functions of Ce-, Nd- and Pd-contents in design. Data of Ce- and Nd-substitutions was previously published [9,10].

3.7. Magnetic Susceptibility

The catalytic ability in the surface of MH alloy was previously correlated successfully to the saturated M.S. [143]. After activation, Zr from the alloy forms surface oxides, and the non-corroded Ni atoms conglomerate and form metallic clusters within the oxides [144]. Since the M.S. of metallic Ni is at least seven orders of magnitude larger than that of the alloy, due to the existence of unpaired electrons in metallic Ni [145], the total percentage of metallic Ni can be estimated from the saturated

M.S. (M_S) by measuring the M.S. of the activated MH alloy. The average size of Ni clusters can also be estimated by the strength of the applied magnetic field that corresponds to half of the M_S value ($H_{1/2}$) [7]. The magnetic properties of several key MH alloys were compared in an earlier publication [1]. Both the M_S and $H_{1/2}$ of alloys in this study are listed in the last two rows in Table 7. The M_S values of the Pd-containing alloys are much lower than that of the Pd-free alloy Pd0. Since the percentage of reduction in M_S of the Pd-containing alloys is much larger than that of the increase in the TiNi phase abundance, Pd in the main C14 phase must also contribute to the reduction in M_S. Moreover, the improved HRD, achieved by adding Pd, is certainly not related to the amount of metallic clusters embedded in the surface oxide. The $H_{1/2}$ values for the alloys indicate that the Ni cluster size is relatively unchanged with the addition of Pd.

M_S and I_o measured at room temperature vs. R measured at −40 °C for several C14-based alloys with 1 at % of various additives are plotted in Figure 15. Except for Pd, M_S and I_o from the same alloy correlate very closely. In other words, surface electrochemical reaction is dominated by the amount of metallic Ni in the surface oxide for the majority of modified C14 MH alloys. However, Pd facilitates the electrochemical reaction by acting as a catalyst. Figure 15 also demonstrates that M_S (I_o) is inversely proportional to R, except for the Nd- and Pd-containing alloys. Nd, although it shows zero solubility in the C14 phase, may participate in the catalytic process through another more complicated route (for example, creating a unique surface oxide structure as in the case of the La-addition [146]).

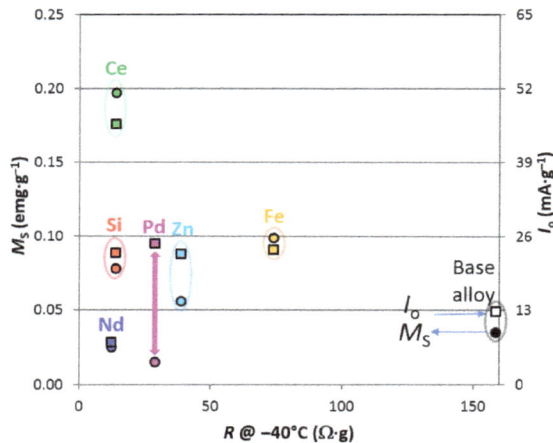

Figure 15. M_S and I_o measured at room temperature vs. surface reaction resistance R measured at −40 °C are plotted for the base alloy ($Ti_{12}Zr_{22.8}V_{7.5}Mn_{8.1}Co_7Ni_{32.2}Al_{0.4}$) and alloys with additions of 1 at % Ce, Si, Pd, Zn, Fe, and Nd. All additives demonstrate a reduction in R measured at −40 °C. The M_S and I_o pair from the Pd-containing alloy shows the largest separation, suggesting that the amount of catalytic Ni embedded in the surface oxide is not the origin of the improvements in I_o and R.

4. Conclusions

Incorporation of Pd in the Zr-based AB_2 multi-phase metal hydride alloy has been systematically studied. XRD analysis results show that Pd occupies the A-site for both the C14 and C15 structures, which results in shrinkage of the unit cells and, consequently, reductions in the gaseous phase and electrochemical capacities. With a strong affinity to Ni, Pd promotes the formation of the Ti(Ni, Pd) phase with a B2 structure as shown by the XRD and SEM/EDS results (where as the Pd-content in the alloy increases, the TiNi abundance and amount of Pd in the phase increase). This secondary phase is beneficial for gaseous phase H-storage, which is indicated by the increase in H-storage capacity despite the decrease in unit cell size of the main storage C14 phase at the point of dramatic increase in TiNi (substitution of 1 at % Pd); however, TiNi is detrimental to various electrochemical properties due to its passivating nature with alkaline electrolytes. Although the reactive surface areas of the Pd-containing alloys are smaller, the completeness of gaseous hydrogen desorption, high-rate dischargeability, and low-temperature performance are all improved with the addition of highly catalytic Pd at only 1 at %. Therefore, combining a small amount of Pd with other substitution elements with the capability of increasing capacity and/or reactive surface area, such as Ce, Y, and Si, is recommend for future modification research.

Acknowledgments: The authors would like to thank the following individuals from BASF-Ovonic for their help: Su Cronogue, Baoquan Huang, Diana F. Wong, David Pawlik, Allen Chan, and Ryan J. Blankenship.

Author Contributions: Kwo-Hsiung Young designed the experiments and analyzed the results. Taihei Ouchi prepared the alloy samples and performed the PCT and XRD analysis. Jean Nei prepared the electrode samples and conducted the magnetic measurements. Shiuan Chang assisted in data analysis and manuscript preparation.

Conflicts of Interest: The authors declare no conflict of interest.

Abbreviations

The following abbreviations are used in this manuscript:

MH	Metal hydride
H-storage	Hydrogen storage
HRD	High-rate dischargeability
AM	Arc melting
RD	Replacement-diffusion
MA	Mechanical alloying
TA	Thermal annealing
IM	Induction melting
MS	Melt spinning
LM	Levitation melting
WI	Wet impregnation
GP	Gaseous phase
EC	Electrochemical
I_o	Surface exchange current
XRD	X-ray diffractometer
SEM	Scanning electron microscope
PCT	Pressure concentration temperature
ICP-OES	Inductively coupled plasma optical emission spectrometer
EDS	Energy dispersive spectroscopy
M.S.	Magnetic susceptibility
ΔH_h	Heat of hydride formation
hcp	Hexagonal close-packed
fcc	Face-centered-cubic
bcc	Body-centered-cubic
IMC	Intermetallic compound
e/a	Average electron density
V_{C14}	Unit cell volume of the C14 phase
FWHM	Full width at half maximum
BEI	Back-scattering electron image
ΔS_h	Change in entropy
T	Absolute temperature
R	Ideal gas constant
D	Bulk diffusion coefficient
R	Surface charge-transfer resistance
C	Surface double-layer capacitance
M_S	Saturated magnetic susceptibility
$H_{1/2}$	Applied magnetic field strength corresponding to half of saturated magnetic susceptibility

References

1. Young, K.; Nei, J. The current status of hydrogen storage alloy development for electrochemical applications. *Materials* **2013**, *6*, 4574–4608. [CrossRef] [PubMed]
2. Chang, S.; Young, K.; Nei, J.; Fierro, C. Reviews on the U.S. Patents regarding nickel/metal hydride batteries. *Batteries* **2016**, *2*. [CrossRef]
3. Shaltiel, D.; Jacob, I.; Davidov, D. Hydrogen absorption and desorption properties of AB_2 Laves-phase pseudobinary compounds. *J. Less Common Met.* **1977**, *53*, 117–131. [CrossRef]
4. Jacob, I.; Shaltiel, D.; Davidov, D.; Miloslavski, I. A phenomenological model for the hydrogen absorption capacity in pseudobinary Laves phase compounds. *Solid State Commun.* **1977**, *23*, 669–671. [CrossRef]
5. Shaltiel, D. Hydride properties of AB_2 Laves-phase compounds. *J. Less Common Met.* **1978**, *62*, 407–416. [CrossRef]
6. Mendelsohn, M.H.; Gruen, D.M. The pseudo-binary system $Zr(V_{1-x}Cr_x)_2$: Hydrogen absorption and stability considerations. *J. Less Common Met.* **1981**, *78*, 275–280. [CrossRef]
7. Chang, S.; Young, K.; Ouchi, T.; Meng, T.; Nei, J.; Wu, X. Studies on incorporation of Mg in Zr-based AB_2 metal hydride alloys. *Batteries* **2016**, *2*. [CrossRef]
8. Young, K.; Wong, D.F.; Ouchi, T.; Huang, B.; Reichman, B. Effects of La-addition to the structure, hydrogen storage, and electrochemical properties of C14 metal hydride alloys. *Electrochim. Acta* **2015**, *174*, 815–825. [CrossRef]
9. Young, K.; Ouchi, T.; Nei, J.; Moghe, D. The importance of rare-earth additions in Zr-based AB_2 metal hydride alloys. *Batteries* **2016**, *2*. [CrossRef]
10. Wong, D.F.; Young, K.; Nei, J.; Wang, L.; Ng, K.Y.S. Effects of Nd-addition on the structural, hydrogen storage, and electrochemical properties of C14 metal hydride alloys. *J. Alloys Compd.* **2015**, *647*, 507–518. [CrossRef]
11. Griessen, R.; Riesterer, T. Heat of Formation Models. In *Hydrogen in Intermetallic Compounds I*; Schlapbach, L., Ed.; Springer: Berlin/Heidelberg, Germany, 1988.
12. Osumi, Y. *Suiso Kyuzou Goukin*; Agune Co. Ltd.: Tokyo, Japan, 1993. (In Japanese)
13. Graham, T. On the absorption and dislytic separation of gases by colloid septa. *Philos. Trans. R. Soc. Lond.* **1866**, *156*, 399–439. [CrossRef]
14. Smith, D.P. *Hydrogen in Metals*; The University of Chicago Press: Chicago, IL, USA, 1947.
15. Mackay, K.M. *Hydrogen Compounds of the Metallic Elements*; E.&F.N. Spon Ltd.: London, UK, 1966.
16. Muetterties, E.L. The transition metal-hydrogen interaction. In *Transition Metal Hydrides*; Muetterties, E.L., Ed.; Marcel Dekker, Inc.: New York, NY, USA, 1971.
17. Sakamoto, Y.; Yuwasa, K.; Hirayama, H. X-ray investigation of the absorption of hydrogen by several palladium and nickel solid solution alloys. *J. Less Common Met.* **1982**, *88*, 115–124. [CrossRef]
18. Żurowski, A.; Łukaszewski, M.; Czerwiński, A. Electrosorption of hydrogen into palladium-rhodium alloys Part 2. Pd-rich electrodes of various thickness. *Electrochim. Acta* **2008**, *53*, 7812–7816. [CrossRef]
19. Rousselot, S.; Bichat, M.-P.; Guay, D.; Roué, L. Structure and electrochemical behavior of metastable $Mg_{50}Ti_{50}$ alloy prepared by ball milling. *J. Power Sources* **2008**, *175*, 621–624. [CrossRef]
20. Łukaszewski, M.; Hubkowska, K.; Koss, U.; Czerwiński, A. On the nature of voltammetric signals originating from hydrogen electrosorption into palladium-noble metal alloys. *Materials* **2013**, *6*, 4817–4835. [CrossRef] [PubMed]
21. Badri, V.; Hermann, A.M. Metal hydride batteries: Pd nanotube incorporation into the negative electrode. *Int. J. Hydrogen Energy* **2000**, *25*, 249–253. [CrossRef]
22. Zaluska, A.; Zaluski, L.; Ström-Olsen, J.O. Nanocrystalline magnesium for hydrogen storage. *J. Alloys Compd.* **1999**, *288*, 217–225. [CrossRef]
23. Kohno, T.; Yamamoto, M.; Kanda, M. Electrochemical properties of mechanically ground Mg_2Ni alloy. *J. Alloys Compd.* **1999**, *293–295*, 643–647. [CrossRef]
24. Janot, R.; Rougier, A.; Aymard, L.; Lenain, C.; Herrera-Urbina, R.; Narzi, G.A.; Tarascon, J.M. Enhancement of hydrogen storage in MgNi by Pd-coating. *J. Alloys Compd.* **2003**, *356–357*, 438–441. [CrossRef]
25. Rivera, M.A.; Pal, U.; Wang, X.; Gonzalez-Rodriguez, J.G.; Gamboa, S.A. Rapid activation of $MmNi_{5-x}M_x$ based MH alloy through Pd Nanoparticle impregnation. *J. Power Sources* **2006**, *155*, 470–474. [CrossRef]

26. Visintin, A.; Castro, E.B.; Real, S.G.; Trica, W.E.; Wang, C.; Soriaga, M.P. Electrochemical activation and electrocatalytic enhancement of a hydride-forming metal alloy modified with palladium, platinum and nickel. *Electrochim. Acta* **2006**, *51*, 3658–3667. [CrossRef]

27. Shan, X.; Payer, J.H.; Jennings, W.D. Mechanism of increased performance and durability of Pd-treated metal hydriding alloys. *Int. J. Hydrogen Energy* **2009**, *34*, 363–369. [CrossRef]

28. Uchida, H.H.; Wulz, H.-G.; Fromm, E. Catalytic effect of nickel, iron and palladium on hydriding titanium and storage materials. *J. Less Common Met.* **1991**, *172–174*, 1076–1083. [CrossRef]

29. Matsuoka, M.; Kohno, T.; Iwakura, C. Electrochemical properties of hydrogen storage alloys modified with foreign metals. *Electrochim. Acta* **1993**, *38*, 789–791. [CrossRef]

30. Hjort, P.; Krozer, A.; Kasemo, B. Hydrogen sorption kinetics in partly oxidized Mg films. *J. Alloys Compd.* **1996**, *237*, 74–80. [CrossRef]

31. Visintin, A.; Tori, C.A.; Garaventta, G.; Triaca, W.E. The electrochemical performance of Pd-coated metal hydride electrodes with different binding additives in alkaline solution. *J. Electrochem. Soc.* **1998**, *145*, 4169–4172. [CrossRef]

32. Cuevas, F.; Hirscher, M. The hydrogen desorption kinetics of Pd-coated LaNi$_5$-type films. *J. Alloys Compd.* **2000**, *313*, 269–275. [CrossRef]

33. Hara, M.; Hatano, Y.; Abe, T.; Watanabe, K.; Naitoh, T.; Ikeno, S.; Honda, Y. Hydrogen absorption by Pd-coated ZrNi prepared by using Barrel-sputtering system. *J. Nucl. Mat.* **2003**, *320*, 265–271. [CrossRef]

34. Park, H.J.; Goo, N.H.; Lee, K.S. In situ Pd deposition on Mg$_2$Ni electrode for Ni/MH secondary batteries during charge cycles. *J. Electrochem. Soc.* **2003**, *150*, A1328–A1332. [CrossRef]

35. Barsellini, D.B.; Visintin, A.; Triaca, W.E.; Soriaga, M.P. Electrochemical characterization of a hydride-forming metal alloy surface-modified with palladium. *J. Power Sources* **2003**, *124*, 309–313. [CrossRef]

36. Parimala, R.; Ananth, M.V.; Ramaprabhu, S.; Raju, M. Effect of electroless coating of Cu, Ni and Pd on ZrMn$_{0.2}$V$_{0.8}$Fe$_{0.8}$Ni$_{0.8}$ alloy used as anodes in Ni-MH batteries. *Int. J. Hydrogen Energy* **2004**, *29*, 509–513. [CrossRef]

37. Yoshimura, K.; Yamada, Y.; Okada, M. Hydrogenation of Pd capped Mg thin films at room temperature. *Surf. Sci.* **2004**, *566–568*, 751–754. [CrossRef]

38. Souza, E.C.; Ticianelli, E.A. Structural and electrochemical properties of MgNi-based alloys with Ti, Pt and Pd additives. *Int. J. Hydrogen Energy* **2000**, *25*, 249–253. [CrossRef]

39. Xin, G.; Yang, J.; Fu, H.; Zheng, J.; Li, X. Pd capped Mg$_x$Ti$_{1-x}$ films: Promising anode materials for alkaline secondary batteries with superior discharge capacities and cyclic stabilities. *Int. J. Hydrogen Energy* **2013**, *38*, 10625–10629. [CrossRef]

40. Jung, H.; Cho, S.; Lee, W. A catalytic effect on hydrogen absorption kinetics in Pd/Ti/Mg/Ti multilayer thin films. *J. Alloys Compd.* **2015**, *635*, 203–206. [CrossRef]

41. Zhu, M.; Lu, Y.; Ouyang, L.; Wang, H. Thermodynamic tuning of Mg-based hydrogen storage alloys: A review. *Materials* **2013**, *6*, 4654–4674. [CrossRef] [PubMed]

42. Zhang, M.; Hu, R.; Zhang, T.; Kou, H.; Li, J.; Xue, X. Hydrogenation properties of Pd-coated Zr-based Laves phase compounds. *Vacuum* **2014**, *109*, 191–196. [CrossRef]

43. Geng, M. Electrochemical characteristics of Ni-Pd-coated MmNi$_5$-based alloy powder for nickel-metal hydride batteries. *J. Alloys Compd.* **1995**, *217*, 90–93. [CrossRef]

44. Geng, M. Electrochemical characterization of MmNi$_5$-based alloy powder coated with palladium and nickel-palladium. *J. Alloys Compd.* **1994**, *215*, 151–153. [CrossRef]

45. Williams, M.; Lototsky, M.V.; Davids, M.W.; Linkov, V.; Yartyes, V.A.; Solberg, J.K. Chemical surface modification for the improvement of the hydrogenation kinetics and poisoning resistance of TiFe. *J. Alloys Compd.* **2011**, *509*, S770–S774. [CrossRef]

46. Williams, M.; Lototsky, M.; Nechaev, A.; Linkov, V.; Vartys, V.; Li, Q. Surface-modified AB$_5$ alloys with enhanced hydrogen absorption kinetics. In *Carbon Nanomaterials in Clean Energy Hydrogen Systems*; Barabowski, B., Zaginaichenko, S.Y., Schur, D.V., Skorokhod, V.V., Veziroglu, A., Eds.; Springer: Dordrecht, The Netherlands, 2008; pp. 625–636.

47. Van Mal, H.H.; Buschow, K.H.J.; Miedema, A.R. Hydrogen absorption in LaNi$_5$ and related compounds: Experimental observations and their explanation. *J. Less Common Met.* **1974**, *35*, 65–76. [CrossRef]

48. Zhang, Y.; Ji, J.; Yuan, H.; Chen, S.; Wang, D.; Zang, T. Synthesis of hydrogen storage compound Mg$_2$Ni$_{0.75}$Pd$_{0.25}$ and studies on hydriding-dehydriding properties. *Acta Sci. Nat. Univ. Nan Kaiensis* **1991**, *1*, 93–98. (In Chinese)

49. Zaluski, L.; Zaluska, A.; Ström-Olsen, J.O. Hydrogen absorption in nanocrystalline Mg$_2$Ni formed by mechanical alloying. *J. Alloys Compd.* **1995**, *217*, 245–249. [CrossRef]

50. Zaluski, L.; Zaluska, A.; Tessier, P.; Ström-Olsen, J.O.; Schulz, R. Effects of relaxation on hydrogen absorption in Fe-Yi produced by ball-milling. *J. Alloys Compd.* **1995**, *227*, 53–57. [CrossRef]

51. Tsukahara, M.; Takahashi, K.; Mishima, T.; Isomura, A.; Sakai, T. Influence of various additives in vanadium-based alloys V$_3$TiNi$_{0.56}$ on secondary phase formation, hydrogen storage properties and electrode properties. *J. Alloys Compd.* **1996**, *245*, 59–65. [CrossRef]

52. Yamashita, I.; Tanaka, H.; Takeshita, H.; Kuriyama, N.; Sakai, T.; Uehara, I. Hydrogenation characteristics of TiFe$_{1-x}$Pd$_x$ (0.05 $\leq$ $x\leq$ 0.30) alloys. *J. Alloys Compd.* **1997**, *253–254*, 238–240. [CrossRef]

53. Wang, C.S.; Lei, Y.Q.; Wang, Q.D. Effects of Nb and Pd on the electrochemical properties of a Ti-Ni hydrogen-storage electrode. *J. Power Sources* **1998**, *70*, 222–227. [CrossRef]

54. Yang, X.G.; Zhang, W.K.; Lei, Y.Q.; Wang, Q.D. Electrochemical properties of Zr-V-Ni system hydrogen storage alloys. *J. Electrochem. Soc.* **1999**, *146*, 1245–1250. [CrossRef]

55. Zeppelin, F.; Reule, H.; Hirscher, M. Hydrogen desorption kinetics of nanostructured MgH$_2$ composite materials. *J. Alloys Compd.* **2002**, *330–332*, 723–726. [CrossRef]

56. Ovshinsky, S.R.; Young, R. High Power Nickel-Metal Hydride Batteries and High Power Alloys/Electrodes for Use Therein. U.S. Patent 6,413,670 B1, 2 July 2002.

57. Yamaura, S.; Kim, H.; Kimura, H.; Inoue, A.; Arata, Y. Thermal stabilities and discharge capacities of melt-spun Mg-Ni-based amorphous alloys. *J. Alloys Compd.* **2002**, *339*, 230–235. [CrossRef]

58. Yamaura, S.; Kim, H.; Kimura, H.; Inoue, A.; Arata, Y. Electrode properties of rapidly solidified Mg$_{67}$Ni$_{23}$Pd$_{10}$ amorphous alloy. *J. Alloys Compd.* **2002**, *347*, 239–243. [CrossRef]

59. Yamaura, S.; Kimura, H.; Inoue, A. Discharge capacities of melt-spun Mg-Ni-Pd amorphous alloys. *J. Alloys Compd.* **2003**, *358*, 173–176. [CrossRef]

60. Ma, J.; Hatano, Y.; Abe, T.; Watanabe, K. Effects of Pd addition on electrochemical properties of MgNi. *J. Alloys Compd.* **2004**, *372*, 251–258. [CrossRef]

61. Ma, T.; Hatano, Y.; Abe, T.; Watanabe, K. Effects of bulk modification by Pd on electrochemical properties of MgNi. *J. Alloys Compd.* **2005**, *391*, 313–317. [CrossRef]

62. Miyamura, H.; Takada, M.; Kikuchi, S. Characteristics of hydride electrode using Ti-Fe-Pd-X alloys. *J. Alloys Compd.* **2005**, *404–406*, 675–678. [CrossRef]

63. Tian, Q.; Zhang, Y.; Sun, L.; Xu, F.; Tan, Z.; Yuan, H.; Zhang, T. Effects of Pd substitution on the electrochemical properties of Mg$_{0.9-x}$Ti$_{0.1}$Pd$_x$Ni (x = 0.04–0.1) hydrogen storage electrode alloys. *J. Power Sources* **2006**, *158*, 1463–1471. [CrossRef]

64. Tian, Q.; Zhang, Y.; Chu, H.; Sun, L.; Xu, F.; Tan, Z.; Yuan, H.; Zhang, T. The electrochemical performances of Mg$_{0.9}$Ti$_{0.1}$Ni$_{1-x}$Pd$_x$ (x = 0–0.15) hydrogen storage electrode alloys. *J. Power Sources* **2006**, *159*, 155–158. [CrossRef]

65. Tian, Q.; Zhang, Y.; Tan, Z.; Xu, F.; Sun, L.; Zhang, T.; Yuan, H. Effects of Pd substitution for Ni on the corrosion performances of Mg$_{0.9}$Ti$_{0.1}$Ni$_{1-x}$Pd$_x$ hydrogen storage electrode alloys. *Trans. Nonferrous Met. Soc. China* **2006**, *16*, 497–501. [CrossRef]

66. Tian, Q.; Zhang, Y.; Sun, L.; Xu, F.; Yuan, H. The hydrogen desorption kinetics of Mg$_{0.9-x}$Ti$_{0.1}$Pd$_x$Ni (x = 0.04, 0.06, 0.08, 0.1) electrode alloys. *J. Alloys Compd.* **2007**, *446–447*, 121–123. [CrossRef]

67. Pinkerton, F.E.; Balogh, M.P.; Meyer, M.S.; Meisner, G.P. Hydrogen Generation Material. U.S. Patent Application 2006/0,057,049 A1, 16 March 2006.

68. Yermakov, A.Y.; Mushnikov, N.V.; Uimin, M.A.; Gaviko, V.S.; Tankeev, A.P.; Skripov, A.V.; Soloninin, A.V.; Buzlukov, A.L. Hydrogen reaction kinetics of Mg-based alloys synthesized by mechanical milling. *J. Alloys Compd.* **2006**, *425*, 367–372. [CrossRef]

69. Berlouis, L.E.A.; Honnor, P.; Hall, P.J.; Morris, S.; Dodd, S.B. An investigation of the effect of Ti, Pd, and Zr on the dehydriding kinetics of MgH$_2$. *J. Mater. Sci.* **2006**, *41*, 6403–6408. [CrossRef]

70. Kalisvaart, W.P.; Wondergem, H.J.; Bakker, F.; Notten, P.H.L. Mg-Ti based materials for electrochemical hydrogen storage. *J. Mater. Res.* **2007**, *22*, 1640–1649. [CrossRef]

71. Spassov, T.; Todorova, S.; Jung, W.; Borissova, A. Hydrogen sorption properties of ternary intermetallic Mg-(Ir,Rh,Pd)-Si compounds. *J. Alloys Compd.* **2007**, *429*, 306–310. [CrossRef]
72. Yvon, K.; Rapin, J.-Ph.; Penin, N.; Ma, Z.; Chou, M.Y. LaMg$_2$PdH$_7$, a new complex metal hydride containing tetrahedral [PdH$_4$]$^{4-}$ anion. *J. Alloys Compd.* **2007**, *446–447*, 34–38. [CrossRef]
73. Jeng, R.; Lee, S.; Hsu, C.; Wu, Y.; Lin, J. Effects of the addition of Pd on the hydrogen absorption-desorption characteristics of Ti$_{33}$V$_{33}$Cr$_{34}$ alloys. *J. Alloys Compd.* **2008**, *464*, 467–471. [CrossRef]
74. Liu, Y.; Zhang, S.; Li, R.; Gao, M.; Zhong, K.; Miao, H.; Pan, H. Electrochemical performances of the Pd-added Ti-V-based hydrogen storage alloys. *Int. J. Hydrogen Energy* **2008**, *33*, 728–734. [CrossRef]
75. Liu, B.; Zhang, Y.; Mi, G.; Zhang, Z.; Wang, L. Crystallographic and electrochemical characteristics of Ti-Zr-Ni-Pd quasicrystalline alloys. *Int. J. Hydrog. Energy* **1008**, *34*, 6925–6929. [CrossRef]
76. Tian, Q.; Zhang, Y.; Chu, H.; Ding, Y.; Wu, Y. Electrochemical impedance study of discharge characteristics of Pd substituted MgNi-based hydrogen storage electrode alloys. *J. Alloys Compd.* **2009**, *481*, 826–829. [CrossRef]
77. Gao, L.; Yao, E.; Nakamura, J.; Zhang, W.; Chua, H. Hydrogen storage in Pd-Ni doped defective carbon nanotubes through the formation of CH$_x$ (x = 1, 2). *Carbon* **2010**, *48*, 3250–3255. [CrossRef]
78. Rousselot, S.; Gazeau, A.; Guay, D.; Roué, L. Influence of Pd on the structure and electrochemical hydrogen storage properties of Mg$_{50}$Ti$_{50}$ alloy prepared by ball milling. *Electrochim. Acta* **2010**, *55*, 611–619. [CrossRef]
79. Ruiz, F.C.; Peretti, H.A.; Visintin, A. Electrochemical hydrogen storage in ZrCrNiPd$_x$ alloys. *Int. J. Hydrogen Energy* **2010**, *35*, 5963–5967. [CrossRef]
80. Okonska, I.; Jurczyk, M. Hydriding properties of Mg-3d/M-type nanocomposites (3d = Cu, Ni; M = Ni, Cu, Pd). *Phys. Status Solidi A* **2010**, *207*, 1139–1143. [CrossRef]
81. Emami, H.; Cuevas, F. Hydrogenation properties of shape memory Ti(Ni,Pd) compounds. *Intermetallics* **2011**, *19*, 876–886. [CrossRef]
82. Anik, M.; Özdemir, G.; Küçükdeveci, N. Electrochemical hydrogen storage characteristics of Mg-Pd-Ni ternary alloys. *Int. J. Hydrogen Energy* **2011**, *36*, 6744–6750. [CrossRef]
83. Prigent, J.; Joubert, J.-M.; Gupta, M. Investigation of modification of hydrogenation and structural properties of LaNi$_5$ intermetallic compound induced by substitution of Ni by Pd. *J. Solid State Chem.* **2011**, *184*, 123–133. [CrossRef]
84. Anik, M. Improvement of the electrochemical hydrogen storage performance of magnesium based alloys by various additive elements. *Int. J. Hydrogen Energy* **2012**, *37*, 1905–1911. [CrossRef]
85. Lin, K.; Mai, Y.; Chiu, W.; Yang, J.; Chan, S.L.I. Synthesis and characterization of metal hydride/carbon aerogel composites for hydrogen storage. *J. Nanomater.* **2012**, *20154*. [CrossRef]
86. Etiemble, A.; Rousselot, S.; Guo, W.; Idrissi, H.; Roué, L. Influence of Pd addition on the electrochemical performance of Mg-Ni-Ti-Al-based metal hydride for Ni-MH batteries. *Int. J. Hydrogen Energy* **2013**, *38*, 10625–10629. [CrossRef]
87. Santos, S.F.; Castro, J.F.R.; Ticianelli, E.A. Microstructures and electrode performances of Mg$_{50}$Ni$_{(50-x)}$Pd$_x$ alloys. *Cent. Eur. Chem.* **2013**, *11*, 485–491. [CrossRef]
88. Williams, M.; Sibanyoni, J.M.; Lototskyy, M.; Pollet, B.G. Hydrogen absorption study of high-energy reactive ball milled Mg composites with palladium additives. *J. Alloys Compd.* **2013** *580*, S144–S148. [CrossRef]
89. Nikkuni, F.R.; Santos, S.F.; Ticianelli, E.A. Microstructures and electrochemical properties of Mg$_{49}$Ti$_6$Ni$_{(45-x)}$M$_x$ (M = Pd and Pt) alloy electrodes. *Int. J. Energy Res.* **2013**, *37*, 706–712. [CrossRef]
90. Teresiak, A.; Uhlemann, M.; Thomas, J.; Eckert, J.; Gebert, A. Influence of Co and Pd on the formation of nanostructured LaMg$_2$Ni and its hydrogen reactivity. *J. Alloys Compd.* **2014** *582*, 647–658. [CrossRef]
91. Balcerzak, M.; Nowak, M.; Jakubowicz, J.; Jurczyk, M. Electrochemical behavior of Nanocrystalline TiNi doped by MWCNTs and Pd. *Renew. Energy* **2014**, *62*, 432–438. [CrossRef]
92. Zhang, Y.; Zhuang, X.; Zhu, Y.; Zhan, L.; Pu, Z.; Wan, N.; Li, L. Effects of additive Pd on the structures and electrochemical hydrogen storage properties of Mg$_{67}$Co$_{33}$-based composited or alloys with BCC phase. *J. Alloys Compd.* **2015**, *622*, 580–586. [CrossRef]
93. Zhan, L.; Zhang, Y.; Zhu, Y.; Zhuang, X.; Dong, J.; Guo, X.; Chen, J.; Wang, Z.; Li, L. The electrochemical hydrogen storage properties of Mg$_{67-x}$Pd$_x$Co$_{33}$ (x = 1, 3, 5, 7) electrodes with BCC phase. *J. Alloys Compd.* **2016**, *662*, 396–403. [CrossRef]

94. Banerjee, S.; Dasgupta, K.; Kumar, A.; Ruz, P.; Vishwanadh, B.; Joshi, J.B.; Sudarsan, V. Comparative evaluation of hydrogen storage behavior of Pd dopes carbon nanotubes prepared by wet impregnation and polyol methods. *Int. J. Hydrogen Energy* **2015**, *40*, 3268–3276. [CrossRef]

95. Dündar-Tekkaya, E.; Yürüm, Y. Effect of loading bimetallic mixture of Ni and Pd on hydrogen storage capacity of MCM-41. *Int. J. Hydrog. Energy* **2015**, *40*, 7636–7643. [CrossRef]

96. Ismail, N.; Madian, M.; El-Shall, M.S. Reduced graphene oxide doped with Ni/Pd nanoparticles for hydrogen storage application. *J. Ind. Eng. Chem.* **2015**, *30*, 328–335. [CrossRef]

97. Balcerzak, M. Electrochemical and structural studies on Ti-Zr-Ni and Ti-Zr-Ni-Pd alloys and composites. *J. Alloys Compd.* **2016**, *658*, 576–587. [CrossRef]

98. Giasafaki, D.; Charalambopoulou, G.; Tampaxis, Ch.; Dimos, K.; Gournis, D.; Stubos, A. Comparing hydrogen sorption in different Pd-doped pristine and surface-modified nanoporous carbons. *Carbon* **2016**, *98*, 1–14. [CrossRef]

99. Crivello, J.-C.; Denys, R.V.; Dornheim, M.; Felderhoff, M.; Grant, D.M.; Huot, J.; Jensen, T.R.; Jongh, P.; Latroche, M.; Walker, G.S.; et al. Mg-based compounds for hydrogen and energy storage. *Appl. Phys. A* **2016**, *122*, 85. [CrossRef]

100. Abdul, J.M. Development of titanium alloys for hydrogen storage. Ph.D. Thesis, University of the Witwatersrand, Johannesburg, South Africa, December 2015. Available online: http://wiredspace.wits.ac.za/handle/10539/21151 (accessed on 11 August 2017).

101. Zhan, L.; Zhang, Y.; Zhu, Y.; Zhuang, X.; Wan, N.; Qu, Y.; Guo, X.; Chen, J.; Wang, Z.; Li, L. Electrochemical performances of $Mg_{45}M_5Co_{50}$ (M = Pd, Zr) ternary hydrogen storage electrodes. *Trans. Nonferrous Met. Soc. China* **2016**, *26*, 1388–1395. [CrossRef]

102. Tosques, J.; Guerreiro, B.H.; Martin, M.H.; Roué, L.; Guay, D. Hydrogen solubility of bcc PdCu and PdCuAg alloys prepared by mechanical alloys. *J. Alloys Compd.* **2017**, *698*, 725–730. [CrossRef]

103. Young, K.; Fetcenko, M.A.; Li, F.; Ouchi, T. Structural, thermodynamics, and electrochemical properties of $Ti_xZr_{1-x}(VNiCrMnCoAl)_2$ C14 Laves phase alloys. *J. Alloys Compd.* **2008**, *464*, 238–247. [CrossRef]

104. Young, K.; Fetcenko, M.A.; Koch, J.; Morii, K.; Shimizu, T. Studies of Sn, Co, Al, and Fe additives in C14/C15 Laves alloys for NiMH battery application by orthogonal arrays. *J. Alloys Compd.* **2009**, *486*, 559–569. [CrossRef]

105. Price of Palladium. Available online: http://www.apmex.com/spotprices/palladium-price (accessed on 18 October 2016).

106. Price of Nickel. Available online: www.infomine.com/investment/metal-prices/nickel/ (accessed on 18 October 2016).

107. Rennert, P.; Radwan, A.M. Structural investigation of the Laves phase $MgZn_2$ with model potential calculations. *Phys. Status Solidi B* **1977**, *79*, 167–173. [CrossRef]

108. Douglas, B.E.; Ho, S.M. *Structure and Chemistry of Crystalline Solids*; Springer Science + Business, Inc.: New York, NY, USA, 2006.

109. Yakoubi, A.; Baraka, O.; Bouhafs, B. Structural and electronic properties of the Laves phase based on rare earth type BaM_2 (M = Rh, Pd, Pt). *Results Phys.* **2012**, *2*, 58–65. [CrossRef]

110. Palladium Hydride. Available online: https://en.wikipedia.org/wiki/Palladium_hydride (accessed on 18 October 2016).

111. Adams, B.D.; Chen, A. The role of palladium in a hydrogen economy. *Mater. Today* **2011**, *14*, 282–289. [CrossRef]

112. Nihon Kinzoku Gakkai. *Hi Kagaku Ryouronteki Kinzoku Kagobutsu*; Maruzen: Tokyo, Japan, 1975; p. 296.

113. Crystal Structure of the Elements. Available online: http://www.periodictable.com/Properties/A/CrystalStructure.html (accessed on 13 October 2016).

114. Lide, D.R. *CRC Handbook of Chemistry and Physics*, 74th ed.; CRC Press Inc.: Boca Raton, FL, USA, 1993.

115. Murray, J.L. Ti-Zr binary phase diagram. In *ASM Handbook, Vol. 3 Alloy Phase Diagram*; Baker, H., Ed.; ASM International: Materials Park, OH, USA, 1992.

116. Zhu, J.H.; Liu, C.T.; Pike, L.M.; Liaw, P.K. Enthalpies of formation of binary Laves phases. *Intermetallics* **2002**, *10*, 579–595. [CrossRef]

117. Liu, C.T.; Zhu, J.H.; Brady, M.P.; McKamey, C.G.; Pike, L.M. Physical metallurgy and mechanical properties of transition-metal Laves phase alloys. *Intermetallics* **2000**, *8*, 1119–1129. [CrossRef]

118. Johnston, R.L.; Hoffmann, R. Structure-bonding relationships in the Laves phases. *Z. Anorg. Allg. Chem.* **1992**, *616*, 105–120. [CrossRef]
119. Young, K.; Ouchi, T.; Fetcenko, M.A. Pressure-composition-temperature hysteresis in C14 Laves phase alloys: Part 1. Simple ternary alloys. *J. Alloys Compd.* **2009**, *480*, 428–433. [CrossRef]
120. Nei, J.; Young, K.; Salley, S.O.; Ng, K.Y.S. Determination of C14/C15 phase abundance in Laves phase alloys. *Mater. Chem. Phys.* **2012**, *135*, 520–527. [CrossRef]
121. Young, K.; Fetcenko, M.A.; Li, F.; Ouchi, T.; Koch, J. Effect of vanadium substitution in C14 Laves phase alloys for NiMH battery application. *J. Alloys Compd.* **2009**, *468*, 482–492. [CrossRef]
122. Young, K.; Regmi, R.; Lawes, G.; Ouchi, T.; Fetcenko, M.A.; Wu, A. Effect of aluminum substitution in C14-rich multi-component alloys for NiMH battery application. *J. Alloys Compd.* **2010**, *490*, 282–292. [CrossRef]
123. Young, K.; Fetcenko, M.A.; Ouchi, T.; Li, F.; Koch, J. Effect of Sn-substitution in C14 Laves phase alloys for NiMH battery application. *J. Alloys Compd.* **2009**, *469*, 406–416. [CrossRef]
124. Boriskina, N.G.; Kenina, E.M. Phase equilibria in the Ti-TiPd-TiNi system alloys. In *Proceedings of the 4th International Conference on Titanium 80, Science & Technology, Kyoto, Japan, 19–22 May 1980*; Kimura, H., Izumi, O., Eds.; The Metallurgical Society of AIME: Warrendale, PA, USA, 1980; pp. 2917–2927.
125. Ghost, G. Nickel-Palladium-Titanium. In *Light Metal System. Part 4 Selected Systems from Al-Si-Ti to Ni-Si-Ti*; Effenberg, D., Ilyenko, S., Eds.; Springer: Berlin/Heidelberg, Germany, 2006; pp. 425–434.
126. Thoma, D.J.; Perepezko, J.H. A geometric analysis of solubility ranges in Laves phases. *J. Alloys Compd.* **1995**, *224*, 330–341. [CrossRef]
127. Young, K.; Ouchi, T.; Nei, J.; Wang, L. Annealing effects on Laves phase-related body-centered-cubic solid solution metal hydride alloys. *J. Alloys Compd.* **2016**, *654*, 216–225. [CrossRef]
128. Bououdina, M.; Soubeyroux, J.L.; De Rango, P.; Fruchart, D. Phase stability and neutron diffraction studies of the laves phase compounds $Zr(Cr_{1-x}Mo_x)_2$ with $0.0 \leq x \leq 0.5$ and their hydrides. *Int. J. Hydrogen Energy* **2000**, *25*, 1059–1068. [CrossRef]
129. Young, K.; Ouchi, T.; Huang, B.; Reichman, B.; Fetcenko, M.A. Effect of molybdenum content on structural, gaseous storage, and electrochemical properties of C14-predominated AB_2 metal hydride alloys. *J. Power Sources* **2011**, *196*, 8815–8821. [CrossRef]
130. Young, K.; Ouchi, T.; Lin, X.; Reichman, B. Effects of Zn-addition to C14 metal hydride alloys and comparisons to Si, Fe, Cu, Y, and Mo-additives. *J. Alloys Compd.* **2016**, *655*, 50–59. [CrossRef]
131. Boettinger, W.J.; Newbury, D.E.; Wang, K.; Bendersky, L.A.; Chiu, C.; Kattner, U.R.; Young, K.; Chao, B. Examination of multiphase (Zr, Ti) (V, Cr, Mn, Ni)$_2$ Ni-MH electrode alloys: Part I. Dendritic solidification structure. *Metall. Mater. Trans.* **2010**, *41A*, 2033–2047. [CrossRef]
132. Bendersky, L.A.; Wang, K.; Boettinger, W.J.; Newbury, D.E.; Young, K.; Chao, B. Examination of multiphase (Zr, Ti) (V, Cr, Mn, Ni)$_2$ Ni-MH electrode alloys: Part II. Solid-state transformation of the interdendric B2 phase. *Metall. Mater. Trans.* **2010**, *41A*, 1891–1906. [CrossRef]
133. Liu, Y.; Young, K. Microstructure investigation on metal hydride alloys by electron backscatter diffraction technique. *Batteries* **2016**, *2*. [CrossRef]
134. Wong, D.F.; Young, K.; Ouchi, T.; Ng, K.Y.S. First-principles point defect models for Zr_7Ni_{10} and Zr_2Ni_7 phases. *Batteries* **2016**, *2*. [CrossRef]
135. Young, K.; Ouchi, T.; Koch, J.; Fetcenko, M.A. Compositional optimization of vanadium-free hypo-stoichiometric AB_2 metal hydride for Ni/MH battery application. *J. Alloys Compd.* **2012**, *510*, 97–106. [CrossRef]
136. Young, K.; Ouchi, T.; Huang, B.; Reichman, B.; Blankenship, R. Improvement in $-40\,^{\circ}C$ electrochemical properties of AB_2 metal hydride alloy by silicon incorporation. *J. Alloys Compd.* **2013**, *575*, 65–72. [CrossRef]
137. Young, K.; Ouchi, T.; Meng, T.; Wong, D.F. Studies on the synergetic effects in multi-phase metal hydride alloys. *Batteries* **2016**, *2*. [CrossRef]
138. Young, K.; Wong, D.F.; Nei, J. Effects of vanadium/nickel contents in Laves phase-related body-centered-cubic solid solution metal hydride alloys. *Batteries* **2015**, *1*, 34–53. [CrossRef]
139. Sinha, V.K.; Pourarian, F.; Wallace, W.E. Hydrogenation characteristics of $Zr_{1-x}Ti_xMnFe$ alloys. *J. Less Common Met.* **1982**, *87*, 283–296. [CrossRef]
140. Lee, H.; Lee, K.; Lee, J. The hydrogenation characteristics of Ti-Zr-V-Mn-Ni C14 type Laves phase alloys for metal hydride electrodes. *J. Alloys Compd.* **1997**, *253–254*, 601–604. [CrossRef]

141. Pourbaix, M.J.N.; Van Muylder, J.; De Zoubov, N. Electrochemical properties of the platinum metals. *Platinum Met. Rev.* **1959**, *3*, 100–106.

142. Young, K.; Reichman, B.; Fetcenko, M.A. Electrochemical performance of AB$_2$ metal hydride alloys measured at −40 °C. *J. Alloys Compd.* **2013**, *580*, S349–S352. [CrossRef]

143. Young, K.; Huang, B.; Regmi, R.K.; Lawes, G.; Liu, Y. Comparisons of metallic clusters imbedded in the surface of AB$_2$, AB$_5$, and A$_2$B$_7$ alloys. *J. Alloys Compd.* **2010**, *506*, 831–840. [CrossRef]

144. Young, K.; Chao, B.; Liu, Y.; Nei, J. Microstructures of the oxides on the activated AB$_2$ and AB$_5$ metal hydride alloys surface. *J. Alloys Compd.* **2014**, *606*, 97–104. [CrossRef]

145. Stucki, F.; Schlapbach, L. Magnetic properties of LaNi$_5$, FeTi, Mg$_2$Ni and their hydrides. *J. Less Common Met.* **1980**, *74*, 143–151. [CrossRef]

146. Young, K.; Chao, B.; Pawlik, D.; Shen, H.T. Transmission electron microscope studies in the surface oxide on the La-containing AB$_2$ metal hydride alloy. *J. Alloys Compd.* **2016**, *672*, 356–365. [CrossRef]

batteries

MDPI

Article

Effects of Boron-Incorporation in a V-Containing Zr-Based AB$_2$ Metal Hydride Alloy

Shiuan Chang [1], Kwo-Hsiung Young [2,3,*], Taihei Ouchi [2], Jean Nei [2] and Xin Wu [1]

[1] Department of Mechanical Engineering, Wayne State University, Detroit, MI 48202, USA; ShiuanC@wayne.edu (S.C.); xin.wu@wayne.edu (X.W.)

[2] BASF/Battery Materials—Ovonic, 2983 Waterview Drive, Rochester Hills, MI 48309, USA; taihei.ouchi@basf.com (T.O.); jean.nei@basf.com (J.N.)

[3] Department of Chemical Engineering and Materials Science, Wayne State University, Detroit, MI 48202, USA

* Correspondence: kwo.young@basf.com; Tel.: +1-248-293-7000

Received: 23 September 2017; Accepted: 25 October 2017; Published: 14 November 2017

Abstract: In this study, boron, a metalloid element commonly used in semiconductor applications, was added in a V-containing Zr-based AB$_2$ metal hydride alloy. In general, as the boron content in the alloy increased, the high-rate dischargeability, surface exchange current, and double-layer capacitance first decreased and then increased whereas charge-transfer resistance and dot product of charge-transfer resistance and double-layer capacitance changed in the opposite direction. Electrochemical and gaseous phase characteristics of two boron-containing alloys, with the same boron content detected by the inductively coupled plasma optical emission spectrometer, showed significant variations in performances due to the difference in phase abundance of a newly formed tetragonal V$_3$B$_2$ phase. This new phase contributes to the increases in electrochemical high-rate dischargeability, surface exchange current, charge-transfer resistances at room, and low temperatures. However, the V$_3$B$_2$ phase does not contribute to the hydrogen storage capacities in either gaseous phase and electrochemical environment.

Keywords: metal hydride; nickel/metal hydride battery; Laves phase alloy; boron; electrochemistry; pressure concentration isotherm

---

1. Introduction

As energy storage devices, nickel/metal hydride (Ni/MH) batteries are important in consumer battery, transportation, and alternative energy-related stationary applications, and they are superior in operation temperature range and cycle life compared to the competing Li-ion batteries [1]. However, the specific energy of Ni/MH battery is lower than that of Li-ion battery, which has become the main research direction in the field [2]. Two commonly used stoichiometries in Ni/MH batteries are AB$_2$ and AB$_5$, where the A-site is occupied by larger elements with stronger affinities with hydrogen, and the B-site elements are relatively smaller and have the tendency to weaken the alloy's average metal-hydrogen bond strength [3]. Nevertheless, flexibility in stoichiometry [4,5] and abundantly available secondary phases [6] in the AB$_2$ metal hydride (MH) alloy allow for fine tailoring in chemical composition to fulfill the stringent demands from different applications. Among various types of AB$_2$ MH alloys, one based on the C14 Laves phase—an intermetallic phase with an AB$_2$ stoichiometry—is commonly used as the negative electrode active material of the Ni/MH battery [6]. Laves phases, named after Fritz Laves (1906–1978), have three different classes: a cubic MgCu$_2$ (C15), a hexagonal MgZn$_2$ (C14), and a dihexagonal MgNi$_2$ (C36) [7]. Although the theoretical hydrogen storage of C15 phase is higher than that in the C14 phase (6.33 vs. 7 per AB$_2$ formula) [3], C14 alloy was used in the first commercially built electric vehicle (EV-1 by General Motor, Detroit, MI, USA) due to its large capacity and superior cycle stability over C15 phase. Typical A-site elements in the C14 MH alloy are Ti and Zr, while B-site elements are mainly Ni with a small percentage of V, Cr, Mn, and Co. Contribution of each element can be found in a recently published review article [6]. C14 Laves phase-based AB$_2$ MH alloy showed a reversible hydrogen storage (H-storage) capacity up to 2.1 wt % [9] and an electrochemical discharge capacity reaching 436 mAh·g^{-1} [10], which provides a pathway to narrow the gap in specific energy between Ni/MH and the rival Li-ion battery technologies. While most of the substitution studies on the AB$_2$ MH alloys were performed with transition metals with metallic natures similar to other constituent elements (Zr and Ni), the number of studies with non-transition metals, especially non-metals, is very limited [6].

Boron, a non-transition metal, was doped into a body-centered-cubic MH alloy and demonstrated the capability of decreasing the alloy's hydrogen absorption kinetics [11]. Moreover, a study in which potassium boride was added to a Ti-based AB$_2$ MH alloy revealed that boron, not potassium, is responsible for the increase in electrochemical capacity [12]. In terms of electrochemical cycling stability, boron was a beneficial additive in the La–Mg–Ni-based superlattice MH alloy system [13]. In a Zr-based AB$_2$ MH alloy, boron reduced the unit cell volume and decreased the reaction kinetics of hydrogen absorption with the increase in incorporation [14]. However, adding boron in a Zr-based V-free AB$_2$ alloy expanded the unit cell volume and lowered the gaseous phase desorption pressure whereas the gaseous phase and electrochemical capacities were both reduced [15]. Therefore, as an additive, boron's functions in the AB$_2$ MH alloys are not clear and warrant further evaluation. In this study, boron is added in a Zr-based V-containing AB$_2$ alloy to investigate its effects on structural, gaseous phase, and electrochemical properties.

2. Experimental Setup

Sample preparation began with a conventional melt-and-cast by placing the designated atomic proportionated raw materials into an Al$_2$O$_3$ crucible under a helium protective atmosphere with the induction melting furnace heating up and holding at 1500 °C for 5 min; an iron plate was later used for holding the melt for cooling. A hydriding/dehydriding process followed to pulverize the ingot into powder smaller than 200 mesh. The actual chemical composition of each sample was analyzed with a Varian Liberty 100 inductively coupled plasma optical emission spectrometer (ICP-OES, Agilent Technologies, Santa Clara, CA, USA). A Philips X'Pert Pro X-ray diffractometer (XRD, Philips, Amsterdam, The Netherlands) and a JEOL-JSM6320F scanning electron microscope (SEM, JEOL, Tokyo, Japan) with energy dispersive spectroscopy (EDS) capability were used for the phase identification, phase distribution, and phase composition studies. A Suzuki–Shokan multi-channel pressure–concentration–temperature (PCT, Suzuki Shokan, Tokyo, Japan) system was used for the gaseous phase H-storage study. A 2-h thermal cycle activation between room temperature (RT) and 300 °C under 2.5 MPa H$_2$ pressure was conducted before the PCT measurements at 30, 60 and 90 °C. For the electrochemical measurements, −200 mesh alloy powder was compacted onto an expanded nickel mesh by applying 2 GPa pressure with a hydraulic press to form a negative electrode that was 1 cm^2 in size and 0.2 mm thick. Half-cell testing performed with an Arbin BT2000 battery tester (Arbin, College Station, TX, USA) was used to measure electrochemical properties at RT with a 30 wt % KOH aqueous solution as the electrolyte and two pieces of sintered Ni(OH)$_2$/NiOOH as the counter positive electrode. The system was charged at a current density of 50 mA·g^{-1} for 10 h and then discharged at a current density of 50 mA·g^{-1} until a cutoff voltage of 0.9 V was reached. Next, the system was discharged at a current density of 12 mA·g^{-1} until a cutoff voltage of 0.9 V was reached and again discharged at a current density of 4 mA·g^{-1} until a cutoff voltage at 0.9 V was reached [16,17]. AC impedance measurements were also performed to characterize the RT and low temperature (LT, −40 °C) electrochemical properties. Each negative electrode was first subjected to three full charge/discharge activation cycles, charged fully, and discharged to 80% state-of-charge at a 0.1 C rate with an Arbin BT2000 battery tester. A Solartron 1250 Frequency Response Analyzer (Solartron Analytical, Leicester, England) with a sine wave amplitude of 10 mV and a frequency range of 0.5 mHz to 10 kHz was then used to conduct the AC impedance measurements. Magnetic susceptibility (M) of the activated alloy powder surface (by immersing the alloy powder in a 30 wt % KOH solution at 100 °C for 4 h) was measured by a Digital Measurement Systems Model 880 vibrating sample magnetometer (MicroSense, Lowell, MA, USA).

3. Results and Discussion

3.1. Chemical Composition

A 2 kg induction melting furnace with an Al_2O_3 crucible was used to prepare six alloys (B0 to B5) with the design compositions of $Ti_{12}Zr_{21.5}V_{10-0.4x}Cr_{7.5}Mn_{8.1}Co_8Ni_{32.2-0.6x}Al_{0.4}Sn_{0.3}B_x$, where x is from 0 to 5. The base alloy (B0) was chosen because of its balanced performances in capacity, high-rate dischargeability (HRD), charge retention, and cycle stability, and was previously used as the base material in other substitution studies [18–22]. ICP results are compared with the design compositions in Table 1. For each design composition, the vanadium- and nickel-contents were reduced by 40% and 60% of the added boron content, respectively, to maintain the B/A ratio. Therefore, while 60% of boron was designed to occupy the A-site, 40% of boron was designed to occupy the B-site. Boron, with the highest melting point among the constituent elements (2076 °C), was dissolved (instead of melted) in the liquid during melting. The participation rate (ratio between the content in the final alloy and content in the raw materials) of boron decreases with the increase in the boron content, which results in the lower boron contents and B/A ratios in alloys B3 to B5 compared to the corresponding design values.

Average electron density (e/a), as calculated by averaging the numbers of outer-shell electrons of constituent elements, is a parameter used to predict the C14/C15 ratio in MH alloys [23–25]. In this study, e/a of 6.57 to 6.82 is designed to promote the formation of C14 over C15 [25]. The design e/a value decreases with the increase in the boron content due to boron's relatively low number of outer-shell electrons (3) compared to the replaced vanadium (5) and nickel (10). Even with the increased deviation between the design and ICP result, as the boron content increases, the e/a values calculated from the ICP results do not deviate much from the design values since boron shows the largest deviation between the design and ICP result but has only three outer-shell electrons (two $2s$ and one $2p$).

Table 1. Design compositions (in **bold**) and inductively coupled plasma (ICP) results in at %. e/a is the average electron density. B/A is the atomic ratio of the B-atoms (elements other than Ti and Zr) to the A-atoms (Ti and Zr).

Alloy	Source	Ti	Zr	V	Cr	Mn	Co	Ni	Al	Sn	B	e/a	B/A
B0	Design	12.0	21.5	10.0	7.5	8.1	8.0	32.2	0.4	0.3	0.0	6.82	1.99
	ICP	12.0	21.5	10.0	7.5	8.1	8.0	32.2	0.3	0.4	0.0	6.82	1.99
B1	Design	12.0	21.5	9.6	7.5	8.1	8.0	31.6	0.4	0.3	1.0	6.77	1.99
	ICP	12.1	21.5	9.7	7.3	7.9	8.0	31.6	0.7	0.4	0.9	6.76	1.98
B2	Design	12.0	21.5	9.2	7.5	8.1	8.0	31.0	0.4	0.3	2.0	6.72	1.99
	ICP	12.0	21.4	9.7	7.2	8.0	8.0	31.1	0.7	0.3	1.6	6.72	2.00
B3	Design	12.0	21.5	8.8	7.5	8.1	8.0	30.4	0.4	0.3	3.0	6.67	1.99
	ICP	12.1	21.7	9.2	7.1	8.0	8.1	31.0	0.7	0.4	1.8	6.72	1.96
B4	Design	12.0	21.5	8.4	7.5	8.1	8.0	29.8	0.4	0.3	4.0	6.62	1.99
	ICP	12.2	22.1	8.6	6.9	8.1	8.3	31.0	0.7	0.4	1.8	6.72	1.92
B5	Design	12.0	21.5	8.0	7.5	8.1	8.0	29.2	0.4	0.3	5.0	6.57	1.99
	ICP	12.1	22.8	8.0	7.0	8.7	8.1	29.5	0.7	0.3	2.7	6.62	1.86

3.2. XRD Analysis

XRD was used to analyze the alloys' structural characteristics, and the resulting patterns are shown in Figure 1. For the Laves phases, peaks from the hexagonal C14 and cubic C15 phases overlapped at several angles. A peak at around 41.5° is observed in each pattern and identified as the TiNi phase with a B2 cubic structure. Furthermore, a tetragonal phase is detected in most boron-containing alloys and assigned to the V_3B_2 phase with a tetragonal $tP10$ structure (Figure 2). Structure characteristics, such as lattice constants and phase abundances, were analyzed by the Jade 9.0 software (MDI, Livermore,

CA, USA), and the results are listed in Table 2 and Table S1. With the increased boron concentration, the main peaks of the C14, C15, and TiNi phases shift to lower angles, indicating increases in lattice constants according to Bragg's Law. For the C14 phase, both lattice constants a and c increase, and the a/c ratio decreases slightly, suggesting an anisotropic expansion in unit cell as the boron content increases. Although boron is the smallest among all constituent elements, the C14 unit cell expansion due to the addition of boron has been reported previously and attributed to the dumbbell model of two smaller atoms occupying one A-site, which causes an anisotropic unit cell expansion (lower a/c ratio) [4]. The C14 unit cell expansion in the current study is compared with that from a previous study on the boron substitution in a V-free AB$_2$ alloy [15] in Figure 3, and it demonstrates that more boron can be incorporated in C14 of the V-free MH alloy (as shown by the higher rate of increase in the C14 unit cell vs. the boron content in the alloy) than that of the V-containing alloy because boron forms a new V$_3$B$_2$ secondary phase in the V-containing MH alloy. Moreover, lattice constants of the C15 and TiNi phases follow the increasing trend observed in the C14 phase as the boron content increases. It is possible that boron always enters a structure in pairs. Differently to the Laves and cubic TiNi phases, the V$_3$B$_2$ unit cell volume does not have a linear dependency with the boron content, and all are smaller than the reported value (100.17 Å^3 from the JCPDF file [26]) because the mixed boron configuration is suspected to be in play, where a reduced unit cell indicates a single boron atom in the B-site, and an enlarged unit cell implies double boron atoms occupying the A-site in a dumbbell manner [27]. Rietveld refinement was used to obtain the constituent phase abundances from the XRD patterns, and the results are listed in Table S1 and illustrated in Figure 4. As the concentration of boron increases, the TiNi phase abundance increases at the expense of the Laves phase abundance monotonically, and the V$_3$B$_2$ phase starts to form.

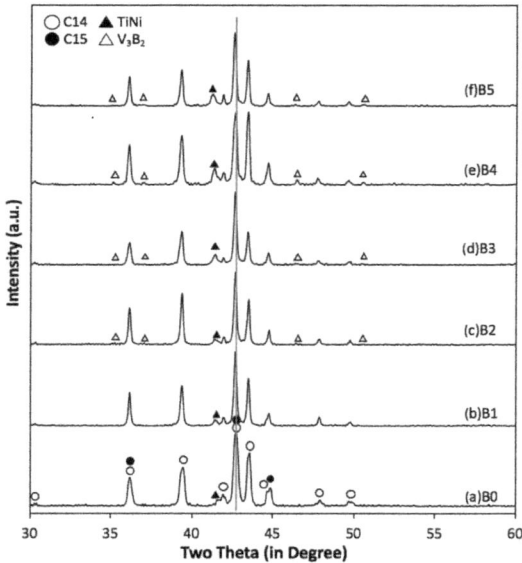

Figure 1. XRD patterns using Cu-K$_\alpha$ as the radiation source for alloys (**a**) B0; (**b**) B1; (**c**) B2; (**d**) B3; (**e**) B4; and (**f**) B5. Besides the two Laves phases, a cubic phase (TiNi) and a tetragonal phase (V$_3$B$_2$) can also be identified. Vertical line indicates the main C14/C15 peak shifting into a lower angle with the increase in the boron content.

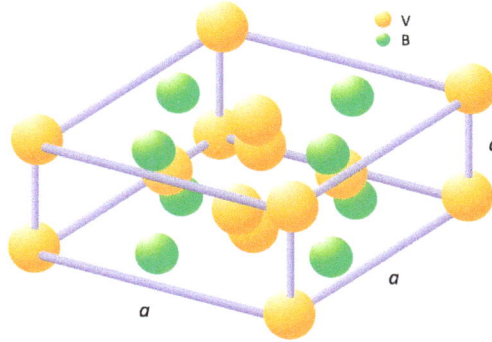

Figure 2. Schematic of the V_3B_2 crystal structure (*tP*10). Orange and green spheres represent the vanadium and boron atoms, respectively.

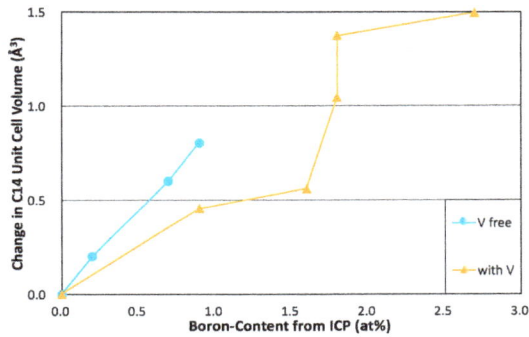

Figure 3. C14 unit cell volume change vs. boron content in the V-free [15] and V-containing AB_2 alloys.

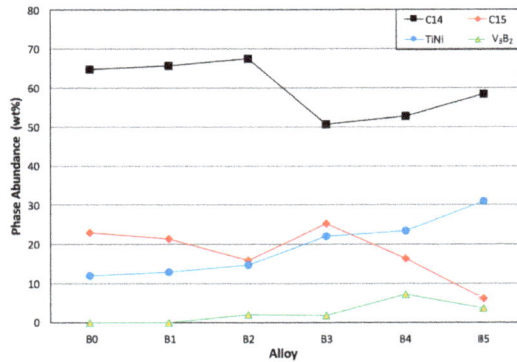

Figure 4. Evolutions of the C14, C15, TiNi, and V_3B_2 phase abundances vs. alloy number.

Table 2. Lattice constants *a* and *c*, *a*/*c* ratio, unit cell volume, and crystallite size of the main C14 phase of alloys B0 to B5 from the XRD analysis. Lattice information of the secondary phases are also included. N.D. denotes non-detectable.

Alloy	C14 a (Å)	C14 c (Å)	C14 a/c	V_{C14} (Å³)	C14 Crystallite Size (Å)	C15 a (Å)	TiNi a (Å)	V_3B_2 a (Å)	V_3B_2 c (Å)	V_3B_2 a/c	$V_{V_3B_2}$ (Å³)
B0	4.9635	8.0906	0.6135	172.62	260	7.0002	3.0637	N.D.	N.D.	N.D.	N.D.
B1	4.9655	8.0959	0.6133	172.87	591	7.0077	3.0770	N.D.	N.D.	N.D.	N.D.
B2	4.9667	8.0961	0.6135	172.96	643	7.0087	3.0775	5.7125	3.0232	1.8896	98.66
B3	4.9702	8.1072	0.6131	173.44	487	7.0148	3.0813	5.7089	3.0486	1.8726	99.36
B4	4.9744	8.1110	0.6133	173.81	496	7.0154	3.0841	5.7193	3.0330	1.8857	99.21
B5	4.9845	8.1330	0.6129	174.99	571	7.0293	3.0916	5.7287	3.0397	1.8846	99.76

3.3. SEM/EDS Analysis

Figure 5 shows the SEM back-scattering electron image (BEI) of each boron-containing alloy. Table 3 lists the composition information of selected and numbered spots with different contrasts (indicated in Figure 5) analyzed by EDS in each BEI. The microstructure of alloy B0 was studied and reported previously [18–22]. For the B/A ratio calculation, vanadium is assumed to occupy the B-site in the Laves phases [28] and the A-site in the non-Laves phases [29]. For the boron-containing alloys B1 to B5, the brightest phase with the B/A ratio between 1.38 and 1.44 is identified as Zr_7Ni_{10}. Three phases, TiNi, C15, and C14, are detected next to each other. The TiNi phase with the second brightest contrast and a low vanadium-content exhibits the lowest B/A ratio (1.18 to 1.21) among the three phases. The phase with an e/a value (7.07 to 7.10) higher than the C14/C15 threshold [25] and a hypo-stochiometric composition (1.82 to 1.88) is identified as C15. The C15 phase solidifies between the C14 and TiNi phases in the cooling sequence [30]. The C14 phase with the darkest contrast among the three phases has an e/a value below 7.0 and a hyper-stoichiometry of 2.02 to 2.12. Furthermore, EDS shows a high oxygen-content in a relatively dark area (second darkest among all phases due to the light weight of oxygen), and this area is identified as ZrO_2. By examining both the EDS and XRD results, the phase with the darkest contrast is the tetragonal V_3B_2 phase. Boron is too light to be detected by EDS, but the low BEI contrast indicates the existence of an element lighter than carbon. The calculated B/A ratio for the V_3B_2 phase is not the real B/A ratio since boron is not detectable by EDS; instead, it is the ratio of sum of the chromium-, manganese-, cobalt-, nickel-, aluminum-, and tin-contents to that of the zirconium-, titanium-, and vanadium-contents. This B/A ratio increases from 0.32 to 0.48 with the increase in the boron content from alloys B1 to B4, which suggests increases in the amounts of smaller transition metals in the V_3B_2 phase.

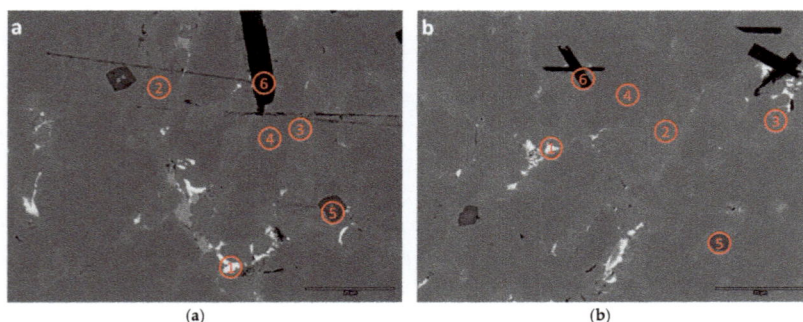

(a)

(b)

Figure 5. *Cont.*

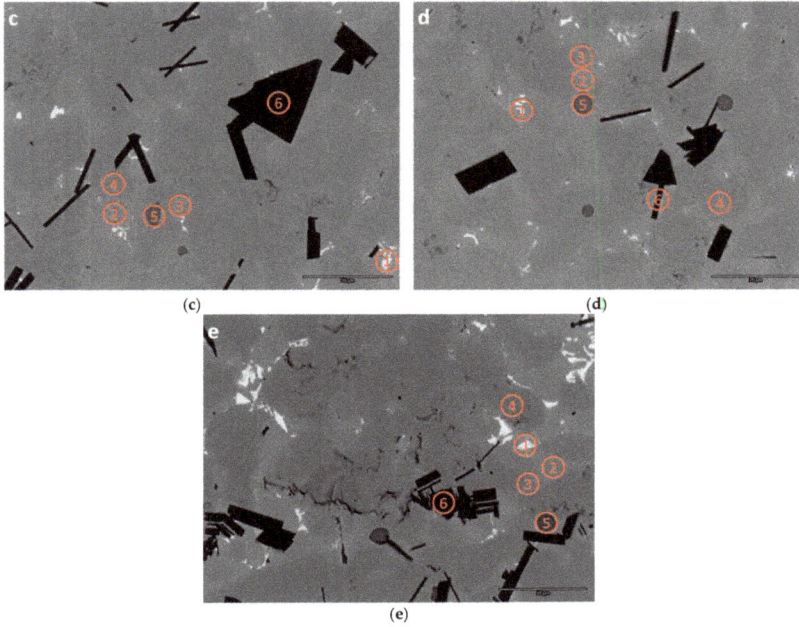

Figure 5. SEM back-scattering electron images (BEIs) of alloys (**a**) B1; (**b**) B2; (**c**) B3; (**d**) B4; and (**e**) B5. Compositions of the numbered areas were analyzed by energy dispersive spectroscopy (EDS), and the results are shown in Table 4. Areas 1–6 are identified as the Zr_7Ni_{10}, TiNi, C15, C14, Zr_xO_2, and V_3B_2 phases, respectively.

Table 3. Summary of the EDS results. All compositions are in at%. Compositions of the main **C14** and *C15* phases are in **bold** and *italic*, respectively.

Location	Ti	Zr	V	Cr	Mn	Co	Ni	Al	Sn	O	e/a	B/A	Phase
B1-1	4.8	35.9	0.4	0.2	1.6	1.4	40.3	0.4	15.3	0.0	-	1.44	Zr_7Ni_{10}
B1-2	25.3	18.9	1.2	0.4	2.3	7.5	43.8	0.4	0.4	0.0	-	1.21	TiNi
B1-3	*11.8*	*22.9*	*6.8*	*3.4*	*7.0*	*6.5*	*40.6*	*0.6*	*0.5*	*0.0*	*7.10*	*1.88*	*C15*
B1-4	**9.6**	**22.4**	**11.9**	**9.7**	**9.6**	**8.7**	**27.2**	**0.5**	**0.1**	**0.0**	**6.65**	**2.12**	**C14**
B1-5	0.2	33.8	0.1	0.1	0.1	0.2	0.6	0.1	0.0	64.9	-	1.94	ZrO_2
B1-6	18.5	0.5	56.9	16.0	4.8	1.6	1.6	0.1	0.0	0.0	-	-	V_3B_2
B2-1	6.8	34.6	0.7	0.3	1.8	2.2	39.6	0.4	13.7	0.0	-	1.38	Zr_7Ni_{10}
B2-2	24.7	19.7	1.1	0.4	2.3	7.3	43.8	0.3	0.3	0.0	-	1.20	TiNi
B2-3	*11.5*	*23.3*	*6.5*	*3.7*	*7.1*	*6.6*	*40.2*	*0.6*	*0.4*	*0.0*	*7.08*	*1.87*	*C15*
B2-4	**9.6**	**22.8**	**11.3**	**10.9**	**10.0**	**9.0**	**25.9**	**0.5**	**0.2**	**0.0**	**6.64**	**2.09**	**C14**
B2-5	0.1	34.2	0.1	0.1	0.1	0.2	0.7	0.1	0.0	64.5	-	1.91	ZrO_2
B2-6	10.4	0.4	62.2	21.1	3.6	1.0	1.3	0.1	0.0	0.0	-	-	V_3B_2
B3-1	6.2	34.4	0.5	0.3	1.9	2.3	38.5	0.4	13.4	1.9	-	1.38	Zr_7Ni_{10}
B3-2	23.7	21.0	1.0	0.3	2.2	7.3	43.5	0.3	0.4	0.0	-	1.18	TiNi
B3-3	*11.4*	*24.0*	*6.0*	*3.8*	*7.4*	*6.8*	*39.7*	*0.7*	*0.4*	*0.0*	*7.08*	*1.83*	*C15*
B3-4	**10.1**	**23.1**	**10.4**	**9.8**	**10.0**	**9.1**	**27.0**	**0.5**	**0.1**	**0.0**	**6.67**	**2.02**	**C14**
B3-5	1.2	31.5	0.2	0.1	0.2	0.4	2.2	0.1	0.0	64.1	-	2.04	ZrO_2
B3-6	11.4	1.2	59.5	21.2	3.5	0.9	2.0	0.1	0.0	0.0	-	-	V_3B_2
B4-1	7.1	34.2	0.8	0.5	2.3	2.5	40.0	0.4	12.4	0.0	-	1.38	Zr_7Ni_{10}
B4-2	22.8	21.9	0.9	0.4	2.4	7.6	43.4	0.4	0.4	0.0	-	1.20	TiNi
B4-3	*11.0*	*24.3*	*5.7*	*3.6*	*8.1*	*7.0*	*39.3*	*0.6*	*0.4*	*0.0*	*7.07*	*1.83*	*C15*
B4-4	**9.7**	**23.2**	**10.3**	**10.1**	**10.7**	**9.4**	**26.0**	**0.5**	**0.2**	**0.0**	**6.66**	**2.04**	**C14**
B4-5	0.1	32.6	0.1	0.1	0.1	0.1	0.7	0.1	0.0	66.1	-	2.04	ZrO_2
B4-6	14.8	1.8	52.0	23.4	4.1	1.2	2.6	0.1	0.1	0.0	-	-	V_3B_2

Table 3. *Cont.*

Location	Ti	Zr	V	Cr	Mn	Co	Ni	Al	Sn	O	e/a	B/A	Phase
B5-1	4.5	36.5	0.5	0.3	2.4	2.3	39.2	0.6	13.9	0.0	-	1.42	Zr_7Ni_{10}
B5-2	21.3	23.5	0.8	0.4	2.5	7.2	43.6	0.4	0.5	0.0	-	1.20	TiNi
B5-3	10.9	24.6	5.1	3.5	9.1	7.6	38.4	0.7	0.4	0.0	7.08	1.82	C15
B5-4	**9.7**	**23.4**	**9.3**	**9.6**	**11.9**	**9.7**	**25.9**	**0.6**	**0.1**	**0.0**	**6.68**	**2.03**	**C14**
B5-5	0.1	32.3	0.1	0.1	0.1	0.2	0.6	0.1	0.0	66.5	-	2.08	ZrO_2
B5-6	17.5	2.0	48.2	22.8	4.6	1.5	3.2	0.1	0.2	0.0	-	-	V_3B_2

3.4. PCT Analysis

PCT measurements were used to study the gaseous phase H-storage characteristics of alloys in this study. Isotherms from the PCT results measured at 30 and 60 °C are plotted in Figure 6. At the same pressure, hydrogen storage at lower temperature (30 °C) is higher than that at a higher temperature (30 °C). Compared to the AB_5 MH alloy, the highly disordered AB_2 MH alloy has a relatively slanted isotherm without an easily defined plateau region [31–36]. Instead, the desorption pressure at 0.75 wt % is introduced as an indicator for the plateau pressure and used in the calculations of hysteresis, heat of hydride formation (ΔH_h), and entropy change (ΔS_h) [37]. Table 4 shows the gaseous phase H-storage properties obtained from the PCT isotherms. In general, the boron-containing alloys have higher maximum capacities and lower reversible capacities when compared to those from the boron-free alloy.

Desorption pressure measured at 0.75 wt % H-storage decreases with the increase in the boron content and can be attributed to the enlarged C14 unit cell revealed by the XRD analysis. Slope factor (SF), defined as the ratio of desorption capacity between 0.01 and 0.5 MPa to total reversible capacity [6,17,22], is an indicator of the degree of disorder (DOD) in the MH alloy [38] and smaller in a more slanted isotherm. SF measured at 30 °C is listed in the fifth column in Table 4 and shows a decreasing trend with the increase in the boron content. This finding suggests a decrease in uniformity with the addition of boron and is consistent with the decrease in the main C14 phase abundance. Hysteresis of PCT isotherm, defined as the natural log of ratio between the absorption and desorption equilibrium pressures [39–41], can be used to predict the degree of pulverization through cycling [28,37,42,43]. In the current study, hysteresis is listed in the sixth column in Table 4 and does not change significantly as the boron content increases. Therefore, boron is not likely to have a major impact on the pulverization rate during cycling.

ΔH_h and ΔS_h can be calculated from the desorption equilibrium pressures at 0.75 wt % H-storage at 30, 60, and 90 °C by the Van't Hoff equation,

$$\Delta G = \Delta H_h - T\Delta S_h = RT\ln P, \tag{1}$$

where T and R are the absolute temperature and ideal gas constant, respectively. ΔH_h is negative since the hydrogen absorbing process is exothermic. The different between ΔS_h and -130.7 J·mol H_2^{-1}·K^{-1} for H_2 gas indicates DOD of hydrogen in the MH alloy [42]. Both $-\Delta H_h$ and $-\Delta S_h$ are listed in Table 4 and share a similar trend with the gaseous phase H-storage capacities due to the existence of two competing driving forces—an increase in the C14 unit cell volume and a decrease in the C14 phase abundance as the boron content increases.

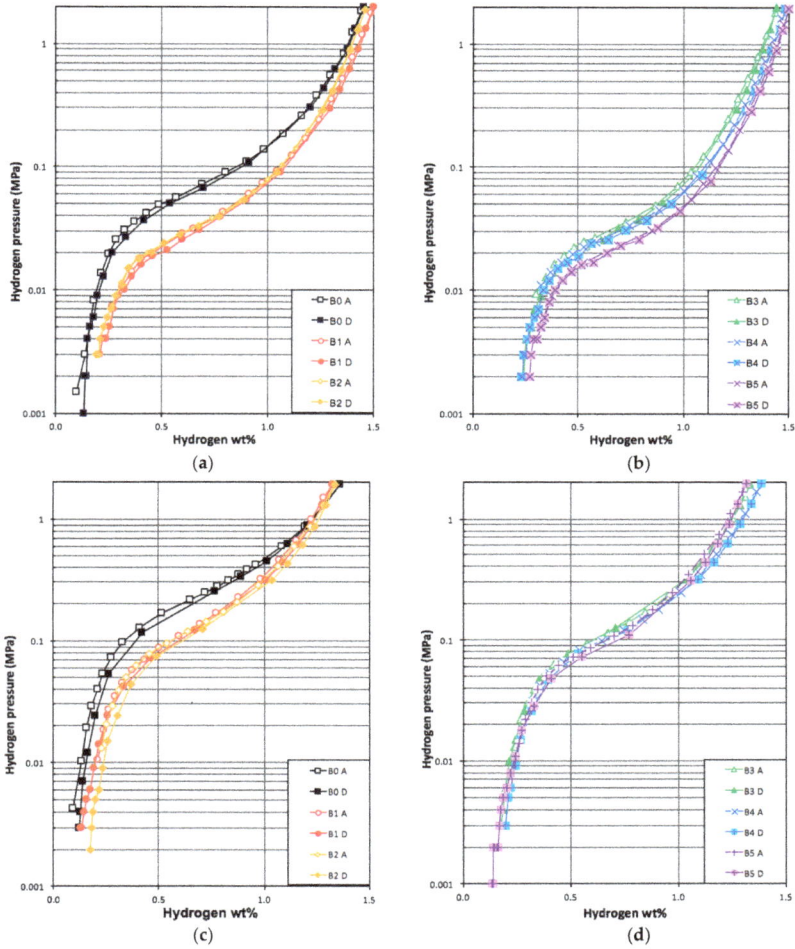

Figure 6. 30 °C PCT isotherms of alloys (**a**) B0, B1, and B2 and (**b**) B3, B4, and B5 and 60 °C PCT isotherms of alloys (**c**) B0, B1, and B2 and (**d**) B3, B4, and B5. Open and solid symbols are for absorption and desorption curves, respectively.

Table 4. Summary of the gaseous phase properties.

Alloy	Maximum Capacity at 30 °C wt %	Reversible Capacity at 30 °C wt %	Desorption PRESSURE at 0.75 wt %, 30 °C MPa	Slope Factor at 30 °C %	Hysteresis at 0.75 wt %, 30 °C	$-\Delta H_h$ kJ·mol H_2^{-1}	$-\Delta S_h$ J·mol H_2^{-1}·K^{-1}
B0	1.45	1.32	0.078	60	0.04	32.8	107
B1	1.50	1.29	0.037	54	0.08	38.9	120
B2	1.46	1.27	0.038	53	0.06	37.5	116
B3	1.44	1.21	0.036	55	0.07	35.3	108
B4	1.46	1.23	0.032	58	0.05	35.5	108
B5	1.50	1.22	0.025	50	0.05	42.9	130

3.5. Electrochemical Analysis

Half-cell full discharge capacity is obtained by summing the discharge capacities at three different rates, and half-cell HRD is defined as the ratio between the high-rate and full capacities. Electrochemical activation behaviors of full capacity and HRD in the first 13 cycles are illustrated in Figure 7. All alloys require the same number of activation cycles (2) to reach the stabilized capacity. As the boron content increases, capacity decrease (Figure 7a), and more activation cycles are needed to reach 85% of maximum HRD (Figure 7b). HRD starts to drop at the eighth cycle in alloys with higher boron contents. HRD at the 10th cycle for each alloy is listed in Table 5 and shows a first-decrease-then-increase trend with the increase in the boron content. Therefore, boron in the Laves phase-based MH alloy is suspected to deteriorate HRD [4], but the increase in the V_3B_2 phase abundance at higher boron content benefits it. In alloy B5, although the V_3B_2 phase abundance is smaller than that in alloy B4, the abundance of another beneficial phase, TiNi [44], is higher, which results in the positive net effect on HRD.

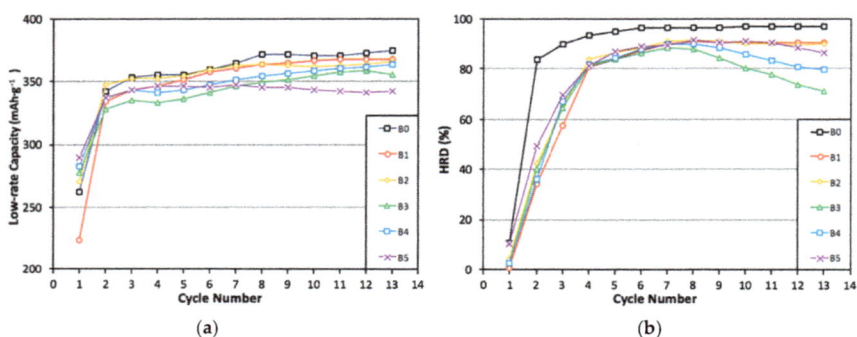

Figure 7. Activation behaviors observed from (**a**) half-cell capacity measured at 4 mA·g^{-1} and (**b**) half-cell high-rate dischargeability (HRD) for the first 13 electrochemical cycles.

Table 5. Summary of the RT electrochemical half-cell results.

Alloy	10th Cycle High-Rate Discharge Capacity mAh·g^{-1}	10th Cycle Full Discharge Capacity mAh·g^{-1}	10th Cycle HRD %	Number of Activation Cycles to Reach 85% HRD
B0	359	371	97	3
B1	330	366	90	4
B2	325	362	90	5
B3	284	354	80	6
B4	307	359	86	6
B5	311	343	91	6

Both surface exchange current (I_o) and bulk diffusion coefficient (D) are used to trace the cause of change in HRD, and the results are listed in Table 6. While I_o reflects the reaction kinetics at the electrode surface, D shows the ability of hydrogen diffusing from the particle core to surface. Details of these measurements can be found in previous studies [45]. Other temperature- and hydrogen content-dependent measurements of hydrogen diffusion constant, such as nuclear magnetic resonance [46] and quasi-elastic neutron scattering [47], in Laves phase alloys were also conducted before. In this study, both I_o and D first decrease and then increase with the increase in alloy number as shown in Figure 8a. However, since the variation in D is on a smaller scale compared to that in I_o, I_o is considered the main factor in affecting HRD. In other words, the HRD performances of the boron-containing alloys are dominated by the speed of surface reaction. According to our previous study, the TiNi phase contributes more to D than I_o [44], which is confirmed by comparing HRDs of alloys B4 and B5 in this study.

Table 6. Summary of the room temperature (RT) and low temperature (LT) electrochemical properties (D: bulk diffusion coefficient, I_o: surface exchange current, R: charge-transfer resistance, C: double-layer capacitance, and RC product).

Alloy	D at RT 10^{-10} cm$^2 \cdot$s^{-1}	I_o at RT mA·g^{-1}	R at RT Ω·g	C at RT F·g^{-1}	RC Product at RT s	R at LT Ω·g	C at LT F·g^{-1}	RC Product at LT s
B0	1.3	38	0.44	0.24	0.11	16.5	0.21	3.5
B1	1.3	26	0.71	0.22	0.16	27.4	0.28	7.7
B2	1.2	22	1.04	0.21	0.22	83.9	0.18	15.1
B3	1.0	18	2.74	0.18	0.49	171.5	0.22	37.7
B4	1.2	25	0.99	0.20	0.20	70.0	0.16	11.2
B5	1.3	21	0.59	0.24	0.15	47.1	0.23	10.8

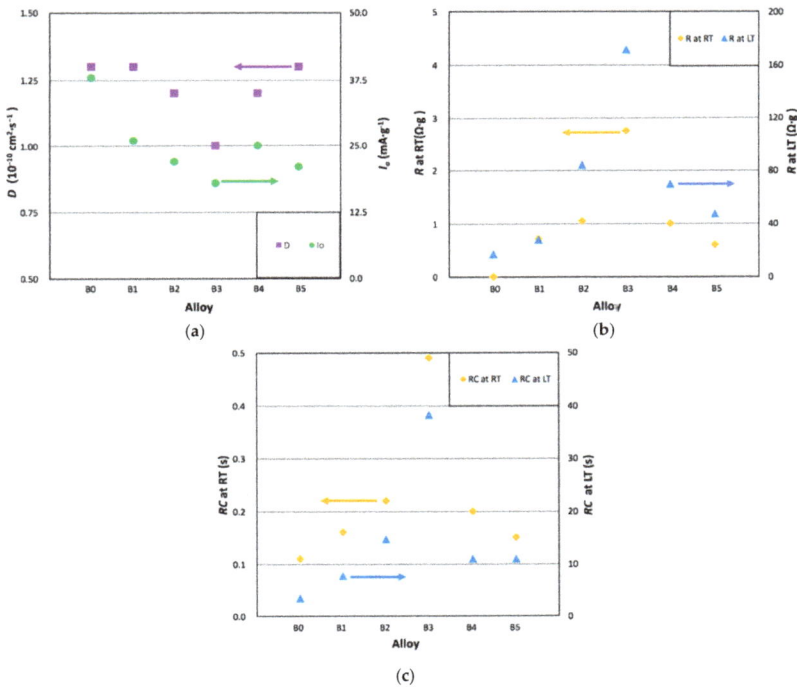

Figure 8. Correlations of (**a**) D and I_o vs. alloy number, (**b**) R at RT and LT vs. alloy number, and (**c**) RC at RT and LT vs. alloy number.

To further investigate the connection between HRD and the surface reaction, AC impedance measured at both RT and LT were conducted, and the resulting charge-transfer resistances (R) and double-layer capacitances (C) from the Cole–Cole plots are summarized in Table 6. While R is closely related to the speed of electrochemical reaction, C is proportional to the reactive surface area, and their product (RC) represents the surface catalytic ability without the contribution from surface area. From the results shown in Table 6 and plotted in Figure 8b,c, we found that R and RC obtained at both RT and LT follow the same trend with the half-cell HRD (Table 6). Changes in surface reactive area (proportional to C) at RT and LT are not as significant as those in R and RC at RT and LT. Therefore, we conclude that the improvement in HRD by the introduction of a new V_3B_2 secondary phase is from the increase in surface catalytic ability.

3.6. Magnetic Susceptibility

To continue the study of surface catalytic ability, M measurements on the activated alloys were conducted to quantify the amount of metallic nickel embedded in the surface oxide. Metallic nickel in the surface acts as a catalyst for water splitting and recombination in the electrochemical reaction and contributes directly to the surface exchange current. Saturated M (M_S) from the ferromagnetic nickel in the surface oxide can be obtained by deducting the paramagnetic part of M curve [48]. Strength of the applied magnetic field corresponding to half of the M_S value ($H_{1/2}$) is associated with the magnetic field domain size and can be used to estimate the average size of nickel clusters. The M vs. applied magnetic field curves are plotted in Figure 9, and the calculated M_S and $H_{1/2}$ are listed in Table 7. M_S first decreases slightly and then increases with the increase in the boron content. The initial decrease in M_S is due to the reduction in the nickel-content in the alloy, and the later increase is caused by the increase in the nickel-content in the catalytic V_3B_2 phase. As the boron content increases, not only the V_3B_2 phase abundance increases, but the nickel content in the V_3B_2 phase also increases (Table 4). The $H_{1/2}$ values listed in Table 7 indicate a decrease in size and an increase in surface area of nickel clusters with the increase in the boron content.

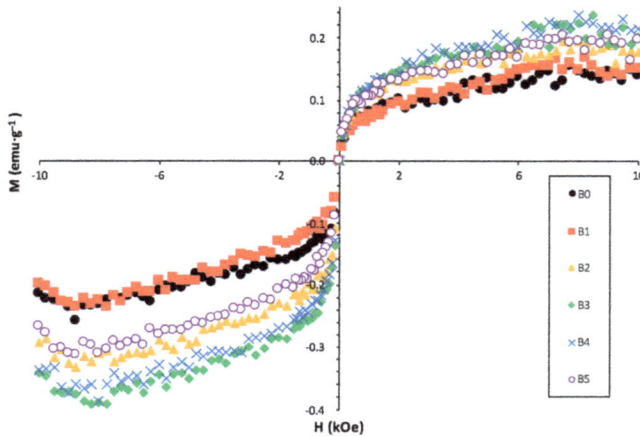

Figure 9. Magnetic susceptibilities of alloys in this study.

Table 7. Summary of the RT magnetic properties (M_S: saturated magnetic susceptibility and $H_{1/2}$: applied field at half of the M_S value).

Alloy	M_S emu·g^{-1}	$H_{1/2}$ kOe
B0	0.064	0.064
B1	0.059	0.120
B2	0.093	0.138
B3	0.099	0.139
B4	0.101	0.106
B5	0.090	0.139

Batteries **2017**, *3*, 36

4. Conclusions

The influences of boron addition on the microstructural, gaseous phase, and electrochemical properties of a V-containing Laves phase-based metal hydride alloy have been studied. A small amount of boron (<2 at %) decreases the hydrogen equilibrium pressure because of the enlargement in unit cell of the C14 main phase, decreases the discharge capacities, and impedes the high-rate performance. However, a further increase in the boron content promotes the formation of a new V_3B_2 secondary phase and increase in the abundance of the beneficial TiNi phase, which contribute positively to the high-rate and low-temperature performances. Future studies will focus on optimizing the V_3B_2 phase abundance and introducing this phase in other metal hydride alloy families.

Acknowledgments: The authors would like to thank the following individuals from BASF-Ovonic for their help: Su Cronogue, Diana F. Wong, Shuli Yan, Tiejun Meng, David Pawlik, Allen Chen, Ryan J. Blankenship, Chaolan Hu, and Reichman Benjamin.

Author Contributions: Shiuan Chang designed the experiments, prepared alloy samples, and performed electrochemical and structural analysis. Taihei Ouchi performed the PCT analysis. Kwo-Hsiung Young, Jean Nei, and Xin Wu mentored the work and helped in manuscript preparation.

Conflicts of Interest: The authors declare no conflict of interest.

Abbreviations

The following abbreviations are used in this manuscript:

Ni/MH	Nickel/metal hydride
MH	Metal hydride
H-storage	Hydrogen-storage
ICP-OES	Inductively coupled plasma-optical emission spectrometer
XRD	X-ray diffractometer
SEM	Scanning electron microscope
EDS	Energy dispersive spectroscopy
PCT	Pressure concentration temperature
RT	Room temperature
LT	Low temperature at $-40\,^\circ$C
M	Magnetic susceptibility
HRD	High-rate dischargeability
e/a	Average electron density
V_{C14}	Unit cell volume of the C14 phase
$V_{V_3B_2}$	Unit cell volume of the V_3B_2 phase
BEI	Back-scattering electron image
ΔH_h	Heat of hydride formation
ΔS_h	Change in entropy
SF	Slope factor
DOD	Degree of disorder
T	Absolute temperature
R	Ideal gas constant
I_o	Surface exchange current
D	Bulk diffusion coefficient
R	Surface charge-transfer resistance
C	Surface double-layer capacitance
M_S	Saturated magnetic susceptibility
$H_{1/2}$	Applied magnetic field strength corresponding to half of saturated magnetic susceptibility

References

1. Zelinsky, M.A.; Koch, J.M.; Young, K. Performance comparison of rechargeable batteries for stationary applications (Ni/MH vs. Ni-Cd and VRLA). *Batteries*. submitted.

2. Young, K.; Ng, K.Y.S.; Bendersky, L.A. A technical report of the Robust Affordable Next Generation Energy Storage System-BASF program. *Batteries* **2016**, *2*, 2. [CrossRef]

3. Young, K.; Nei, J. The current status of hydrogen storage alloy development for electrochemical applications. *Materials* **2013**, *6*, 4574–4608. [CrossRef] [PubMed]

4. Huang, T.; Wu, Z.; Xia, B.; Xu, N. Influence of stoichiometry and alloying elements on the crystallography and hydrogen sorption properties of TiCr based alloys. *Mater. Sci. Eng. A* **2005**, *397*, 284–287.

5. Kim, D.M.; Jeon, S.W.; Lee, J.Y. A study of the development of a high capacity and high performance Zr–Ti–Mn–V–Ni hydrogen storage alloy for Ni–MH rechargeable batteries. *J. Alloys Compd.* **1998**, *279*, 209–214. [CrossRef]

6. Young, K.; Chang, S.; Lin, X. C14 Laves phase metal hydride alloys for Ni/MH batteries applications. *Batteries* **2017**, *3*, 27. [CrossRef]

7. Wilipedia. Laves Phase. Available online: https://en.wikipedia.org/wiki/Laves_phase (accessed on 20 October 2017).

8. Shoemaker, D.P.; Shoemaker, C.B. Concerning atomic sites and capacities for hydrogen absorption in the AB$_2$ Friauf-Laves phases. *J. Less-Common Met.* **1979**, *68*, 43–58. [CrossRef]

9. Jacob, I.; Stern, A.; Moran, A.; Shaltiel, D.; Davidov, D. Hydrogen absorption in $(Zr_xTi_{1-x})B_2$ (B = Cr, Mn) and the phenomenological model for the absorption capacity in pseudo-binary Laves-phase compounds. *J. Less-Common Met.* **1980**, *73*, 1369–1376. [CrossRef]

10. Young, K.; Ouchi, T.; Koch, J.; Fetcenko, M.A. The role of Mn in C14 Laves phase multi-component alloys for NiMH battery application. *J. Alloys Compd.* **2009**, *477*, 749–758. [CrossRef]

11. Sehn, C.-C.; Chou, J.C.-P.; Li, H.-C.; Wu, Y.-P.; Perng, T.-P. Effect of interstitial boron and carbon on the hydrogenation properties of $Ti_{25}V_{35}Cr_{40}$ alloy. *Int. J. Hydrogen Energy* **2010**, *35*, 11975–11980. [CrossRef]

12. Luan, B.; Chu, N.; Zhao, H.J.; Liu, H.K.; Dou, S.X. Effects of potassium-boron addition on the performance of titanium based hydrogen storage alloy electrodes. *Int. J. Hydrogen Energy* **1996**, *21*, 373–379. [CrossRef]

13. Zhang, Y.-H.; Dong, X.-P.; Wang, G.-Q.; Guo, S.-H.; Ren, J.-Y.; Wang, X.-L. Effect of boron additive on electrochemical cycling life of La-Mg-Ni alloys prepared by casting and rapid quenching. *Int. J. Hydrogen Energy* **2007**, *32*, 594–599. [CrossRef]

14. Leela, A.M.R.; Ramaprabhu, S. Hydrogen diffusion studies in Zr-based Laves phase AB$_2$ alloys. *J. Alloys Compd.* **2008**, *460*, 268–271.

15. Young, K.; Ouchi, T.; Huang, B.; Fetcenko, M.A. Effects of B, Fe, Gd, Mg, and C on the structure, hydrogen storage, and electrochemical properties of vanadium-free AB$_2$ metal hydride alloy. *J. Alloys Compd.* **2012**, *511*, 242–250. [CrossRef]

16. Young, K.; Fetcenko, M.A.; Li, F.; Ouchi, T. Structural, thermodynamics, and electrochemical properties of $Ti_xZr_{11-x}(VNiCrMnCoAl)_2$ C14 Laves phase alloys. *J. Alloys Compd.* **2008**, *464*, 238–247. [CrossRef]

17. Young, K.; Fetcenko, M.A.; Koch, J.; Morii, K.; Shimizu, T. Studies of Sn, Co, Al, and Fe additives in C14/C15 Laves alloys for NiMH battery application by orthogonal arrays. *J. Alloys Compd.* **2009**, *486*, 559–569. [CrossRef]

18. Young, K.; Ouchi, T.; Huang, B.; Reichman, B.; Fetcenko, M.A. Studies of copper as a modifier in C14-predominant AB$_2$ metal hydride alloys. *J. Power Sources* **2012**, *204*, 205–212. [CrossRef]

19. Young, K.; Ouchi, T.; Huang, B.; Reichman, B.; Fetcenko, M.A. The structure, hydrogen storage, and electrochemical properties of Fe-doped C14-predominating AB$_2$ metal hydride alloys. *Int. J. Hydrogen Energy* **2011**, *36*, 12296–12304. [CrossRef]

20. Young, K.; Ouchi, T.; Huang, B.; Reichman, B. Effect of molybdenum content on structural, gaseous storage, and electrochemical properties of C14-predominant AB$_2$ metal hydride alloys. *J. Power Sources* **2011**, *196*, 8815–8821. [CrossRef]

21. Erika, T.; Ricardo, F.; Fabricio, R.; Fernando, Z.; Verónica, D. Electrochemical and metallurgical characterization of $ZrCr_{1-x}NiMo_x$ AB$_2$ metal hydride alloys. *J. Alloys Compd.* **2015**, *649*, 267–274. [CrossRef]

22. Young, K.; Ouchi, T.; Lin, X.; Reichman, B. Effects of Zn-addition to C14 metal hydride alloys and comparisons to Si, Fe, Cu, Y, and Mo-additives. *J. Alloys Compd.* **2016**, *655*, 50–59. [CrossRef]

23. Johnson, R.L.; Hoffmann, R. Z. Structure-bonding relationships in the Laves Phases. *Z. Anorg. Allg. Chem.* **1992**, *616*, 105–120. [CrossRef]

24. Liu, C.T.; Zhu, J.H.; Brady, M.P.; McKamey, C.G.; Pike, L.M. Physical metallurgy and mechanical properties of transition-metal Laves phase alloys. *Intermetallics* **2000**, *8*, 1119–1129. [CrossRef]

25. Nei, J.; Young, K.; Salley, S.O.; Ng, K.Y.S. Determination of C14/C15 phase abundance in Laves phase alloys. *Mat. Chem. Phys.* **2012**, *136*, 520–527. [CrossRef]

26. *Power Diffraction File (PDF) Database*; MSDS No. 04-003-6123; International Centre for Diffraction Data: Newtown Square, PA, USA, 2011.

27. Notten, P.H.L.; Einerhand, R.E.F.; Daams, J.L.C. How to achieve long-term electrochemical cycling stability with hydride-forming electrode materials. *J. Alloys Compd.* **1995**, *231*, 604–610. [CrossRef]

28. Young, K.; Fetcenko, M.A.; Li, F.; Ouchi, T.; Koch, J. Effect of vanadium substitution in C14 Laves phase alloys for NiMH battery application. *J. Alloys Compd.* **2009**, *468*, 482–492. [CrossRef]

29. Young, K.; Ouchi, T.; Fetcenko, M.A.; Mays, W.; Reichman, B. Structural and electrochemical properties of $Ti_{1.5}Zr_{5.5}V_xNi_{10-x}$. *Int. J. Hydrogen Energy* **2009**, *34*, 8695–8706. [CrossRef]

30. Liu, Y.; Young, K. Microstructure investigation on metal hydride alloys by electron backscatter Diffraction Technique. *Batteries* **2016**, *2*, 26. [CrossRef]

31. Ovshinsky, S.R.; Fetcenko, M.A. Electrochemical Hydrogen Storage Alloys and Batteries Fabricated from Mg Containing Base Alloys. U.S. Patent 5,506,069, 9 April 1996.

32. Young, K.; Huang, B.; Ouchi, T. Studies of Co, Al, and Mn substitutions in $NdNi_5$ metal hydride alloys. *J. Alloys Compd.* **2012**, *543*, 90–98. [CrossRef]

33. Schlapbach, L.; Züttel, A. Hydrogen-storage materials for mobile applications. *Nature* **2001**, *414*, 353–358. [CrossRef] [PubMed]

34. Züttel, A. Materials for hydrogen storage. *Mater. Today* **2003**, *6*, 24–33. [CrossRef]

35. Sastri, M.V.C. Introduction to metal hydrides: basic chemistry and thermodynamics of their formation. In *Metal Hydride*; Sastri, M.V.C., Viswanathan, B., Murthy, S.S., Eds.; Springer-Verlag: Berlin, Germany, 1998; p. 5.

36. Young, K.; Ouchi, T.; Meng, T.; Wong, D.F. Studies on the synergetic effects in multi-phase metal hydride alloys. *Batteries* **2016**, *2*, 15. [CrossRef]

37. Young, K.; Ouchi, T.; Fetcenko, M.A. Pressure-composition-temperature hysteresis in C14 Laves phase alloys: Part 1. Simple ternary alloys. *J. Alloys Compd.* **2009**, *480*, 428–433. [CrossRef]

38. Wong, D.F.; Young, K.; Nei, J.; Wang, L.; Ng, K.Y.S. Effects of Nd-addition on the structural, hydrogen storage, and electrochemical properties of C14 metal hydride alloys. *J. Alloys Compd.* **2015**, *647*, 507–518. [CrossRef]

39. Scholtus, N.A.; Hall, W.K. Hysteresis in the palladium-hydrogen system. *J. Chem. Phys.* **1963**, *39*, 868–870. [CrossRef]

40. Makenas, B.J.; Birnbaum, H.K. Phase changes in the niobium-hydrogen system 1: Accommodation effects during hydride precipitation. *Acta Metall. Mater.* **1980**, *28*, 979–988. [CrossRef]

41. Balasubramaniam, R. Accommodation effects during room temperature hydrogen transformations in the niobium-hydrogen system. *Acta Metall. Mater.* **1993**, *41*, 3341–3349. [CrossRef]

42. Lide, D.R. *CRC Handbook of Chemistry and Physics*, 74th ed.; CRC Press: Boca Raton, FL, USA, 1993; pp. 6–22.

43. Jeng, R.; Lee, S.; Hsu, C.; Wu, Y.; Lin, J. Effects of the addition of Pd on the hydrogen absorption-desorption characteristics of $Ti_{33}V_{33}Cr_{34}$ alloys. *J. Alloys Compd.* **2008**, *464*, 467–471. [CrossRef]

44. Young, K.; Ouchi, T.; Nei, J.; Moghe, D. The importance of rare-earth additions in Zr-based AB$_2$ metal hydride alloys. *Batteries* **2016**, *2*, 25. [CrossRef]

45. Young, K.; Wong, D.F.; Ouchi, T.; Huang, B.; Reichman, B. Effects of La-addition to the structure, hydrogen storage, and electrochemical properties of C14 metal hydride alloys. *Electrochim. Acta* **2015**, *174*, 815–825. [CrossRef]

46. Renz, W.; Majer, G.; Skripov, A.V.; Seeher, A. A pulsed-field-gradient NMR study of hydrogen diffusion in the Laves-phase compounds $ZrCr_2H_x$. *J. Phys. Condens. Matter.* **1994**, *6*, 6367–6474. [CrossRef]

47. Campbell, S.I.; Kemali, M.; Ross, D.K.; Bull, D.J.; Fernandez, J.F.; Johnson, M.R. Quasi-elastic neutron scattering study of the hydrogen diffusion in the C15 Laves structure, $TiCr_{1.85}$. *J. Alloys Compd.* **1999**, *293–295*, 351–355. [CrossRef]

48. Stucki, F.; Schlapbach, L. Magnetic properties of $LaNi_5$, FeTi, Mg_2Ni and their hydrides. *J. Less-Common Met.* **1980**, *74*, 143–151. [CrossRef]

batteries

MDPI

Article

Comparison of C14- and C15-Predomiated AB$_2$ Metal Hydride Alloys for Electrochemical Applications

Kwo-Hsiung Young [1,2,*], Jean Nei [2], Chubin Wan [3,4], Roman V. Denys [3] and Volodymyr A. Yartys [3,5]

1   Department of Chemical Engineering and Materials Science, Wayne State University, Detroit, MI 48202, USA
2   BASF/Battery Materials—Ovonic, 2983 Waterview Drive, Rochester Hills, MI 48309, USA; jean.nei@basf.com
3   Institute for Energy Technology, P.O. Box 40, NO-2027 Kjeller, Norway; cbinwan@gmail.com (C.W.);
    roman.v.denys@gmail.com (R.V.D.); volodymyr.yartys@ife.no (V.A.Y.)
4   Department of Physics, University of Science and Technology Beijing, Beijing 100083, China
5   Department of Materials Science and Engineering, Norwegian University of Science and Technology,
    NO-7491 Trondheim, Norway
*   Correspondence: kwo.young@basf.com; Tel.: +1-248-293-7000

Academic Editor: Catia Arbizzani
Received: 24 May 2017; Accepted: 11 July 2017; Published: 28 July 2017

Abstract: Herein, we present a comparison of the electrochemical hydrogen-storage characteristics of two state-of-art Laves phase-based metal hydride alloys ($Zr_{21.5}Ti_{12.0}V_{10.0}Cr_{7.5}Mn_{8.1}Co_{8.0}Ni_{32.2}Sn_{0.3}Al_{0.4}$ vs. $Zr_{25.0}Ti_{6.5}V_{3.9}Mn_{22.2}Fe_{3.8}Ni_{38.0}La_{0.3}$) prepared by induction melting and hydrogen decrepitation. The relatively high contents of lighter transition metals (V and Cr) in the first composition results in an average electron density below the C14/C15 threshold ($e/a \sim 6.9$) and produces a C14-predomiated structure, while the average electron density of the second composition is above the C14/C15 threshold and results in a C15-predomiated structure. From a combination of variations in composition, main phase structure, and degree of homogeneity, the C14-predomiated alloy exhibits higher storage capacities (in both the gaseous phase and electrochemical environment), a slower activation, inferior high-rate discharge, and low-temperature performances, and a better cycle stability compared to the C15-predomiated alloy. The superiority in high-rate dischargeability in the C15-predomiated alloy is mainly due to its larger reactive surface area. Annealing of the C15-predomiated alloy eliminates the ZrNi secondary phase completely and changes the composition of the La-containing secondary phase. While the former change sacrifices the synergetic effects, and degrades the hydrogen storage performance, the latter may contribute to the unchanged surface catalytic ability, even with a reduction in total volume of metallic nickel clusters embedded in the activated surface oxide layer. In general, the C14-predomiated alloy is more suitable for high-capacity and long cycle life applications, and the C15-predomiated alloy can be used in areas requiring easy activation, and better high-rate and low-temperature performances.

Keywords: metal hydride; nickel metal hydride battery; Laves phase alloy; electrochemistry; synergetic effect

1. Introduction

Nickel/metal hydride (Ni/MH) rechargeable batteries are widely used in today's consumer electronics, stationary power storage, and transportation applications. One of the major factors limiting the performance of Ni/MH batteries is a relatively low gravimetric energy density, compared to the rival lithium-ion battery technology [1]. For the active materials in the negative electrode of Ni/MH battery, Laves phase-based AB$_2$ metal hydride (MH) alloy containing 1.85 wt % H with a potential capacity of 434 mAh·g^{-1} [2] has commonly been a high-energy alternative to the conventional rare earth-based AB$_5$ alloys, which have a capacity of approximately 330 mAh·g^{-1}. Other performance comparisons between these two MH alloy families are available in an earlier review article [3]. Different from the single CaCu$_5$ crystal structure in the AB$_5$ MH alloys, the main phase in the AB$_2$ MH alloys can be C14, C15, or a mixture of two, which provides additional freedom in composition design to address various requirements, such as ultra-low temperature performance, high-temperature storage, and overcharge performance [4]. C14 and C15 are two Laves structures and form the largest intermetallic compound group [5].

The difference between these two structures originates from the different types of packings in two types of metal nets, Kagome 6363 nets formed by B atoms and containing hexagons and triangles, and A$_2$B buckled nets formed by both A and B atoms [6]. There are 6 types of these nets, depending of their orientation along the [001] direction of the hexagonal/trigonal unit cells, as shown in Figure 1; A, B, and C nets for the Kagome 6363 nets and a, b, and c nets for the A$_2$B buckled nets. The packing of these nets creates AcBc 2-layer stacking, resulting in a hexagonal C14 type Laves type structure, or 3-layer stacking (AcBaCb), resulting in a face-centered cubic (fcc) C15 Laves type structure, both with AB$_2$ stoichiometry. As shown in Figure 1, atoms in the A layer form a triangular net and there are two possible arrangements for the next layer—atoms in the B or C position. If the stacking of the triangular nets follows the sequence A-B-A-B, as shown in Figure 1b, a hexagonal crystal structure is formed. In the case of another stacking sequence, A-B-C-A-B-C, the structure is fcc with the same packing density as for the hexagonal one (Figure 1c). For the Laves phases, the triangular net is replaced by an A$_4$B$_8$ slab with an A$_2$B-B$_3$-A$_2$B-B$_3$ structure, and C14 and C15 are formed following the A-B-A-B and A-B-C-A-B-C stacking sequences, respectively. Another member of the Laves phases, hexagonal C36, has the same building slabs, but they are stacked in a different sequence, AbCaBaCb. However, the C36 type of structure is much less abundant than C14 and C15 [7], and we will not discuss it further in this work.

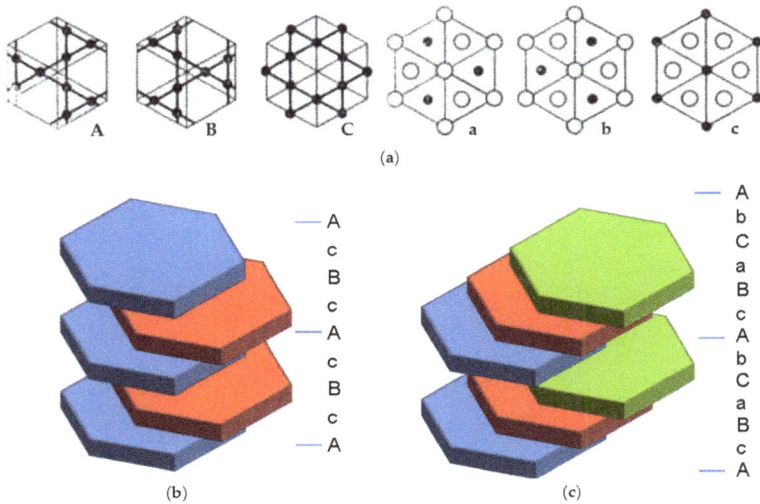

Figure 1. Stacking (**a**) units of each layer, (**b**) C14, and (**c**) C15 Laves type structures.

Figure 2 shows the crystal structures of C14 and C15 type alloys, and Table 1 summarizes the crystallographic data for both structures. Ideally, the lattice parameters are closely related in each structure and between structures. However, in the actual C14-predominated MH alloys, the c/a ratio is slightly lower than the theoretical value ($2\sqrt{\frac{2}{3}} \cong 1.633$) [8,9]. Three types of positions are available for hydrogen occupation tetrahedral sites (A_2B_2, AB_3, and B_4) in both C14 and C15 structures, as shown in Figure 2. In the Laves phases, octahedral sites are not present at all, therefore the following discussion will only concentrate on tetrahedral sites.

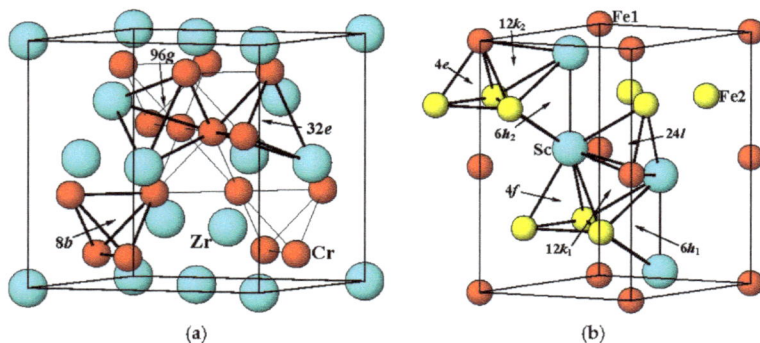

(a) (b)

Figure 2. Unit cells for (a) ZrCr$_2$ (C15) and (b) ScFe$_2$ (C14) structures. Various tetrahedral hydrogen occupation sites (A_2B_2, AB_3, and B_4) are indicated by arrows.

Table 1. Basic physical parameters of C14 and C15. H-site denotes hydrogen occupying site.

Parameter	C14	C15
Crystal symmetry	Hexagonal (*hP12*)	Face-centered-cubic (*cF24*)
Space group	$P6_3/mmc$	$Fd\bar{3}m$
Ideal lattice constant a	a_{C14}	$\sqrt{2}a_{C14}$
Ideal lattice constant c	$2\sqrt{\frac{2}{3}}a_{C14}$	-
Number of A_2B_2/Full unit cell AB_2 tetrahedral H-sites per formula [10]	12 ($6h_1 + 6h_2 + 12k_1 + 24l$)	12 (96g)
Number of AB_3 tetrahedral H-site per formula [10]	4 (4f + 12k_2)	4 (32e)
Number of B_4 tetrahedral H-site per formula [10]	1 (4e)	1 (8b)
Maximum H-storage	Up to 5.4 at. H/AB$_2$ [11]	Up to 7 at. H/AB$_2$ [12]
Theoretical maximum H-storage	6.33 per AB$_2$ [13]	7 per AB$_2$ [13]

Hydrogen occupation occurs first at the A_2B_2 site, next at the AB_3 site, and finally at the B_4 site [14–17]. Furthermore, the ratio between the tetrahedral sites is 12 A_2B_2, 4 AB_3, and 1 B_4 for both C14 and C15 structures. However, not all these sites can be occupied at the same time as the sites with a common triangular face are blocked from simultaneous occupancy. This is because the distance between their centers appears to be well below 0.2 nm, the minimum allowed H-H distance in the structures of metal hydrides [18].

Storages of up to 5.4 and 4.25 H atoms per AB$_2$ formula unit for the C14 and C15 structures, respectively, have been demonstrated at room temperature and in a reasonable pressure range, although their theoretical hydrogen storage (H-storage) capacities are higher (6.33 and 6 H atoms per AB$_2$ formula unit for C14 and C15, respectively).

The choice of the dominating Laves structure at room temperature in the alloy is not random. While several determining factors, such as atomic size ratio, difference in electronegativity between the A-site and B-site atoms [19], and stoichiometry [20] have been discussed in literature, average electron density (e/a) was found to be the most crucial parameter and can be directly correlated to the ratio of C14/C15 at room temperature [21]. An empirical model for predicating the C14/C15 formation was originally supported by a set of tight-binding calculations [22] and recently confirmed by a density function theory calculation [23]. In addition, the model for e/a at the C14/C15 threshold (C14:C15

= 1:1) was further improved to include the contribution from the A-site atoms by incorporating the average chemical potential of the A-site atoms [24].

While the initial studies of the AB_2 MH alloys in the gaseous phase [25] and electrochemical environment [26] started and were later commercialized with the C14 phase [3], studies on the C15-predominated AB_2 alloys for Ni/MH application were common from 1991 to 2004. The major accomplishments during this period are summarized in Table 2. In order to increase the stability of the C15 phase, these alloys are generally designed to have a higher Zr-content (which moves e/a at the C14/C15 threshold to a lower value [27]), lower the V- and Cr-content (which increases e/a to a value above the e/a of the C14/C15 threshold [27]), and have a hyperstoichiometry (B/A > 2), compared to the C14-predominated MH alloys used in Ni/MH applications. Although the C15 alloys that exhibit a high capacity and excellent high-rate dischargeability (HRD) have been successfully developed, they are not as popular as the C14 alloys. Through the years, the performance of the C14 and C15 phases in both the gaseous phase and electrochemical environment have been compared over a dozen times either in the alloys with the same composition but different preparation processes, or in a series of alloys with close compositions. Unfortunately, the findings are inconsistent (Table 3). It is very difficult to determine which phase has better performance with regard to capacity, HRD, and cycle life. In this paper, we provide a different approach to compare these two Laves structures. To this end, two state-of-art C14 and C15-based MH alloys were selected and their gaseous and electrochemical H-storages were compared. We hope this work would illuminate future AB_2 MH alloy research.

Table 2. Summary of previous studies on the hydrogen storage properties of C15-predominated MH alloys in an electrochemical environment. AM, Ann, PM, LM, IM denote arc melting, annealing, plasma melting, levitation melting, and induction melting preparation methods, respectively. Cm is the discharge capacity obtained with an m mA·g^{-1} discharge current. HRD and EC are abbreviations for high-rate dischargeability and electrochemistry, respectively.

Basic Composition	Preparation Method	Major Achievements	References	Year
$ZrCr_{0.4}Mn_{0.4}Ni_{1.2}$	AM + Ann	C30 of ~320 mAh·g^{-1}	[28]	1991
$ZrV_{0.8}Mn_{0.4}Ni_{1.2}$	Ar PM	C10 of ~366 mAh·g^{-1}	[29]	1991
$ZrV_{0.5}Ni_{1.5}$	LM	C100 of ~365 mAh·g^{-1}	[30,31]	1994
$ZrV_{0.05}Cr_{0.25}Mn_{0.6}Ni_{1.3}$	AM	C50 of 343 mAh·g^{-1}	[32]	1995
$ZrV_{0.5}Mn_{0.5}NiMo_{0.15}$	AM	C50 of 339 mAh·g^{-1}	[33]	1995
$ScCr_{0.2}Mn_{0.5}Co_{0.2}Ni_{1.1}$	AM	C70 of 400 mAh·g^{-1}	[34]	1995
$ZrV_{0.33}Mn_{0.86}Co_{0.11}Ni_{0.9}$	AM	C17 of 440 mAh·g^{-1}	[35]	1995
$ZrV_{1.5}Ni_{1.5}$	LM	C2 of 800 mAh·g^{-1}	[36]	1997
$ZrV_{0.2}Cr_{0.1}Mn_{0.6}Ni_{1.2}$	IM + Ann	C80 of 330 mAh·g^{-1}	[37]	1997
$ZrV_{0.5}Mn_{0.7}Ni_{1.2}$	AM	C100 of 330 mAh·g^{-1}	[38]	1998
$ZrMn_{1-x}V_xNi_{1.4+y}$	AM	Surface area dominates EC performance	[39]	1998
$ZrV_{0.2}Cr_{0.05}Mn_{0.6}Co_{0.05}Ni_{1.2}$	IM + Ann	C70 of 370 mAh·g^{-1}	[40,41]	1998
$ZrV_{1.5}Ni_{1.5}$	IM	C160 of 356 mAh·g^{-1}	[42]	1999
$Zr(VMnCoNi)_{2+\alpha}$	IM	300 cycle with stable capacity C60 = 342 mAh·g^{-1}	[43]	1999
$Zr_{0.5}Ti_{0.5}V_{0.6}Mn_{0.2}Pd_{0.1}Ni_{0.8}Fe_{0.2}$	AM	C50 of 372 mAh·g^{-1}	[44]	1999
$ZrV_{0.2}Mn_{0.6}Cr_{0.1}Ni_{1.2}$	AM	F-treatment with Ni improves cycle life	[45]	1999
$Zr_{0.4}Ti_{0.6}V_{1.2}Cr_{0.3}Ni_{1.5}$	AM	200 cycle with stable capacity	[46]	2000
$ZrV_{0.2}Mn_{0.6}Co_{0.1}Ni_{1.2}$	AM + Ann	C50 of ~350 mAh·g^{-1}	[47]	2000
$ZrV_{0.4}Mn_{0.5}Co_{0.05}Ni_{1.1}$	AM	Co improves HRD, cycle stability, and self-discharge	[48]	2001
$ZrTi_{0.1}V_{0.2}Cr_{0.1}Mn_{0.6}Co_{0.1}Ni_{1.2}$	AM + Ann	C100 of 390 mAh·g^{-1}	[49]	2001
$Zr_{0.9}Ti_{0.1}V_{0.2}Mn_{0.56}Co_{0.1}Ni_{1.14}$	IM	C60 of 350 mAh·g^{-1}	[50]	2002
$Zr(NiVMnCoSn_x)_{2+\alpha}$	IM	Sn has detrimental effects to EC performance.	[51]	2006

Table 3. Summary of previous comparative studies on the hydrogen storage properties of C14 and C15 in gaseous phase (GP) or electrochemical (EC) environment. ΔH_h denotes heat of hydride formation.

Basic Composition	Preparation Method	Application	Major Findings	References
$(TiZr)V_{0.5}Mn_{0.2}Fe_{0.2}Ni_{1.1}$	AM	EC	C14 has a better HRD	[52]
$Zr(CrNi)_2$	AM + Ann	GP	No difference if composition is the same	[53]
$(ZrTi)(VMnNi)x$	AM	EC	C14 has a higher discharge capacity ($x < 2$) C15 has a better HRD ($x > 2$)	[35,54]
$(ZrTi)(NiMnM)_x$, where M = Cr, V, Co, Al	IM + Ann	EC	C15 has a better cycle life but slower activation	[40]
$(ZrTi)(VMnCoNi)_2$	LM	EC	C14 has a higher capacity and HRD	[55]
$(ZrTi)(VMnNi)_2$	AM	EC	With no Ti, C15 has more desirable capacities, with Ti, C14 has a high capacity	[44]
$Zr(VMnNi)_2$	AM	EC	C15 is better with regards to capacity, HRD, and activation	[56]
$(ZrTi)(VCrMnNi)_2$	AM	EC	C14 has a higher capacity and HRD	[57]
$(ZrTi)(VMnNi)_x$	IM + Ann	GP	C15 has a longer cycle life	[58]
$Zr(VFe)_x$	AM	GP	C14 has a higher H-storage capacity	[59]
$(ZrTi)(VAl)_2$	AM	GP	C14 has a lower ΔH_h	[60]
$(ZrTi)(VCrMnCoNiAl)_2$	IM	EC	C15 has a better HRD and low-temperature performance, but shorter cycle life	[61]
$(ZrTi)(VCrMnNil)_2$	IM	EC	C14 has a better charge retention and cycle life, but lower capacity and HRD	[62]
$(ZrTi)(VCrMnCoNiAl)_2$	IM	EC	C15 phase improves both activation and HRD	[63]

2. Experimental Setup

Each ingot sample was prepared by an induction melting process under a 0.08 MPa Ar protection atmosphere and elemental raw materials with a purity of >99.9% (except for Zr, where Sn-containing (1%) zircaloy was used). An MgO crucible, an alumina tundish, and a steel mold were used for melting. Annealing was performed in vacuum (achieved with a diffusion pump) for 6 h at 960 °C with a 3 h temperature ramp-up period. The ingot was then cooled naturally to room temperature. For powder fabrication, the ingot underwent a hydriding/dehydriding process, which introduced initial volume expansion/contraction to create internal stress before it was crushed and ground to a −200 mesh powder. A Varian *Liberty* 100 inductively coupled plasma optical emission spectrometer (ICP-OES, Agilent Technologies, Santa Clara, CA, USA) was employed to study the chemical composition. A Philips *X'Pert Pro* XRD (X-ray diffractometer, Philips, Amsterdam, The Netherlands) was used to perform the phase analysis, and a JEOL-*JSM6320F* scanning electron microscope (SEM, JEOL, Tokyo, Japan) with energy dispersive spectroscopy (EDS) was also used to investigate the phase distribution and composition. A Suzuki Shokan multi-channel pressure-concentration-temperature system (PCT, Suzuki Shokan, Tokyo, Japan) was used to measure the gaseous phase H-storage characteristics. PCT measurements at 30, 60, and 90 °C were performed after activation, which consisted of a 2 h thermal cycle between room temperature and 300 °C under 2.5 MPa H_2 pressure. MH alloy electrodes were prepared by directly pressing the MH alloy powder onto an expanded Ni substrate (1 cm × 1 cm) with a 10-ton press without the use of any metallic or organic binder. Electrochemical measurements, including capacities at various rates, bulk diffusion coefficient (D), and surface exchange current (I_o) were performed on an Arbin Instruments BT-2143 Battery Test Equipment (Arbin Instruments, College Station, TX, USA). A Solartron 1250 Frequency Response Analyzer (Solartron Analytical, Leicester, UK) with a sine wave amplitude of 10 mV and a frequency range of 0.5 mHz to 10 kHz was used to conduct the alternating current (AC) impedance measurements. A Digital Measurement Systems Model 880 vibrating sample magnetometer (MicroSense, Lowell, MA, USA) was used to measure the magnetic susceptibility of the activated alloy surfaces (etched for 4 h in 30 wt % KOH at 100 °C).

3. Results and Discussion

Two compositions, $Zr_{21.5}Ti_{12.0}V_{10.0}Cr_{7.5}Mn_{8.1}Co_{8.0}Ni_{32.2}Sn_{0.3}Al_{0.4}$ and $Zr_{25.0}Ti_{6.5}V_{3.9}Mn_{22.2}Fe_{3.8}Ni_{38.0}La_{0.3}$, were selected for this comparative study. Their target compositions and ICP results are summarized in

Table 4. The first composition is a stoichiometric C14 composition and was used as the base alloy for a number of comparative studies [8,63–66] due to its overall balanced performance with regard to activation, HRD, and cycle stability. The e/a of the first composition is below the C14/C15 threshold ($e/a \sim$ 6.9 [24]), and therefore a C14-predominated structure occurs. The second composition was chosen based on a series of refinements targeting high-rate Ni/MH applications, and further by containing an optimized Ti and Zr ratio with Ni, Mn, V, and Fe, with a minor amount of La additive [67]. The half-cell capacity for the alloy with the second composition mixed with 80% carbonyl nickel approached 460 mAh·g^{-1} at a discharge current density of 10 mA·g^{-1} [68]. Compared to the first composition, the second composition is hyperstoichiometric and has a higher Zr-content, lower V-content, no Cr, and higher Ni-content, which contribute to a higher e/a value and result in a C15-predominated alloy. A small amount of La was added in the C15-predominated alloy to facilitate the activation process [37,69,70]. While only the un-annealed C14 alloy was used for this comparative work, two versions of the C15 alloys were assessed: pristine (C15) and annealed alloys (C15A). Since the effects of annealing on the multi-phase C14-predominated AB$_2$ MH alloys have been well studied (elimination/reduction in secondary phase abundance results in reduction of synergetic effects, leading to deterioration of electrochemical properties) [55,61,71,72], only the impacts of annealing on the C15 AB$_2$ MH alloy will be verified in this work. ICP results of the three alloys (C14, C15, and C15A) are in excellent agreement with the corresponding design values.

Table 4. Design compositions (in bold) and ICP results in at %. e/a is the average electron density. B/A is the atomic ratio of B-atom (elements other than Ti and Zr, and La) to A-atom (Ti, Zr, and La).

Alloy	Source	Zr	Ti	V	Cr	Mn	Fe	Co	Ni	Sn	Al	La	e/a	B/A
C14	Design	21.5	12.0	10.0	7.5	8.1	-	8.0	32.2	0.3	0.4	-	6.82	1.99
	ICP	21.5	12.0	10.0	7.5	8.1	-	8.0	32.2	0.4	0.3	-	6.82	1.99
C15	Design	25.0	6.5	3.9	-	22.2	3.8	-	38.0	0.3	0.0	0.3	7.13	2.14
	ICP	24.7	6.5	3.9	-	21.9	4.2	-	38.3	0.3	0.1	0.2	7.16	2.18
C15A	Design	25.0	6.5	3.9	-	22.2	3.8	-	38.0	0.3	0.0	0.3	7.13	2.14
	ICP	24.9	6.5	3.9	-	21.8	4.2	-	38.2	0.3	0.1	0.2	7.15	2.16

3.1. X-Ray Diffractometer Analysis

XRD analysis was used to study the constituent phases occurring in the alloys. The obtained XRD patterns are shown in Figure 3. The XRD pattern from the C14 alloy demonstrates a C14-predominated structure with overlapping C15 peaks and a minor TiNi peak. Both XRD patterns from C15 and C15A alloys show a C15 structure with a small ZrNi peak in the pristine alloy. Results from full XRD pattern fitting with Jade 9.0 software (MDI, Livermore, CA, USA) are summarized in Table 5. The c/a ratio obtained for the C14 alloy (1.629) is only slightly lower than the ideal ratio (1.633), and this deviation is commonly seen in C14 alloys for Ni/MH application. The atomic size ratio, R_A/R_B (where R_A and R_B represent the average atomic radii of the A-site and B-site atoms, respectively), in the C14 alloy (1.216) is slightly lower than the ideal ratio of $\sqrt{\frac{3}{2}} \cong 1.225$ [73], which causes a deviation in the c/a ratio from the ideal value. Moreover, the secondary phases found in the C14 and C15 alloys belong to TiNi and ZrNi structures, respectively. After annealing, the ZrNi secondary phase in the C15 alloy becomes undetectable. This reduction/diminishing of the secondary phase after annealing also occurs in the C14 AB$_2$ MH alloys [55,61,71,72]. In addition to C14 and TiNi, there is also a 5.2 wt % of C15 found in the C14 alloy since the alloy's e/a (6.82) is close to the e/a at the C14/C15 threshold for Zr/Ti ($\cong$1.8 (6.91)) [24]. The C15 phase is usually located between the C14 main matrix and other Zr$_x$Ni$_y$ secondary phases [74]. Therefore, due to the mixed nature of the C14 and C15 phases in the C14 alloy, the crystallites in the C14 alloy are smaller than those in the C15 alloy. Furthermore, the annealed C15 (C15A) has even larger crystallites. The increase in crystallite size after annealing is a common observation in Laves phase-based MH alloys [61,75].

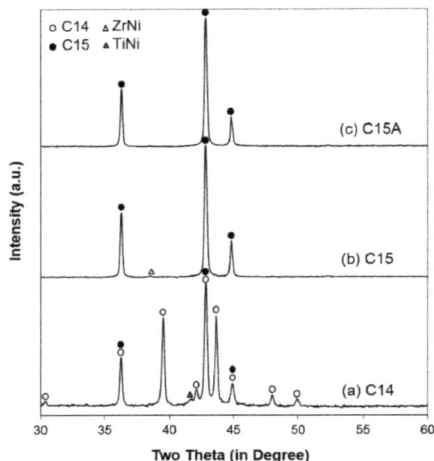

Figure 3. X-ray diffractometer patterns using Cu-K$_\alpha$ as the radiation source for the various alloys. (**a**) C14, (**b**) C15, and (**c**) C15A.

Table 5. Lattice constants, abundance, crystallite size (CS) of the C14 and C15 phases of the C14, C15, and C15A alloys. Abundances of TiNi and ZrNi secondary phases are also included.

Alloy	C14	C15	C15A
C14 Lattice constant *a*, nm	0.49545	-	-
C14 Lattice constant *c*, nm	0.80733	-	-
C14 Abundance, wt%	93.7	-	-
C14 CS, nm	68	-	-
C15 Lattice constant *a*, nm	0.69932	0.70061	0.70047
C15 Abundance, %	5.2	99.3	100
C15 CS, nm	54	96	>100
TiNi Abundance, wt%	1.2	-	-
ZrNi Abundance, wt%	-	0.7	-

3.2. Scanning Electron Microscope/Energy Dispersive Spectroscopy Analysis

SEM back-scattering electron images (BEI) from the alloys are presented in Figure 4. The composition of several representative areas (identified by Roman numerals) in the SEM micrographs were studied by EDS, and the results are summarized in Table 6. SEM micrographs of the C14 alloy shows a very typical multi-phase C14-C15-Zr$_x$Ni$_y$ microstructure, which has been extensively studied with transmission electron microscopy (TEM) [76,77] and electron backscattering diffraction (EBSD) [75]. Occasional ZrO$_2$ inclusions are also seen in the C14 alloy and act as oxygen scavengers [78], which may contribute positively to the bulk diffusion of hydrogen and provide surface protection against oxidation by the electrolyte [79]. In the SEM micrographs of the C15 and C15A alloys, a LaNi or La-rich phase with a high contrast is observed, suggesting segregation of La from the main phase. Since La does not precipitate into the Zr-based Laves phase, it segregates into a LaNi secondary phase, as in the cases of other rare earth element substitutions [9]. The relatively high solubility of the LaNi phase in the KOH electrolyte results in the facilitation of an initial formation process in alkaline solution [70]. The La-content and Ni-content of the La-rich secondary phase in the C15 alloy increases and decreases after annealing, respectively. It should be noted that the XRD analysis does not detect any La-containing phase, due to its small overall abundance. In addition, the SEM micrographs shown in Figure 4 are not typical, but exhibit the most features and therefore reveal all phases of the alloys. Additionally, the measured Sn-content in the LaNi phase before annealing is quite high (15.7 at %) and becomes even higher (21.5 at %) after annealing.

In the Laves phase MH alloys, Sn dissolves into the main C14 Laves phase and the ZrNi secondary phase without forming any Sn-rich secondary phase [8,78,80–84], and more Sn migrates into the ZrNi secondary phase after annealing [61]. The presence of Sn in the composition of the Zr-containing MH alloy is due to a cost saving consideration—the market price of Sn-containing zircaloy scrap, which is used as one of the raw materials in the current study, at one time was only one tenth of the cost of pure Zr scrap. In general, a small percentage of Sn (approximately 0.2 to 0.4 at %), if dissolved fully into the main phase, facilitates hydride formation but reduces HRD and cycle life [79]. Moreover, a phase with a slightly brighter contrast (Spot 2 in Figure 4b) and a composition close to (Zr,Ti)Ni can be found in the C15 alloy. It is eliminated during the annealing process and disappears in the SEM micrograph taken from the C15A alloy (Figure 4c).

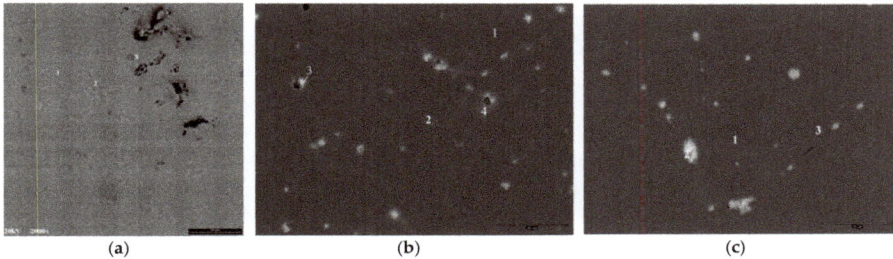

(a)   (b)   (c)

Figure 4. SEM BEI micrographs from the (a) C14, (b) C15, and (c) C15A alloys. The composition of the numbered areas was analyzed by EDS and the results are shown in Table 6. The bar at the lower right corner in each micrograph represents 25 μm.

Table 6. Summary of the EDS results. All compositions are in %. Compositions of the main AB_2 phase are in bold.

Location	Zr	Ti	V	Cr	Mn	Fe	Co	Ni	La	Sn	B/A	e/a	Phase
C14-1	**19.4**	**14.6**	**8.6**	**5.1**	**7.5**	**0.0**	**6.4**	**38.3**	**0.0**	**0.1**	**1.94**	**6.46**	**AB_2**
C14-2	18.0	23.3	3.0	1.4	3.5	0.0	5.1	45.6	0.0	0.1	1.42	-	TiNi
C14-3	80.4	3.9	3.4	2.2	2.1	0.0	1.3	6.2	0.0	0.5	0.19	-	ZrO_2
C15-1	**24.6**	**6.5**	**3.7**	**0.0**	**22.2**	**4.4**	**0.0**	**38.6**	**0.0**	**0.0**	**2.22**	**7.20**	**AB_2**
C15-2	23.6	15.2	0.9	0.0	8.4	1.4	0.0	50.2	0.3	0.0	1.56	-	ZrNi
C15-3	5.0	1.7	0.7	0.0	4.7	0.8	0.0	20.9	50.5	15.7	0.75	-	LaNi
C15-4	15.6	5.3	1.3	0.0	7.5	1.2	0.0	26.4	40.9	1.8	0.62	-	Oxide
C15A-1	**24.8**	**6.5**	**3.8**	**0.0**	**22.6**	**4.2**	**0.0**	**37.9**	**0.1**	**0.0**	**2.18**	**7.15**	**AB_2**
C15A-2	0.6	0.2	0.5	0.0	0.7	0.1	0.0	11.1	61.3	25.5	0.61	-	La-rich
C15A-3	58.7	5.6	2.1	0.0	12.2	2.4	0.0	19.0	0.0	0.0	0.56	-	ZrO_2

3.3. Pressure-Concentration-Temperature Analysis

The PCT isotherms were measured at 30, 60, and 90 °C, and the results from the first two temperatures are shown in Figure 5. PCT isotherms measured at 90 °C are not complete due to an increase in plateau pressure (out of range for the testing apparatus), and therefore are not shown. Gaseous phase H-storage characteristics obtained from the PCT analysis are summarized in Table 7. Compared to the C14 alloy, the PCT isotherms of both the C15 and C15A alloys show a very steep takeoff from the α (metal)-to-β (metal hydride) region, which is similar to the observations seen in Nd-based AB_5 [85] and A_2B_7 [86] MH alloys, and a lower self-discharge is expected. Moreover, the C15 and C15A alloys show very flat plateaus, which are extremely uncommon in multi-phase MH alloys [8,87]. In order to quantify the plateau flatness, slope factor (as previously defined in [8]: the ratio of storage capacity between 0.01 MPa and 0.5 MPa to total capacity in the desorption isotherm) of each alloy was calculated. The increase in slope factor (plateau flatness) from the C14 alloy (0.60) to the C15 and C15A alloys (0.87 and 0.90, respectively) is a direct result of the elimination of multi-phase features and the accompanying synergetic mode [88]. Annealing of the C15 alloy decreases the storage capacity, slightly decreases the plateau pressure, and increases the absorption/desorption hysteresis (defined as $\ln\left(\frac{\text{absorption plateau pressure}}{\text{desorption plateau pressure}}\right)$) in the middle of the pressure plateau due to the improvement in the homogeneity and complete removal of the ZrNi secondary phase, which is very critical for supplying the synergetic effects [89,90]. Several speculations have been proposed for the possible origin of PCT hysteresis [91–94]. The energy required for elastic lattice deformation in the metal/MH interface area during absorption [95] is currently the most accepted explanation. The reduction in PCT hysteresis in the multi-phase system has been explained previously and is caused by the remaining hydrogenated phase (from activation or previous hydrogenation) at the grain boundary between phases (Figure 14 in [89]). Cleanness at the interface (free of amorphous and impurity phase) between phases removes a possible source for dissipation of stresses at a boundary between the major and the secondary phases and is important for the occurrence of such phenomenon. It has been confirmed in similar alloys through the use of TEM and EBSD [74,96]. Therefore, the C14 alloy, that has the highest secondary phase abundance (6.4 wt %), also has the smallest PCT hysteresis (0.04); the C15 alloy, that has a lower secondary phase abundance (0.7 wt %), has a larger PCT hysteresis (0.13), while the C15A alloy has no detectable (through XRD analysis) secondary phase and shows the largest PCT hysteresis (0.31). Synergetic effects resulted by the presence of the secondary phase and composition inhomogeneity reduce the hysteresis and make more storage sites accessible, so the plateau region of the PCT isotherm can be extended [89]. Furthermore, both the ΔH_h and difference in entropy (ΔS_h) were estimated using desorption plateau pressures at 30 and 60 °C with the following equation:

$$\Delta H_h - T\Delta S_h = \Re\, T \ln P \qquad (1)$$

where $\Re$ is the ideal gas constant and T is the absolute temperature. Although the C15 alloy has a significantly higher plateau pressures compared to the C14 alloy, they exhibit similar ΔH_h values, which indicates that the current comparative study between C14 and C15 is fair. After annealing, the C15A alloy demonstrates a lower ΔH_h (more stable hydride) and a ΔS_h closer to the ideal value between free hydrogen gas and solid ($-130.7\ \text{J·mol}^{-1}\text{·K}^{-1}$) [97]. The formation of the more ordered hydride from C15A is resulted by the improvement in homogeneity by annealing.

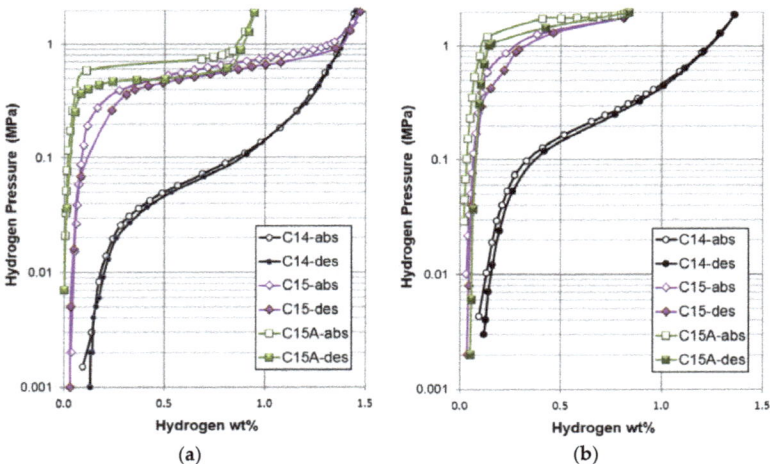

Figure 5. Pressure-concentration-temperature (PCT) isotherms from the C14, C15, and C15a alloys measured at (**a**) 30 and (**b**) 60 °C. Open and solid symbols represent the absorption and desorption curves, respectively.

Table 7. Summary of the gaseous phase properties of the C14, C15, and C15A AB$_2$ alloys.

Alloy	C14	C15	C15A
Maximum Capacity @ 30 °C in wt%	1.45	1.47	0.95
Reversible Capacity @ 30 °C in wt%	1.32	1.44	0.94
Desorption Pressure @ 30 °C in MPa	0.078	0.55	0.50
Slope Factor @ 30 °C	0.60	0.87	0.90
PCT Hysteresis @ 30 °C	0.04	0.13	0.31
$-\Delta H_h$ in kJ·mol^{-1}	32	31.8	35.4
$-\Delta S_h$ in J·mol^{-1}·K^{-1}	104	119	130

3.4. Electrochemical Analysis

The electrochemical capacity and activation characteristics of the alloys were studied using half-cell measurements in a flooded configuration (for details, see [98]). Evolution of full capacity (measured at a discharge rate of 4 mA·g^{-1}) and HRD (the ratio of capacity at a discharge rate of 50 mA·g^{-1} to that at a discharge rate of 4 mA·g^{-1}) for the first 13 cycles are plotted in Figure 6, and the electrochemical properties are summarized in Table 8. Figure 6 shows that the C14 alloy has a higher low-rate capacity, a lower HRD, and is more difficult to activate, compared to the C15 alloys. Since the C15 and C15A alloys' plateaus pressures are higher than 0.1 MPa (one atmosphere) and therefore cannot be fully charged in the open-to-air half-cell configuration, their discharge capacities are lower than the expected values from the conversion of the gaseous phase H-storage capacities (1 wt % = 268 mAh·g^{-1}). If the C15 alloy powder samples are entirely embedded in a soft metallic binder (Ni or Cu), their full capacities can be obtained [68]. However, for our measurements, the MH powder was directly compacted onto a Ni substrate without any binder or metallic fine particles, which results in the easy release of hydrogen gas from the surface and incomplete charge. Moreover, the C15 and C15A alloys show better HRD and activation performances than the C14 alloy, and the HRD of the C15 alloy is slightly higher than that of the C15A alloy, due to the eliminations of the secondary phase and accompanied synergic effect by annealing. We believe that the differences in activation, degradation, and HRD originated from the composition rather than the structure. By comparing the compositions of alloys C14 and C15 (Table 4), Cr, a very important substitution element in the MH alloy for the enhancement of corrosion resistance by forming the V-Cr-based solid solution secondary phase [78,99], is absent in alloy C15. Alloy C15 also has a higher Ni-content, which is known for achieving a better HRD performance [100,101]. While the Cr-containing alloy C14 is more difficult to activate but maintains the discharge capacity in the first 13 cycles, the Cr-free alloy C15 shows some capacity degradation. After annealing, the capacity degradation of alloy C15A is improved but still noticeable in cycles 9 to 13 (Figure 6a).

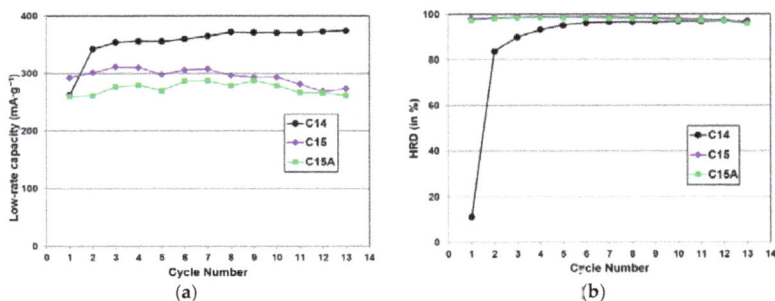

Figure 6. Activation behaviors observed from (**a**) full capacity (measured at low-rate) and (**b**) HRD for the first 13 electrochemical cycles measured at room temperature.

Table 8. Summary of the room temperature electrochemical and magnetic results (capacity, rate, D, I_o, and M_S, $H_{1/2}$,) of C14, C15, and C15A alloys. RT denotes room temperature.

Alloy	C14	C15	C15A
3rd cycle capacity @50 mA·g^{-1} in mAh·g^{-1}	318	307	273
3rd cycle capacity @4 mA·g^{-1} in mAh·g^{-1}	354	311	277
HRD @ 3rd cycle	0.9	0.99	0.98
Activation cycle to reach 95% HRD	5	1	1
Diffusion coefficient, D @ RT in 10^{-10} cm^2·s^{-1}	2.5	2.4	1.6
Exchange current, I_o @ RT in mA·g^{-1}	22.5	46.8	39.4
M_S in emu·g^{-1}	0.037	0.042	0.017
$H_{1/2}$ in kO2	0.11	0.26	0.48

The superiority in HRD of alloys C15 and C15A was further investigated by electrochemically measuring D and I_o. D was measured by a potentiostatic discharge process—the electrode was first fully charged and then discharged at a potential of +0.6 V for 7200 s, and the anodic current response is tracked with respect to time during the process. Figure 7a shows the resulted semi-logarithmic curves of the anodic current response vs. time for the three alloys. D was estimated using the slope of the linear region of the semi-logarithmic response according to the equation [102]:

$$\log i = \log\left(\frac{6FD}{da^2}(C_o - C_s)\right) - \frac{\pi^2}{2.303}\frac{D}{a^2}t \tag{2}$$

where i is the specific diffusion current (A·g^{-1}), F is the Faraday constant, C_o is the initial hydrogen concentration in the alloy bulk (mol·cm^{-3}), C_s is the hydrogen concentration on the surface of the alloy particles (mol·cm^{-3}), d is the density of the H-storage alloy (g·cm^{-3}), a is the alloy particle radius (cm), and t is the discharge time (s). I_o was measured by linear polarization, specifically the electrode was first fully charged, then discharged to 50% depth-of-discharge, and then scanned within a small overpotential range of ±10 mV. In this small overpotential range, the current vs. overpotential shows a linear dependence, as seen in Figure 7b, and I_o can be obtained from the equation [103]:

$$I_o = \frac{i\Re T}{F\eta} \tag{3}$$

where i is the specific current (A·g^{-1}), F is the Faraday constant, and η is the overpotential. Further details for the D and I_o calculations have been previously reported [98], and the D and I_o values for the current set of alloys are listed in Table 8. While the D values of the C14 and C15 alloys are close, it deteriorates after annealing for the C15 alloy, due to the elimination of the secondary phase. The main difference between the C14 and C15 alloys occurs in the surface reaction, where I_o in the C15 alloy is more than double that of the C14 alloy. The I_o value found for the C15 alloy is even higher than that in an AB$_5$ MH alloy that has a higher Ni-content [104]. After annealing, I_o decreases, which confirms the positive contribution of the ZrNi secondary phase to the surface reaction. ZrNi is more susceptible to dissolution in KOH solution, and its existence in the AB$_2$ MH alloys has been shown to improve HRD [62]. In conclusion, the superior HRD of the C15 alloy comes from the higher I_o value, which indicates a faster surface reaction.

In order to investigate the source of the faster surface reaction (higher I_o) found in the C15 alloy, AC impedance measurements were conducted at both room temperature (RT) and -40 °C. The charge-transfer resistance (R) and double-layer capacitance (C) obtained from Cole-Cole plots [88] are listed in Table 9. There are two factors dominating the R values: the amount of reactive surface area and surface reaction catalytic ability. While the former is directly proportional to the capacitance, the latter can be related to the RC product (a higher RC corresponds to a worse catalytic surface) [105]. From the comparisons in Table 9, the C15 alloy has lower R values at both RT and -40 °C, mainly due to its higher amount of reactive surface area (higher *C*), which is closely related to the addition of La [70]. In addition, the surface catalytic abilities of the C15 alloy at RT and −40 °C are the same as and slightly worse, respectively, than those of the C14 alloy. Therefore, we conclude that the higher I_0 of the C15 alloy originates from the higher amount of reactive surface, which is due to the additional La, an absence of Cr, and a higher Mn-content in the composition. As for the annealing effects for the C15 alloy, the data in Table 9 show deteriorated *R*'s at both RT and −40 °C after annealing, which is due to the reduction in reactive surface area. The unchanged *RC* product with annealing requires further investigation and is discussed in the next section.

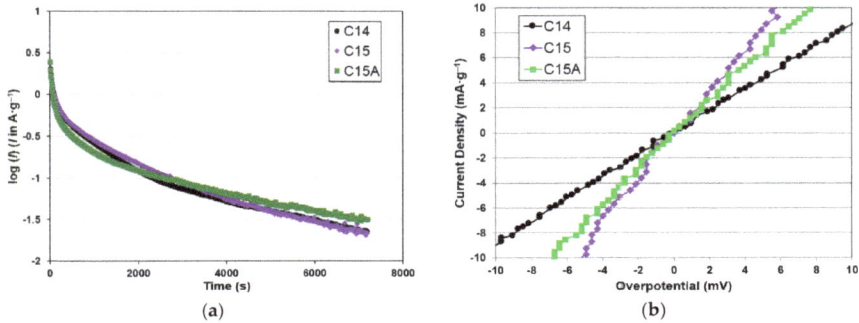

Figure 7. (**a**) Semilogarithmic curves of anodic current vs. time during potentiostatic discharge and (**b**) linear polarization curves measured at 50% depth-of-discharge from the C14, C15, and C15A alloys.

Table 9. A summary of the electrochemical results from AC impedance measurement (*R*—charge transfer resistance, *C*—double-layer capacitance) at room temperature (RT) and −40 °C for the C14, C15, and C15A alloys.

Property	*R* @ RT	*C* @ RT	*RC*	*R* @ −40 °C	*C* @ −40 °C	*RC*
Unit	$\Omega{\cdot}g$	$Farad{\cdot}g^{-1}$	s	$\Omega{\cdot}g$	$Farad{\cdot}g^{-1}$	s
C14	0.32	0.34	0.11	29	0.24	7.0
C15	0.23	0.49	0.11	22	0.49	10.7
C15A	0.32	0.34	0.11	26	0.41	10.7

3.5. Magnetic Susceptibility Analysis

Measuring saturated values of magnetic susceptibility (M_s) is a convenient way to quantify the amount of catalytic metallic Ni clusters embedded in the surface oxide, which has been shown to strongly influence the surface catalytic ability of MH alloys for electrochemical reactions [79,104,106]. However, recent TEM studies revealed that other fine surface structures in the Si- [82] and La- [70] containing AB₂ MH alloys can also affect surface catalytic ability. Furthermore, applied magnetic fields corresponding to half of M_s ($H_{1/2}$) can be used as a parameter to quantify the size of the metallic clusters; more specifically, it is inversely proportional to the size of the magnetic domain of these clusters [8]. Results of M_s and $H_{1/2}$ from the three alloys in this study are listed in the last two columns in Table 8. The C15 alloy has a slightly higher M_s (more catalytic surface) and $H_{1/2}$ (smaller metallic clusters) compared to the C14 alloy, which may be related to the larger surface area of the C15 alloy (higher RT *C* in Table 9). After annealing, M_s is reduced by a large amount, but this change in M_s is not reflected in the *RC* product. Other factors, such as the change in composition of the La-containing phase (increases in La and Sn but reduction in Ni) after annealing, may be the cause of the unchanged catalytic ability, even though M_s is reduced with annealing.

4. Conclusions

The gaseous phase and electrochemical hydrogen storage characteristics of two Laves phase MH alloys are compared. In order to secure the dominance of a single phase, a composition with a higher percentage of transition metals with lower number of valence electrons (V and Cr) was adopted to achieve a low average electron density and the consequent C14-predominated alloy (Alloy C14), and a composition designed oppositely is also adopted to acquire a C15-predominated alloy (Alloy C15). The following performance variations are most likely linked to composition differences, rather than structural difference. Alloy C15 in this study has a higher plateau pressure, lower gaseous phase storage capacities under 2 MPa (both maximum and reversible), and a lower discharge capacity compared to Alloy C14. The flatter PCT isotherm with a larger hysteresis and a smaller change in entropy in Alloy C15 are due to the higher uniformity in the alloy (higher main phase abundance). The increased high-rate performance of Alloy C15 originates from its larger surface exchange current, which is the result of a higher amount of surface area from the addition of La. The effects of annealing on Alloy C15 are identified as similar to those on the C14-predominated MH alloys, specifically elimination/reduction in secondary phase abundance, which causes deterioration in the gaseous phase and electrochemical hydrogen storage performances. However, the surface catalytic ability is unchanged after annealing, even though a reduction in the total volume of surface metallic Ni inclusions is observed. Therefore, other causes, such as a change in composition of the La-rich secondary phase after annealing, may be in play. In summary, from the two compositions used in this study, Alloy C15 is recommended for applications requiring high capacity and long cycle durability, while Alloy C15 is more geared toward those requiring high-rate capability and easy formation.

Acknowledgments: The authors would like to thank the following individuals from BASF-Ovonic for their assistance: Taihei Ouchi, Su Cronogue, Baoquan Huang, Diana F. Wong, David Pawlik, Allen Chan, and Ryan J. Blankenship. The work is related to the collaboration between IFE and BASF on the project MoZEES, funded by Norwegian Research Council.

Author Contributions: Kwo-Hsiung Young designed the experiments. Jean Nei performed the experiments. Chubin Wan, Roman V. Denys and Volodymyr A. Yartys analyzed the results and prepared the manuscript.

Conflicts of Interest: The authors declare no conflict of interest.

Abbreviations

Ni/MH	Nickel/metal hydride
MH	Metal hydride
fcc	Face-centered cubic
H-site	Hydrogen occupying site
H-storage	Hydrogen storage
e/a	Average electron density
HRD	High-rate dischargeability
AM	Arc melting
Ann	Annealing
PM	Plasma melting
LM	Levitation melting
IM	Induction melting
C_m	Discharge capacity obtained at an m mA·g^{-1} discharge current
GP	Gaseous phase
EC	Electrochemical
ΔH_h	Heat of hydride formation
ICP-OES	Inductively coupled plasma optical emission spectrometer
XRD	X-ray diffractometer
SEM	Scanning electron microscope
EDS	Energy dispersive spectroscopy
PCT	Pressure concentration temperature

D	Bulk diffusion coefficient
I_o	Surface exchange current
AC	Alternating current
R_A	Average atomic radius of the A-site atoms
R_B	Average atomic radius of the B-site atoms
CS	Crystallite size
BEI	Back-scattering electron image
TEM	Transmission electron microscope
EBSD	Electron backscattering diffraction
α	Metal
β	Metal hydride
ΔS_h	Change in entropy
R	Ideal gas constant
T	Absolute temperature
i	Specific diffusion current
F	Faraday constant
C_o	Initial hydrogen concentration in alloy bulk
C_s	Hydrogen concentration on alloy particle surface
d	Density of hydrogen storage alloy
a	Alloy particle radius
t	Discharge time
η	Overpotential
RT	Room temperature
R	Surface charge-transfer resistance
C	Surface double-layer capacitance
M_s	Saturated magnetic susceptibility
$H_{1/2}$	Applied magnetic field strength corresponding to half of saturated magnetic susceptibility

References

1. Young, K.; Ng, K.Y.S.; Bendersky, L.A. A technical report of the Robust Affordable Next Generation Energy Storage System-BASF Program. *Batteries* **2016**, *2*. [CrossRef]
2. Young, K.; Ouchi, T.; Koch, J.; Fetcenko, M.A. The role of Mn in C14 Laves phase multi-component alloys for NiMH battery application. *J. Alloys Compd.* **2009**, *477*, 749–758. [CrossRef]
3. Chang, S.; Young, K.; Nei, J.; Fierro, C. Reviews on the U.S. Patents regarding nickel/metal hydride batteries. *Batteries* **2016**, *2*, 10. [CrossRef]
4. Young, K.; Yasuoka, S. Capacity degradation mechanisms in nickel/metal hydride batteries. *Batteries* **2016**, *2*. [CrossRef]
5. Yurchenko, N.; Stepanov, N.; Salishchev, G. Laves-phase formation criterion for high-entropy alloys. *Mater. Sci. Tech.* **2017**, *33*, 17–22. [CrossRef]
6. Pearson, W.B. *The Crystal Chemistry and Physics of Metals and Alloys*; John Wiley & Sons: New York, NY, USA, 1972; p. 24.
7. Aufrecht, J.; Leineweber, A.; Mittemeijer, E.J. Metastable hexagonal modifications of the NbCr$_2$ Laves phase as function of cooling rate. *Mater. Res. Soc. Symp. Proc.* **2009**, *1128*. [CrossRef]
8. Chang, S.; Young, K.; Ouchi, T.; Meng, T.; Nei, J.; Wu, X. Studies on incorporation of Mg in Zr-based AB$_2$ metal hydride alloys. *Batteries* **2016**, *2*. [CrossRef]
9. Young, K.; Ouchi, T.; Nei, J.; Moghe, D. The importance of rare-earth additions in Zr-based AB$_2$ metal hydride alloys. *Batteries* **2016**, *2*. [CrossRef]
10. Ivey, D.; Northwood, D. Hydrogen site occupancy in AB$_2$ Laves phases. *J. Less-Common Met.* **1986**, *115*, 23–33. [CrossRef]
11. Yartys, V.A.; Burnasheva, V.V.; Semmenenko, K.N.; Fadeeva, N.V.; Solov'ev, S.P. Crystal chemistry of RT$_5$H(D)$_x$, RT$_2$H(D)$_x$ and RT$_3$H(D)$_x$ hydrides based on intermetallic compounds of CaCu$_5$, MgCu$_2$, MgZn$_2$ and PuNi$_3$ structure types. *Int. J. Hydrogen Energy* **1982**, *7*, 957–965. [CrossRef]

12. Gingl, F.; Yvon, K.; Vogt, T.; Hewat, A.W. Synthesis and crystal structure of tetragonal $LnMg_2H_7$ (Ln=La, Ce), two Laves phase hydride derivatives having ordered hydrogen distribution. *J. Alloys Compd.* **1997**, *253*, 313–317. [CrossRef]

13. Shoemaker, D.P.; Shoemaker, C.B. Concerning atomic sites and capacities for hydrogen absorption in the AB_2 Friauf-Laves phases. *J. Less-Common Met.* **1979**, *68*, 43–58. [CrossRef]

14. Midden, H.J.P.; Prodan, A.; Zupanič, E.; Žitko, R.; Makridis, S.S.; Stubos, A.K. Structural and electronic properties of the hydrogenated $ZrCr_2$ Laves phase. *J. Phys. Chem. C* **2010**, *114*, 4221–4227. [CrossRef]

15. Hong, S.; Fu, C.L. Hydrogen in Laves phase ZrX_2 (X = V, Cr, Mn, Fe, Co, Ni) compounds: Binding energies and electronic and magnetic structure. *Phys. Rev. B* **2002**, *66*, 094109. [CrossRef]

16. Li, F.; Zhao, J.; Tian, D.; Zhang, H.; Ke, X.; Johansson, B.J. Hydrogen storage behavior in C15 Laves phase compound $TiCr_2$ by first principles. *J. Appl. Phys.* **2009**, *105*, 043707. [CrossRef]

17. Merlino, A.R.; Luna, C.R.; Juan, A.; Pronsato, M.E. A DFT study of hydrogen storage in $Zr(Cr_{0.5}Ni_{0.5})_2$ Laves phase. *Int. J. Hydrogen Energy* **2016**, *41*, 2700–2710. [CrossRef]

18. Westlake, D.G. A geometric model for the stoichiometry and interstitial site occupancy in hydrides (deuterides) of $LaNi_5$, $LaNi_4Al$ and $LaNi_4Mn$. *J. Less-Common Met.* **1983**, *91*, 275–292. [CrossRef]

19. Stein, F.; Palm, M.; Sauthoff, G. Structure and stability of Laves phases. Part I. Critical assessment of factors controlling Laves phase stability. *Intermetallics* **2004**, *12*, 713–720. [CrossRef]

20. Young, K.; Ouchi, T.; Yang, J.; Fetcenko, M.A. Studies of off-stoichiometric AB_2 metal hydride alloy: Part 1. Structural characteristics. *Int. J. Hydrogen Energy* **2011**, *36*, 11137–11145. [CrossRef]

21. Liu, C.T.; Zhu, J.H.; Brady, M.P.; McKamey, C.G.; Pike, L.M. Physical metallurgy and mechanical properties of transition-metal Laves phase alloys. *Intermetallics* **2000**, *8*, 1119–1129. [CrossRef]

22. Johnston, R.L.; Hoffmann, R. Structure-bonding relationships in the Laves phases. *Z. Anorg. Allg. Chem.* **1992**, *616*, 105–120. [CrossRef]

23. Wong, D.F.; Young, K.; Ng, K.Y.S. First-principles study of structure, initial lattice expansion, and pressure-composition-temperature hysteresis for substituted $LaNi_5$ and $TiMn_2$ alloys. *Model. Simul. Mater. Sci. Eng.* **2016**, *24*, 085007. [CrossRef]

24. Nei, J.; Young, K.; Salley, S.O.; Ng, K.Y.S. Determination of C14/C15 phase abundance in Laves phase alloys. *Mater. Chem. Phys.* **2012**, *135*, 520–527. [CrossRef]

25. Shaltiel, D.; Jacob, I.; Davidov, D. Hydrogen absorption and desorption properties of AB_2 Laves-phase pseudobinary compounds. *J. Less-Common Met.* **1977**, *53*, 117–131. [CrossRef]

26. Sapru, K.; Hong, K.; Fetcenko, M.A.; Venkatesan, S. Hydrogen storage materials and methods of sizing and preparing the same for electrochemical applications. U.S. Patent 4551400, 5 November 1985.

27. Zhu, J.H.; Liaw, P.K.; Liu, C.T. Effect of electron concentration on the phase stability of $NbCr_2$-based Laves phase alloys. *Mater. Sci. Eng.* **1997**, *A239-240*, 260–264. [CrossRef]

28. Moriwaki, Y.; Gamo, T.; Seri, H.; Iwaki, T. Electrode characteristics of C15-type Laves phase alloys. *J. Less-Common Met.* **1991**, *172–174*, 1211–1218. [CrossRef]

29. Wakao, S.; Sawa, H.; Furukawa, J. Effects of partial substitution and anodic oxidation treatment of Zr-V-Ni alloys on electrochemical properties. *J. Less-Common Met.* **1991**, *172-174*, 1219–1226. [CrossRef]

30. Züttel, A.; Meli, F.; Schlapbach, L. Electrochemical and surface properties of $Zr(V_xNi_{1-x})_2$ alloys as hydrogen-absorbing electrodes in alkaline electrolyte. *J. Alloys Compd.* **1994**, *203*, 235–241. [CrossRef]

31. Züttel, A.; Meli, F.; Chartouni, D.; Schlapbach, L.; Lichtenberg, F.; Friedrich, B. Properties of $Zr(V_{0.25}Ni_{0.75})_2$ metal hydride as active electrode material. *J. Alloys Compd.* **1996**, *239*, 175–182. [CrossRef]

32. Yu, J.Y.; Lei, Y.Q.; Chen, C.P.; Wu, J.; Wang, Q.D. The electrochemical properties of hydrogen storage Zr-based Laves phase alloys. *J. Alloys Compd.* **1995**, *231*, 578–581. [CrossRef]

33. Gao, X.; Song, D.; Zhang, Y.; Zhou, Z.; Yang, H.; Zhang, W.; Shen, P.; Wang, M. Characteristics of the superstoichiometric C15-type Laves phase alloys and their hydride electrodes. *J. Alloys Compd.* **1995**, *231*, 582–586. [CrossRef]

34. Yoshida, M.; Ishibashi, H.; Susa, K.; Ogura, T.; Akiba, E. Crystal structure, hydrogen absorbing properties and electrode performances of Sc-based Laves phase alloys. *J. Alloys Compd.* **1995**, *226*, 161–165. [CrossRef]

35. Nakano, H.; Wakao, S. Substitution effect of elements in Zr-based alloys with Laves phase of nickel-hydride battery. *J. Alloys Compd.* **1995**, *231*, 587–593. [CrossRef]

36. Züttel, A.; Chartouni, D.; Gross, K.; Bächler, M.; Schlapbach, L. Structural- and hydriding-properties of the $Zr(V_{0.25}Ni_{0.75})\alpha$ ($1 \leq \alpha \leq 4$) alloys system. *J. Alloys Compd.* **1997**, *253-254*, 587–589. [CrossRef]

37. Sun, D.; Latroche, M.; Percheron-Guégan, A. Effects of lanthanum or cerium on the equilibrium of $ZrNi_{1.2}Mn_{0.6}V_{0.2}Cr_{0.1}$ and its related hydrogenation properties. *J. Alloys Compd.* **1997**, *248*, 215–219. [CrossRef]

38. Kim, D.; Lee, S.; Jang, K.; Lee, J. The electrode characteristics of over-stoichiometric $ZrMn_{0.5}V_{0.5}Ni_{1.4+y}$ (y = 0.0, 0.2, 0.4 and 0.6) alloys with C15 Laves phase structure. *J. Alloys Compd.* **1998**, *268*, 241–247. [CrossRef]

39. Kim, D.; Lee, S.; Jung, J.; Jang, K.; Lee, J. Electrochemical properties of over-stoichiometric $ZrMn_{1-x}V_xNi_{1.4+y}$ alloys with C15 Laves phase. *J. Electrochem. Soc.* **1998**, *145*, 93–98. [CrossRef]

40. Knosp, B.; Jordy, C.; Blanchard, P.; Berlureau, T. Evaluation of $Zr(Ni, Mn)_2$ Laves phase alloys as negative active material for Ni-MH electric vehicle batteries. *J. Electrochem. Soc.* **1998**, *145*, 1478–1482. [CrossRef]

41. Knosp, B.; Vallet, L.; Blanchard, P. Performance of an AB_2 alloy in sealed Ni-MH batteries for electric vehicles: quantification of corrosion rate and consequences on the battery performance. *J. Alloys Compd.* **1999**, *293-295*, 770–774. [CrossRef]

42. Lupu, D.; Biris, A.R.; Indrea, E.; Biris, A.S.; Nele, G.; Schlapbach, L.; Züttle, A. Hydrogen absorption and hydride electrode behaviour of the Laves phase $ZrV_{1.5-x}Cr_xNi_{1.5}$. *J. Alloys Compd.* **1999**, *291*, 289–294. [CrossRef]

43. Chen, L.; Wu, F.; Tong, M.; Chen, D.M.; Long, R.B.; Shang, Z.Q.; Liu, H.; Sun, W.S.; Yang, K.; Wang, L.B.; et al. Advanced nanocrystalline Zr-based AB_2 hydrogen storage electrode materials for NiMH EV batteries. *J. Alloys Compd.* **1999**, *293-295*, 508–520. [CrossRef]

44. Yang, X.G.; Zhang, W.K.; Lei, Y.Q.; Wang, Q.D. Electrochemical properties of Zr-V-Ni system hydrogen storage alloys. *J. Electrochem. Soc.* **1999**, *146*, 1245–1250. [CrossRef]

45. Gao, X.; Sun, X.; Toyoda, E.; Higuchi, H.; Nakagima, T.; Suda, S. Deterioration of Laves phase alloy electrode during cycling. *J. Power Sources* **1999**, *833*, 100–107. [CrossRef]

46. Lupu, D.; Biriş, A.S.; Biriş, A.R.; Mişan, I.; Indrea, E. Hydrogen absorption and electrode properties of $Zr_{1-x}Ti_xV_{1.2}Cr_{0.3}Ni_{1.5}$ Laves phases. *J. Alloys Compd.* **2000**, *312*, 302–306. [CrossRef]

47. Hsu, Y.; Chiou, S.; Peng, T. Electrochemical hydrogenation behavior of C15-type $Zr(Mn, Ni)_2$ alloy electrode. *J. Alloys Compd.* **2000**, *313*, 263–268. [CrossRef]

48. Zhang, H.; Lei, Y.; Li, D. Electrochemical performance of $ZrMn_{0.5}V_{0.4}Ni_{1.1}Co_x$ Laves phase alloy electrode. *J. Power Sources* **2001**, *99*, 48–53. [CrossRef]

49. Cao, J.; Gao, X.; Lin, D.; Zhou, X.; Yuan, H.; Song, D.; Shen, P. Activation behavior of the Zr-based Laves phase alloy electrode. *J. Power Sources* **2001**, *93*, 141–144.

50. Yang, K.; Chen, D.; Chen, L.; Guo, Z.X. Microstructure, electrochemical performance and gas-phase hydrogen storage property of $Zr_{0.9}Ti_{0.1}[(Ni,V,Mn)_{0.95}Co_{0.05}]_\alpha$ Laves phase alloys. *J. Alloys Compd.* **2002**, *333*, 184–189. [CrossRef]

51. Liu, H.; Li, R. Effect of Sn content on properties of AB_2 hydrogen storage alloy. *Foundry Technol.* **2006**, *27*, 503–505. (In Chinese)

52. Huot, J.; Akiba, E.; Ogura, T.; Ishido, Y. Crystal structure, phase abundance and electrode performance of Laves phase compounds $(Zr, A)V_{0.5}Ni_{1.1}Mn_{0.2}Fe_{0.2}$ ($A \equiv$ Ti, Nb or Hf). *J. Alloys Compd.* **1995**, *218*, 101–109. [CrossRef]

53. Joubert, J.; Latroche, M.; Percheron-Guégan, A.; Bouet, J. Improvement of the electrochemical activity of Zr-Ni-Cr Laves phase hydride electrode by secondary phase precipitation. *J. Alloys Compd.* **1996**, *240*, 219–228. [CrossRef]

54. Nakano, H.; Wakao, S.; Shimizu, T. Correlation between crystal structure and electrochemical properties of C14 Laves-phase alloys. *J. Alloys Compd.* **1997**, *253-254*, 609–612. [CrossRef]

55. Zhang, Q.A.; Lei, Y.Q.; Yang, X.G.; Ren, K.; Wang, Q.D. Annealing treatment of AB_2-type hydrogen storage alloys: II. Electrochemical properties. *J. Alloys Compd.* **1999**, *292*, 241–246. [CrossRef]

56. Song, X.; Zhang, X.; Leo, Y.; Wang, Q. Effect of microstructure on the properties of Zr-Mn-V-Ni AB_2 type hydride electrode alloys. *Int. J. Hydrogen Energy* **1999**, *24*, 455–459. [CrossRef]

57. Du, Y.L.; Yang, X.G.; Zhang, Q.A.; Lei, Y.Q.; Zhang, M.S. Phase structures and electrochemical properties of the Laves phase hydrogen storage alloys $Zr_{1-x}Ti_x(Ni_{0.6}Mn_{0.3}V_{0.1}Cr_{0.05})_2$. *Int. J. Hydrogen Energy* **2001**, *26*, 333–337. [CrossRef]

58. Iosub, V.; Joubert, J.; Latroche, M.; Cerny, R.; Percheron-Guegan, A. Hydrogen cycling induced diffraction peak broadening in C14 and C15 Laves phases. *J. Solid State Chem.* **2005**, *178*, 1799–1806. [CrossRef]

59. Banerjee, S.; Kumar, A.; Pillai, C.G.S. Improvement on the hydrogen storage properties of $ZrFe_2$ Laves phase alloy by vanadium substitution. *Intermetallics* **2014**, *51*, 30–36. [CrossRef]

60. Wu, T.; Xue, X.; Zhang, T.; Hu, R.; Kou, H.; Li, J. Microstructures and hydrogenation properties of $(ZrTi)(V_{1-x}Al_x)_2$ Laves phase intermetallic compounds. *J. Alloys Compd.* **2015**, *645*, 358–368. [CrossRef]

61. Young, K.; Ouchi, T.; Huang, B.; Chao, B.; Fetcenko, M.A.; Bendersky, L.A.; Wang, K.; Chiu, C. The correlation of C14/C15 phase abundance and electrochemical properties in the AB_2 alloys. *J. Alloys Compd.* **2010**, *506*, 841–848. [CrossRef]

62. Young, K.; Nei, J.; Ouchi, T.; Fetcenko, M.A. Phase abundances in AB_2 metal hydride alloys and their correlations to various properties. *J. Alloys Compd.* **2011**, *509*, 2277–2284. [CrossRef]

63. Young, K.; Ouchi, T.; Lin, X.; Reichman, B. Effects of Zn-addition to C14 metal hydride alloys and comparisons to Si, Fe, Cu, Y, and Mo-additives. *J. Alloys Compd.* **2016**, *655*, 50–59. [CrossRef]

64. Young, K.; Ouchi, T.; Huang, B.; Reichman, B.; Fetcenko, M.A. Studies of copper as a modifier in C14-predominant AB_2 metal hydride alloys. *J. Power Sources* **2012**, *204*, 205–212. [CrossRef]

65. Young, K.; Ouchi, T.; Huang, B.; Reichman, B.; Fetcenko, M.A. The structure, hydrogen storage, and electrochemical properties of Fe-doped C14-predominating AB_2 metal hydride alloys. *Int. J. Hydrogen Energy* **2011**, *36*, 12296–12304. [CrossRef]

66. Young, K.; Ouchi, T.; Huang, B.; Reichman, B. Effect of molybdenum content on structural, gaseous storage, and electrochemical properties of C14-predominant AB_2 metal hydride alloys. *J. Power Sources* **2011**, *196*, 8815–8825. [CrossRef]

67. Yartys, V.A. Ti-Zr Based AB_2 Alloys for High Power Metal Hydride Batteries. In Proceedings of the 15th International Symposium on Metal-Hydrogen System, Interlaken, Switzerland, 7–12 August 2016.

68. Wan, C.; Ju, X.; Wang, Y. EXAFS characterization of TiVCrMn hydrogen storage alloy upon hydrogen absorption-desorption cycles. *Int. J. Hydrogen Energy* **2012**, *37*, 990–994. [CrossRef]

69. Kim, S.R.; Lee, J.Y. Activation behaviour of $ZrCrNiM_{0.05}$ metal hydride electrodes (M = La, Mm (misch metal), Nd). *J. Alloys Compd.* **1992**, *185*, L1–L4. [CrossRef]

70. Young, K.; Wong, D.F.; Ouchi, T.; Huang, B.; Reichman, B. Effects of La-addition to the structure, hydrogen storage and electrochemical properties of C14 metal hydride alloys. *Electrochim. Acta* **2015**, *174*, 815–825. [CrossRef]

71. Visintin, A.; Peretti, A.A.; Fruiz, F.; Corso, H.L.; Triaca, W.E. Effect of additional catalytic phases imposed by sintering on the hydrogen absorption behavior of AB_2 type Zr-based alloys. *J. Alloys Compd.* **2007**, *428*, 244–251. [CrossRef]

72. Zhang, W.K.; Ma, C.A.; Yang, X.G.; Lei, Y.Q.; Wang, Q.D.; Lu, G.L. Influences of annealing heat treatment on phase structure and electrochemical properties of $Zr(MnVNi)_2$ hydrogen storage alloys. *J. Alloys Compd.* **1999**, *293–295*, 691–697. [CrossRef]

73. Rennert, P.; Radwan, A.M. Structural investigation of the Laves phase $MgZn_2$ with model potential calculations. *Phys. Status Solidi (b)* **1977**, *79*, 167–173. [CrossRef]

74. Liu, Y.; Young, K. Microstructure investigation on metal hydride alloys by electron backscatter diffraction technique. *Batteries* **2016**, *2*. [CrossRef]

75. Young, K.; Ouchi, T.; Banik, A.; Koch, J.; Fetcenko, M.A. Improvement in the electrochemical properties of gas atomized AB_2 metal hydride alloys by hydrogen annealing. *Int. J. Hydrogen Energy* **2011**, *36*, 3547–3555. [CrossRef]

76. Boettinger, W.J.; Newbury, D.E.; Wang, K.; Bendersky, L.A.; Chiu, C.; Kattner, U.R.; Young, K.; Chao, B. Examination of multiphase (Zr, Ti)(V, Cr, Mn, Ni)$_2$ Ni-MH electrode alloys: Part I. Dendritic solidification structure. *Metall. Mater. Trans. A* **2010**, *41*, 2033–2047. [CrossRef]

77. Bendersky, L.A.; Wang, K.; Boettinger, W.J.; Newbury, D.E.; Young, K.; Chao, B. Examination of multiphase (Zr, Ti)(V, Cr, Mn, Ni)$_2$ Ni-MH electrode alloys: Part II. Solid-state transformation of the interdendric B2 phase. *Metall. Mater. Trans. A* **2010**, *41*, 1891–1906. [CrossRef]

78. Young, K.H.; Fetcenko, M.A.; Ovshinsly, S.R.; Ouchi, T.; Reichman, B.; Mays, W. Improved surface catalysis of Zr-based Laves phase alloys for NiMH Batteries. In *Hydrogen at Surface and Interface*; Jerkiewicz, G., Feliu, J.M., Popov, B.N., Eds.; The Electrochemical Society, Inc.: New Jersey, USA, 2000; Volume 2000-16, pp. 59–71.

79. Young, K.; Regmi, R.; Lawes, G.; Ouchi, T.; Reichman, B.; Fetcenko, M.A.; Wu, A. Effects of aluminum substitution in C14-rich multi-component alloys for NiMH battery applications. *J. Alloys Compd.* **2010**, *490*, 282–292. [CrossRef]

80. Young, K.; Fetcenko, M.A.; Ouchi, T.; Li, F.; Koch, J. Effect of Sn-substitution in C14 Laves phase alloys for NiMH battery application. *J. Alloys Compd.* **2009**, *469*, 406–416. [CrossRef]

81. Young, K.; Ouchi, T.; Reichman, B.; Mays, W.; Regmi, R.; Lawes, G.; Fetcenko, M.A.; Wu, A. Optimization of Co-content in C14 Laves phase multi-component alloys for NiMH battery application. *J. Alloys Compd.* **2010**, *489*, 202–210. [CrossRef]

82. Young, K.; Ouchi, T.; Huang, B.; Reichman, B.; Blankenship, R. Improvement in -40 °C electrochemical properties of AB$_2$ metal hydride alloy in silicon incorporation. *J. Alloys Compd.* **2013**, *575*, 65–72. [CrossRef]

83. Young, K.; Young, M.; Ouchi, T.; Reichman, B.; Fetcenko, M.A. Improvement in high-rate dischargeability, activation, and low-temperature performance in multi-phase AB$_2$ alloys by partial substitution of Zr with Y. *J. Power Sources* **2012**, *215*, 279–287. [CrossRef]

84. Young, K.; Fetcenko, M.A.; Koch, J.; Morii, K.; Shimizu, T. Studies of Sn, Co, Al, and Fe additives in C14/C15 Laves alloys for NiMH battery application by orthogonal arrays. *J. Alloys Compd.* **2009**, *486*, 559–569. [CrossRef]

85. Young, K.; Huang, B.; Ouchi, T. Studies of Co, Al, and Mn substitutions in NdNi$_5$ metal hydride alloys. *J. Alloys Compd.* **2012**, *543*, 90–98. [CrossRef]

86. Young, K.; Ouchi, T.; Huang, B. Effects of various annealing conditions on (Nd, Mg, Zr)(Ni, Al, Co)$_{3.74}$ metal hydride alloys. *J. Power Sources* **2014**, *248*, 147–153. [CrossRef]

87. Young, K.; Ouchi, T.; Huang, B.; Nei, J. Structure, hydrogen storage, and electrochemical properties of body-centered-cubic Ti$_{40}$V$_{30}$Cr$_{15}$Mn$_{13}$X$_2$ alloys (X = B, Si, Mn, Ni, Zr, Nb, Mo, and La). *Batteries* **2015**, *1*, 74–90. [CrossRef]

88. Wong, D.F.; Young, K.; Nei, J.; Wang, L.; Ng, K.Y.S. Effects of Nd-addition on the structural, hydrogen storage, and electrochemical properties of C14 metal hydride alloys. *J. Alloys Compd.* **2015**, *647*, 507–518. [CrossRef]

89. Young, K.; Ouchi, T.; Meng, T.; Wong, D.F. Studies on the synergetic effects in multi-phase metal hydride alloys. *Batteries* **2016**, *2*. [CrossRef]

90. Mosavati, N.; Young, K.; Meng, T.; Ng, K.Y.S. Electrochemical open-circuit voltage and pressure-concentration-temperature isotherm comparison for metal hydride alloys. *Batteries* **2016**, *2*. [CrossRef]

91. Young, K.; Ouchi, T.; Fetcenko, M.A. Pressure-composition-temperature hysteresis in C14 Laves phase alloys: Part 1. Simple ternary alloys. *J. Alloys Compd.* **2009**, *480*, 428–433. [CrossRef]

92. Trapanese, M.; Franzitta, V.; Viola, A. Description of hysteresis of nickel metal hydride battery. In Proceedings of the 38th Annual Conference on IEEE Industrial Electronics Society, Montreal, QC, Canada, 25–28 October 2012; pp. 967–970.

93. Trapanese, M.; Franzitta, V.; Viola, A. Description of hysteresis in lithium battery by classical Preisach model. *Adv. Mater. Res.* **2013**, *622*, 1099–1103.

94. Trapanese, M.; Franzitta, V.; Viola, A. The Jiles Atherton model for description on hysteresis in lithium battery. In Proceedings of the Twenty-Eighth Annual IEEE Applied Power Electronics Conference and Exposition (APEC), Long Beach, CA, USA, 17–21 March 2013; pp. 2772–2775.

95. Schwarz, R.B.; Khachaturyan, A.G. Thermodynamics of open two-phase systems with coherent interface. *Phys. Rev. Lett.* **1995**, *74*, 2523–2526. [CrossRef]

96. Shen, H.T.; Young, K.H.; Meng, T.; Bendersky, L.A. Clean grain boundary found in C14/body-center-cubic multi-phase metal hydride alloys. *Batteries* **2016**, *2*. [CrossRef]

97. Lide, D.R. *CRC Handbook of Chemistry and Physics*, 74th ed.; CRC Press: Boca Raton, FL, USA, 1993; pp. 6–22.

98. Young, K.; Fetcenko, M.A.; Li, F.; Ouchi, T. Structural, thermodynamic, and electrochemical properties of Ti$_x$Zr$_{1-x}$(VNiCeMnCiAl)$_2$ C14 Laves phase alloys. *J. Alloys Compd.* **2008**, *464*, 238–247. [CrossRef]

99. Young, K.; Ouchi, T.; Fetcenko, M.A. Roles of Ni, Cr, Mn, Sn, Co, and Al in C14 Laves phase alloys for NiMH battery application. *J. Alloys Compd.* **2009**, *476*, 774–781. [CrossRef]

100. Young, K.; Ouchi, T.; Koch, J.; Fetcenko, M.A. Compositional optimization of vanadium-free hypo-stoichiometric AB$_2$ metal hydride alloy for Ni/MH battery application. *J. Alloys Compd.* **2012**, *510*, 97–106. [CrossRef]

101. Young, K.; Wong, D.F.; Nei, J. Effects of vanadium/nickel contents in Laves phase-related body-centered-cubic solid solution metal hydride alloys. *Batteries* **2015**, *1*, 34–53. [CrossRef]

102. Zheng, G.; Popov, B.N.; White, R.E. Electrochemical determination of the diffusion coefficient of hydrogen through an $LaNi_{4.25}Al_{0.75}$ electrode in alkaline aqueous solution. *J. Electrochem. Soc.* **1995**, *142*, 2695–2698.

103. Notten, P.H.L.; Hokkeling, P. Double-phase hydride forming compounds: A new class of highly electrocatalytic materials. *J. Electrochem. Soc.* **1991**, *138*, 1877–1885. [CrossRef]

104. Young, K.; Nei, J. The current status of hydrogen storage alloy development for electrochemical applications. *Materials* **2013**, *6*, 4574–4608. [CrossRef]

105. Young, K.; Reichman, B.; Fetcenko, M.A. Electrochemical properties of AB_2 metal hydride alloys measured at −40 °C. *J. Alloys Compd.* **2013**, *580*, S349–S353. [CrossRef]

106. Young, K.; Huang, B.; Regmi, R.K.; Lawes, G.; Liu, Y. Comparisons of metallic clusters imbedded in the surface of AB_2, AB_5, and A_2B_7 alloys. *J. Alloys Compd.* **2010**, *506*, 831–840. [CrossRef]

batteries

MDPI

Article

Hydrogen Storage Characteristics and Corrosion Behavior of Ti$_{24}$V$_{40}$Cr$_{34}$Fe$_{2}$ Alloy

Jimoh Mohammed Abdul [1,2,*], Lesley Hearth Chown [1], Jamiu Kolawole Odusote [3], Jean Nei [4], Kwo-Hsiung Young [4] and Woli Taiye Olayinka [2]

[1] School of Chemical and Metallurgical Engineering, Faculty of Engineering and Built Environment, University of the Witwatersrand, Johannesburg Private Bag 3, Wits 2050, South Africa; lhchown@gmail.com
[2] Department of Mechanical Engineering, Faculty of Engineering, Federal Polytechnic, Offa 24013, Nigeria; taiyewoli@gmail.com
[3] Department of Materials and Metallurgical Engineering, Faculty of Engineering and Technology, University of Ilorin, Ilorin 240003, Nigeria; jamiukolawole@gmail.com
[4] BASF/Battery Materials-Ovonic, 2983 Waterview Drive, Rochester Hills, MI 48309, USA; jean.nei@basf.com (J.N.); kwo.young@basf.com (K.-H.Y.)
* Correspondence: jmabdul@gmail.com or jmabdul66@yahoo.com; Tel.: +1-234-805-562-9924 or +23490-848-80106

Academic Editor: Hua Kun Liu
Received: 14 February 2017; Accepted: 8 June 2017; Published: 14 June 2017

Abstract: In this work, we investigated the effects of heat treatment on the microstructure, hydrogen storage characteristics and corrosion rate of a Ti$_{34}$V$_{40}$Cr$_{24}$Fe$_{2}$ alloy. The arc melted alloy was divided into three samples, two of which were separately quartz-sealed under vacuum and heated to 1000 °C for 1 h; one of these samples was quenched and the other furnace-cooled to ambient temperature. The crystal structures of the samples were studied via X-ray diffractometry and scanning electron microscopy. Hydrogenation/dehydrogenation characteristics were investigated using a Sievert apparatus. Potentiostat corrosion tests on the alloys were performed using an AutoLab® corrosion test apparatus and electrochemical cell. All samples exhibited a major body-center-cubic (BCC) and some secondary phases. An abundance of Laves phases that were found in the as-cast sample reduced with annealing and disappeared in the quenched sample. Beside suppressing Laves phase, annealing also introduced a Ti-rich phase. The corrosion rate, maximum absorption, and useful capacities increased after both heat treatments. The annealed sample had the highest absorption and reversible capacity. The plateau pressure of the as-cast alloy increased after quenching. The corrosion rate increased from 0.0004 mm/y in the as-cast sample to 0.0009 mm/y after annealing and 0.0017 mm/y after quenching.

Keywords: Ti-V-Cr-Fe alloy; hydrogen storage characteristics; metal corrosion; heat treatment; crystal structure

---

1. Introduction

Ti-V-Cr body-centered-cubic (BCC) solid solution alloys are very promising for the storage of a large quantity of hydrogen at room temperature [1–3]. Some of the identified shortcomings of these alloys include poor pressure-composition-temperature (PCT) plateau characteristics, low hydrogen desorption capacities, and long activation times [4–7]. In an attempt to improve on these shortcomings, controlled quantities of additives such as Fe, Zr, and Mn have been found to be effective in lowering cost and enhancing the overall performance of the alloy [8–11]. The effects of substituting Fe for Cr on the hydrogen storage property of the Ti$_{0.32}$Cr$_{0.43}$V$_{0.25}$ alloy showed that the desorption plateau pressure increased without decrease in effective hydrogen capacity, suggesting the possibility of using ferrovanadium as a substitute for the expensive pure vanadium [12]. Increasing the V content in V_x–$(Ti–Cr–Fe)_{100-x}$ ($Ti/(Cr+Fe) = 1.0$, $Cr/Fe = 2.5$, $x = 20$–55) alloys led to an increase in both the hydrogen absorption capacity and desorption capacity, but decreased the plateau pressure [13]. Miao et al. [14] found that all of the alloys $Ti_{0.8}Zr_{0.2}V_{2.7}Mn_{0.5}Cr_{0.8-x}Ni_{1.25}Fe_x$ ($x = 0.0$–0.8) mainly consisted of two phases, the C14 Laves phase with a three-dimensional network and the dendritic V-based solid solution phase. Further, the lattice parameters of the two phases and the maximum discharge capacity decreased with an increase in Fe content, but the cyclic stability and the high rate dischargeability increased first and then decreased with increasing x. Liu et al. [15] found that Ce addition favored the chemical homogeneity of the BCC phase and, therefore, improved the hydrogen storage properties of the $(Ti_{0.267}Cr_{0.333}V_{0.40})_{93}Fe_7Ce_x$ ($x = 0$, 0.4, 1.1 and 2.0 at%) alloys. To increase the hydrogen storage capacity and the plateau pressure of the $Ti_{0.32}Cr_{0.43}V_{0.25}$ alloy, Yoo et al. [16] replaced a fraction of the Cr with Mn or a combination of Mn and Fe. When Mn was used alone, the effective hydrogen storage capacity increased to about 2.5 wt% though the plateau pressure showed no significant change. However when Fe was added with Mn, both the effective hydrogen storage capacity and the plateau pressure increased.

Further efforts include assessing the effects of heat treatment on hydrogen absorption properties; Okada et al. [1] found that moderate heat treatment, specifically annealing at 1573 K for 1 min, enhanced the storage capacity and flattened the desorption plateau of $Ti_{25}Cr_{40}V_{35}$ alloy. Liu et al. [17] reported that heat treatment effectively improved the flatness of the plateau and improved the hydrogenation capacity of $Ti_{32}Cr_{46}V_{22}$ alloy by lowering the oxygen concentration and homogenizing the composition and microstructure. A hydrogen desorption capacity of 2.3 wt % was achieved when $Ti_{32}Cr_{40}V_{25}$ was annealed at 1653 K for 1 min. [18]. Chuang et al. [19] found that annealing atomized powder of Ti-Zr based alloy at 1123 K for 4 h greatly enhanced the discharge capacity. Hang et al. [20] heat-treated $Ti_{10}V_{77}Cr_6Fe_6Zr$ alloy at a relatively lower temperature, but elongated the soaking time by annealing at 1523 K for 5 min and at 1373 K for 8 h, and found that the sample annealed at 1523 K for 5 min had the best overall hydrogen storage properties, with a desorption capacity of 1.82 wt % and a dehydriding plateau pressure of 0.75 MPa.

Although BCC solid solution alloys have very high gaseous phase hydrogen storage capacities, they suffer from severe capacity degradation during electrochemical applications due to the leaching of Vanadium (V) into the KOH electrolyte [21,22]. The preferential leaching of V in the negative electrode material has been previously identified [23], and V-free Laves phase alloys have been adopted to mitigate the consequent cycle life and self-discharge issues originating from V-corrosion [24,25].

Metal corrosion mainly occurs through electrochemical reactions at the interface between the metal and electrolyte [26]. The basic process of metallic corrosion in an aqueous solution consists of the anodic dissolution of metals and the cathodic reduction of oxidants present in the solution:

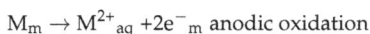

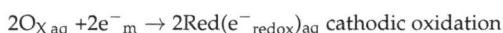

$$M_m \rightarrow M^{2+}{}_{aq} + 2e^-{}_m \text{ anodic oxidation}$$

$$2Ox_{aq} + 2e^-{}_m \rightarrow 2Red(e^-{}_{redox})_{aq} \text{ cathodic oxidation}$$

In the formulae, M_m is the metal in the state of metallic bonding; $M^{2+}{}_{aq}$ is the hydrated metal ion in an aqueous solution; $e^-{}_m$ is the electron in the metal; Ox_{aq} is an oxidant; $Red(e^-{}_{redox})_{aq}$ is a reductant; and $e^-{}_{redox}$ is the redox electron in the reductant. The overall corrosion reaction is then written as follows:

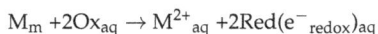

$$M_m + 2Ox_{aq} \rightarrow M^{2+}{}_{aq} + 2Red(e^-{}_{redox})_{aq}$$

These reactions are charge-transfer processes that occur across the interface between the metal and the aqueous solution, hence they are dependent on the interfacial potential that essentially corresponds to what is called the electrode potential of metals in electrochemistry terms. In physics terms, the electrode potential represents the energy level of electrons, called the Fermi level, in an electrode immersed in electrolyte. For normal metallic corrosion, in practice, the cathodic process is carried out by the reduction of hydrogen ions and/or the reduction of oxygen molecules in an aqueous solution.

These two cathodic reductions are electron transfer processes that occur across the metal–solution interface, whereas anodic metal dissolution is an ion transfer process across the interface.

The rate of the reaction is evaluated in terms of the corrosion current. The natural logarithm of the absolute value of the corrosion current versus the potential value is plotted as a Tafel curve. The corrosion current values can be transformed to corrosion rate (*CR*) values (e.g., mm/y) using Equation (1) [27]:

$$CR = K \frac{i_{corr}}{\rho} EW \tag{1}$$

where K is a constant that depends on the unit of corrosion rate; K = 3272 for mm/y (mmpy), or = 1.288×10^5 for milli-inches/y (mpy), i_{corr} = corrosion current density ($\mu A \cdot Cm^{-2}$), ρ = alloy density $(g \cdot Cm^{-3})$, *EW* = Equivalent weight = 1/electron equivalent (*Q*) where:

$$Q = \sum \frac{n_i f_i}{W_i} \tag{2}$$

where n_i = the valence if *i*th element of the alloy, f_i = the mass fraction of the *i*th element in the alloy, Wi = the atomic weight of the *i*th element in the alloy.

By combining Equations (1) and (2), the penetration rate (*CR* and mass loss, *ML*) of an alloy is given by:

$$CR = K_1 \frac{i_{corr}}{\rho \sum \frac{n_i f_i}{W_i}} \tag{3}$$

$$ML = K_2 i_{corr} EW \tag{4}$$

The equations above give values of 0.1288 if the unit of *CR* is m/y and 0.03272 if the unit of *CR* is mm/y [28]. This work investigates the influence of heat treatment on hydrogen storage capacity and corrosion rate of $Ti_{34}V_{40}Cr_{24}Fe_2$ in standard KOH electrolyte. V-based hydrogen storage alloys are often used as the anode in NiMH batteries [29].

Cho et al. [29] identified a composition region in the Ti-V-Cr phase diagram as having the highest hydrogen uptake. This informed the choice of the base alloy $Ti_{25}V_{40}Cr_{35}$, which falls within the region. The choice to use Fe as the additive was made because of its relatively low cost. Literature has shown that in the ternary Ti-V-Cr system [12–16,30], Fe substituted Cr or V and FeV was used in place of expensive V. Thus, research gap is noted, as the effect of substituting Fe at the Ti site on hydrogen storage and corrosion behavior of $Ti_{25}V_{40}Cr_{35}$ has not yet been investigated. The present work on substituting an equal quantity of Fe for Cr and Ti on the hydrogen storage and corrosion behavior of $Ti_{25}V_{40}Cr_{35}$ therefore fills this gap.

2. Experimental Setup

The raw materials for this work were sourced from Metrohm, South Africa, including iron (325–290 mesh, 99% purity, 0.01% C and 0.015% P and S); chromium (<0.3mm, 99.8% purity); vanadium (−325 mesh, 99.5% purity); and titanium (−325 mesh, 99.5% purity). A 10-g sample of $Ti_{34}V_{40}Cr_{24}Fe_2$ alloy was prepared in a water-cooled, copper-crucible arc melting furnace under argon atmosphere. The ingot was turned over and remelted three times to ensure homogeneity. After melting, the ingot was divided into three pieces. Two cut samples were vacuum-sealed in separate silica glass tubes in preparation for heat treatment.

The two quartz-sealed specimens were loaded in a heat treatment furnace and heated to 1000 °C for 1 h. One tube was immediately removed and broken in cold water to quench the alloy, thereby locking the microstructure, while the second sample was slowly furnace-cooled to room temperature.

The crystal structure and lattice parameters in the as-cast and heat-treated samples were determined by X-ray diffraction (XRD) analysis, using a Bruker D2_Phaser X-ray® diffraction instrument (Bruker AXS, Inc., Madison, WI, USA) with Cu-K$_\alpha$ radiation from 2θ = 10° to 80°. Xpert High Score® phase identification software produced by Philips analytical B.V. Almelo Netherlands was used to identify the phases from the XRD data. Elemental compositions of the phases were determined using an FEI Nova NanoSEM 200® scanning electron microscope (SEM) (FEI, Hillsboro, OR, USA) fitted with EDAX® advanced microanalysis solution (EDAX Inc., Mahwah, NJ, USA). The amount of phases was determined by image analysis using the ImageJ freeware (National Institute of Mental Health, Bethesda, MD, USA).

Pottentiostatic corrosion tests on the alloys were performed using an AutoLab® corrosion test apparatus (Metrohm Autolab B.V., Utrecht, The Netherlands) and an electrochemical cell consisting of a tri-electrode (a platinum reference electrode, Ag/AgCl counter electrode, and 0.14 cm^2 test alloy as the working electrode). An aqueous solution of 6 mol·L^{-1} KOH was used as the electrolyte. The alloys were cut into rectangles and a copper wire of suitable length was attached to one side of the specimen with aluminum tape. The sample was then covered in cold-resin for 24 h to enable curing, while leaving only the test surface exposed. When cured, the test surface was ground to 1200 grit. The corrosion experiments were performed at 25 °C and Tafel curves were recorded from -1.4 V to -0.2 V with a scanning rate of 1 mV·sec^{-1}.

Measurement of the Pressure Composition Temperature, PCT isotherms was performed using a Suzuki-Shokan multi-channel PCT (Suzuki Shokan, Tokyo, Japan) system. Samples were crushed into particles <75 μm in size, and 1 g of each alloy was sealed into a stainless steel reactor and heated to 573 K. Next, 3 MPa of hydrogen pressure was introduced into the apparatus for 30 min, followed by slow cooling to room temperature in a hydrogen atmosphere. The alloys absorbed most of the hydrogen and were pulverized in this step. After the hydrogenation process, the samples were heated to 573 K and chamber was evacuated for 1 h with a mechanical pump to completely dehydrogenate the samples for PCT measurements at 303, 333, and 363 K successively.

3. Results and Discussion

3.1. Microstrcuture

The XRD pattern in Figure 1 shows mainly a BCC (V) phase and some minor peaks from secondary phases. Phase analysis indicated that both C14 and C15 phases co-existed in the as-cast sample and transformed into an α-Ti phase after annealing (Table 1). Changes in the lattice parameters of the BCC main phase and the α-Ti secondary phase with the different preparation methods are negligible. The annealed sample showed the least amount of α-Ti.

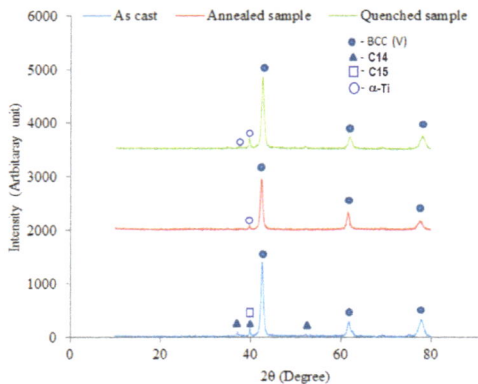

Figure 1. XRD patterns of the as-cast and heat-treated Ti$_{34}$V$_{40}$Cr$_{24}$Fe$_2$ alloys.

Table 1. XRD crystallographic parameters of the as-cast and heat-treated $Ti_{34}V_{40}Cr_{24}Fe_2$.

Sample	Phases	Space Group (No.)	*A*	*c*	Unit Cell Volume ($Å^3$)
			Crystallographic Description		
As-cast	BCC (V)	$Im\overline{3}m$ (229)	3.00		27.08
	C14 (Laves)	$P6_3/mmc$ (194)	4.85	7.94	161.7
	C15 (Laves)	$Fd\overline{3}m$ (227)	6.943		334.69
Annealed	BCC (V)	$Im\overline{3}m$ (229)	3.01		27.37
	α-Ti	$P6_3/mmc$ (194)	2.98	4.73	36.24
Quenched	BCC (V)	$Im\overline{3}m$ (229)	3.01		27.37
	α-Ti	$P6_3/mmc$ (194)	2.98	4.73	36.24

Figure 2 shows the representative back scattering electron (BSE) images of the as-cast and heat-treated (both slow cooled and quenched) $Ti_{34}V_{40}Cr_{24}Fe_2$. The microstructure of the alloy was a primary, light grey phase (C) with some darker intergranular phases (D and E) in all three samples. The primary light grey phase corresponds to the main peak of BCC (V) phase, while the black intergranular and the dark phases were the Laves and Ti-rich secondary phases. X-ray energy dispersive spectroscopy (EDS) was used to measure compositions in the areas of interest, and the results are summarized in Table 2. The EDS technique measures the average composition within 1–2 microns volume (depending on the primary electron energy) due to the nature of electron scattering in the solid. The secondary phases in this study are either below or about that range, and therefore EDS result can only indicate relative change in composition. In the as-cast sample, areas D and E have relatively higher Ti-contents with a smaller atomic weight, which resulted in a darker contrast. Areas D and E are assigned to C14 and C15 from the combined results of the XRD and EDS measurements. In the annealed sample, C15 phase transformed into C14 phase (area D) and a Ti-rich phase (area E) started to appear. We believe this Ti-rich phase can be assigned to the α-Ti phase found in the XRD analysis. The α-Ti phase only has a very small solubility of Cr and V [31]. The high V and Cr contents found in the EDS data in Table 2 can be explained by either the small grain size of Ti-phase, which is below the sampling volume of the EDS technique, or a mixture of microcrystalline Ti/C15 that occurred as the product of a Eutectic solidification at 667 °C [32]. Quenching introduced twinning in the secondary phase, as shown inside the circle (F). The addition of Fe into the TiVCr solid solution is known to promote the secondary Laves phase [33–35], which is considered to be a catalyst that facilitates hydrogenation/dehydrogenation kinetics [36]. The reduction in the Laves phase abundance by thermal annealing was reported before with a Laves phase-related BCC TiZrV-based alloy [37].

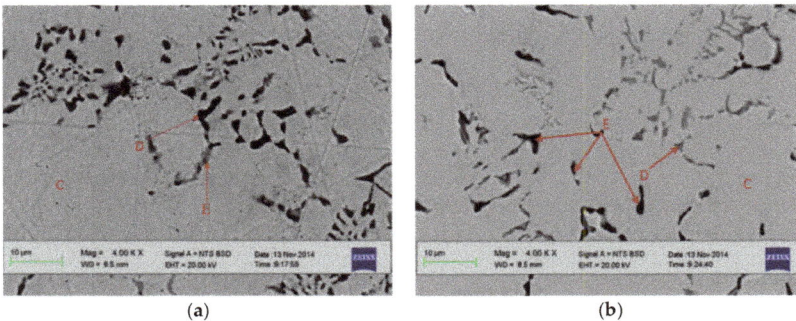

(a)

(b)

Figure 2. *Cont.*

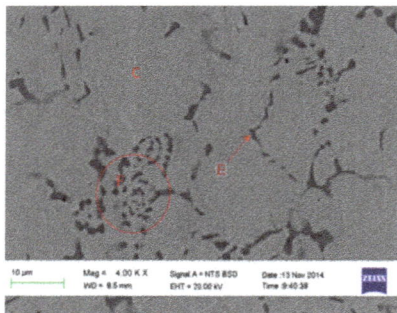

(c)

Figure 2. SEM BSE micrographs of (**a**) the as-cast, (**b**) annealed, and (**c**) quenched $Ti_{34}V_{40}Cr_{24}Fe_2$.

Table 2 shows that the BCC (V) in all three samples contained 17–22 at% Ti, 40–44 at% V, and ~36 at% Cr. The black intergranular E areas in Figure 2b,c contained 23–66 at% Ti, 18–41 at% V, and 14–36 at% Cr. The heat treatment reduced the Laves phase abundance from 21.5 to 18% and the quenching totally removed it. The Ti/Cr ratio in the BCC phase is almost equal in all three samples. However, the ratio of the darker areas (containing Laves phase and Ti-phase) increased from 0.6 to 4.4 after annealing and to ~2 after quenching.

Table 2. EDS results of the as-cast and heat-treated $Ti_{34}V_{40}Cr_{24}Fe_2$.

Sample	Location	Composition (at%)				Ti/Cr Ratio	Phase Identification	Phase Proportion (% Area)
		Ti	V	Cr	Fe			
As-cast	C	17.7 (1.1)	43.5 (1.6)	36.5 (2.2)	2.3 (0.7)	0.5	BCC	78.5
	D	24.7 (0.8)	37.8 (0.7)	34.0 (0.4)	3.5 (0.3)	0.7	C14	1.3
	E	23.0 (3.7)	41.4 (2.4)	35.5 (1.4)	2.0 (0.2)	0.6	C15	20.2
Annealed	C	17.7 (1.9)	43.5 (1.3)	36.5 (2.9)	2.3 (0.2)	0.5	BCC	82.0
	E	65.9 (5.2)	18.6 (2.8)	14.9 (2.2)	0.6 (0.2)	4.4	α-Ti/BCC	12.0
Quenched	C	21.8 (1.5)	40.0 (1.3)	35.6 (1.2)	2.6 (1.7)	0.6	BCC	80.5
	E	47.3 (3.7)	21.6 (2.5)	24.7 (1.4)	6.4 (0.5)	1.9	α-Ti/BCC	19.5

3.2. Gaseous Phase Hydrogen Storage

Table 3 shows that both heat treatment processes increased the reversible hydrogen storage capacity (RHSC); the capacity increased from 0.44 to 1.26 wt % after annealing and to 0.65 wt % in the quenched sample. Literature indicates that annealing increases hydrogen capacity [13,14]. The BCC phase enhances hydrogen capacity [11,22,29]; an increase in its unit cell volume implies the availability of more hydrogen absorption sites or spaces, leading to an increase in storage capacity. The Laves phase is detrimental to storage capacity [10,38–40], and alloys with larger Laves unit cell volumes or high Laves proportions are known to exhibit lower hydrogen capacity. Therefore, the observed increase in useful capacity after heat treatment was due to the increase in the BCC unit cell volume. The abundance of catalytic Laves phase in the as-cast sample reduced after annealing; this suggests another reason for the higher hydrogen capacity observed in the annealed sample.

Table 3 also shows that the desorption plateau pressure reduced after annealing the alloy at 1000 °C for 1 h, while it rose after quenching. Specifically, the pressure of the desorption plateau decreased from 1.32 to 0.68 MPa after annealing and increased to 2.34 MPa after quenching. The plateau properties are affected by both the homogeneity and oxygen content in Ti-V based hydrogen storage alloys [10]. Inhomogeneity of microstructure can be minimized by heat treatment at higher temperatures for a short time [41]. The observed reduction in the plateau pressure after annealing suggests a homogenized microstructure; in addition, the sample was annealed in a vacuum-sealed quartz tube, thereby preventing oxygen intake. The quenched sample was exposed to oxygen in the quenching medium; therefore, the rise in plateau pressure suggests the presence of a higher oxygen content.

Table 3. Effect of heat treatment on H storage properties of $Ti_{34}V_{40}Cr_{24}Fe_2$ alloy.

Sample	Absorption Capacity (wt %)	Capacity Remaining (wt %)	RHSC (wt %)	Plateau Pressure (MPa)
As-cast	0.86	0.42	0.44	1.32
Annealed	1.54	0.28	1.26	0.68
Quenched	1.04	0.39	0.65	2.34

In Figure 3a, the PCT isotherms for both the as-cast and heat-treated alloys were steep, an indication of high plateau pressure. However, the isotherm for the annealed sample showed a flatter and wider plateau, indicating a reduction in plateau pressure, and the wider plateau also indicated higher hydrogen capacity.

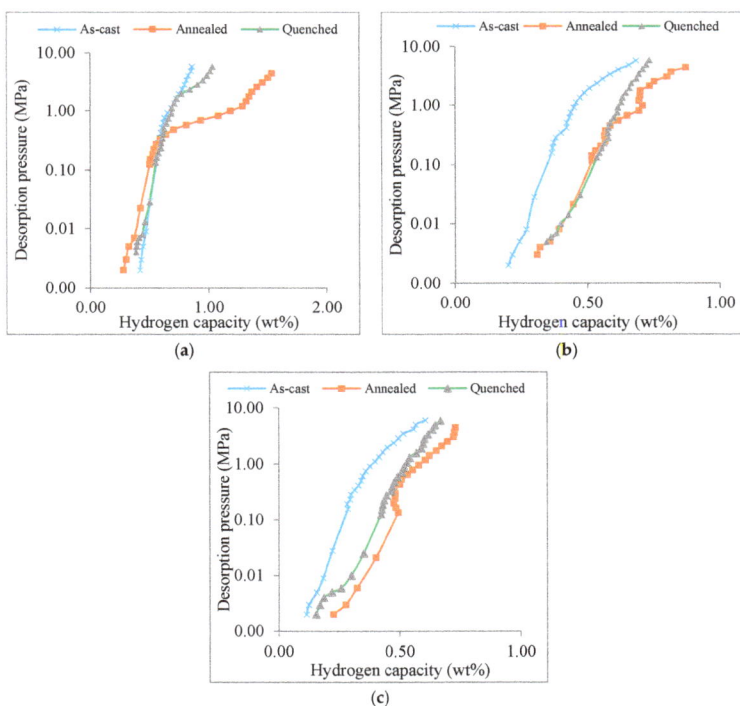

Figure 3. PCT desorption curves of the as-cast and heat-treated $Ti_{34}V_{40}Cr_{24}Fe_2$ alloys at (**a**) 303, (**b**) 333, and (**c**) 363 K.

In Figure 4, the maximum absorption capacity decreased with increasing isotherm temperatures, similar to what has been previously described in the literature [42]. For the as-cast sample, the maximum absorption capacity was 0.86 wt % at 30 °C, and the capacity declined to 0.69 and 0.62 wt % as the temperature rose to 333 and 363 K, respectively. Similar trends were observed for the annealed and quenched samples. The kinetic energy of gas increased with increasing temperatures;

a low kinetic energy is associated with lower temperatures, while increased temperature leads to high kinetic energy. Hydrogen gas atoms with a low kinetic energy are more easily absorbed than those with high kinetic energy because gases with higher kinetic energy move faster, thus requiring additional force to attract to the surface of the adsorbate. This explains the observed higher capacity at low temperatures and lower capacity at high temperatures.

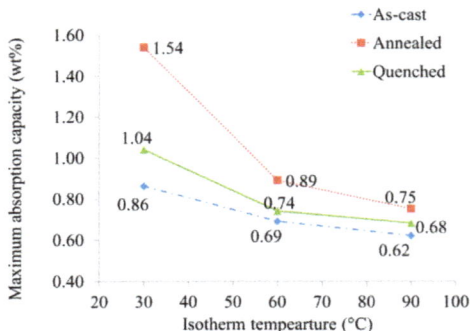

Figure 4. Influence of isotherm temperatures on maximum hydrogen capacity.

The stability of the current alloy can be improved with the addition of Ni to promote other secondary phases, such as TiNi [43] and Ti_2Ni [44]. The synergetic interaction between the main storage phase (BCC) and secondary phases can reduce the equilibrium plateau pressure of the BCC phase and make them available for electrochemical storage purposes [45].

3.3. Corrosion Behavior

The corrosion behavior of $Ti_{34}V_{40}Cr_{24}Fe_2$ alloys in 30% KOH aqueous solution was studied by Tafel curve measurements, and the results are plotted in Figure 5a. No significant change was observed in the corrosion potential (E_{corr}) for the as-cast alloy after annealing. However, a decrease from −0.80 to −0.867 mV was observed after quenching. Both heat treatment processes increased the corrosion rate of the alloy (Figure 5b). The rate increased from 0.0004 in the as-cast alloy to 0.0009 mm/y after annealing and to 0.0017 mm/y after quenching.

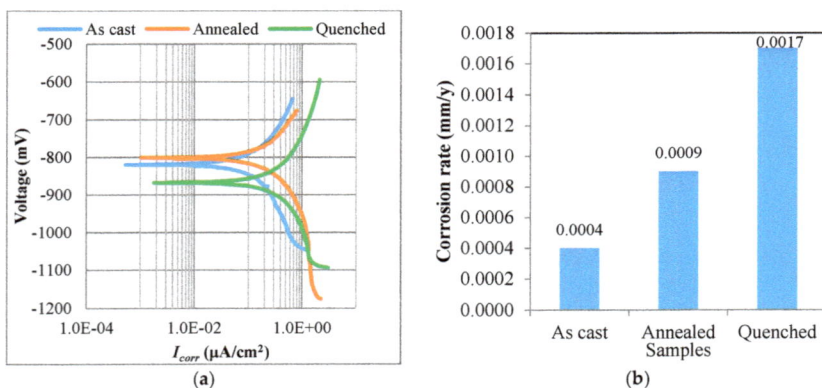

Figure 5. (a) Tafel curves and (b) corrosion rates of the as-cast and heat-treated $Ti_{34}V_{40}Cr_{24}Fe_2$.

Cr is known to improve the corrosion resistance of an alloy [40,41]. Samples containing high at % Cr are expected to have low corrosion rates and vice versa. Therefore, an increase in the corrosion rate of heat-treated samples could be a result of a reduction in Cr content in the secondary phase. The phase structure and oxide layer are among the factors that determine the corrosion rate of an alloy. The Laves structure has been reported to have a thinner oxide layer than the BCC structure [42], and this oxide layer is known for passivation of corrosion [43–45]. It is therefore implied that for a dual phase alloy, such as the alloys being investigated in this research, the corrosion rate will increase when the proportion of the Ti-rich phase with a thinner oxide layer increases. The Ti-rich phase is more easily corroded than the BCC phase [46].

4. Conclusions

The influence of heat treatment on the microstructure and hydrogen storage capacity of $Ti_{34}V_{40}Cr_{24}Fe_2$ at% alloys was investigated. The main phase of all alloys under different preparations is a BCC phase, while the secondary phase shifted from a Laves phase to a Ti-rich phase with annealing. Heat treatment was beneficial with regard to hydrogen capacity, but detrimental to corrosion behavior. Though both heat treatment processes enhanced useful hydrogen capacity, the annealed sample had superior storage characteristics. Although both annealing and quenching increased the corrosion rate of the alloy, the rate of corrosion was found to be highest in the quenched sample with the highest amount of Ti-rich phase.

Acknowledgments: The work is supported by African Material Science and Engineering Network, AMSEN, and the National Research Found NRF Thuthuka programme.

Author Contributions: Jimoh Mohammed Abdul and Lesley Hearth Chown designed the experiments and analyzed the results. Jamiu K. Odusote and Jean Nei conducted the corrosion and PCT measurements, respectively, and Kwo-Hsiung Young and Woli Taiye Olayinka, assisted in data analysis and manuscript preparation.

Conflicts of Interest: The authors declare no conflict of interest.

Abbreviations

BCC	Body-centered-cubic
PCT	Pressure-concentration-temperature
CR	Corrosion rate
i_{corr}	Corrosion current density
EW	Equivalent weight
ML	Mass loss
XRD	X-ray diffractometer
SEM	Scanning electron microscopy
BSE	Back-scattering electron
EDS	Energy dispersive spectroscopy
RHSC	Reversible hydrogen storage capacity
E_{corr}	Corrosion potential

References

1. Okada, M.; Kuriiwa, T.; Tamura, T.; Kamegawa, A. Ti-V-Cr b.c.c. alloys with high protium content. *J. Alloys Compd.* **2002**, *330–332*, 511–516. [CrossRef]
2. Tamura, T.; Kazumi, T.; Kamegawa, A.; Takamura, H.; Okada, M. Protium absorption properties and protide formations of Ti-Cr-V alloys. *J. Alloys Compd.* **2003**, *356–357*, 505–509. [CrossRef]
3. Young, K.; Fetcenko, M.A.; Ouchi, T.; Im, J.; Ovshinsky, S.R.; Li, F.; Reinhout, M. Hydrogen Storage Materials Having Exunit cellent Kinetics, Capacity, and Cycle Stability. U.S. Patent 7.344,676, 18 March 2008.
4. Akiba, E.; Iba, H. Hydrogen absorption by Laves phase related BCC solid solution. *Intermetallics* **1998**, *6*, 461–470. [CrossRef]
5. Cho, S.; Han, C.; Park, C.; Akiba, E. The hydrogen storage characteristics of Ti–Cr–V alloys. *J. Alloys Compd.* **1999**, *288*, 294–298. [CrossRef]

6. Kuriwa, T.; Tamura, T.; Amemiya, T.; Fusa, T.; Kamegawa, A.; Takemura, T.; Okada, M. New V-based alloys with high protium absorption and desorption capacity. *J. Alloys Compd.* **1999**, *293–295*, 433–436. [CrossRef]

7. Seo, C.; Kim, J.; Lee, P.S.; Lee, J. Hydrogen storage properties of vanadium-based b.c.c. solid solution metal hydrides. *J. Alloys Compd.* **2003**, *348*, 252–257. [CrossRef]

8. Tamura, T.; Tominaga, Y.; Matsumoto, K.; Fuda, T.; Kuriiwa, T.; Kamegawa, A.; Takamura, H.; Okada, M. Protium absorption properties of Ti-V-Cr-Mn alloys with a b.c.c. structure. *J. Alloys Compd.* **2002**, *330–332*, 522–525. [CrossRef]

9. Huang, T.; Wu, Z.; Chen, J.; Yu, X.; Xia, B.; Xu, N. Dependence of hydrogen storage capacity of $TiCr_{1.8-x}(VFe)_x$ on V-Fe content. *Mater. Sci. Eng. A* **2004**, *385*, 17–21.

10. Yu, X.B.; Wu, Z.; Xia, B.J.; Xu, N.X. Enhancement of hydrogen storage capacity of Ti–V–Cr–Mn BCC phase alloys. *J. Alloys Compd.* **2004**, *372*, 272–277. [CrossRef]

11. Hang, Z.; Xiao, X.; Tan, D.; He, Z.; Li, W.; Li, S.; Chen, C.; Chen, L. Microstructure and hydrogen storage properties of $Ti_{10}V_{84-x}Fe_6Zr_x$ (x = 1–8) alloys. *Int. J. Hydrog. Energy* **2010**, *35*, 3080–3086. [CrossRef]

12. Yooa, J.; Shim, G.; Cho, S.; Park, C. Effects of desorption temperature and substitution of Fe for Cr on the hydrogen storage properties of $Ti_{0.32}Cr_{0.43}V_{0.25}$ alloy. *Int. J. Hydrog. Energy* **2007**, *32*, 2977–2981. [CrossRef]

13. Yan, Y.; Chen, Y.; Liang, H.; Zhou, X.; Wu, C.; Tao, M.; Pang, L. Hydrogen storage properties of V–Ti–Cr–Fe alloys. *J. Alloys Compd.* **2008**, *454*, 427–431. [CrossRef]

14. Miao, H.; Gao, M.; Liu, Y.; Lin, Y.; Wang, J.; Pan, H. Microstructure and electrochemical properties of Ti–V-based multiphase hydrogen storage electrode alloys. *Int. J. Hydrog. Energy* **2007**, *32*, 3947–3953. [CrossRef]

15. Liu, X.P.; Cuevas, F.; Jiang, L.J.; Latroche, M.; Li, Z.; Wang, S. Improvement of the hydrogen storage properties of Ti–Cr–V–Fe BCC alloy by Ce addition. *J. Alloys Compd.* **2009**, *476*, 403–407. [CrossRef]

16. Yoo, J.; Shim, G.; Park, C.; Kim, W.; Cho, S. Influence of Mn or Mn plus Fe on the hydrogen storage properties of the Ti-Cr-V alloy. *Int. J. Hydrog. Energy* **2009**, *34*, 9116–9121. [CrossRef]

17. Liu, X.; Jiang, L.; Li, Z.; Huang, Z.; Wang, S. Improve plateau property of $Ti_{32}Cr_{46}V_{22}$ BCC alloy with heat treatment and Ce additive. *J. Alloys Compd.* **2009**, *471*, L36–L38. [CrossRef]

18. Chen, X.; Yuan, Q.; Madigan, B.; Xue, W. Long-term corrosion behavior of martensitic steel welds in static molten Pb–17Li alloy at 550 °C. *Corros. Sci.* **2015**, *96*, 178–185. [CrossRef]

19. Chuang, H.J.; Huang, S.S.; Ma, C.Y.; Chan, S.L.I. Effect of annealing heat treatment on an atomized AB_2 hydrogen storage alloy. *J. Alloys Compd.* **1999**, *285*, 284–291. [CrossRef]

20. Hang, Z.; Xiao, X.; Li, S.; Ge, H.; Chen, C.; Chen, L. Influence of heat treatment on the microstructure and hydrogen storage properties of $Ti_{10}V_{77}Cr_6Fe_6Zr$ alloy. *J. Alloys Compd.* **2012**, *529*, 128–133. [CrossRef]

21. Yu, X.B.; Wu, Z.; Xia, B.J.; Xu, N.X. A Ti-V-based bcc phase alloy for use as metal hydride electrode with high discharge capacity. *J. Chem. Phys.* **2004**, *121*, 987–990. [CrossRef] [PubMed]

22. Young, K.; Ouchi, T.; Huang, B.; Nei, J. Structure, hydrogen storage, and electrochemical properties of body-centered-cubic $Ti_{40}V_{30}Cr_{15}Mn_{13}X_2$ alloys (X = B, Si, Mn, Ni, Zr, Nb, Mo, and La). *Batteries* **2015**, *1*, 74–90. [CrossRef]

23. Young, K.; Huang, B.; Regmi, R.K.; Lawes, G.; Liu, Y. Comparisons of metallic clusters imbedded in the surface oxide of AB_2, AB_5, and A_2B_7 alloy. *J. Alloys Compd.* **2010**, *506*, 831–840. [CrossRef]

24. Young, K.; Ouchi, T.; Koch, J.; Fetcenko, M.A. Compositional optimization of vanadium-free hypo-stoichiometric AB_2 metal hydride alloy for Ni/MH battery application. *J. Alloys Compd.* **2012**, *510*, 97–106. [CrossRef]

25. Young, K.; Ouchi, T.; Huang, B.; Fetcenko, M.A. Effects of B, Fe, Gd, Mg, and C on the structure, hydrogen storage, and electrochemical properties of vanadium-free AB_2 metal hydride alloy. *J. Alloys Compd.* **2012**, *511*, 242–250. [CrossRef]

26. American Society for Testing and Materials, ASTM International. *Standard Practice for Calculation of Corrosion Rates and Related Information from Electrochemical Measurements*; ASTM: West Conshohocken, PA, USA, 1999.

27. Handzlik, P.; Fitzner, K. Corrosion resistance of Ti and Ti–Pd alloy in phosphate buffered saline solutions with and without H_2O_2 addition. *Trans. Nonferrous Met. Soc. China* **2013**, *23*, 866–875. [CrossRef]

28. Young, K.; Nei, J. The current status of hydrogen storage alloy development for electrochemical application. *Materials* **2013**, *6*, 4574–4608. [CrossRef]

29. Cho, S.; Enoki, H.; Akiba, E. Effect of Fe addition on hydrogen storage characteristics of $Ti_{0.16}Zr_{0.05}Cr_{0.22}V_{0.57}$ alloy. *J. Alloys Compd.* **2000**, *307*, 304–310. [CrossRef]

30. Enomoto, M. The Cr-Ti-V system. *J. Phase Equilibria* **1992**, *13*, 195–200. [CrossRef]
31. Dou, T.; Wu, Z.; Mao, J.; Xu, N. Application of commercial ferrovanadium to reduce cost of Ti–V-based BCC phase hydrogen storage alloys. *Mater. Sci. Eng. A* **2008**, *476*, 34–38. [CrossRef]
32. Santos, S.F.; Huot, J. Hydrogen storage in TiCr$_{1.2}$(FeV)x BCC solid solutions. *J. Alloys Compd.* **2009**, *472*, 247–251. [CrossRef]
33. Chen, N.; Li, R.; Zhu, Y.; Liu, Y.; Pan, H. Electrochemical hydrogenation and dehydrogenation mechanisms of the Ti–V base multiphase hydrogen storage electrode alloy. *Acta Metall. Sin.* **2004**, *40*, 1200–1204.
34. Young, K.; Ouchi, T.; Nei, J.; Wang, L. Annealing effects on Laves phase-related body-centered-cubic solid solution metal hydride alloys. *J. Alloys Compd.* **2016**, *654*, 216–225. [CrossRef]
35. Itoh, H.; Arashima, H.; Kubo, K.; Kabutomori, T. The influence of microstructure on hydrogen absorption properties of Ti–Cr–V alloys. *J. Alloys Compd.* **2002**, *330*, 287–291. [CrossRef]
36. Towata, S.; Noritake, T.; Itoh, A.; Aoki, M.; Miwa, K. Effect of partial niobium and iron substitution on short-term cycle durability of hydrogen storage Ti–Cr–V alloys. *Int. J. Hydrog. Energy* **2013**, *38*, 3024–3029. [CrossRef]
37. Ashworth, M.A.; Davenport, A.J.; Ward, R.M.; Hamilton, H.G.C. Microstructure and corrosion of Pd-modified Ti alloys produced by powder metallurgy. *Corros. Sci.* **2010**, *52*, 2413–2421. [CrossRef]
38. Young, K.; Wong, D.F.; Nei, J. Effects of vanadium/nickel contents in Laves phase-related body-centered-cubic solid solution metal hydride alloys. *Batteries* **2015**, *1*, 34–53. [CrossRef]
39. Young, K.; Ouchi, T.; Meng, T.; Wong, D.F. Studies on the synergetic effects in multi-phase metal hydride alloys. *Batteries* **2016**, *2*, 15. [CrossRef]
40. Zhou, Y.; Chen, J.; Xu, Y.; Liu, Z. Effects of Cr, Ni and Cu on the Corrosion Behavior of Low Carbon Microalloying Steel in a Cl$^-$ Containing Environment. *J. Mater. Sci. Technol.* **2013**, *29*, 168–174. [CrossRef]
41. Kamimura, T.; Stratmann, M. The influence of chromium on the atmospheric corrosion of steel. *Corros. Sci.* **2001**, *43*, 429–447. [CrossRef]
42. Shih, C.; Shih, C.; Su, Y.; Su, L.H.L.; Chang, M.; Lin, S. Effect of surface oxide properties on corrosion resistance of 316L stainless steel for biomedical applications. *Corros. Sci.* **2004**, *46*, 427–441. [CrossRef]
43. Güleryüz, H.; Çimenoğlu, H. Effect of thermal oxidation on corrosion and corrosion–wear behaviour of a Ti–6Al–4V alloy. *Biomaterials* **2004**, *25*, 3325–3333. [CrossRef] [PubMed]
44. Shi, P.; Ng, W.F.; Wong, M.H.; Cheng, F.T. Improvement of corrosion resistance of pure magnesium in Hanks' solution by microarc oxidation with sol–gel TiO$_2$ sealing. *J. Alloys Compd.* **2009**, *469*, 286–292. [CrossRef]
45. Iba, H.; Akiba, E. The relation between microstructure and hydrogen absorbing property in Laves phase-solid solution multiphase alloys. *J. Alloys Compd.* **1995**, *231*, 508–512. [CrossRef]
46. Tsukahara, M.; Takahashi, K.; Mishima, T.; Isomura, A.; Sakai, T. V-based solid solution alloys with Laves phase network: Hydrogen absorption properties and microstructure. *J. Alloys Compd.* **1996**, *236*, 151–155. [CrossRef]

batteries

MDPI

Article

Comparison among Constituent Phases in Superlattice Metal Hydride Alloys for Battery Applications

Kwo-Hsiung Young [1,2,*], Taihei Ouchi [2], Jean Nei [2], John M. Koch [2] and Yu-Ling Lien [3]

[1]   Department of Chemical Engineering and Materials Science, Wayne State University, Detroit, MI 48202, USA
[2]   BASF/Battery Materials—Ovonic, 2983 Waterview Drive, Rochester Hills, MI 48309, USA;
      taihei.ouchi@basf.com (T.O.); jean.nei@basf.com (J.N.); john.m.koch@basf.com (J.M.K.)
[3]   Department of Chemistry, Michigan State University, East Lansing, MI 48824, USA; yulinglien@gmail.com
*   Correspondence: kwo.young@basf.com; Tel.: +1-248-293-7000

Academic Editor: Andreas Jossen
Received: 14 September 2017; Accepted: 18 October 2017; Published: 31 October 2017

Abstract: The effects of seven constituent phases—$CeNi_3$, $NdNi_3$, Nd_2Ni_7, Pr_2Ni_7, Sm_5Ni_{19}, Nd_5Co_{19}, and $CaCu_5$—on the gaseous phase and electrochemical characteristics of a superlattice metal hydride alloy made by induction melting with a composition of $Sm_{14}La_{5.7}Mg_{4.0}Ni_{73}Al_{3.3}$ were studied through a series of annealing experiments. With an increase in annealing temperature, the abundance of non-superlattice $CaCu_5$ phase first decreases and then increases, which is opposite to the phase abundance evolution of Nd_2Ni_7—the phase with the best electrochemical performance. The optimal annealing condition for the composition in this study is 920 °C for 5 h. Extensive correlation studies reveal that the A_2B_7 phase demonstrates higher gaseous phase hydrogen storage and electrochemical discharge capacities and better battery performance in high-rate dischargeability, charge retention, and cycle life. Moreover, the hexagonal stacking structure is found to be more favorable than the rhombohedral structure.

Keywords: metal hydride (MH); nickel/metal hydride (Ni/MH) battery; hydrogen absorbing alloy; electrochemistry; superlattice alloy

---

1. Introduction

Misch metal (Mm, a mixture of more than one rare earth element)-based superlattice metal hydride (MH) alloys are very important for today's nickel/metal hydride (Ni/MH) batteries, because of their higher hydrogen storage (H-storage) capacity, better high-rate dischargeability (HRD) capability, superior low-temperature and charge retention performances, and improved cycle stability [1–8]. The three main components in the superlattice alloy can be classified by chemical stoichiometry—or more precisely, the B/A ratio—and they are AB_3, A_2B_7, and A_5B_{19}. In each component, there are two different types of structures—hexagonal and rhombohedral, depending on the stacking sequence for the A_2B_4 slabs (illustrated in pink in Figure 1). Each A_2B_4 slab shifts on the *ab*-plane by 1/3 $\vec{a}$ and 1/3 $\vec{b}$ from its neighbor. Studies on superlattice alloys began with the structure [9] and gaseous phase (GP) H-storage characteristics of the single rare earth element (RE)-based AB_3 alloys [10–12] and soon shifted to the Mm-based A_2B_7 chemistry for battery applications [13–16]. Some researchers continued to work on RE-based AB_3 chemistry for its basic electrochemical (EC) properties [17–22]. In the meantime, RE-based A_5B_{19} was also highly promoted for battery applications [23–27]. Several comparative works on the EC performances of various superlattice phases were previously reported and are summarized as follows. In a $(LaMg)Ni_x$ (x = 3, 3.5, and 3.8) system, the capacity decreased, and both HRD and cycle stability increased with the increase of x from 3 to 3.5 and finally 3.8 [28]. In a $(LaY)(NiMnAl)_x$

(x = 3, 3.5, and 3.8) system, both capacity and cycle stability reached the maximum at x = 3.5, and HRD increased with the increase in x [29]. The EC properties of the Mm-based AB_x (where A is Mm, B is a combination of several transition metals and Al, and x = 2, 3, 3.5, 3.8, and 5) were compared, and A_2B_7 showed the best overall performance [30,31], which prompted many investigations in the A_2B_7 superlattice alloys [32–37]. However, no data were provided to support the comparative work. Therefore, in this work we detail the correlations between the constituent phase abundances and various properties.

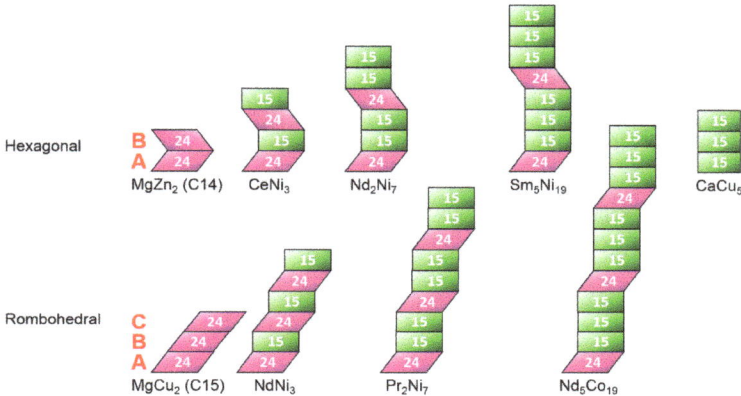

Figure 1. Stacking sequences along the *c*-axis direction of various hexagonal and rhombohedral structures available for the superlattice metal hydride alloys.

Annealing has been used to effectively alter the phase components in the AB [38,39], AB_2 [40–43], AB_3 [44–47], A_2B_7 [48–51], A_5B_{19} [52], AB_5 [53–55], A_8B_{21} [56,57], and body-centered-cubic (bcc) [58] MH alloys. According to first-principle calculations on superlattice MH alloy systems, the preferable phase abundance can be influenced by the starting composition and annealing condition [59,60]. For example, the A_2B_7 phase is more stable than the mixture of the AB_3 and AB_5 phases in the La_3MgNi_{14} alloy, however, the opposite is true in the $La_2CeMgNi_{14}$ alloy [60]. Therefore, while annealing increased the A_2B_7 phase abundance in the La-based superlattice alloy [61], it promoted phase segregation in the LaPrNd-based alloy [62]. Moreover, phase segregation is more prominent in the Mm-based (A is more than one RE element) superlattice alloys than in the RE-based (A is only one RE element) alloys. In this work, annealing is applied to engineer the abundances of constituent components in a SmLa-based superlattice alloy, and the correlations of phase abundances with the GP and EC properties are reported.

2. Experimental Setup

First, Eutectix (Troy, MI, USA) prepared a 250 kg ingot using the conventional induction melting method [31]. Five 2 kg ingot pieces were annealed in 1 atm atmosphere of Ar for 5 h at different temperatures. Each ingot was hydrided and then crushed and ground into the size of −200 mesh. Chemical compositions of ingots before and after annealing were verified with a Varian Liberty 100 inductively coupled plasma-optical emission spectrometer (ICP-OES, Agilent Technologies, Santa Clara, CA, USA). A Philips X'Pert Pro X-ray diffractometer (XRD, Amsterdam, The Netherlands) and a JEOL-JSM6320F scanning electron microscope (SEM, Tokyo, Japan) with energy dispersive spectroscopy (EDS) was used to conduct microstructure analysis. GP H-storage characteristics were evaluated with a Suzuki-Shokan multi-channel pressure-concentration-temperature system (PCT, Tokyo, Japan). Negative electrodes were fabricated by compacting the alloy powder onto an expanded nickel substrate through a roll mill without any binder. Half-cell measurements were performed using a CTE MCL2 Mini cell testing system (Chen Tech Electric MFG. Co., Ltd., New Taipei, Taiwan) with a partially pre-charged $Ni(OH)_2$ positive electrode and a 30% KOH electrolyte. The electrode was first charged with a current density of 100 mA·g^{-1} for 4 h and then discharged with a current density of 400 mA·g^{-1} until a cutoff voltage of 0.9 V was reached. More discharges at smaller current densities (300, 200, 100, 50, and 5 mA·g^{-1}) with the same cutoff voltage were performed afterward with a 2 min rest in between. The sum of capacities from six discharge stages was used as the full discharge capacity. The HRD was defined as the ratio of the capacity obtained at the highest rate (400 mA·g^{-1}) vs. full discharge capacity. A Solartron 1250 Frequency Response Analyzer (Solartron Analytical, Leicester, UK) with a sine wave amplitude of 10 mV and a frequency range of 0.5 mHz to 10 kHz was used for the AC impedance measurement. A Digital Measurement Systems Model 880 vibrating sample magnetometer (MicroSense, Lowell, MA, USA) was used to measure the magnetic susceptibility of the alloy powder surface after activation, which was performed by immersing the powder in 30 wt % KOH solution at 100 °C for 4 h.

For the sealed cell testing, a C-size cylindrical cell was chosen. While the negative electrode was fabricated by dry compacting the alloy powder onto nickel mesh current collectors, the counter positive electrode was fabricated by pasting a mixture of 89% standard AP50 [63] with the composition of $Ni_{0.91}Co_{0.045}Zn_{0.045}(OH)_2$ (BASF—Ovonic, Rochester Hills, MI, USA), 5 wt % Co powder, and 6% CoO powder onto nickel foam substrates. Scimat 700/79 acrylic acid grafted polypropylene/polyethylene separators were used (Freudenberg Group, Weinheim, Germany). A 1.5 to 1.7 negative-to-positive capacity ratio cell design was used to maintain a good balance between the over-charge and over-discharge reservoirs [64]. A 30 wt % KOH solution with LiOH (1.5 wt %) additive was used as the electrolyte. Formation was performed with a six-cycle process using a Maccor Battery Cycler (Maccor, Tulsa, OK, USA). Details of cell testing can be found in an earlier publication [65].

3. Results and Discussion

The design compositions for this study are presented in Table 1. Sm was chosen as the main RE element because it is relatively inexpensive (like La and Ce) and less oxidable (for a comparison to other RE elements, see Table 7 in [31]). However, the metal–hydrogen bond strength of the Sm-based superlattice MH alloy is too weak (Sm_2Ni_7 has a discharge capacity of 170 mAh·g^{-1} [66]), and thus adding La is necessary to increase the storage capacity [67]. LaSm-based superlattice MH alloys with the La/Sm ratio above 1 were previously reported. While the AB$_5$ phase cannot be removed completely by annealing in those alloys, the AB$_5$ phase abundance still decreased with the increase in Sm-content [27,68–72]. In this experiment, a La/Sm ratio of 0.4 was adopted to attempt to suppress the AB$_5$ phase abundance. Two common constituent elements used in the AB$_5$ MH alloy, Mn and Co, are excluded in this study for better charge retention and cycle stability [1,73]. Al is included to prevent the hydrogenation-induced-amorphization in the superlattice MH alloys [36].

Table 1. Designed composition and ICP results in at%. B/A is the ratio of the B-atom (Ni and Al) to the A-atom (La, Sm, and Mg).

Alloy	Annealing Temperature	Source	La	Sm	Mg	Ni	Al	B/A
-	-	Design	5.7	14.0	4.0	73.0	3.3	3.2
A0	-	ICP	5.7	14.0	3.9	73.0	3.4	3.2
A1	880 °C	ICP	6.0	14.5	4.2	72.1	3.2	3.0
A2	900 °C	ICP	5.8	13.9	4.0	72.9	3.4	3.2
A3	920 °C	ICP	5.8	13.9	4.0	73.0	3.3	3.2
A4	940 °C	ICP	5.8	14.3	4.0	72.7	3.2	3.1
A5	960 °C	ICP	5.8	14.5	4.0	72.5	3.2	3.1

Five pieces of ingot from the induction melting were annealed at 880, 900, 920, 940, and 960 °C for 5 h in an Ar environment. Compositions of the as-cast (alloy A0) and annealed ingots (alloys A1 to A5) were measured by ICP, and the results are summarized in Table 1. ICP results reveal that the annealed ingots have similar Mg-content, but are slightly rich in La and lean in Ni.

3.1. Microstructure Analysis

XRD patterns of alloys in this study are presented in Figure 2. These patterns show the typical multi-phase superlattice structures. By using the Jade 9.0 software (MDI, Livermore, CA, USA), we were able to deconvolute each pattern into its constituent phases, and the results are summarized in Table 2. One example of such deconvolution is plotted in Figure 3. There are two stacking structures for each stoichiometry (AB_3, A_2B_7, and A_5B_{19})—hexagonal (H) and rhombohedral (R). Most of the phase transformations in the superlattice phases are through the peritectic reaction and are very sensitive to the annealing conditions (temperature and duration). From Table 2, it is clear that annealing initially suppresses the unwanted non-superlattice AB_5 phase, which decreases both the capacity and HRD [72,73]. However, further increases in annealing temperature promotes the formation of AB_5. Evolutions in phase stoichiometry (x value in AB_x) and stacking structure (hexagonal vs. rhombohedral) with different annealing temperatures are plotted in Figure 4. Annealing first increases the A_2B_7 abundance at the expense the AB_3 abundance, but further increases in annealing temperature decreases the A_2B_7 abundance. In comparison, the changes in A_5B_{19} abundance are less obvious. For the stacking structure evolution, annealing first increases the hexagonal structure and then slowly decreases it as the annealing temperature increases (Figure 4b).

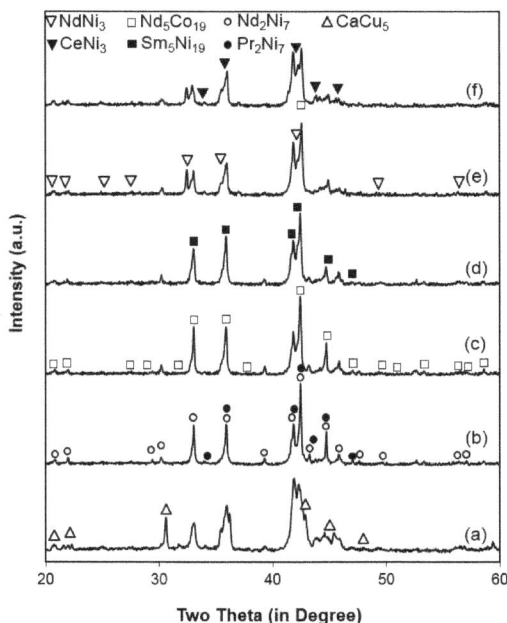

Figure 2. XRD patterns using Cu-K$_\alpha$ as the radiation source for alloys (**a**) A0, (**b**) A1, (**c**) A2, (**d**) A3, (**e**) A4, and (**f**) A5.

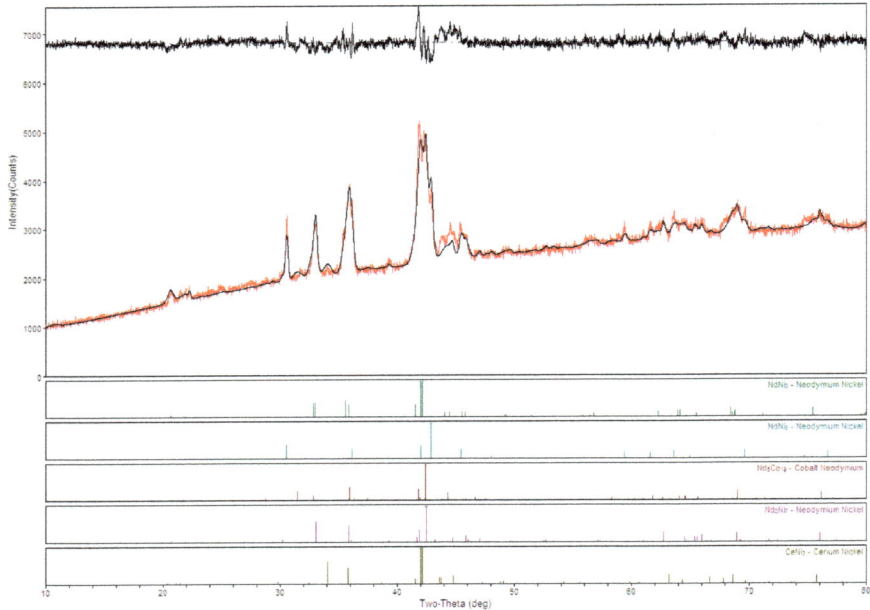

Figure 3. Phase deconvolution performed by Jade 9 software on XRD pattern from alloy A0. The red and black lines are the raw data and fitting curve, respectively. The residue of fitting is plotted in the top of the figure.

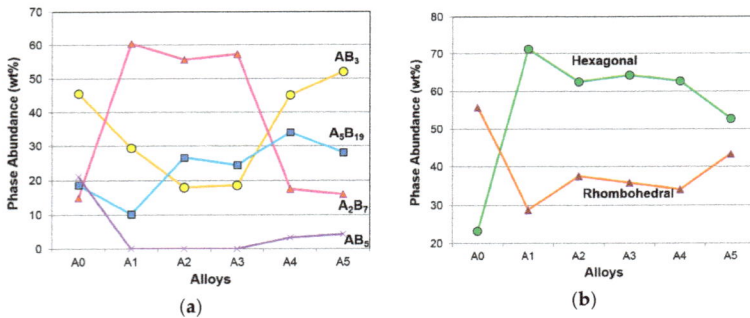

Figure 4. Phase abundance distributions classified by (**a**) stoichiometry and (**b**) structure family.

Table 2. Abundances (in wt %) of the constituent phases obtained from the XRD data for hexagonal (H) and rhombohedral (R) systems.

Alloy	$CeNi_3$ (H)	$NdNi_3$ (R)	Nd_2Ni_7 (H)	Pr_2Ni_7 (R)	Sm_5Ni_{19} (H)	Nd_5Co_{19} (R)	$CaCu_5$ (H)
A0	8.4	37.1	14.9	-	-	18.6	21.0
A1	20.8	8.6	50.4	10.0	-	10.2	-
A2	-	17.9	48.6	7.0	13.9	12.6	-
A3	-	18.5	52.4	4.8	11.9	12.5	-
A4	24.3	20.8	17.5	-	20.9	13.2	3.3
A5	24.3	27.7	15.9	-	12.5	15.5	4.2

Representative SEM backscattering electron image (BEI) micrographs from alloys A0 (as-cast) and A3 (annealed at 920 °C) are compared in Figure 5. While obvious phase segregation can be observed in alloy A0 (Figure 5a), alloy A3 appears to be more uniform (Figure 5b). EDS was used to measure chemical compositions of several spots in each micrograph, and the results are summarized in Table 3. In alloy A0 (Figure 5a), undissolved La (spot 1), the SmNi (spot 2), AB_3 (spot 4), AB_2 (spot 6) and AB_5 (spot 7) phases can be identified. The main phase (spot 5) is a mixture of the AB_3, A_2B_7, and A_5B_{19} phases. In the annealed alloy A3, only occasional undissolved La and Sm (spot 1) and the AB_2 phase (spot 4) can be identified, and the majority is composed of a fine mixture of the AB_3, A_2B_7, and A_5B_{17} phases (Figure 5b).

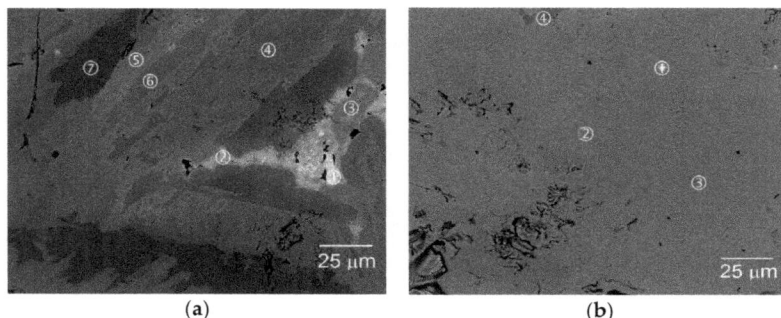

Figure 5. SEM backscattering electron image (BEI) micrographs from alloys (**a**) A0 and (**b**) A3. Compositions of the numbered areas were analyzed by EDS, and the results are shown in Table 3.

Table 3. Summary of EDS results from several selective spots in the SEM–BEI micrographs of alloys A0 and A3 shown in Figure 5a,b.

Sample	Location	La	Sm	Mg	Ni	Al	B/A	Phase
	3a-1	90.8	5.1	1.5	1.9	0.7	0.03	La
	3a-2	2.5	42.9	7.3	47.0	0.3	0.90	AB
	3a-3	9.3	13.6	5.2	65.7	6.2	2.56	AB_2/AB_3
A0	3a-4	4.1	15.3	5.4	71.6	3.6	3.03	AB_3
	3a-5	3.4	15.0	4.4	73.6	3.6	3.39	$AB_3/A_2B_7/A_5B_{19}$
	3a-6	7.2	14.3	13.5	63.7	1.2	1.85	AB_2
	3a-7	3.5	14.0	0.6	78.3	3.6	4.52	AB_5
	3b-1	27.9	61.0	2.7	7.3	1.1	0.09	La/Sm
A3	3b-2	5.1	14.4	5.1	71.1	4.3	3.07	AB_3/A_2B_7
	3b-3	3.1	13.9	4.0	75.2	3.8	3.76	A_2B_7/A_5B_{19}
	3b-4	4.8	14.5	15.6	64.0	1.0	1.86	AB_2

3.2. Gaseous Phase Hydrogen Storage

In this study, PCT analysis was used to examine hydrogen absorption/desorption characteristics of alloys, and the isotherms measured at 30 and 45 °C are plotted in Figure 6. While no obvious plateau is observed in the isotherms of alloy A0, isotherms of the annealed alloys show well-defined plateau regions. GP H-storage properties obtained from the PCT analysis are summarized in Table 4. Both maximum and reversible capacities first increase and then decrease with an increase in annealing temperature, and the annealed alloys have higher storage capacities compared to the as-cast alloy A0. The desorption pressure at 0.75 wt % H-storage increases monotonically with an increase in annealing temperature. For the annealed alloys, the slope factor (ratio of the H-storage capacity between 0.02 and 0.5 MPa to the total reversible capacity) first decreases (more slanted, less uniform in composition [74]) and then increases (flatter and more uniform in composition) as the annealing temperature increases, but all are higher than that of the as-cast alloy A0. PCT hysteresis, defined as

$$\text{PCT hysteresis} = \ln\left(\frac{\text{absorption pressure at 0.75 wt \% H} - \text{storage}}{\text{desorption pressure at 0.75 wt \% H} - \text{storage}}\right) \tag{1}$$

of alloys A4 and A5 are relatively smaller, which suggest a better resistance to pulverization during charge/discharge cycling [75]. Changes in enthalpy (ΔH) and entropy (ΔS) were calculated by using the desorption pressures (P) at 0.75 wt % H-storage measured at different temperatures in the Van't Hoff equation,

$$\Delta G = \Delta H - T\Delta S = \Re\, T \ln P, \tag{2}$$

where T and $\Re$ are the absolute temperature and ideal gas constant, respectively. Calculation results are listed in the last two rows of Table 4. Both ΔH and ΔS of the annealed alloys are lower than those of the as-cast alloy A0. Among the annealed alloys, both ΔH and ΔS increase (except for alloy A5) with an increase in annealing temperature. The lowest ΔS value obtained from alloy A5 is the closest to the ΔS value for H_2 gas ($-135\ \text{J·mol H}_2^{-1}\ \text{K}^{-1}$), indicating the highest degree of order of hydrogen in the MH alloy [76].

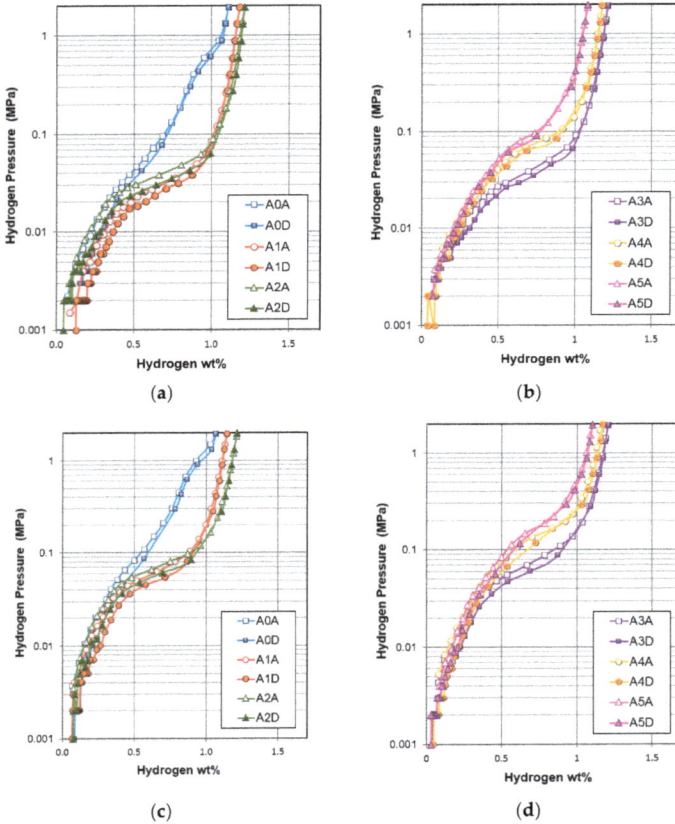

Figure 6. PCT isotherms measured at (**a**,**b**) 30 and (**c**,**d**) 45 °C for alloys A0–A5. Open and solid symbols represent absorption and desorption curves, respectively.

Table 4. Summary of gaseous phase properties measured at 30 °C.

Gaseous Phase Properties	Unit	A0	A1	A2	A3	A4	A5
Capacity at 2 MPa	wt %	1.11	1.18	1.21	1.21	1.18	1.08
Reversible capacity	wt %	0.95	0.98	1.17	1.17	1.13	1.00
Desorption pressure	MPa	0.16	0.029	0.035	0.037	0.071	0.091
Slope factor	%	69	78	76	72	77	79
Hysteresis		0.17	0.16	0.22	0.17	0.08	0.09
$-\Delta H$	kJ·mol H_2^{-1}	25.8	38.8	35.2	33.6	31.8	39.3
$-\Delta S$	J·mol H_2^{-1}·K^{-1}	89	118	107	102	102	129

3.3. Electrochemical and Magnetic Susceptibility Measurements

Before the half-cell capacity measurement, the compacted negative electrode went through a 4 h activation process in 30 wt % KOH at 100 °C. The second cycle capacities obtained at different discharge currents for each alloy are plotted in Figure 7a together with those for a standard and commercially available AB$_5$ alloy (Alloy B with a chemical composition of $La_{10.5}Ce_{4.3}Pr_{0.5}Nd_{1.4}Ni_{60}Co_{12.7}Mn_{5.9}Al_{4.7}$ supplied by Eutectix). Discharge capacity decreases with the increase in discharge rate in a roughly linear manner. The slope of capacity vs. discharge rate for the superlattice alloys (dashed red line) is lower than the slope for the AB$_5$ alloy (solid green line), which confirms the superiority of superlattice alloys in HRD, as previously reported [30]. EC test results of alloys in this study are summarized in Table 5. Annealing improves the capacity substantially. Both the capacity and half-cell HRD (the ratio between the capacities obtained at 400 and 5 mA·g^{-1}) increase first and then decrease with the increase in annealing temperature and peak at alloy A2 (Figure 7b). EC discharge capacities of alloys A1 to A3 (323 to 326 mAh·g^{-1}) are very close to the maximum H-storage capacities in GP (1.18 to 1.21 wt %, which is equivalent to 316 to 324 mAh·g^{-1}).

Evolution in half-cell HRD was further investigated by the bulk hydrogen diffusion constant (D) and surface reaction current (I_o), and the results are listed in Table 5. Details of both measurements were previously reported [33]. While the D value remains approximately the same, the I_o value peaks at alloy A2. Therefore, we conclude the half-cell HRD in this series of annealed superlattice alloys is related more to the surface catalytic ability, which agrees with our previous findings from the Co-substituted Mm-based superlattice MH alloys [33]. With the −40 °C AC impedance measurement, both charge-transfer resistance (R) and double-layer capacitance (C) were obtained, and are summarized in Table 5. With an increase in annealing temperature, both R and RC product (a measure of surface catalytic ability [77]) first decrease and then increase; this trend is similar to that observed in HRD. Moreover, alloys A2, A3, and A4 demonstrate the best low-temperature performance (the lowest Rs), which is dominated by the surface catalytic ability (the lowest RC products).

Magnetic susceptibility was measured to study the evolution in surface metallic Ni with the annealing temperature. Details of this measurement and the connection between magnetic susceptibility and HRD were previously published [78,79]. Saturated magnetic susceptibility (M_S, closely related to the surface catalytic ability) first decreases and then increases with an increase in annealing temperature, suggesting that the surface's catalytic ability first decreases and then increases. This result is contradictory to the conclusion drawn from the RC product result. Therefore, the lowest Rs and RC products observed in alloys A2, A3, and A4 do not correlate with the amount of surface metallic nickel. This discrepancy is rare, but there is example when the surface oxide microstructure, rather than the metallic nickel, played an important role in affecting the HRD and low temperature performances [80]. Further study of the surface oxide microstructure by transmission microscope is necessary to explain the source of highly catalytic surface of alloys A2, A3, and A4. Lastly, the applied field strength corresponding to half of M_S ($H_{1/2}$) of these alloys are similar, indicating the size of metallic nickel inclusion in the surface oxide remains constant despite the change in annealing condition.

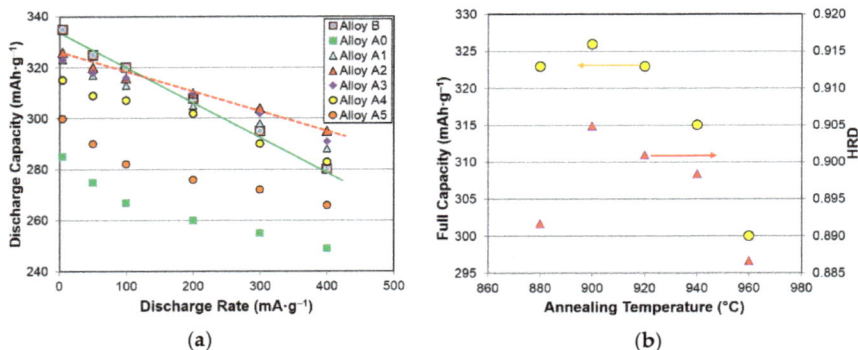

Figure 7. Electrochemical discharge capacities as functions of (**a**) discharge rate and (**b**) annealing temperature. Alloy B is a commercially available La-rich AB$_5$ MH alloy.

Table 5. Summary of electrochemical half-cells.

Electrochemical and Magnetics Properties	Unit	A0	A1	A2	A3	A4	A5
High-rate discharge capacity	mAh·g^{-1}	249	288	295	291	283	266
Full discharge capacity	mAh·g^{-1}	285	323	326	323	315	300
Half-cell HRD	%	87.5	89.2	90.4	90.1	89.8	88.7
Diffusion coefficient, D	10^{-10} cm^2·s^{-1}	4.0	4.0	4.2	4.4	4.2	4.1
Surface reaction current, I_o	mA·g^{-1}	24.2	24.0	33.1	23.8	21.9	17.6
Charge-transfer resistance at −40 °C, R	Ω·g	4.9	8.8	4.0	3.7	3.4	5.3
Double-layer capacitance at −40 °C, C	F·g^{-1}	1.6	0.86	1.02	0.88	1.12	1.29
RC product at −40 °C	s	7.7	7.6	4.1	3.3	3.8	6.8
Total saturated magnetic susceptibility, M_S	emu·g^{-1}	1.45	1.36	0.96	0.60	1.05	1.12
Applied field where M.S. = $\frac{1}{2}$ M_S, $H_{1/2}$	kOe	0.11	0.10	0.11	0.12	0.10	0.10

3.4. Sealed Cell Performance

Five annealed alloys (alloys A1 to A5) were incorporated into cylindrical C-size cells (20 cells for each alloy), and each cell has a nominal capacity of 5.0 Ah. After a standard formation process [65], cells were distributed to various tests, and the results are discussed in the following sections.

3.4.1. High-Rate Performance

Four different discharge rates (C/5, C/2, C, and 2C) were used to obtain the room temperature (RT) discharge voltage curves, and the results from cells made with alloys A1 and A3 are shown in Figure 8. As the discharge rate increases, voltage is suppressed by the internal resistivity in the cell [65], and the discharge capacity obtained at the fixed cutoff voltage (0.8 V) decreases. Cells made with alloy A3 show a smaller voltage reduction with the increase in discharge rate, which indicates a lower cell internal resistance compared to the internal resistance in the cell made with alloy A1. The normalized discharge capacity obtained at a 2C rate for cells made with the five annealed alloys are listed in Table 6. The data displays an initial increasing and later decreasing trend with an increase in alloy annealing temperature. Cells made with alloy A3 yield the best RT high-rate (2C) performance in this test.

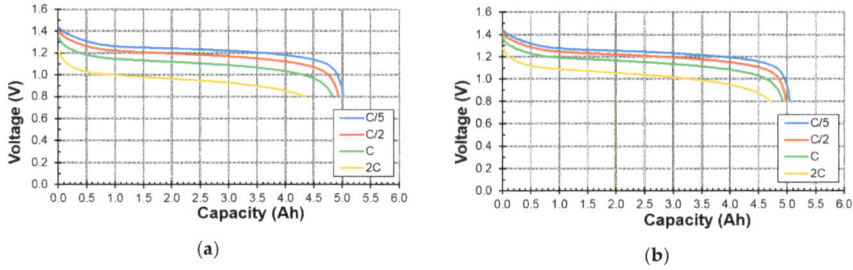

Figure 8. Room temperature discharge voltage curves at different discharge rates (C/5, C/2, C, and 2C) for cells made with alloys (**a**) A1 and (**b**) A3.

Table 6. Summary of C-cell test results (RT stands for room temperature).

C-Cell Results	Unit	A1	A2	A3	A4	A5
2C at RT capacity/0.2C at RT capacity	%	87	90	93	90	87
1C at −10 °C capacity/0.2C at RT capacity	%	92	94	95	94	91
14-day charge retention	%	85.3	84.9	84.8	76.1	85.3
28-day 45 °C voltage stand	V	1.213	1.222	1.218	1.218	1.220
Peak power at RT (20th cycle)	W·kg^{-1}	183	198	206	200	194
0.5C/0.5C cycle life (before reaching 3 Ah)	Number of cycles	220	255	365	340	210
C/C cycle life (before reaching 3 Ah)	Number of cycles	110	185	205	130	120

3.4.2. Low-Temperature Performance

Low-temperature performance was evaluated at −10 °C and a 1C discharge rate, and the results are summarized in Table 6. Like the trend in high-rate performance, a trend of initial increase followed by decrease is observed for the low-temperature performance with an increase in alloy annealing temperature. The cell made with alloy A3 shows the best low-temperature result, which can be attributed to its surface catalytic ability being the best (lowest *RC* product in Table 5).

3.4.3. Charge Retention

RT charge retention test results are shown in Figure 9a. All cells (except for the one made with alloy A4) exhibit similar charge retention characteristics. Three cells made with alloy A4 were tested, and all show inferior charge retention performance compared to the cells made with the other annealed alloys. 45 °C voltage stand test results are plotted in Figure 9b and summarized in Table 6. In this test, the cells made with alloy A2 demonstrate the best result.

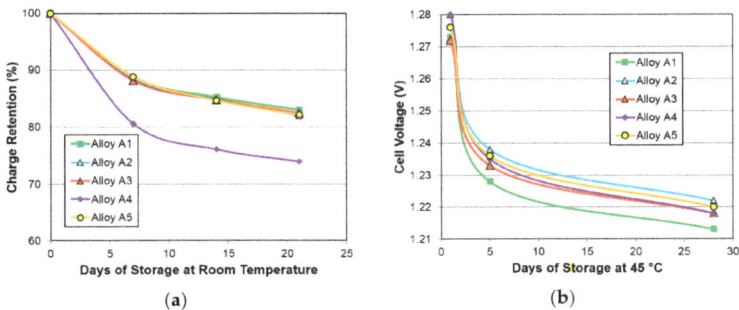

Figure 9. (**a**) Room temperature charge retentions and (**b**) 45 °C voltage stands for cells made with alloys A1 to A5.

3.4.4. Peak Power

Peak power was measured at the 20th cycle and RT, and the results are compared in Table 6. Peak power follows the same trend as those in high-rate and low-temperature performances. Cells made with alloy A3 show the best balance among peak power, best high-rate, and low-temperature performances.

3.4.5. Cycle Life

The RT cycle life performance was evaluated in two different configurations: a regular configuration charged at a C/2 rate and discharged at a C/2 rate, and an accelerated configuration charged at a C rate and discharged at a C rate. The results are plotted in Figure 10 and summarized in Table 6. The number of cycles before reaching a capacity of 3 Ah first increases and then decreases with an increase in alloy annealing temperature. Cells with alloy A3 show the best cycle stability in both C/2-C/2 and C-C cycling tests. As reported previously, the main degradation mechanism for superlattice alloy free of Mn and Co is the combination of particle pulverization and surface oxidation [34,35].

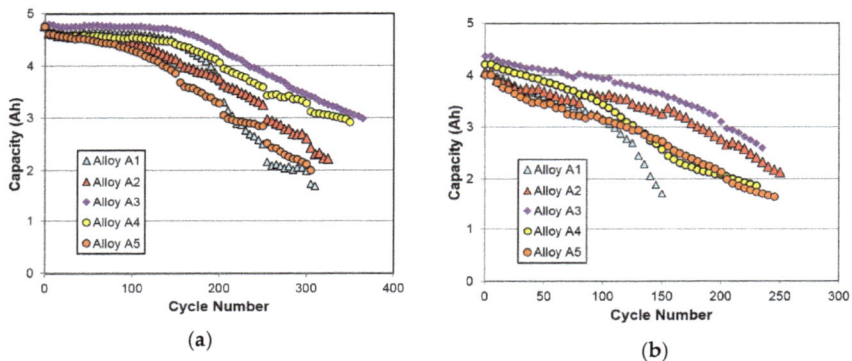

Figure 10. Room temperature cycle life from cells made with alloys A1 to A5 (**a**) in the configuration of a C/2 rate charge, a C/2 rate discharge, and a $-\Delta V$ cutoff voltage of 3 mV and (**b**) in the configuration of a C rate charge, a C rate discharge, and a $-\Delta V$ cutoff voltage of 5 mV.

3.5. Performance Correlation with Individual Phase

Correlations between the seven constituent phase abundances ($CeNi_3$, $NdNi_3$, Nd_2Ni_7, Pr_2Ni_7, Sm_5Ni_{19}, Nd_5Co_{19}, and $CaCu_5$) and seven GP, eight half-cell EC, and five sealed cell properties were studied by the linear regression method, and the resulting correlation factors (R^2) are summarized in Table 7. Although these phase abundances were obtained from the XRD analysis of the un-activated alloys, the XRD performed on the activated alloys only showed additional minute rare-earth oxide (hydride) peaks and no changes in the main phase peaks were found [81]. For the maximum H-storage capacity, Nd_2Ni_7 (Figure 11a) and Pr_2Ni_7 are considered beneficial, whereas $NdNi_3$ and Nd_5Co_{19} are detrimental. Reversible H-storage capacity does not correlate well with any phase. $NdNi_3$, Nd_5Co_{19} (Figure 11b), and $CaCu_5$ (Figure 11c) are effective in increasing the PCT plateau pressure; however, Nd_2Ni_7 and Pr_2Ni_7 have an opposite effect. $CeNi_3$ increases the flatness of PCT isotherm, but $CaCu_5$ decreases it. While $CeNi_3$ reduces the PCT hysteresis, Nd_2Ni_7 and Pr_2Ni_7 increase it (Figure 11d). The only strong influence on the change in enthalpy is from $CaCu_5$, which contributes to a much less negative ΔH value (weaker metal-hydrogen bond strength). $NdNi_3$ and Nd_5Co_{19} may also increase ΔH. None of the phases have a significant correlation with the change in entropy (which is essentially the change of entropy between hydrogen gas and hydrogen in an ordered solid).

Batteries **2017**, *3*, 34

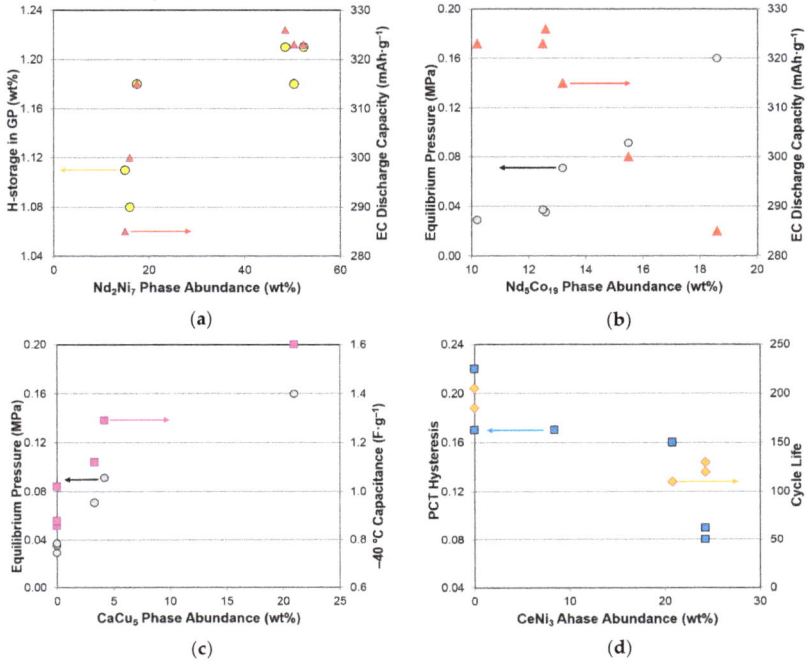

(a)

(b)

(c)

(d)

Figure 11. Examples of correlations between the constituent phase abundances and several gaseous phase/electrochemical properties: (**a**) Nd_2Ni_7 (**b**) Nd_5Co_{19} (**c**) $CaCu_5$ (**d**) $CeNi_3$.

Table 7. Correlation factors (R^2) between the constituent phase abundances and various gaseous phase, half-cell, and sealed cell properties. Plus (+) and minus (−) signs after the numbers indicate positive and negative correlations, respectively. Significant correlations with $R^2 \geq 0.6$ are highlighted in red. EC full capacity is the discharge capacity measured with a 5 mA·g^{-1} current density.

Properties	CeNi₃	NdNi₃	Nd₂Ni₇	Pr₂Ni₇	Sm₅Ni₁₉	Nd₅Co₁₉	CaCu₅
GP maximum capacity	0.26−	0.51−	0.63+	0.42+	0.07+	0.55−	0.38−
GP reversible capacity	0.24−	0.14−	0.21+	0.03+	0.57+	0.20−	0.37−
Equilibrium pressure	0.02+	0.89+	0.68−	0.58−	0.09−	0.92+	0.92+
PCT slope factor	0.42+	0.31−	0.00	0.05+	0.14+	0.31−	0.46−
PCT hysteresis	0.73−	0.03−	0.46+	0.39+	0.17−	0.02−	0.00
ΔH	0.12−	0.43+	0.16−	0.24−	0.02−	0.42+	0.61+
ΔS	0.28−	0.16+	0.01−	0.05−	0.01−	0.15+	0.33+
EC high-rate capacity	0.06−	0.79−	0.65+	0.49+	0.17+	0.84−	0.84−
EC full capacity	0.05−	0.84−	0.69+	0.56+	0.13+	0.89−	0.84−
HRD	0.09−	0.52−	0.43+	0.22+	0.42+	0.59−	0.76−
Diffusion constant, D	0.25−	0.05−	0.16+	0.00	0.42+	0.09−	0.27−
Exchange Current, I_o	0.52−	0.08−	0.33+	0.31+	0.00	0.07−	0.02−
−40 °C resistivity, R	0.15+	0.17−	0.06+	0.32+	0.53−	0.11−	0.01−
−40 °C capacitance, C	0.02+	0.91+	0.71−	0.59−	0.05+	0.94+	0.85+
RC product	0.17+	0.08+	0.11−	0.00	0.65−	0.13+	0.29+
High rate	0.54−	0.00	0.14+	0.00	0.20+	0.00	0.15−
Low temperature	0.50−	0.02−	0.18+	0.01+	0.19+	0.07−	0.23−
Charge retention	0.17−	0.03−	0.31+	0.30+	0.48−	0.01−	0.21−
Peak power	0.29−	0.25+	0.00	0.23−	0.56+	0.17−	0.00
Cycle life	0.90−	0.00	0.31+	0.02+	0.07+	0.00	0.30−

195

For the EC properties, Nd_2Ni_7 (Figure 11a) and Pr_2Ni_7 make positive contributions to the high-rate and low-rate discharge capacities, while $NdNi_3$, Nd_5Co_{19} (Figure 11b) and $CaCu_5$ have an opposite effect. Nd_2Ni_7 and Sm_5Ni_{19} increase HRD, but $NdNi_3$, Nd_5Co_{19}, and $CaCu_5$ decrease it. The increase in HRD comes from different sources: improvement in surface reaction for Nd_2Ni_7 (indicated by its positive correlation with I_o) and enhancement of bulk diffusion for Sm_5Ni_{19} (indicated by its positive correlation with D). For the low-temperature characteristics determined by the AC impedance measured at $-40\ °C$, Sm_5Ni_{19} decreases the charge-transfer resistance by increasing the surface's catalytic ability (indicated by its negative correlation with the RC product), but Pr_2Ni_7 increases it by reducing the surface's reactive area (indicated by its negative correlation with C). Although $NdNi_3$, Nd_5Co_{19}, and $CaCu_5$ (Figure 11c) can increase the surface reactive area, they have a marginal effect on the charge-transfer resistance, due to their relatively low catalytic abilities.

Judging from the ratio between the number of significant correlations ($R^2 \geq 0.60$) found and number of properties (7/7 for the GP properties, 14/8 for the half-cell EC properties, and 2/5 for the sealed cell properties), correlations with the sealed cell properties are the weakest. There are two explanations for the difficulty in setting up correlations for the sealed cell properties: manual assembly operation and limited cell number. While the former adds inconsistency, the latter limits the accuracy of sampling. Therefore, further confirmation on a larger scale (hundreds of cells) is needed. Nevertheless, correlations obtained from the 100 cells (20 for each annealed alloy) in this study are shown in the bottom five rows of Table 7. Only $CeNi_3$ shows a significant and detrimental effect on the high-rate performance. For the low-temperature performance, only Sm_5Ni_{19} exhibits a positive influence. Both Nd_2Ni_7 and Pr_2Ni_7 show a positive impact on the charge retention performance, but Sm_5Ni_{19} influences it negatively. Among all phases, Sm_5Ni_{19} most effectively increases the peak power. Only $NdNi_3$ contributes to the increase in cycle stability, whereas $CeNi_3$ (Figure 11d) and $CaCu_5$ deteriorate it.

As the exact chemical composition in each superlattice phase cannot be quantified by SEM–EDS, the correlations found in this session may come from the non-uniform distribution of A-site and B-site elements in these phases, but this cannot be verified.

3.6. Performance Correlation with Phase Stoichiometry

In earlier superlattice MH alloy studies of various phases' contributions to the EC performances, phases are grouped according to the phase stoichiometry, such as AB_3 ($CeNi_3$ and $NdNi_3$), A_2B_7 (Nd_2Ni_7 and Pr_2Ni_7), and A_5B_{19} (Sm_5Ni_{19} and Nd_5Co_{19}) [28–30]. In this study, we correlated the properties with the phase stoichiometry, and the resulted R^2s are summarized and shown in the first three columns of Table 8. AB_3 decreases the GP and EC capacities (Figure 12a), increases the plateau pressure, decreases HRD because of the reductions in D and I_o, which also contribute to an inferior high-rate performance. Moreover, although AB_3 decreases the PCT hysteresis, it still deteriorates the cycle stability. A_2B_7 increases the GP and EC capacities (Figure 12b), lowers the equilibrium pressure, increases HRD because of the enhanced D and I_o, and improves both the charge retention and cycle life. A_5B_{19} increases the GP reversible H-storage, improves the $-40\ °C$ performance via the increase in surface catalytic ability, increases the peak power, and deteriorates the charge retention. By comparing these correlations, A_2B_7 appears to be the most desirable stoichiometry for battery applications among all stoichiometries for the superlattice MH alloys. This conclusion is in complete agreement with previous reports [29–31].

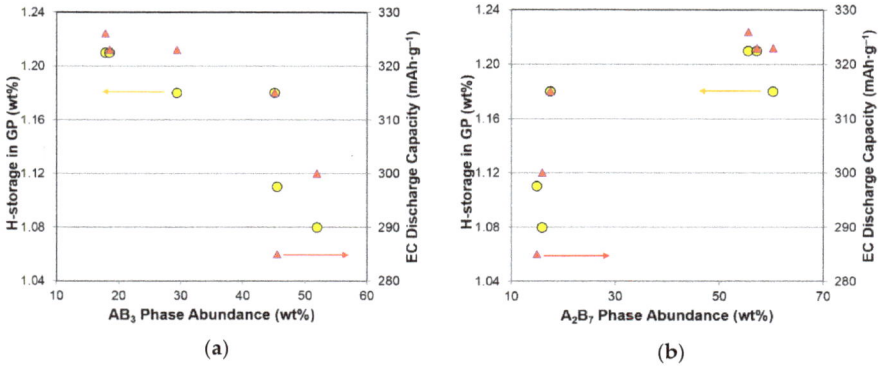

Figure 12. Examples of correlations between the abundances of phases with (**a**) an AB$_3$ and (**b**) an A$_2$B$_7$ stoichiometry and several gaseous phase/electrochemical properties.

Table 8. Correlation factors (R^2) between the phase stoichiometry/structure and various gaseous phase, half-cell, and sealed cell properties and. Plus (+) and minus (−) signs after the numbers indicate positive and negative correlations, respectively. Significant correlations with $R^2 \geq 0.6$ are highlighted in red. EC full capacity is the discharge capacity measured with a 5 mA·g^{-1} current density.

Properties	AB$_3$	A$_2$B$_7$	A$_5$B$_{19}$	Hexagonal	Rhombohedral
GP maximum capacity	0.75−	0.60+	0.00	0.44+	0.46−
GP reversible capacity	0.40−	0.17+	0.36+	0.29+	0.21−
Equilibrium pressure	0.52+	0.68−	0.00	0.93−	0.85+
PCT slope factor	0.02+	0.01+	0.03+	0.45+	0.41−
PCT hysteresis	0.61−	0.46+	0.21−	0.00	0.00
ΔH	0.03+	0.17−	0.01+	0.54−	0.43+
ΔS	0.02−	0.01−	0.00	0.26−	0.18+
EC high-rate capacity	0.60−	0.63+	0.01+	0.84+	0.78−
EC full capacity	0.61−	0.68+	0.00	0.86+	0.81−
HRD	0.49−	0.40+	0.14+	0.70+	0.59−
Diffusion constant, D	0.29−	0.11+	0.29+	0.18+	0.11−
Exchange Current, I_o	0.56−	0.34+	0.01−	0.02+	0.01−
−40 °C resistivity, R	0.00	0.10+	0.72−	0.04+	0.08−
−40 °C capacitance, C	0.53+	0.71−	0.01+	0.90−	0.88+
RC product	0.26+	0.08−	0.46−	0.23−	0.16+
High rate	0.34−	0.09+	0.13+	0.02+	0.00
Low temperature	0.42−	0.13+	0.10+	0.11+	0.09−
Charge retention	0.17−	0.32+	0.38−	0.00	0.04+
Peak power	0.05−	0.02−	0.53+	0.11−	0.16+
Cycle life	0.58−	0.24+	0.05+	0.00	0.03+

3.7. Performance Correlation with Phase Structure

Another way to classify the various phases in the superlattice MH alloys is by structure symmetry. Stacking of the A$_2$B$_4$ slabs in the CeNi$_3$, Nd$_2$Ni$_7$, and Sm$_5$Ni$_{19}$ phases is the same as that in the C14 crystal structure and is considered as the hexagonal group, and the rhombohedral group containing the NdNi$_3$, Pr$_2$Ni$_7$, and Nd$_5$Co$_{19}$ phases shares the same A$_2$B$_4$ stacking as the C15 structure. The C14- and C15-based MH alloys behave differently in the EC environment, and we have reported the comparison previously [82] and concluded that the C14-predominated alloy is more suitable for high-capacity and long-cycle life applications, whereas the C15-predominated alloy is preferable in applications requiring easy activation and good high-rate and low-temperature performances. Therefore, it will be interesting to determine whether similar conclusions can be drawn from the superlattice MH alloy study.

R^2s obtained by correlating the total abundances of the hexagonal and rhombohedral phase groups with various properties are listed in the last two columns in Table 8. Compared to those in the rhombohedral group, the hexagonal group shows higher GP and EC capacities (Figure 13a), a lower equilibrium pressure (Figure 13a), a flatter PCT isotherm, a better HRD (Figure 13b), and a reduced surface reactive area at $-40\ ^\circ$C (Figure 13b), but without affecting the charge-transfer resistance. Based on the results, a hexagonal phase will be more desirable for battery applications. Furthermore, combining the preferences in stoichiometry and structure yields the champion of this study—the Nd_2Ni_7 phase.

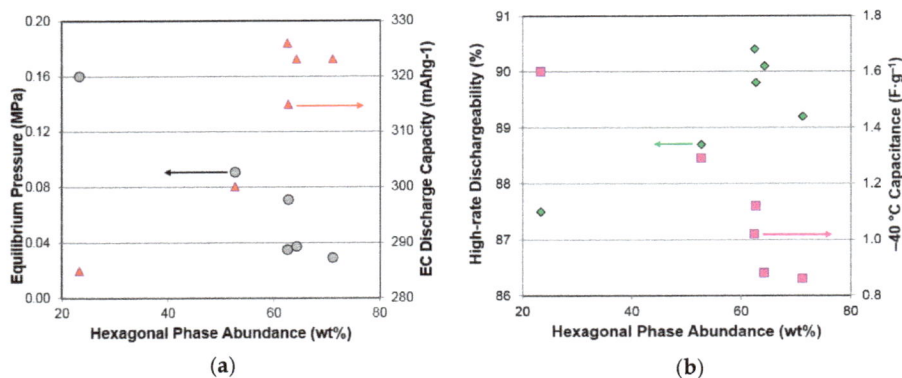

Figure 13. Plots of (**a**) PCT H_2 equilibrium pressure and half-cell electrochemical discharge capacity and (**b**) half-cell high-rate dischargeability and capacitance measured at $-40\ ^\circ$C vs. the hexagonal superlattice structure abundance.

4. Conclusions

Constituent phase abundances of an SmLa-based superlattice alloy were engineered by changing the annealing temperature. Various gaseous phase, half-cell electrochemical, and sealed cell properties and performances were correlated to the individual phase, phase stoichiometry, and type of structure. Many significant correlations were identified; however, because of the complication of phase structure and limited sampling size used in this study, all correlations need further confirmation. In general, an A_2B_7 stoichiometry and a hexagonal structure are more favorable for battery applications. Among the six superlattice phases in this study, Nd_2Ni_7 is the most desirable. As a result, establishment of a single target of maximizing Nd_2Ni_7 phase abundance simplifies the composition/process optimization task. In this study, an annealing condition of 920 $^\circ$C for 5 h for the superlattice $Sm_{14}La_{5.7}Mg_{4.0}Ni_{73}Al_{3.3}$ MH alloy renders the highest Nd_2Ni_7 phase abundance and the best overall electrochemical performance.

Acknowledgments: The authors would like to thank the following individuals from BASF—Ovonic for their help: Su Cronogue, Baoquan Huang, Diana F. Wong, S. Chang, Chaolan Hu, Benjamin Reichman, Nathan English, Sui-ling Chen, Cheryl Setterington, David Pawlik, Allen Chan, and Ryan J. Blankenship.

Author Contributions: Kwo-Hsiung Young designed the experiments and analyzed the results. Taihei Ouchi prepared the alloy samples and performed the PCT and XRD analyses. Jean Nei prepared the electrode samples and conducted the magnetic measurements. Yu-Ling Lien organized the data and prepared the manuscript.

Conflicts of Interest: The authors declare no conflict of interest.

Abbreviations

Mm	Misch metal
MH	Metal hydride alloy
Ni/MH	Nickel/metal hydride
H-storage	Hydrogen-storage
HRD	High-rate dischargeability
GP	Gaseous phase
RE	Rare earth
EC	Electrochemical
bcc	Body-centered-cubic
ICP-OES	Inductively coupled plasma-optical emission spectrometer
XRD	X-ray diffractometer
SEM	Scanning electron microscope
EDS	Energy dispersive spectroscopy
PCT	Pressure-concentration-temperature
BEI	Backscattering electron image
ΔH	Change in enthalpy or heat of hydride formation
ΔS	Change in entropy
P	Desorption pressure
T	Temperature
R	Ideal gas constant
D	Bulk hydrogen diffusion coefficient
I_o	Surface exchange current
R	Surface charge-transfer resistance
C	Surface double-layer capacitance
M_S	Saturated magnetic susceptibility
$H_{1/2}$	Applied field strength corresponding to half of M_S
RT	Room temperature
R^2	Correlation factor

References

1. Yasuoka, S.; Magari, Y.; Murata, T.; Tanaka, T.; Ishida, J.; Nakamura, H.; Nohma, T.; Kihara, M.; Baba, Y.; Teraoka, H. Development of high-capacity nickel-metal hydride batteries using superlattice hydrogen-absorbing alloys. *J. Power Sources* **2006**, *156*, 662–666. [CrossRef]
2. Teraoka, H. Development of Low Self-Discharge Nickel-Metal Hydride Battery. Available online: http://www.scribd.com/doc/9704685/Teraoka-Article-En (accessed on 9 April 2016).
3. Kai, T.; Ishida, J.; Yasuoka, S.; Takeno, K. The effect of nickel-metal hydride battery's characteristics with structure of the alloy. In Proceedings of the 54th Battery Symposium in Japan, Osaka, Japan, 7–9 October 2013; p. 210.
4. Takasaki, T.; Nishimura, K.; Saito, M.; Fukunaga, H.; Iwaki, T.; Sakai, T. Cobalt-free nickel-metal hydride battery for industrial applications. *J. Alloys Compd.* **2013**, *580*, S378–S381. [CrossRef]
5. Teraoka, H. Development of Ni-MH EThSS with Lifetime and Performance Estimation Technology. Presented at the 34th International Battery Seminar & Exhibit, Fort Lauderdale, FL, USA, 20–23 March 2017.
6. Teraoka, H. Ni-MH Stationary Energy Storage: Extreme Temperature & Long Life Developments. Presented at the 33th International Battery Seminar & Exhibit, Fort Lauderdale, FL, USA, 21–24 March 2016.
7. Teraoka, H. Development of Highly Durable and Long Life Ni-MH Batteries for Energy Storage Systems. Presented at the 32th International Battery Seminar & Exhibit, Fort Lauderdale, FL, USA, 9–12 March 2015.
8. Ouchi, T.; Young, K.; Moghe, D. Reviews on the Japanese Patent Applications regarding nickel/metal hydride batteries. *Batteries* **2016**, *2*, 21. [CrossRef]
9. Kadir, K.; Sakai, T.; Uehara, I. Synthesis and structure determination of a new series of hydrogen storage alloys; RMg$_2$Ni$_9$ (R = La, Ce, Pr, Nd, Sm and Gd) built from MgNi$_2$ Laves-type layers alternating with AB$_5$ layers. *J. Alloys Compd.* **1997**, *257*, 115–121. [CrossRef]

10. Kadir, K.; Kuriyama, N.; Sakai, T.; Uehara, I.; Eriksson, L. Structural investigation and hydrogen capacity of CaMg$_2$Ni$_9$: A new phase in the AB$_2$C$_9$ system isostructural with LaMg$_2$Ni$_9$. *J. Alloys Compd.* **1999**, *284*, 145–154. [CrossRef]

11. Kadir, K.; Sakai, T.; Uehara, I. Structural investigation and hydrogen storage capacity of LaMg$_2$Ni$_9$ and (La$_{0.65}$Ca$_{0.35}$)(Mg$_{1.32}$Ca$_{0.68}$)Ni$_9$ of the AB$_2$C$_9$ type structure. *J. Alloys Compd.* **2000**, *302*, 112–117.

12. Hayakawa, H.; Akiba, E.; Gotho, M.; Kohno, T. Crystal structure of hydrogen storage alloys, La-Mg-Ni$_x$ (*x* = 3–4) system. *Jpn. Inst. Met.* **2005**, *69*, 170–178. (In Japanese) [CrossRef]

13. Kohno, T.; Yoshida, H.; Kawashima, F.; Inaba, T.; Sakai, I.; Yamamoto, M.; Kanda, M. Hydrogen storage properties of new ternary system alloys: La$_2$MgNi$_9$, La$_5$Mg$_2$Ni$_{23}$, La$_3$MgNi$_{14}$. *J. Alloys Compd.* **2000**, *311*, L5–L7. [CrossRef]

14. Yoshida, H.; Yamamoto, M.; Sakai, I.; Inaba, T.; Takabayashi, J.; Irie, S.; Suzuki, H.; Takeno, K. Hydrogen Storage Alloy, Alkali Secondary Battery, Hybrid Car and Electric Vehicle. Jpn. Patent 069554, 2002.

15. Inaba, T.; Sakai, I.; Yoshida, H.; Takabayashi, J.; Yamamoto, M.; Suzuki, H.; Irie, S.; Takeno, K. Nickel Hydrogen Secondary Battery, Hybrid Car and Electric Vehicle. Jpn. Patent 083593, 2002.

16. Kawashima, F.; Sakamoto, T.; Arai, T. Hydrogen Storage Alloy and Nickel-Hydrogen Secondary Battery Using the Same. Jpn. Patent 105563, 2002.

17. Liao, B.; Lei, Y.; Chen, L.; Lu, G.; Pan, H.; Wang, Q. A study on the structure and electrochemical properties of La$_2$Mg(Ni$_{0.95}$M$_{0.05}$)$_9$ (M = Co, Mn, Fe, Al, Cu, Sn) hydrogen storage electrode alloys. *J. Alloys Compd.* **2004**, *376*, 186–195. [CrossRef]

18. Pan, H.; Liu, Y.; Gao, M.; Lei, Y.; Wang, Q. Electrochemical properties of the La$_{0.7}$Mg$_{0.3}$Ni$_{2.65-x}$Mn$_{0.1}$Co$_{0.75}$Al$_x$ (*x* = 0–5) hydrogen storage alloy electrode. *J. Electrochem. Soc.* **2005**, *152*, A326–A332. [CrossRef]

19. Liao, B.; Lei, Y.; Chen, L.; Lu, G.; Pan, H.; Wang, Q. The effect of Al substitution for Ni on the structure and electrochemical properties of AB$_3$-type La$_2$Mg(Ni$_{1-x}$Al$_x$)$_9$ (*x* = 0–0.05) alloys. *J. Alloys Compd.* **2005**, *404–406*, 665–668. [CrossRef]

20. Qiu, S.; Chu, H.; Zhang, Y.; Qi, Y.; Sun, L.; Xu, F. Investigation on the structure and electrochemical properties of AB$_3$-type La-Mg-Ni-Co-based hydrogen storage composites. *J. Alloys Compd.* **2008**, *462*, 392–397. [CrossRef]

21. Dong, Z.; Ma, L.; Shen, X.; Wang, L.; Wu, Y.; Wang, L. Cooperative effect of Co and Al on the microstructure and electrochemical properties of AB$_3$-type hydrogen storage electrode alloys for advanced MH/Ni secondary battery. *J. Alloys Compd.* **2011**, *36*, 893–900. [CrossRef]

22. Belgacem, Y.B.; Khaldi, C.; Lamloumi, J. The effect of the discharge rate on the electrochemical properties of AB$_3$-type hydrogen storage alloy as anode in nickel-metal hydride batteries. *Int. J. Hydrogen Energy* **2017**, *42*, 12797–12807. [CrossRef]

23. Liu, Z.; Yan, X.; Wang, N.; Chai, Y.; Hou, D. Cyclic stability and high rate discharge performance of (La,Mg)$_5$Ni$_{19}$ multiphase alloy. *Int. J. Hydrogen Energy* **2011**, *36*, 4370–4374. [CrossRef]

24. Guo, X.; Luo, Y.; Gao, Z.; Zhang, G.; Kang, L. The effect of Mg on the microstructure and electrochemical properties of La$_{0.8-x}$Gd$_{0.2}$Mg$_x$Ni$_{3.3}$Co$_{0.3}$Al$_{0.1}$ (*x* = 0–0.4) hydrogen storage alloys. *Funct. Mater.* **2012**, *43*, 2450–2455. (In Chinese)

25. Zhao, Y.; Han, S.; Li, Y.; Liu, J.; Zhang, L.; Yang, S.; Ke, D. Characterization and improvement of electrochemical properties of Pr$_5$Co$_{19}$-type single-phase La$_{0.84}$Mg$_{0.16}$Ni$_{3.80}$ alloy. *Electrochim. Acta* **2015**, *152*, 265–273. [CrossRef]

26. Li, Y.; Zhang, Y.; Ren, H.; Liu, Z.; Sun, H. Mechanism of distinct high rate dischargeability of La$_4$MgNi$_{19}$ electrode alloys prepared by casting and rapid quenching followed by annealing treatment. *Int. J. Hydrogen Energy* **2016**, *41*, 18571–18581. [CrossRef]

27. Xue, C.; Zhang, L.; Fan, Y.; Fan, G.; Liu, B.; Han, S. Phase transformation and electrochemical hydrogen storage performance of La$_3$RMgNi$_{19}$ (R = La, Pr, Nd, Sm, Gd, and Y) alloys. *Int. J. Hydrogen Energy* **2017**, *42*, 6051–6064. [CrossRef]

28. Liu, J.; Han, S.; Li, Y.; Zhang, L.; Zhao, Y.; Yang, S.; Liu, B. Phase structures and electrochemical properties of La–Mg–Ni-based hydrogen storage alloys with superlattice structure. *Int. J. Hydrogen Energy* **2016**, *41*, 20261–20275. [CrossRef]

29. Yan, H.; Xiong, W.; Wang, L.; Li, B.; Li, J.; Zhao, X. Investigations on AB$_3$-, A$_2$B$_7$- and A$_5$B$_{19}$-type La-Y-Ni system hydrogen storage alloys. *Int. J. Hydrogen Energy* **2017**, *42*, 2257–2264. [CrossRef]

30. Young, K.; Yasuoka, S. Past, present, and future of metal hydride alloys in nickel-metal hydride batteries. In Proceedings of the 14th International Symposium on Metal-Hydrogen Systems, Manchester, UK, 21–25 July 2014.

31. Young, K.; Chang, S.; Lin, X. C14 Laves phase metal hydride alloys for Ni/MH batteries applications. *Batteries* **2017**, *3*, 27. [CrossRef]

32. Young, K.; Wong, D.F.; Wang, L.; Nei, J.; Ouchi, T.; Yasuoka, S. Mn in misch-metal based superlattice metal hydride alloy—Part 1 Structural, hydrogen storage and electrochemical properties. *J. Power Sources* **2015**, *277*, 426–432. [CrossRef]

33. Wang, L.; Young, K.; Meng, T.; Ouchi, T.; Yasuoka, S. Partial substitution of cobalt for nickel in mixed rare earth metal based superlattice hydrogen absorbing alloy—Part 1 structural, hydrogen storage and electrochemical properties. *J. Alloys Compd.* **2016**, *660*, 407–415. [CrossRef]

34. Young, K.; Wong, D.F.; Wang, L.; Nei, J.; Ouchi, T.; Yasuoka, S. Mn in misch-metal based superlattice metal hydride alloy—Part 2 Ni/MH battery performance and failure mechanism. *J. Power Sources* **2015**, *277*, 433–442. [CrossRef]

35. Wang, L.; Young, K.; Meng, T.; English, N.; Yasuoka, S. Partial substitution of cobalt for nickel in mixed rare earth metal based superlattice hydrogen absorbing alloy—Part 2 battery performance and failure mechanism. *J. Alloys Compd.* **2016**, *664*, 417–427. [CrossRef]

36. Yasuoka, S.; Ishida, J.; Kai, T.; Kajiwara, T.; Doi, S.; Yamazaki, T.; Kishida, K.; Inui, H. Function of aluminum in crystal structure of rare earth-Mg-Ni hydrogen-absorbing alloy and deterioration mechanism of $Nd_{0.9}Mg_{0.1}Ni_{3.4}$ and $Nd_{0.9}Mg_{0.1}Ni_{3.3}Al_{0.2}$ alloys. *Int. J. Hydrogen Energy* **2017**, *42*, 11574–11583. [CrossRef]

37. Yasuoka, S.; Ishida, J.; Kishida, K.; Inui, H. Effects of cerium on the hydrogen absorption-desorption properties of rare earth-Mg-Ni hydrogen-absorbing alloys. *J. Power Sources* **2017**, *346*, 56–62. [CrossRef]

38. Zhang, Q.A.; Lei, Y.Q.; Wang, C.S.; Wang, F.S.; Wang, Q.D. Structure of the secondary phase and its effects on hydrogen-storage properties in a $Ti_{0.7}Zr_{0.2}V_{0.1}Ni$ alloy. *J. Power Sources* **1998**, *75*, 288–291. [CrossRef]

39. Jurczyk, M.; Jankowska, E.; Makowiecka, M.; Wieczorek, I. Electrode characteristics of nanocrystalline TiFe-type alloys. *J. Alloys Compd.* **2003**, *354*, L1–L4. [CrossRef]

40. Zhang, Q.A.; Lei, Y.Q.; Yang, X.G.; Ren, K.; Wang, Q.D. Annealing treatment of AB_2-type hydrogen storage alloys: I. crystal structures. *J. Alloys Compd.* **1999**, *292*, 236–240. [CrossRef]

41. Zhang, Q.A.; Lei, Y.Q.; Yang, X.G.; Du, Y.L.; Wang, Q.D. Effects of annealing treatment on phase structures, hydrogen absorption–desorption characteristics and electrochemical properties of a $V_3TiNi_{0.56}Hf_{0.24}Mn_{0.15}Cr_{0.1}$ alloy. *J. Alloys Compd.* **2000**, *305*, 125–129. [CrossRef]

42. Yang, X.G.; Zhang, Q.A.; Shu, K.Y.; Du, Y.L.; Lei, Y.Q.; Wang, Q.D.; Zhang, W.K. The effect of annealing on the electrochemical; properties of $Zr_{0.5}Ti_{0.5}Mn_{0.5}V_{0.3}Co_{0.2}Ni_{1.1}$ alloy electrode. *J. Power Sources* **2000**, *90*, 170–175. [CrossRef]

43. Young, K.; Ouchi, T.; Huang, B.; Chao, B.; Fetcenko, M.A.; Bendersky, L.A.; Wang, K.; Chiu, C. The correlation of C14/C15 phase abundance and electrochemical properties in the AB_2 alloys. *J. Alloys Compd.* **2010**, *506*, 841–848. [CrossRef]

44. Pan, H.; Liu, Y.; Gao, M.; Zhu, Y.; Lei, Y.; Wang, Q. A study on the effect of annealing treatment on the electrochemical properties of $La_{0.67}Mg_{0.33}Ni_{2.5}Co_{0.5}$ alloy electrode. *Int. J. Hydrogen Energy* **2003**, *28*, 113–117. [CrossRef]

45. Yang, Z.P.; Li, Q.; Zhao, X.J. Influence of magnetic annealing on electrochemical performance of $La_{0.67}Mg_{0.33}Ni_{3.0}$ hydride electrode. *J. Alloys Compd.* **2013**, *558*, 99–104. [CrossRef]

46. Hu, W.; Denys, R.V.; Nwakwuo, C.C.; Holm, T.; Maehlen, J.P.; Solberg, J.K.; Yartys, V.A. Annealing effect on phase composition and electrochemical properties of the Co-free La_2MgNi_9 anode for Ni-metal hydride batteries. *Electrochim. Acta* **2013**, *96*, 27–133. [CrossRef]

47. Li, P.; Zhang, J.; Zhai, F.; Ma, G.; Xu, L.; Qu, X. Effect of annealing treatment on the anti-pulverization and anti-corrosion properties of $La_{0.67}Mg_{0.33}Ni_{2.5}Co_{0.5}$ hydrogen storage alloy. *J. Rare Earths* **2015**, *33*, 417–424. [CrossRef]

48. Young, K.; Ouchi, T.; Huang, B. Effects of annealing and stoichiometry to $(Nd, Mg)(Ni, Al)_{3.5}$ metal hydride alloys. *J. Power Sources* **2012**, *215*, 152–159. [CrossRef]

49. Balcerzak, M.; Nowak, M.; Jurczyk, M. Hydrogenation and electrochemical studies of La-Mg-Ni alloy. *Int. J. Hydrogen Energy* **2017**, *42*, 1436–1443. [CrossRef]

50. Xiong, W.; Uan, H.; Wang, L.; Zhao, X.; Li, J.; Li, B.; Wang, Y. Effects of annealing temperature on the structure and properties of the $LaY_2Ni_{10}Mn_{0.5}$ hydrogen storage alloy. *Int. J. Hydrogen Energy* **2017**, *42*, 15319–15327. [CrossRef]

51. Zhang, Y.; Yang, T.; Zhai, T.; Yuan, Z.; Zhang, G.; Guo, S. Effects of stoichiometric ratio La/Mg on structures and electrochemical performances of as-cast and annealed La-Mg-Ni-based A_2B_7-type electrode alloy. *Trans. Nonferrous Met. Soc. China* **2015**, *25*, 1968–1977. [CrossRef]

52. Hayakawa, H.; Enoki, H.; Akiba, E. Annealing conditions with Mg vapor-pressure control and hydrogen storage characteristic of La_4MgNi_{19} hydrogen storage alloy. *Jpn. Inst. Met.* **2006**, *70*, 158–161. (In Japanese) [CrossRef]

53. Hu, W.K.; Kim, D.M.; Jeon, S.W.; Lee, J.Y. Effect of annealing treatment on electrochemical properties of Mm-based hydrogen storage alloys for Ni/MH batteries. *J. Alloys Compd.* **1998**, *270*, 255–264. [CrossRef]

54. Ma, Z.; Qiu, J.; Chen, L.; Lei, Y. Effects of annealing on microstructure and electrochemical properties of the low Co-containing alloy MI(NiCoMnAlFe)$_5$ for Ni/MH battery electrode. *J. Power Sources* **2004**, *125*, 267–272. [CrossRef]

55. Zhou, Z.; Song, Y.; Cui, S.; Huang, C.; Qian, W.; Lin, C.; Zhang, Y.; Lin, Y. Effect of annealing treatment on structure and electrochemical performance of quenched $MmNi_{4.2}Co_{0.3}Mn_{0.4}Al_{0.3}Mg_{0.03}$ hydrogen storage alloy. *J. Alloys Compd.* **2010**, *501*, 47–53. [CrossRef]

56. Nei, J.; Young, K.; Salley, S.O.; Ng, K.Y.S. Effects of annealing on $Zr_8Ni_{19}X_2$ (X = Ni, Mg, Al, Sc, V, Mn, Co, Sn, La and Hf): Structural characteristics. *J. Alloys Compd.* **2012**, *516*, 144–152. [CrossRef]

57. Nei, J.; Young, K.; Salley, S.O.; NG, K.Y.S. Effects of annealing on $Zr_8Ni_{19}X_2$ (X = Ni, Mg, Al, Sc, V, Mn, Co, Sn, La and Hf): Hydrogen storage and electrochemical properties. *Int. J. Hydrogen Energy* **2012**, *37*, 8418–8427. [CrossRef]

58. Young, K.; Ouchi, T.; Nei, J.; Wang, L. Annealing effects on Laves phase-related body-centered-cubic solid solution metal hydride alloys. *J. Alloys Compd.* **2016**, *654*, 216–225. [CrossRef]

59. Crivello, J.-C.; Zhang, J.; Latroche, M. Structural stability of AB_y phases in the (La,Mg)-Ni system obtained by density functional theory calculations. *J. Phys. Chem.* **2011**, *115*, 25470–25478.

60. Wong, D.F.; Young, K. Phase stability of superlattice metal hydride alloy estimated by first principle calculation. Unpublished work, 2017.

61. Zhang, L.; Zhang, J.; Han, S.; Li, Y.; Yang, S.; Liu, J. Phase transformation and electrochemical properties of $La_{0.70}Mg_{0.30}Ni_{3.3}$ super-stacking metal hydride alloy. *Intermetallics* **2015**, *58*, 65–70.

62. Li, F.; Young, K.; Ouchi, T.; Fetcenko, M.A. Annealing effects on structural and electrochemical properties of $(LaPrNdZr)_{0.83}Mg_{0.17}(NiCoAlMn)_{3.3}$ alloy. *J. Alloys Compd.* **2009**, *471*, 371–377. [CrossRef]

63. Young, K.; Wang, L.; Yan, S.; Liao, X.; Meng, T.; Shen, H.; May, W.C. Fabrications of high-capacity alpha-$Ni(OH)_2$. *Batteries* **2017**, *3*, 6. [CrossRef]

64. Young, K.; Wu, A.; Qiu, Z.; Tan, J.; Mays, W. Effects of H_2O_2 addition to the cell balance and self-discharge of Ni/MH batteries with AB_5 and A_2B_7 alloys. *Int. J. Hydrogen Energy* **2012**, *37*, 9882–9891. [CrossRef]

65. Young, K.; Koch, J.M.; Wan, C.; Denys, R.V.; Yartys, V.A. Cell performance comparison between C14- and C15-predominated AB_2 metal hydride alloys. *Batteries* **2017**, *3*, 29. [CrossRef]

66. Charbonnier, V.; Monnier, J.; Zhang, J.; Paul-Boucour, V.; Joiret, S.; Puga, B.; Goubault, L.; Bernard, P.; Latroche, M. Relationship between H_2 sorption properties and aqueous corrosion mechanisms in A_2B_7 hydride forming alloys (A = Y, Gd or Sm). *J. Power Sources* **2016**, *326*, 146–155. [CrossRef]

67. Tang, R.; Wei, X.; Liu, Y.; Zhu, C.; Zhu, J.; Yu, G. Effect of the Sm content on the structure and electrochemical properties of $La_{1.3-x}Sm_xCaMg_{0.7}Ni_9$ (x = 0–0.3) hydrogen storage alloys. *J. Power Sources* **2006**, *155*, 456–460.

68. Ping, L.; Hou, Z.; Yang, T.; Shang, H.; Qu, X.; Zhang, Y. Structure and electrochemical hydrogen storage characteristics of the as-cast and annealed $La_{0.8-x}Sm_xMg_{0.2}Ni_{3.15}Co_{0.2}Al_{0.1}Si_{0.05}$ (x = 0–0.4) alloys. *J. Rare Earths* **2012**, *30*, 696–704.

69. Zhang, Y.; Hou, Z.; Li, B.; Ren, H.; Zhang, G.; Zhao, D. An investigation on electrochemical hydrogen storage performances of the as-cast and -annealed $La_{0.8-x}Sm_xMg_{0.2}Ni_{3.35}Al_{0.1}Si_{0.05}$ (x = 0–0.4) alloys. *J. Alloys Compd.* **2012**, *537*, 175–182. [CrossRef]

70. Zhang, Y.; Li, P.; Yang, T.; Zhai, T.; Yuan, Z.; Guo, S. Effects of substituting La with M (M = Sm, Nd, Pr) on electrochemical hydrogen storage characteristics of A_2B_7-type electrode alloys. *Trans. Nonferrous Met. Soc. China* **2014**, *24*, 4012–4022. [CrossRef]

71. Liu, J.; Han, S.; Li, Y.; Zhao, X.; Yang, S.; Zhao, Y. Cooperative effects of Sm and Mg on electrochemical performance of La–Mg–Ni-based alloys with A_2B_7- and A_5B_{19}-type super-stacking structure. *Int. J. Hydrogen Energy* **2015**, *40*, 1116–1127. [CrossRef]

72. Young, K.; Ouchi, T.; Wang, L.; Wong, D.F. The effects of Al substitution on the phase abundance, structure and electrochemical performance of $La_{0.7}Mg_{0.3}Ni_{2.8}Co_{0.5-x}Al_x$ (x = 0, 0.1, 0.2) alloys. *J. Power Sources* **2015**, *279*, 172–179. [CrossRef]

73. Zhang, L.; Ding, Y.; Zhao, Y.; Du, W.; Li, Y.; Yang, S.; Han, S. Phase structure and cycling stability of A_2B_7 superlattice $La_{0.60}Sm_{0.15}Mg_{0.25}Ni_{3.4}$ metal hydride alloy. *Int. J. Hydrogen Energy* **2016**, *41*, 1791–1800. [CrossRef]

74. Young, K.; Ouchi, T.; Koch, J.; Fetcenko, M.A. The role of Mn in C14 Laves phase multi-component alloys for NiMH battery application. *J. Alloys Compd.* **2009**, *477*, 749–758. [CrossRef]

75. Osumi, Y. *Suiso Kyuzou Goukin*; Agune Technology Center: Tokyo, Japan. 1999; p. 218. (In Japanese)

76. Schlapbach, L.; Züttel, A. Hydrogen-storage materials for mobile applications. *Nature* **2001**, *414*, 353–358. [CrossRef] [PubMed]

77. Young, K.; Ouchi, T.; Nei, J.; Moghe, D. The importance of rare-earth additions in Zr-based AB_2 metal hydride alloys. *Batteries* **2016**, *2*, 25. [CrossRef]

78. Stucki, F.; Schlapbach, L. Magnetic properties of $LaNi_5$, FeTi, Mg_2Ni and their hydrides. *J. Less-Comm. Met.* **1980**, *74*, 143–151. [CrossRef]

79. Young, K.; Huang, B.; Regmi, R.K.; Lawes, G.; Liu, Y. Comparisons of metallic clusters imbedded in the surface of AB_2, AB_5, and A_2B_7 alloys. *J. Alloys Compd.* **2010**, *506*, 831–840. [CrossRef]

80. Young, K.; Chao, B.; Pawlik, D.; Shen, H.T. Transmission electron microscope studies in the surface oxide on the La-containing AB_2 metal hydride alloy. *J. Alloys Compd.* **2016**, *672*, 355–365. [CrossRef]

81. Meng, T.; Young, K.; Hu, C.; Reichman, B. Effects of alkaline pre-etching to metal hydride alloys. *Batteries* **2017**, *3*, 30. [CrossRef]

82. Young, K.; Nei, J.; Wan, C.; Denys, R.V.; Yartys, V.A. Comparison of C15- and C15-predominated AB_2 metal hydride alloys for electrochemical applications. *Batteries* **2017**, *3*, 22. [CrossRef]

batteries

MDPI

Article

Fe-Substitution for Ni in Misch Metal-Based Superlattice Hydrogen Absorbing Alloys—Part 1. Structural, Hydrogen Storage, and Electrochemical Properties

Kwo-Hsiung Young [1,2,*], Taihei Ouchi [2], Jean Nei [2] and Shigekazu Yasuoka [3]

[1] Department of Chemical Engineering and Materials Science, Wayne State University, Detroit, MI 48202, USA
[2] BASF/Battery Materials-Ovonic, 2983 Waterview Drive, Rochester Hills, MI 48309, USA;
    taihei.ouchi@basf.com (T.O.); jean.nei@basf.com (J.N.)
[3] Engineering Division, Ni-MH Group, FDK Corporation, 307-2, Koyagi-Machi, Takasaki, Gunma 370-0042,
    Japan; shigekazu.yasuoka@fdk.co.jp
* Correspondence: kwo.young@basf.com; Tel.: +1-248-293-7000

Academic Editor: Andreas Jossen
Received: 19 October 2016; Accepted: 11 November 2016; Published: 21 November 2016

Abstract: The effects of Fe partially replacing Ni in a misch metal-based superlattice hydrogen absorbing alloy (HAA) were studied. Addition of Fe increases the lattice constants and abundance of the main Ce_2Ni_7 phase, decreases the $NdNi_3$ phase abundance, and increases the $CaCu_5$ phase when the Fe content is above 2.3 at%. For the gaseous phase hydrogen storage (H-storage), Fe incorporation does not change the storage capacity or equilibrium pressure, but it does decrease the change in both entropy and enthalpy. With regard to electrochemistry, >2.3 at% Fe decreases both the full and high-rate discharge capacities due to the deterioration in both bulk transport (caused by decreased secondary phase abundance and consequent lower synergetic effect) and surface electrochemical reaction (caused by the lower volume of the surface metallic Ni inclusions). In a low-temperature environment ($-40\ ^\circ$C), although Fe increases the reactive surface area, it also severely hinders the ability of the surface catalytic, leading to a net increase in surface charge-transfer resistance. Even though Fe increases the abundance of the beneficial Ce_2Ni_7 phase with a trade-off for the relatively unfavorable $NdNi_3$ phase, it also deteriorates the electrochemical performance due to a less active surface. Therefore, further surface treatment methods that are able to increase the surface catalytic ability in Fe-containing superlattice alloys and potentially reveal the positive contributions that Fe provides structurally are worth investigating in the future.

Keywords: metal hydride (MH); nickel/metal hydride (Ni/MH) battery; hydrogen absorbing alloy (HAA); electrochemistry; superlattice alloy

---

1. Introduction

Misch metal-based superlattice hydrogen absorbing alloy (HAA) has become the mainstream negative electrode active material for commercial nickel/metal hydride (Ni/MH) batteries due to its higher capacity, improved high-rate capability, wider operating temperature range, and lower self-discharge compared to the conventional AB_5 HAA [1–4]. Superlattice HAAs belong to a family of alloys mainly composed of AB_3, A_2B_7, and A_5B_{19} structures, which are constructed with various numbers of AB_5 building slabs (one, two, and three, respectively) between the A_2B_4 building slabs [5–7]. Other non-superlattice phases, such as AB_2, $MgLaNi_4$, and AB_5, may also be present in these multi-phase superlattice alloys [8–11]. While the basic formula of the commercial superlattice HAAs contains only rare earth metals (La, Pr, Nd, and/or Sm), Ni, or Al, we have previously reported the effects of adding Ce [12], Mn [8,13], and Co [9,14]. In general, Ce promotes AB_2 phase formation and deteriorates battery performance, Mn improves the high-rate performance but creates micro-shorts in the separator and consequently causes severe self-discharge, and Co improves the low-temperature performance with the sacrifices in self-discharge and high-temperature performance. In this paper, we investigate the effects of Fe, another transition metal with an atomic number between Mn and Co, partially replacing Ni on the structural, gaseous phase storage, and electrochemical properties of HAA.

Fe has been used as a low-cost supplement in AB_2 [15–22], AB_5 [23–34], and body-centered-cubic (bcc) [35–40] HAAs. However, the results of Fe incorporation in these HAAs have so far been inconsistent. For example, the addition of Fe to V-containing AB_2 HAA deteriorated the low-temperature performance but increased the discharge capacity [21] and cycle stability [18]. Contradictory results from adding Fe to V-free AB_2 HAA were also reported (improved low-temperature performance but degraded capacity and cycle performance) [22]. Moreover, Fe incorporation in AB_5 HAA enhanced the low-temperature performance due to increases in surface area [23,28] and surface catalytic ability [34], but lowered the capacity [23,27,28,34] and cycle stability [29,34]. However, negative impact on low-temperature performance [30] and positive contribution to cycle stability [32] from Fe addition in AB_5 HAA have also been previously reported. In the Laves phase-related bcc solid solution HAA, Fe impeded high-rate dischargeability (HRD) [40] and decreased the bcc phase abundance [35,36], but the opposite results have also been reported (improved HRD and increased bcc phase abundance) [37]. In addition, ferrovanadium was used as an alternative and inexpensive source of vanadium in the bcc HAA [38,39]. Very few results regarding Fe addition in the superlattice HAA have been published, and all published articles concerned the La-only alloy, which has very limited use in practical Ni/MH battery applications due to its easily oxidized nature and thus short life cycle [41]. Wang et al. [42] reported that Fe addition in the $(La,Mg)Co_{0.45}Ni_{2.55}$ HAA decreased the discharge capacity and HRD, but improved the charge stability. Wu et al. [43] reported that Fe substitution in the $La_{0.7}Ng_{0.3}Ni_3$ superlattice HAA resulted in a lower capacity and HRD. Since any investigation on the effects of Fe incorporation in the misch metal-based superlattice HAA was absent, such results are very desirable in order to further improve the electrochemical properties of the superlattice HAA. While this paper (Part 1) summarizes the results of structural, gaseous phase, and electrochemical (in half-cell configuration) studies on the Fe-substituted misch metal-based superlattice alloys, the performance and failure analysis of the Ni/MH batteries made using these alloys will be discussed in another publication (Part 2, [44]).

2. Experimental Setup

First, −200 mesh HAA powder (2 kg per composition) was prepared by Japan Metals & Chemicals Company (Tokyo, Japan) using the induction melting method. The chemical composition of the HAA powder was verified with a Varian Liberty 100 inductively coupled plasma-optical emission spectrometer (ICP-OES, Agilent Technologies, Santa Clara, CA, USA). Microstructure analysis was performed with a Philips X'Pert Pro X-ray diffractometer (XRD, Amsterdam, The Netherlands) and a JEOL-JSM6320F scanning electron microscope (SEM, Tokyo, Japan) with energy dispersive spectroscopy (EDS). A Suzuki-Shokan multi-channel pressure-concentration-temperature system (PCT, Tokyo, Japan) was used to measure the gaseous phase hydrogen storage (H-storage) characteristics. PCT measurements at 30 and 45 °C were performed after activation. A negative electrode sample was made by compacting the HAA powder onto an expanded nickel substrate through a roll mill without the use of a binder. The half-cell experiment, which employed a pre-activated sintered $Ni(OH)_2$ counter electrode and 30% KOH as the electrolyte, was performed using a CTE MCL2 Mini cell testing system (Chen Tech Electric Mfg. Co., Ltd., New Taipei, Taiwan). A Solartron 1250 Frequency Response Analyzer (Solartron Analytical, Leicester, UK) with a sine wave amplitude of 10 mV and a frequency range of 0.5 mHz–10 kHz was used for the alternative current (AC) impedance measurements. A Digital Measurement Systems Model 880 vibrating sample magnetometer (MicroSense, Lowell, MA, USA) was used to measure the magnetic susceptibility of the surfaces of the activated alloy powder (activation was performed by immersing the powder in 30 wt % KOH solution at 100 °C for 4 h).

3. Results and Discussion

Five superlattice HAAs (Fe1–Fe5) with a general composition of $Mm_{0.83}Mg_{0.17}Ni_{2.94-x}Al_{0.17}Co_{0.2}Fe_x$ ($x = 0, 0.05, 0.1, 0.15, 0.2$) were prepared for this study. Design compositions and ICP results in at% are listed in Table 1. The B/A ratio was set to 3.31, which is the same as in previous matrices for Mn [8] and Co substitutions [9], and the stoichiometry for the main Ce_2Ni_7 phase was independent of the overall composition [10]. The base alloy (Fe1) was chosen from a previous matrix of Co-substituted superlattice HAAs (C3) due to its superior low-temperature performance [9]. ICP results show that the composition of each alloy is very close to its designed value (Table 1).

Table 1. Designed compositions (in **bold**) and inductively coupled plasma (ICP) results in at%. Mm stands for misch metal (mixture of rare earth elements).

Alloy	Source	Mm	Mg	Ni	Al	Co	Fe	B/A
Fe1	**Design**	**19.3**	**3.9**	**68.2**	**3.9**	**4.6**	**0.0**	**3.31**
	ICP	19.2	4.0	68.0	4.0	4.7	0.1	3.32
Fe2	**Design**	**19.3**	**3.9**	**67.1**	**3.9**	**4.6**	**1.2**	**3.31**
	ICP	19.1	3.9	67.0	4.0	4.8	1.2	3.35
Fe3	**Design**	**19.3**	**3.9**	**65.9**	**3.9**	**4.6**	**2.3**	**3.31**
	ICP	19.1	4.0	65.8	4.0	4.7	2.4	3.33
Fe4	**Design**	**19.3**	**3.9**	**64.7**	**3.9**	**4.6**	**3.5**	**3.31**
	ICP	19.2	4.0	64.6	4.0	4.7	3.6	3.32
Fe5	**Design**	**19.3**	**3.9**	**63.6**	**3.9**	**4.6**	**4.6**	**3.31**
	ICP	19.1	4.0	63.4	4.0	4.7	4.7	3.31

3.1. X-Ray Diffraction Analysis

The powder XRD patterns of the five alloys in this study are shown in Figure 1. Peaks from the Ce_2Ni_7, $NdNi_3$, and $CaCu_5$ phases can easily be identified. The lattice parameters of the main phase (Ce_2Ni_7) and phase abundances of all three phases were calculated using the Rietveld refinement method with the Jade 9 software (Materials Data Inc. (MDI), Livermore, CA, USA), and the results are summarized in Table 2.

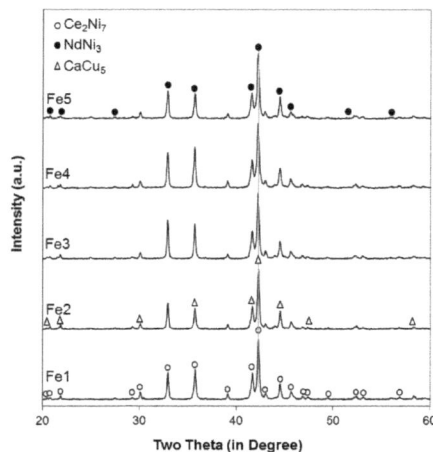

Figure 1. X-ray diffraction (XRD) patterns using a Cu–Kα radiation source for alloys Fe1–Fe5. The vertical line shows the shift in the main peak to lower angles (the unit cell becomes larger) with increasing Fe content.

Table 2. Lattice constants a and c, c/a ratio, unit cell volume, and crystallite size of the main Ce_2Ni_7 phase from a full-pattern fitting, and phase abundances calculated from XRD analysis of alloys Fe1–Fe5. Error ranges are indicated in parentheses after the lattice parameters. R: rhombohedral; and H: hexagonal.

Alloy	Fe1	Fe2	Fe3	Fe4	Fe5
Ce_2Ni_7 a (Å)	5.0198(3)	5.0228(4)	5.0282(5)	5.0306(4)	5.0321(3)
Ce_2Ni_7 c (Å)	24.3951(8)	24.4085(9)	24.4166(7)	24.4310(8)	24.4340(12)
Ce_2Ni_7 c/a ratio	4.860	4.860	4.856	4.856	4.856
Ce_2Ni_7 unit cell volume (Å^3)	532.36	533.29	534.61	535.44	535.83
Ce_2Ni_7 crystallite size (Å)	>1000	>1000	880	>1000	844
Ce_2Ni_7 (H) abundance (wt %)	58.0	62.3	66.4	66.6	68.0
$NdNi_3$ a (Å)	5.0265(3)	5.0363(4)	4.9782(5)	5.0399(5)	4.9539(3)
$NdNi_3$ c (Å)	24.4471(11)	24.4575(9)	24.7684(7)	24.7587(8)	23.9907(10)
$NdNi_3$ (R) abundance (wt %)	20.1	17.5	13.7	10.2	1.5
$CaCu_5$ a (Å)	5.0203(4)	5.0263(5)	5.0293(4)	5.0209(3)	5.0322(6)
$CaCu_5$ c (Å)	4.0693(3)	4.0692(4)	4.0706(6)	4.0729(5)	4.0716(6)
$CaCu_5$ (H) abundance (wt %)	21.9	20.2	19.8	23.2	30.4

As seen in Figure 2, both lattice parameters a and c of the main Ce_2Ni_7 phase increase with increasing Fe content (larger size compared to that of Ni), resulting in the unit cell expansions and, consequently, a shift in the main XRD peak (around 42.2°) to lower angles (larger interplanar distance), as indicated by the vertical line in Figure 1. Moreover, the roughly unchanged c/a ratios shown in Table 2 suggest that the Ce_2Ni_7 unit cell expansion is isotropic, which differs from the faster growth in the a-direction reported in the Mn-substitution study [8]. Changes in the Ce_2Ni_7 unit cell volume with Mn, Fe, and Co substitutions are compared in Figure 3. While the slopes of graphs representing increase versus substitution amount for the Mn and Fe substitutions are similar, no change is observed for the Co substitution due to the similarity in atomic radii between Co and the replaced Ni. Additionally, no particular trend in Ce_2Ni_7 crystallite size (determined by the full width at half-maximum of the (109) peak) with increasing Fe content can be established, which differs from the Mn (decreasing) [8] and Co substitutions (increasing) [9].

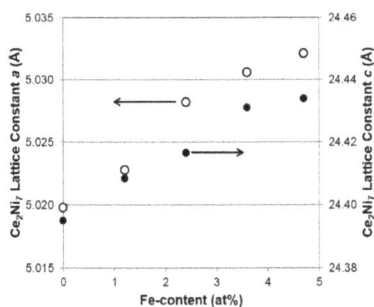

Figure 2. Changes in the Ce_2Ni_7 phase lattice constants a and c with the designed change in Fe content.

Phase abundances of the three constituent phases from the Mn, Fe, and Co substitution studies are plotted in Figure 4. It should be noted that phases with the same composition are merged into one category. For example, calculations for the A_2B_7 phase abundance include both the hexagonal (Ce_2Ni_7) and rhombohedral (Pr_2Ni_7) phase structures. Compared to the large fluctuations in A_2B_7 phase abundance for the Mn and Co substitutions, the Fe substitution increases the amount of the main A_2B_7 phase (Figure 4a). The electrochemical performance of the A_2B_7 phase was believed to be superior to that of the AB_3 phase [45]. However, while additives, including Ce, Nd, Pr, Sm, Zr, Ti, Si, and Cr, decreased the A_2B_7 phase abundance, only Fe and Cu were found to increase the abundance of this phase [46]. It is clear from the comparison shown in Figure 4b that Mn promotes and Fe decreases the AB_3 phase formation. Meanwhile, the effect of Co in this case is not obvious. Trends in AB_5 phase abundance are similar for the increases in all three substitutions, except for the alloy with the least amount of Mn in the Mn-series, and exhibit the behavior of first decreasing and then increasing. Among the three substitutions, the average amount of the AB_5 phase abundance increases in the following trend: Mn < Co < Fe.

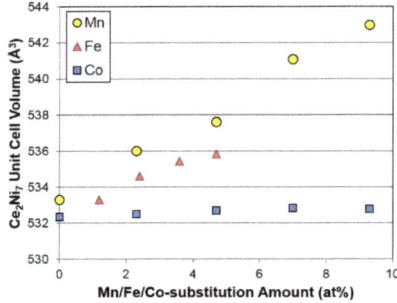

Figure 3. Comparison of Ce_2Ni_7 unit cell volumes with different substitutions.

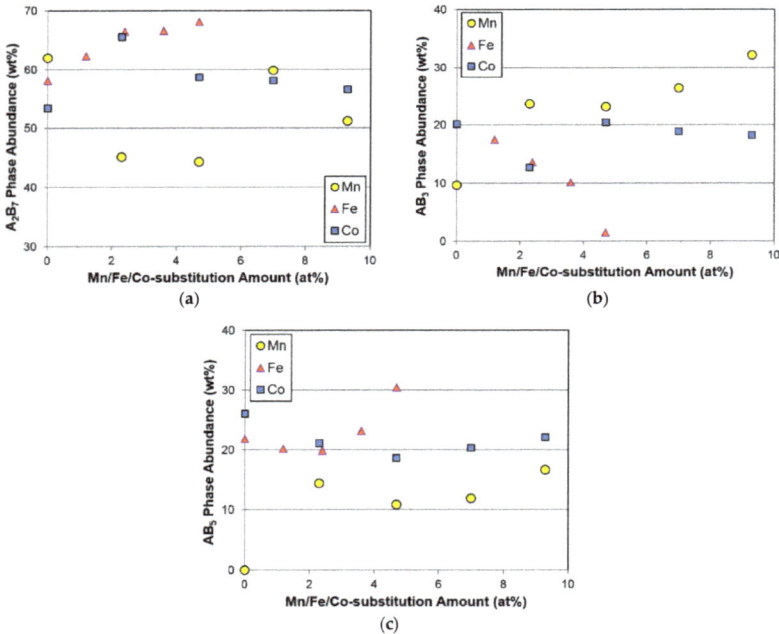

Figure 4. Changes in the (**a**) A_2B_7; (**b**) AB_3; and (**c**) AB_5 phase abundances with increasingdf substitution amounts.

3.2. Scanning Electron Microscope/Energy Dispersive Spectroscopy Analysis

SEM analysis was performed on the polished powder sample surface, and the backscattered electron (BSE) micrographs of the five alloys are shown in Figure 5. In general, the color intensity of each image is very uniform, with some occasional brighter spots and/or areas with slightly darker contrasts. The chemical composition of the numbered areas in each BSE micrograph were studied by EDS, and the results are summarized in Table 3. Brighter spots (Figure 5a-2,b-2) are an excess of rare earth metals, and the darker areas are either the AB_5 (Figure 5b-3,d-2) or $LaMgNi_4$ (Figure 5c-2,e-2) phase. The AB_3 and A_2B_7 phases cannot be separated due to their similarity in composition. The B/A ratios in the main AB_3/A_2B_7 phase of alloys Fe1–Fe5 are in the range of 3.31–3.41, which is between the B/A ratios of AB_3 (3.0) and A_2B_7 (3.5) and closer to the B/A ratio of A_2B_7 (3.5). This finding is in agreement of the XRD results, where the A_2B_7 phase abundance was found to be higher than the AB_3 phase abundance. Similar to the cases of Mn and Co substitutions, Fe does not form any separated secondary phase and dissolves completely in the main AB_3/A_2B_7 phase. Therefore, changes in the lattice constant of the main phase, as revealed by XRD, were the result of Fe incorporation. EDS analysis also shows that the $LaMgNi_4$ phase is deficient in Al, Co, and Fe, while the AB_5 phase is rich in Al, Co, and Fe.

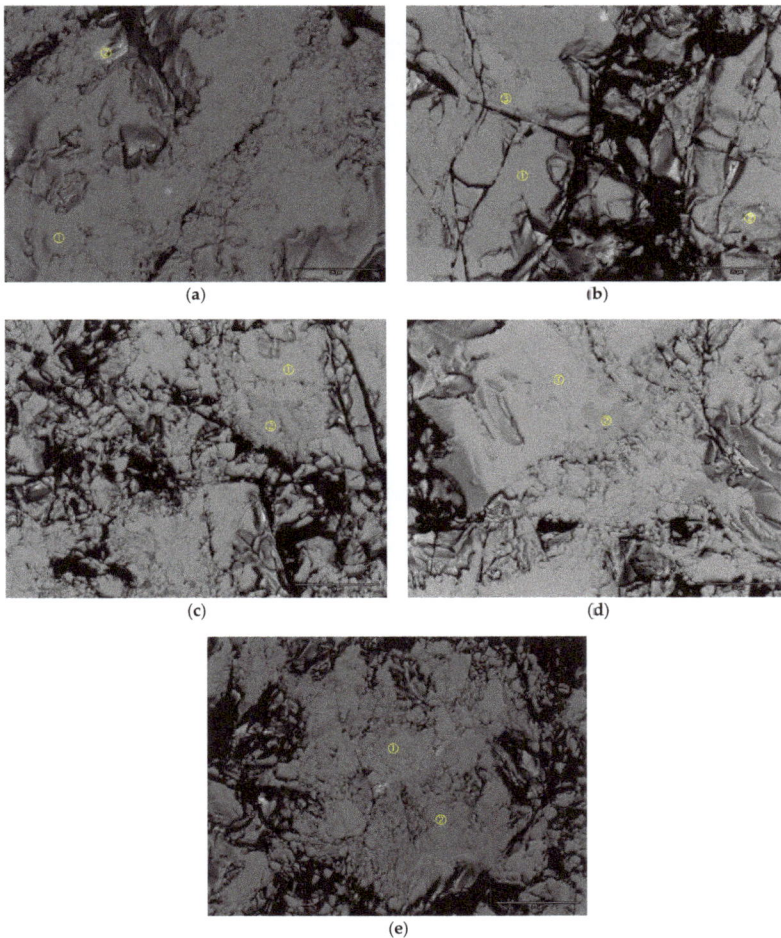

Figure 5. Scanning electron microscope-backscattered electron (SEM-BSE) micrographs from alloys (**a**) Fe1; (**b**) Fe2; (**c**) Fe3; (**d**) Fe4; and (**e**) Fe5. The bar on the lower right corner represents 25 microns. The composition of the numbered areas was analyzed by energy dispersive spectroscopy (EDS), and results are available in Table 3.

Table 3. Summary of EDS results. All compositions are in at%. The composition of the main phase is in **bold**.

Sample	Location	La	Pr	Nd	Mg	Ni	Al	Co	Fe	Mg/A	B/A	Phase
Fe1	**Figure 5a-1**	**3.9**	**8.0**	**8.0**	**3.0**	**68.6**	**3.8**	**4.7**	**0.0**	**0.13**	**3.37**	**AB_3/A_2B_7**
	Figure 5a-2	41.6	27.3	26.7	0.0	3.9	0.5	0.0	0.0	0.00	0.05	Rare Earth
Fe2	**Figure 5b-1**	**4.0**	**7.9**	**7.9**	**3.0**	**66.2**	**4.7**	**5.2**	**1.1**	**0.13**	**3.39**	**AB_3/A_2B_7**
	Figure 5b-2	43.4	22.7	21.1	0.0	11.6	0.8	0.4	0.0	0.00	0.15	Rare Earth
	Figure 5b-3	2.8	7.2	7.8	0.6	69.0	5.2	6.0	1.4	0.03	4.43	AB_5
Fe3	**Figure 5c-1**	**4.2**	**8.0**	**8.0**	**3.0**	**65.5**	**4.3**	**4.8**	**2.2**	**0.13**	**3.31**	**AB_3/A_2B_7**
	Figure 5c-2	3.1	8.1	8.3	14.3	61.7	0.8	2.7	1.0	0.42	1.96	$LaMgNi_4$
Fe4	**Figure 5d-1**	**3.8**	**7.8**	**7.9**	**3.3**	**64.7**	**4.8**	**4.4**	**3.3**	**0.14**	**3.39**	**AB_3/A_2B_7**
	Figure 5d-2	3.2	6.8	6.9	0.4	64.3	8.3	5.3	4.8	0.02	4.78	AB_5
Fe5	**Figure 5e-1**	**4.4**	**8.0**	**7.9**	**2.4**	**62.8**	**4.4**	**5.1**	**5.0**	**0.11**	**3.41**	**AB_3/A_2B_7**
	Figure 5e-2	3.0	8.0	7.9	15.0	60.7	0.8	2.8	1.8	0.44	1.95	$LaMgNi_4$

3.3. Gaseous Phase Hydrogen Storage

The gaseous phase H-storage characteristics of the alloys were studied using PCT measurements performed at 30 and 45 °C. The PCT isotherms are plotted in Figure 6. In general, the changes in shape and position of the isotherms from the Fe-incorporated alloys are small. The isotherms are similar to those from the Co-incorporated alloys [9], but different from the Mn-incorporated alloys [8].

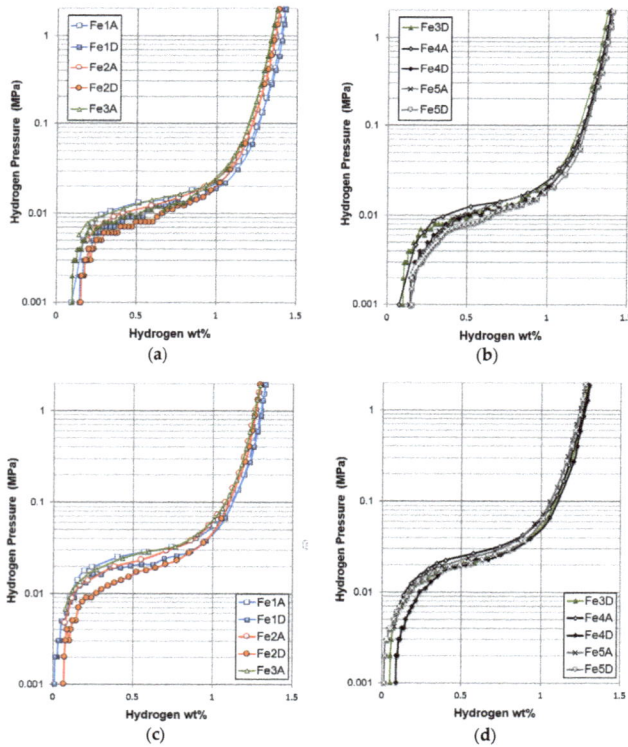

Figure 6. Pressure-concentration-temperature (PCT) isotherms measured at (**a,b**) 30 and (**c,d**) 45 °C for alloys Fe1–Fe5. A and D denote absorption and desorption, respectively. The plateau region has been expanded in the inset.

Several gaseous phase parameters obtained from the PCT measurements are listed in Table 4. Maximum H-storage at 30 °C decreases marginally and then increases slightly with increasing Fe content. Furthermore, changes in maximum and reversible H-storage with the Fe substitution are small, which again reflects similar results to Co substitution, but is unlike the results obtained with Mn substitution (Figure 7a,b). Unexpectedly, increases in unit cell volume (Figure 3) and abundance (Figure 4a) of the main Ce_2Ni_7 phase with increasing Fe content do not increase the gaseous phase H-storage, suggesting that a synergetic effect between the main and secondary phases may dominate the H-storage performance in the Fe-substituted alloys. However, the exact mechanism is not clear at the present time.

Table 4. Summary of gaseous phase hydrogen storage (H-storage) properties. Error ranges are indicated in parentheses after the quantity. PCT: pressure-concentration-temperature; and SF: slope factor.

H-storage Properties	Alloy				
	Fe1	Fe2	Fe3	Fe4	Fe5
Maximum Capacity @ 30 °C (wt %)	1.43	1.39	1.37	1.39	1.41
Reversible Capacity @ 30 °C (wt %)	1.28	1.23	1.27	1.25	1.25
Desorption Pressure@ 0.75%, 30 °C (MPa)	0.0138	0.0116	0.0130	0.0123	0.0128
Desorption Pressure @ 0.75%, 45 °C (MPa)	0.0254	0.0222	0.266	0.261	0.0273
PCT Hysteresis @ 0.75%, 30 °C	0.14	0.23	0.21	0.21	0.14
PCT Hysteresis @ 0.75%, 45 °C	0.24	0.35	0.19	0.22	0.12
PCT SF @ 30 °C	0.83	0.81	0.83	0.79	0.77
PCT SF @ 45 °C	0.84	0.82	0.85	0.82	0.82
$-\Delta H$ (kJ·mol^{-1})	33 (2)	35 (2)	38 (2)	40 (2)	41 (2)
$-\Delta S$ (J·mol^{-1}·K^{-1})	91 (3)	97 (3)	109 (3)	115 (3)	116 (4)

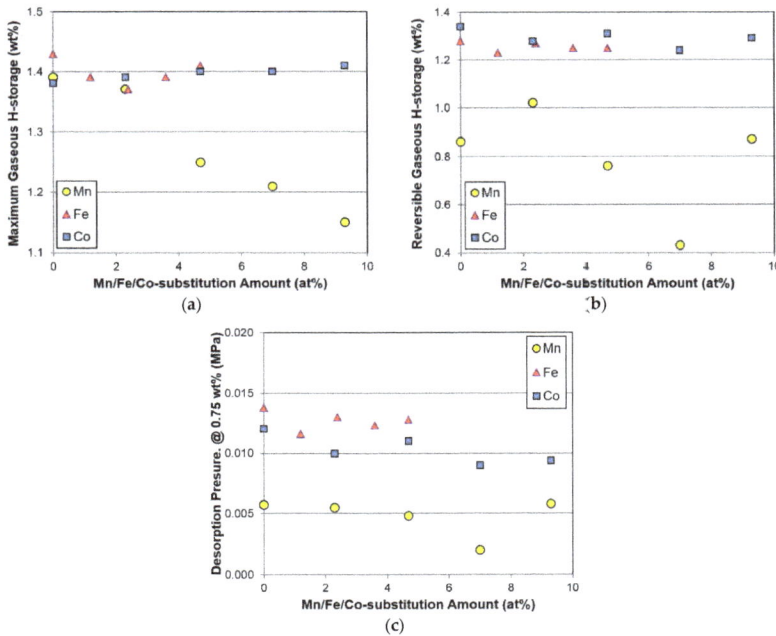

Figure 7. Changes in the (**a**) maximum and (**b**) reversible H-storage capacities; and (**c**) desorption plateau pressure at 0.75 wt % H-storage obtained from the 30 °C PCT isotherms with increasing substitution amount.

Due to the multi-phase nature of the superlattice HAA, the plateau region of the PCT isotherm is not well defined. Therefore, the desorption pressure at 0.75 wt % H-storage is used instead of the plateau pressure in this study. Changes in desorption pressure at 0.75 wt % H-storage for the Fe-substituted alloys were very small, which is similar to what occurs with Co and Mn substitutions (Figure 7c). Composition modifications with Al [11] and La [12] are more effective in changing the equilibrium pressure.

PCT hysteresis is defined as ln(absorption pressure at 0.75 wt % H-storage/desorption pressure at 0.75 wt %). PCT hysteresis is an indication of the elastic deformation energy needed to overcome the lattice expansion at the metal (α)-hydride (β) interface and has been used to predict the pulverization rate during hydride/dehydride cycling [47]. Both PCT hystereses measured at 30 and 45 °C increase and then decrease with increasing Fe content. In addition, Fe-substituted alloys show smaller PCT hysteresis compared to the Mn- and Co-substituted alloys.

Slope factor (SF), defined as the ratio of the H-storage capacity between 0.005 and 0.2 MPa and the reversible H-storage, is related to the degree of homogeneity in the alloy. More specifically, a higher SF value corresponds to a higher uniformity and lesser degree of disorder in the alloy. For the Fe-substituted alloys, SF is independent of Fe content, and the SFs measured at 30 °C are lower than those at 45 °C. Addition of Fe does not alter the alloy homogeneity, which is similar to the case of Co incorporation [9], but different from Mn incorporation, where a reduction in homogeneity was observed [8].

Two thermodynamic parameters, the changes in enthalpy (ΔH) and entropy (ΔS), were estimated using the Van't Hoff equation and the results are listed in Table 4. With increasing Fe content, both ΔH and ΔS become more negative, indicating the formation of a more stable (involving stronger metal–hydrogen (M–H) bonding) and ordered hydride. Compared to the results obtained previously [8,9], the influences on both ΔH and ΔS from Fe are stronger than those from Co and Mn. If we only focus on the Fe-substituted alloys, we may reach the conclusion that with increasing Fe content, the unit cell volume of the main Ce_2Ni_7 phase increases, resulting in a more stable hydride. However, compared to the Fe-substituted alloys, the Mn-substituted alloys have larger Ce_2Ni_7 unit cells, but higher ΔHs. Therefore, the correlation between the structural properties and gaseous phase H-storage characteristics may be too convoluted to establish due to the complex and multi-phase nature of the superlattice alloys.

3.4. Electrochemical Analysis

The open-to-atmosphere half-cell configuration was used to measure the discharge capacities of the five alloys in this study. The dry-compacted electrode was charged with a current density of 100 mA·g^{-1} for 5 h, and it was then discharged initially with the same current density and followed by two pulls at 24 and 8 mA·g^{-1}. These three discharge capacities were added, and the sum considered as the full discharge capacity measured at 8 mA·g^{-1}. In order to examine the activation behaviors, full discharge capacities and HRDs (the ratio of the high-rate discharge capacity to the full discharge capacity) from the first 13 cycles for the five alloys in this study are plotted in Figure 8. Compared to the capacity and HRD curves from the Mn- and Co-substituted alloys, Fe does not facilitate the formation process in the same manner as Mn and Co, which is possibly due to the relatively low solubility of Fe in alkaline solutions, as shown by the following comparison [48]:

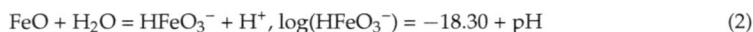

$$MnO + H_2O = HMnO_3^- + H^+, \log(HMnO_3^-) = -16.57 + pH \tag{1}$$

$$FeO + H_2O = HFeO_3^- + H^+, \log(HFeO_3^-) = -18.30 + pH \tag{2}$$

$$CoO + H_2O = HCoO_3^- + H^+, \log(HCoO_3^-) = -16.67 + pH \tag{3}$$

Achievable concentrations of soluble Mn and Co ions are higher than for soluble Fe ion, indicating that Mn and Co are more soluble compared to Fe and therefore contribute positively to the activation performance. Both the full and high-rate discharge capacities, together with HRD, are listed in Table 5. A lower amount of Fe ($\leq$2.3 at%) does not affect the capacity or HRD significantly, but a larger amount of Fe deteriorates both the capacities and HRD. Moreover, the reduction in discharge capacity does not correlate well with the relatively unchanged gaseous phase H-storage values (Figure 7a,b).

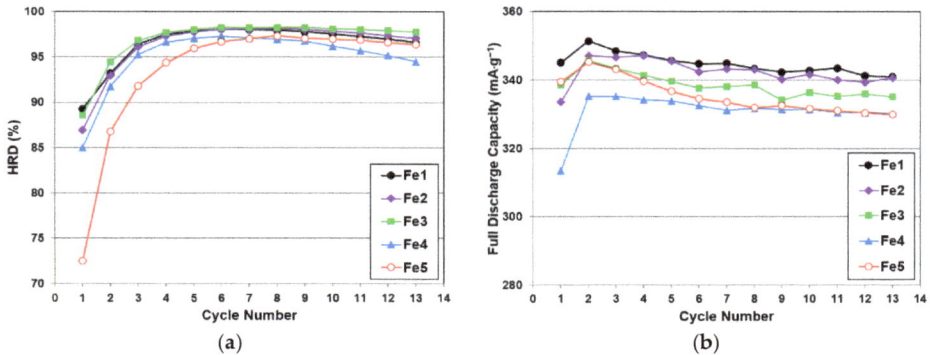

(a)   (b)

Figure 8. Activation characteristics over the first 13 cycles demonstrating (a) full discharge capacity with the lowest discharge current (8 mA·g^{-1}); and (**b**) high-rate dischargeability (HRD) for alloys Fe1–Fe5.

Table 5. Summary of electrochemical properties: capacity, rate, bulk diffusion coefficient (D), and surface exchange current (I_o). RT stands for room temperature.

Alloy	2nd Cycle High-Rate Capacity (mAh·g^{-1})	2nd Cycle Full Capacity (mAh·g^{-1})	HRD (%)	D @ RT (10^{-10} cm^2·s^{-1})	I_0 @ RT (mA·g^{-1})
Fe1	327	351	0.93	5.2	37.5
Fe2	322	347	0.93	4.7	37.2
Fe3	326	346	0.94	3.5	25.7
Fe4	308	335	0.92	3.4	23.6
Fe5	299	345	0.87	2.7	20.5

Changes in full capacity, high-rate capacity, and HRD with different substitution elements are compared in Figure 9a–c, respectively. While Co substitution slightly improves both the full and high-rate capacities, Mn and Fe substitutions (>2.3 at%) decrease the capacities and HRD.

In order to trace the source of the degradations in electrochemical capacities and HRD, both the bulk hydrogen diffusion constant (D) and surface exchange current (I_o) were measured electrochemically (a detailed methodology were reported in our earlier publications [9,49]), and the results are summarized in Table 5. Since D decreases monotonically from the initial measurements, while I_o decreases when the Fe content is greater than 1.2 at%, both bulk hydrogen diffusion and surface catalytic ability deteriorate with Fe incorporation. The decrease in D may be due to a decrease in abundance of the AB$_3$ secondary phase, which decreases the synergetic effects and is an indispensable element in electrochemical H-storage in a multi-phase HAA system [50,51]. Similar to the case in the Laves phase-related bcc solid solution HAA system, where C14 (with stronger M–H bonding) serves as the catalytic phase for the bcc storage phase (with weaker M–H bonding) [51], the AB$_3$ phase can also serve as the catalytic phase that contributes to the synergetic effects despite the fact that it has stronger M–H bonding (judging from its lower B/A ratio compared to the main A$_2$B$_7$ phase). By reducing the AB$_3$ phase abundance through Fe incorporation, the synergetic effects are lowered and, consequently, the hydrogen diffusion in the alloy bulk is impeded. Furthermore, the observed decrease in I_o may be caused by the differences in alloy surface created by addition of Fe, which will be further investigated in the next section. From the observations of D and I_o, it can be concluded that both poor bulk diffusion and surface reaction properties are responsible for the decrease in HRD seen with Fe incorporation. Changes in D and I_o with different substitution elements are compared in Figure 10, and both Mn and Fe result in deterioration, but Co improves the bulk hydrogen diffusion and surface electrochemical reaction.

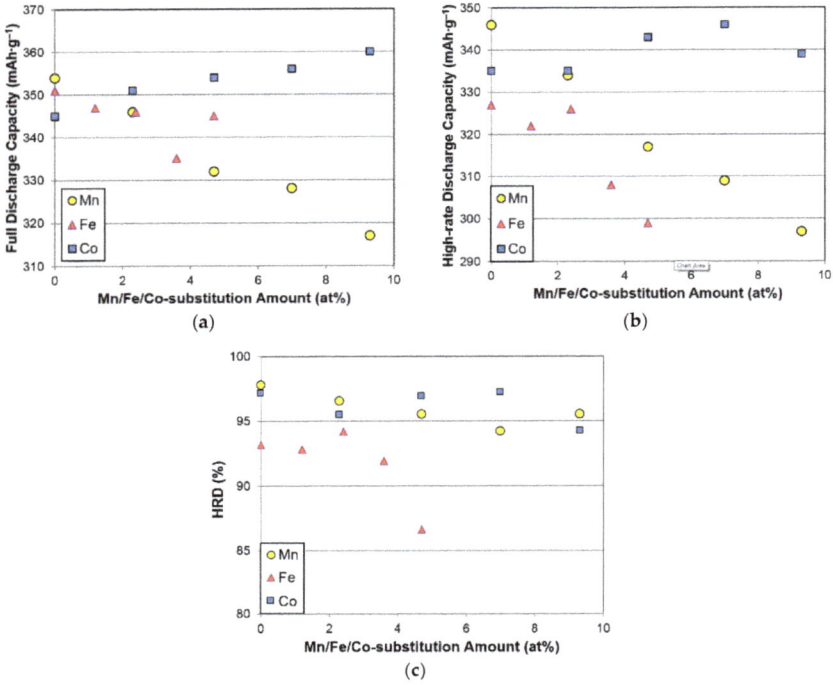

Figure 9. Changes in (a) full and (b) high-rate (100 mA·g^{-1}) capacities; and (c) their ratios (HRD) with increasing substitution amounts.

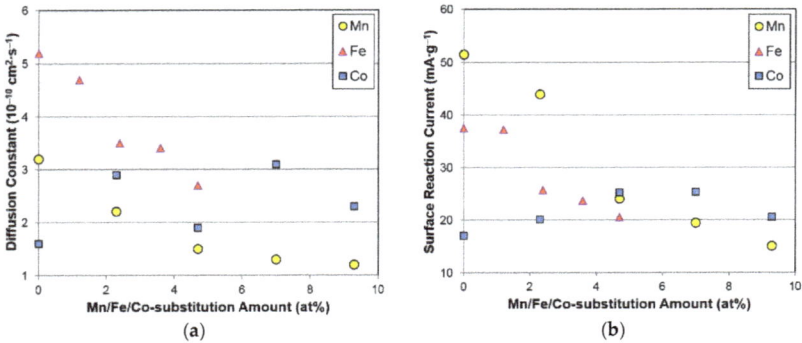

Figure 10. Changes in (a) bulk diffusion constant (D) and (b) surface reaction current (I_o) with increasing substitution amounts.

Overcoming low-temperature performance (especially at $-40\ ^\circ$C) in Ni/MH batteries has always been a very challenging task [52]. We have consistently chosen the AC impedance measurement as the main tool for investigating the $-40\ ^\circ$C electrochemical reaction [53,54]. Details regarding the experimental setup can be found in our earlier publications [55,56]. Surface charge-transfer resistances (R) and surface double-layer capacitances (C) obtained from the Cole–Cole plots measured at both room temperature and −40 °C are summarized in Table 6. In general, R_s and C_s measured at room temperature and −40 °C increase with increasing Fe content. An increase in the reactive surface area (proportional to C) with Fe incorporation was also reported for the AB_2 [21] and AB_5 HAAs [29]. However, the contributions of Fe to R vary for the AB_2 [21] and AB_5 HAAs; R decreases in AB_2 [21], but increases in AB_5 [34]. The RC product has been previously used to characterize the surface catalytic ability [29,34]. With increasing Fe content, the RC product increases in the current study and for the AB_5 HAA [34], but it decreases with the AB_2 HAA [29], which suggests that Fe incorporation impedes the surface electrochemical reaction of A_2B_7 and AB_5, but facilitates that of AB_2. R, C, and RC measured at −40 °C for the Fe- and Mn-substituted superlattice alloys are compared in Figure 11. Although Fe promotes an increase in the surface area more effectively than Mn, the resistances of Fe-containing alloys are much higher than those of the Mn-containing alloys, due to the loss of surface catalytic ability with Fe incorporation, as indicated by the RC plot in Figure 11c.

Table 6. Summary of alternative current (AC) impedance and magnetic susceptibility measurement results. R, C, M_S, and $H_{1/2}$ represent the charge transfer resistance, double-layer capacitance, saturated magnetic susceptibility, and magnetic field at half of M_S, respectively.

Alloy	R @ RT ($\Omega \cdot$g)	C @ RT (Farad$\cdot$g^{-1})	RC @ RT (s)	R @ −40 °C ($\Omega \cdot$g)	C @ −40 °C (Farad$\cdot$g^{-1})	RC @ −40 °C (s)	M_S (memu$\cdot$g^{-1})	$H_{1/2}$ (kOe)
Fe1	0.11	0.30	0.03	5.1	0.58	2.9	1016	0.128
Fe2	0.10	0.42	0.04	5.8	0.75	4.3	835	0.109
Fe3	0.15	0.37	0.06	7.4	0.81	6.0	718	0.106
Fe4	0.13	0.71	0.09	9.1	0.95	8.7	481	0.091
Fe5	0.13	0.82	0.11	10.4	0.99	10.3	341	0.060

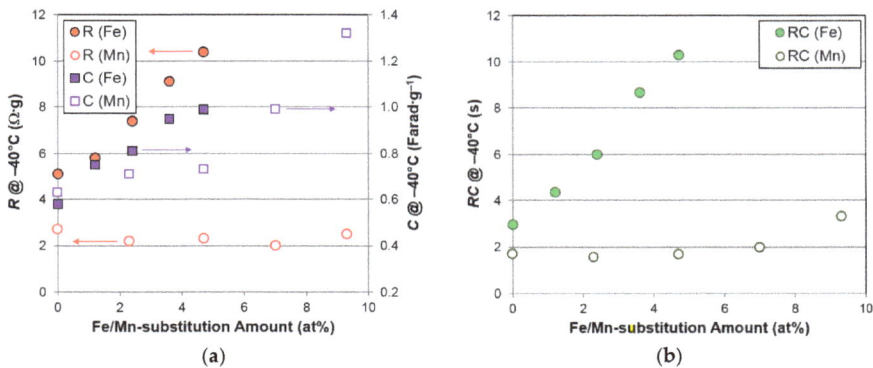

Figure 11. Changes in (**a**) charge transfer resistance (R) and double-layer capacitance (C) calculated from the AC impedance measured at −40 °C; and (**b**) their product (RC) with increasing substitution amounts. A larger RC product corresponds to a slower surface electrochemical reaction and a less catalytic surface.

3.5. Magnetic Properties

The source of degradation in the surface catalytic ability for the Fe-containing superlattice HAAs was further investigated using magnetic susceptibility measurements. Due to the large difference (more than seven orders of magnitude) in the saturated magnetic susceptibilities (M_S) of elemental Ni and Ni in HAA, M_S has been used to estimate the total volume of the metallic Ni clusters imbedded in the surface oxide formed during activation [57]. This measurement has been successfully correlated to HRD of HAA [58]. Moreover, the strength of the applied magnetic field corresponding to half of the M_S

value ($H_{1/2}$) is inversely proportional to the magnetic domain size and used as an indicator of the size of metallic clusters in the surface oxide [59]. Both the M_S and $H_{1/2}$ values obtained from the five alloys in this study are listed in Table 6 and plotted with the data from Mn- and Co-substitution studies [8,9] in Figure 12. With increasing Fe content, both M_S and $H_{1/2}$ decrease, suggesting that Fe incorporation decreases the total volume and surface area of catalytic Ni clusters, judging from the increase in cluster size indicated by the reduction in $H_{1/2}$. The reduction in the amount of catalytic Ni clusters in the surface oxide for the Fe-containing alloys explains the observed decreases in surface catalytic ability, as indicated by the increase in RC in Figure 11c and the surface reaction current (Figure 10b), causing the deterioration in HRD (Figure 9). In addition, the M_Ss of the Mn- and Co-containing alloys are slightly lower than those seen in the base alloy (free of Mn and Co), which suggests that the amount of Ni in the alloy composition correlates closely to the amount of metallic Ni in the surface oxide after activation. Partial replacement of Mn, Fe, and Co for Ni may benefit several electrochemical properties, but certainly deteriorate the high-rate and low-temperature performances due to the reduction in amount of surface catalytic Ni clusters. Interestingly, it was also observed that both $H_{1/2}$ and the AB_3 phase abundance demonstrate similar trends (Figures 4b and 12), and such correlations will be verified in the future.

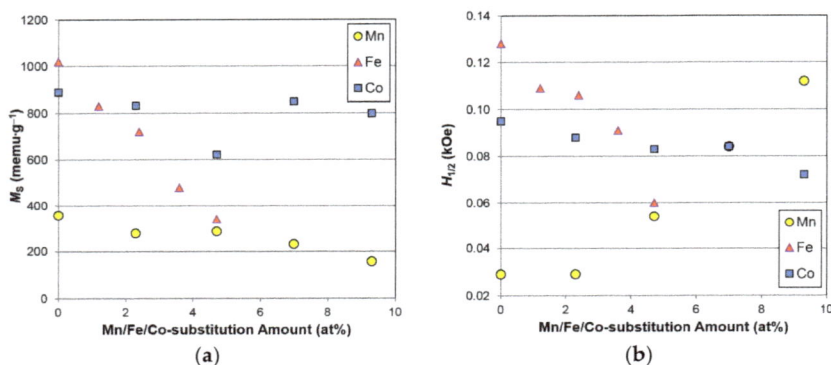

Figure 12. Changes in (**a**) saturated magnetic susceptibility (M_S); and (**b**) strength of the applied magnetic field corresponding to half of M_S ($H_{1/2}$) with increasing substitution amounts.

4. Conclusions

The effects of partial replacement of Ni with Fe in the misch metal-based superlattice HAA were studied and compared with those Mn and Co substitutions. An increase in abundance of the favorable Ce_2Ni_7 phase with Fe incorporation does not significantly affect the H-storage capacity, but the electrochemical properties degraded. Partial replacement of Ni with Fe results in a reduction in total volume of surface metallic Ni inclusions and, consequently, lowers the surface electrochemical reactivity. Even with an increase in the reactive surface area, the Fe-containing alloys exhibit higher surface resistivity at both room temperature and −40 °C, due to severely deteriorated surface catalytic ability. Among various studied substitutions, Co is a better substituting element with regards to general electrochemical performance. However, the Fe-substituting-Ni studied in this work exhibits an increase in the beneficial Ce_2Ni_7 phase abundance and decreases the relatively unfavorable $NdNi_3$ phase among all the AB_3-type phases. Therefore, although Fe incorporation deteriorates most electrochemical properties, further surface treatment may be needed to improve the surface catalytic ability and reveal the positive contribution that Fe structurally provides. Moreover, alternative substitutions targeted to improve the low-temperature performance in Ni/MH battery, such as Mo and Cu, will also be studied in the near future.

Acknowledgments: The authors would like to thank the following individuals from BASF-Ovonic for their help: Su Cronogue, Baoquan Huang, Diana F. Wong, David Pawlik, Allen Chan, and Ryan J. Blankenship.

Author Contributions: Kwo-Hsiung Young designed the experiments and analyzed the results. Taihei Ouchi prepared the alloy samples and performed the PCT and XRD analyses. Jean Nei prepared the electrode samples and conducted the magnetic measurements. Shigekazu Yasuoka designed and obtained the test samples.

Conflicts of Interest: The authors declare no conflict of interest.

Abbreviations

HAA	Hydrogen absorbing alloy
Ni/MH	Nickel/metal hydride
bcc	Body-centered-cubic
HRD	High-rate dischargeability
ICP-OES	Inductively coupled plasma-optical emission spectrometer
XRD	X-ray diffractometer
SEM	Scanning electron microscope
EDS	Energy dispersive spectroscopy
PCT	Pressure-concentration-temperature
AC	Alternative current
H-storage	Hydrogen storage
BSE	Backscattered electron
α	Metal phase or alloy matrix
β	Hydride phase
SF	Slope factor
M–H	Metal–hydrogen
ΔH	Change in enthalpy or heat of hydride formation
ΔS	Change in entropy
D	Bulk hydrogen diffusion coefficient
I_{o}	Surface exchange current
R	Surface charge-transfer resistance
C	Surface double-layer capacitance
RT	Room temperature
M_{S}	Saturated magnetic susceptibility
$H_{1/2}$	Applied magnetic field strength corresponding to half of saturated magnetic susceptibility

References

1. Yasuoka, S.; Magari, Y.; Murata, T.; Tanaka, T.; Ishida, J.; Nakamura, H.; Nohma, T.; Kihara, M.; Baba, Y.; Teraoka, H. Development of high-capacity nickel-metal hydride batteries using superlattice hydrogen-absorbing alloys. *J. Power Sources* **2006**, *156*, 662–666. [CrossRef]
2. Teraoka, H. Development of Low Self-Discharge Nickel-Metal Hydride Battery. 2007. Available online: http://www.scribd.com/doc/9704685/Teraoka-Article-En (accessed on 9 April 2016).
3. Kai, T.; Ishida, J.; Yasuoka, S.; Takeno, K. The Effect of Nickel-Metal Hydride Battery's Characteristics with Structure of the Alloy. In Proceedings of the 54th Battery Symposium, Osaka, Japan, 7–9 October 2013; p. 210.
4. Takasaki, T.; Nishimura, K.; Saito, M.; Fukunaga, H.; Iwaki, T.; Sakai, T. Cobalt-free nickel-metal hydride battery for industrial applications. *J. Alloy. Compd.* **2013**, *580*, S378–S381. [CrossRef]
5. Crivello, J.C.; Zhang, J.; Latroche, M. Structural stability of AB_{y} phases in the (La,Mg)–Ni system obtained by density functional theory calculations. *J. Phys. Chem. C* **2011**, *115*, 25470–25478. [CrossRef]
6. Young, K.; Nei, J. The current status of hydrogen storage alloy development for electrochemical applications. *Materials* **2013**, *6*, 4574–4608. [CrossRef]
7. Verbovyts'kyi, Y.V.; Zavalii, I.Y. New metal-hydride electrode materials based on $R_{1-x}Mg_{x}Ni_{3-4}$ alloys for chemical current sources. *Mater. Sci.* **2016**, *51*, 443–456. [CrossRef]

8. Young, K.; Wong, D.F.; Wang, L.; Nei, J.; Ouchi, T.; Yasuoka, S. Mn in misch-metal based superlattice metal hydride alloy—Part 1 Structural, hydrogen storage and electrochemical properties. *J. Power Sources* **2015**, *277*, 426–432. [CrossRef]

9. Wang, L.; Young, K.; Meng, T.; Ouchi, T.; Yasuoka, S. Partial substitution of cobalt for nickel in mixed rare earth metal based superlattice hydrogen absorbing alloy—Part 1 Structural, hydrogen storage and electrochemical properties. *J. Alloy. Compd.* **2016**, *660*, 407–415. [CrossRef]

10. Young, K.; Ouchi, T.; Huang, B. Effects of annealing and stoichiometry to (Nd, Mg)(Ni, Al)$_{3.5}$ metal hydride alloys. *J. Power Sources* **2012**, *215*, 152–159. [CrossRef]

11. Young, K.; Ouchi, T.; Wang, L.; Wong, D.F. The effected of Al substitution on the phase abundance, structure and electrochemical performance of La$_{0.7}$Mg$_{0.3}$Ni$_{2.8}$Co$_{0.5-x}$Al$_x$ (x = 0, 0.1, 0.2) alloys. *J. Power Sources* **2015**, *279*, 172–179. [CrossRef]

12. Yasuoka, S.; Ishida, J. Effects of cerium (Ce) on the hydrogen absorption-desorption characteristics of RE-Mg-Ni hydrogen absorbing alloy. Unpublished work. 2016.

13. Young, K.; Wong, D.F.; Wang, L.; Nei, J.; Ouchi, T.; Yasuoka, S. Mn in misch-metal based superlattice metal hydride alloy—Part 2 Ni/MH battery performance and failure mechanism. *J. Power Sources* **2015**, *277*, 433–442. [CrossRef]

14. Wang, L.; Young, K.; Meng, T.; English, N.; Yasuoka, S. Partial substitution of cobalt for nickel in mixed rare earth metal based superlattice hydrogen absorbing alloy—Part 2 Battery performance and failure mechanism. *J. Alloy. Compd.* **2016**, *664*, 417–427. [CrossRef]

15. Huang, T.; Wu, Z.; Huang, T.; Ni, J. Influence of V and Fe on the performance of TiMn$_2$ hydrogen storage alloy. *J. Funct. Mat. Devices* **2003**, *9*, 83–86.

16. Li, S.L.; Cheng, H.H.; Deng, X.X.; Chen, W.; Chen, D.M.; Yang, K. Investigation on hydrogen absorption/desorption properties of ZrMn$_{0.85-x}$Fe$_{1+x}$ alloys. *J. Alloy. Compd.* **2008**, *460*, 186–190. [CrossRef]

17. Jain, A.; Jain, R.K.; Agarwal, S.; Ganesan, V.; Lalla, N.P.; Phase, D.M.; Jain, I.P. Synthesis, characterization and hydrogenation of ZrFe$_{2-x}$Ni$_x$ (x = 0.2, 0.4, 0.6, 0.8) alloys. *Int. J. Hydrog. Energy* **2007**, *32*, 3965–3971. [CrossRef]

18. Song, M.Y.; Ahn, S.; Kwon, I.H.; Lee, R.; Rim, H. Development of AB$_2$-type Zr–Ti–Mn–V–Ni–Fe hydride electrodes for Ni–MH secondary batteries. *J. Alloy. Compd.* **2000**, *298*, 254–260. [CrossRef]

19. Song, M.Y.; Kwon, I.H.; Ahn, D.S.; Sohn, M.S. Improvement in the electrochemical properties of ZrMn$_2$ hydrides by substitution of elements. *Met. Mater. Int.* **2001**, *7*, 257–263. [CrossRef]

20. Young, K.; Fetcenko, M.A.; Koch, J.; Morii, K.; Shimizu, T. Studies of Sn, Co, Al, and Fe additives in C14/C15 Laves alloys for NiMH battery application by orthogonal arrays. *J. Alloy. Compd.* **2009**, *486*, 559–569. [CrossRef]

21. Young, K.; Ouchi, T.; Huang, B.; Reichman, B.; Fetcenko, M.A. The structure, hydrogen storage, and electrochemical properties of Fe-doped C14-predominating AB$_2$ metal hydride alloys. *Int. J. Hydrog. Energy* **2011**, *36*, 12296–12304. [CrossRef]

22. Young, K.; Ouchi, T.; Huang, B.; Ftecenko, M.A. Effects of B, Fe, Gd, Mg, and C on the structure, hydrogen storage, and electrochemical properties of vanadium-free AB$_2$ metal hydride alloy. *J. Alloy. Compd.* **2012**, *511*, 242–250. [CrossRef]

23. Ayari, M.; Paul-Boncour, V.; Lamloumi, J.; Percheron-Guégan, A. Magnetic properties of LaNi$_{3.55}$Mn$_{0.4}$Al$_{0.3}$Co$_{0.75-x}$Fe$_x$ (x = 0, 0.35) compounds before and after electrochemical cycles. *J. Mag. Mag. Mat.* **2002**, *242–245*, 850–853. [CrossRef]

24. Singh, R.K.; Gupta, B.K.; Lototsky, M.V.; Srivastava, O.N. On the synthesis and hydrogenation behavior of MmNi$_{5-x}$Fe$_x$ alloys and computer simulation of their *P–C–T* curves. *J. Alloy. Compd.* **2004**, *373*, 208–213. [CrossRef]

25. Khaldi, C.; Mathlouthi, H.; Lamloumi, J.; Percheron-Guégan, A. Electrochemical impedance spectroscopy and constant potential discharge studies of LaNi$_{3.55}$Mn$_{0.4}$Al$_{0.3}$Co$_{0.75-x}$Fe$_x$ hydrides alloy electrodes. *J. Alloy. Compd.* **2004**, *384*, 249–253. [CrossRef]

26. Singh, R.K.; Gupta, B.K.; Lototsky, M.V.; Srivastava, O.N. Thermodynamical, structural, hydrogen storage properties and simulation studied of P–C isotherms of (La, Mm)Ni$_{5-y}$Fe$_y$. *Int. J. Hydrog. Energy* **2007**, *32*, 2971–2976. [CrossRef]

27. Yang, K.; Liu, X.; Dai, C. Mischmetal-Nickel Hydrogen Storage Material by Adding a Third Element to Control the Plateau Pressure. In *Hydrogen Systems*; Veziroglu, T.N.; Zhu, Y., Bao, D., Eds.; Pergamon: New York, NY, USA, 1986; Volume 1, pp. 383–387.

28. Chen, L.; Lei, Y.; Zhu, G.; Wang, Q. The effects of doping Fe to the electrochemical properties of MI(NiCoMnTi)$_5$ metal hydride alloy. *Rare Met. Mater. Eng.* **1998**, *27*, 135–138.

29. Wei, X.; Liu, S.; Dong, H.; Zhang, P.; Liu, Y.; Zhu, J.; Yu, G. Microstructures and electrochemical properties of Co-free AB$_5$-type hydrogen storage alloys through substitution of Ni by Fe. *Electrochim. Acta* **2007**, *52*, 2423–2428. [CrossRef]

30. Yang, H.; Chen, Y.; Tao, M.; Wu, C.; Shao, J.; Deng, G. Low temperature electrochemical properties of LaNi$_{4.6-x}$Mn$_{0.4}$M$_x$ (M = Fe or Co) and effect of oxide layer on EIS response in metal hydride electrodes. *Electrochim. Acta* **2010**, *55*, 648–655. [CrossRef]

31. Liu, B.; Li, A.; Fan, Y.; Hu, M.; Zhang, B. Phase structure and electrochemical properties of La$_{0.7}$Ce$_{0.3}$Ni$_{3.75}$Mn$_{0.35}$Al$_{0.15}$Cu$_{0.75-x}$Fe$_x$ hydrogen storage alloys. *Trans. Nonferrous Met. Soc. China* **2012**, *22*, 2730–2735. [CrossRef]

32. Chao, D.; Zhong, C.; Ma, Z.; Yang, F.; Wu, Y.; Zhu, D.; Wu, C.; Chen, Y. Improvement in high-temperature performance of Co-free high-Fe AB$_5$-type hydrogen storage alloys. *Int. J. Hydrog. Energy* **2012**, *37*, 12375–12383.

33. Kaabi, A.; Khaldi, C.; Lamloumi, J. Thermodynamic and kinetic parameters and high rate discharge-ability of the AB$_5$-type metal hydride anode. *Int. J. Hydrog. Energy* **2016**, *41*, 9914–9923. [CrossRef]

34. Young, K.; Ouchi, T.; Reichman, B.; Koch, J.; Fetcenko, M.A. Improvement in the low-temperature performance of AB$_5$ metal hydride alloys by Fe-addition. *J. Alloy. Compd.* **2011**, *509*, 7611–7617. [CrossRef]

35. Cho, S.; Enoki, H.; Akiba, E. Effect of Fe addition on hydrogen storage characteristics of Ti$_{0.16}$Zr$_{0.05}$Cr$_{0.22}$V$_{0.57}$ alloy. *J. Alloy. Compd.* **2000**, *307*, 304–310. [CrossRef]

36. Young, K.; Nei, J.; Wong, D.F.; Wang, L. Structural, hydrogen storage, and electrochemical properties of Laves phase-related body-centered-cubic solid solution metal hydride alloys. *Int. J. Hydrog. Energy* **2014**, *39*, 21489–21499. [CrossRef]

37. Pan, H.; Li, R.; Liu, Y.; Gao, M.; Miao, H.; Lei, Y.; Wang, Q. Structure and electrochemical properties of the Fe substituted Ti-V-based hydrogen storage alloys. *J. Alloy. Compd.* **2008**, *463*, 189–195. [CrossRef]

38. Dou, T.; Wu, Z.; Mao, J.; Xu, N. Application of commercial ferrovanadium to reduce cost of Ti–V-based BCC phase hydrogen storage alloys. *Mater. Sci. Eng. A* **2008**, *476*, 34–38. [CrossRef]

39. Santos, S.F.; Huot, J. Hydrogen storage in TiCr$_{1.2}$(FeV)$_x$ BCC solid solutions. *J. Alloy. Compd.* **2009**, *472*, 247–251. [CrossRef]

40. Young, K.; Ouchi, T.; Nei, J.; Meng, T. Effects of Cr, Zr, V, Mn, Fe, and Co to the hydride properties of Laves phase-related body-centered-cubic solid solution alloys. *J. Power Sources* **2015**, *281*, 164–172. [CrossRef]

41. Liu, Y.; Cao, Y.; Huang, L.; Gao, M.; Pan, H. Rare earth-Mg-Ni-based hydrogen storage alloys as negative electrode materials for Ni/MH batteries. *J. Alloy. Compd.* **2011**, *509*, 675–686. [CrossRef]

42. Zhang, Y.; Li, B.; Ren, H.; Wu, Z.; Dong, X.; Wang, X. Influences of the substitution of Fe for Ni on structures and electrochemical performances of the as-cast and quenched La$_{0.7}$Mg$_{0.3}$Co$_{0.45}$Ni$_{2.55-x}$Fe$_x$ (x = 0–0.4) electrode alloys. *J. Alloy. Compd.* **2008**, *460*, 414–420. [CrossRef]

43. Wu, F.; Zhang, M.; Mu, D. Effect of B and Fe substitution on structure of AB$_3$-type Co-free hydrogen storage alloy. *Trans. Nonferrous Met. Soc. China* **2010**, *20*, 1885–1891. [CrossRef]

44. Meng, T.; Young, K.; Yasuoka, S. Fe-substitution for Ni in misch metal-based superlattice hydrogen absorbing alloys—Part 2. Performance and failure mechanism in sealed cell. *Batteries* **2016**. to be submitted for publication.

45. Young, K.; Yasuoka, S. Past, Present, and Future of Metal Hydride Alloys in Nickel-Metal Hydride Batteries. In Proceedings of the 14th International Symposium on Metal-Hydrogen Systems, Manchester, UK, 21–25 July 2014.

46. Liu, J.; Han, S.; Li, Y.; Zhang, L.; Zhao, Y.; Yang, S.; Liu, B. Phase structures and electrochemical properties of La–Mg–Ni-based hydrogen storage alloys with superlattice structure. *Int. J. Hydrog. Energy* **2016**. [CrossRef]

47. Young, K.; Ouchi, T.; Fetcenko, M.A. Pressure-composition-temperature hysteresis in C14 Laves phase alloys: Part 1. Simple ternary alloys. *J. Alloy. Compd.* **2009**, *480*, 428–433. [CrossRef]

48. Pourbaix, M. *Atlas of Electrochemical Equilibrium in Aqueous Solutions*; National Association of Corrosion Engineers: Houston, TX, USA, 1974.

49. Li, F.; Young, K.; Ouchi, T.; Fetcenko, M.A. Annealing effects on structural and electrochemical properties of (LaPrNdZr)$_{0.83}$Mg$_{0.17}$(NiCoAlMn)$_{3.3}$ alloy. *J. Alloy. Compd.* **2009**, *471*, 371–377. [CrossRef]

50. Mosavati, N.; Young, K.; Meng, T.; Ng, K.Y.S. Electrochemical open-circuit voltage and pressure-concentration-temperature isotherm comparison for metal hydride alloys. *Batteries* **2016**, *2*. [CrossRef]

51. Young, K.; Ouchi, T.; Meng, T.; Wong, D.F. Studies on the synergetic effects in multi-phase metal hydride alloys. *Batteries* **2016**, *2*. [CrossRef]

52. Young, K. Electrochemical Applications of Metal Hydrides. In *Compendium of Hydrogen Energy*; Barbir, F., Basile, A., Veziroğlu, N., Eds.; Woodhead Publishing Ltd.: Waltham, MA, USA, 2016; Volume 3.

53. Young, K.; Wong, D.F.; Nei, J.; Reichman, B. Electrochemical properties of hypo-stoichiometric Y-doped AB$_2$ metal hydride alloys at ultra-low temperature. *J. Alloy. Compd.* **2015**, *643*, 17–27. [CrossRef]

54. Young, K.; Ouchi, T.; Nei, J.; Moghe, D. The importance of rare-earth additions in Zr-based AB$_2$ metal hydride alloys. *Batteries* **2016**, *2*. [CrossRef]

55. Young, K.; Wong, D.F.; Ouchi, T.; Huang, B.; Reichman, B. Effects of La-addition to the structure, hydrogen storage, and electrochemical properties of C14 metal hydride alloys. *Electrochim. Acta* **2015**, *174*, 815–825. [CrossRef]

56. Wong, D.F.; Young, K.; Nei, J.; Wang, L.; Ng, K.Y.S. Effects of Nd-addition on the structural, hydrogen storage, and electrochemical properties of C14 metal hydride alloys. *J. Alloy. Compd.* **2015**, *647*, 507–518. [CrossRef]

57. Stucki, F.; Schlapbach, L. Magnetic properties of LaNi$_5$, FeTi, Mg$_2$Ni and their hydrides. *J. Less Commun. Met.* **1980**, *74*, 143–151. [CrossRef]

58. Young, K.; Huang, B.; Regmi, R.K.; Lawes, G.; Liu, Y. Comparisons of metallic clusters imbedded in the surface of AB$_2$, AB$_5$, and A$_2$B$_7$ alloys. *J. Alloy. Compd.* **2010**, *506*, 831–840. [CrossRef]

59. Chang, S.; Young, K.; Ouchi, T.; Meng, T.; Nei, J.; Wu, X. Studies on incorporation of Mg in Zr-based AB$_2$ metal hydride alloys. *Batteries* **2016**, *2*. [CrossRef]

batteries

MDPI

Article

Effects of Alkaline Pre-Etching to Metal Hydride Alloys

Tiejun Meng [1], Kwo-Hsiung Young [1,2,*], Chaolan Hu [1] and Benjamin Reichman [1]

[1]   BASF/Battery Materials–Ovonic, 2983 Waterview Drive, Rochester Hills, MI 48309, USA;
     pkumeng@hotmail.com (T.M.); sherry.hu@basf.com (C.H.); Benjamin.reichman@basf.com (B.R.)
[2]   Department of Chemical Engineering and Materials Science, Wayne State University, Detroit, MI 48202, USA
*   Correspondence: kwo.young@basf.com; Tel.: +1-248-293-7000

Received: 10 September 2017; Accepted: 29 September 2017; Published: 5 October 2017

Abstract: The responses of one AB_5, two AB_2, four A_2B_7, and one C14-related body-centered-cubic (BCC) metal hydrides to an alkaline-etch (45% KOH at 110 °C for 2 h) were studied by internal resistance, X-ray diffraction, scanning electron microscope, inductively coupled plasma, and AC impedance measurements. Results show that while the etched rare earth–based AB_5 and A_2B_7 alloys surfaces are covered with hydroxide/oxide (weight gain), the transition metal–based AB_2 and BCC-C14 alloys surfaces are corroded and leach into electrolyte (weight loss). The C14-predominated AB_2, La-only A_2B_7, and Sm-based A_2B_7 showed the most reduction in the internal resistance with the alkaline-etch process. Etched A_2B_7 alloys with high La-contents exhibited the lowest internal resistance and are suggested for use in the high-power application of nickel/metal hydride batteries.

Keywords: metal hydride alloy; nickel metal hydride battery; alkaline bath; high rate performance; superlattice alloys; surface morphology

---

1. Introduction

Metal hydride (MH) alloy, or hydrogen storage alloy, is a group of intermetallic alloys (IMCs) capable of storing hydrogen in the solid form [1]. One of its key applications is the rechargeable nickel/metal hydride (Ni/MH) battery, which is used widely in consumer electronics, as well as stationary and transportation energy storage areas. A large variety of IMCs have been used/proposed as the active materials in the negative electrode of Ni/MH battery, such as A_2B [2], AB [3,4], AB_2 [5], AB_3 [6], A_2B_7 [7], A_5B_{19} [8], AB_5 [9], body-centered-cubic (BCC) solid solution [10], and their combinations [11,12]. These IMCs are composed of mostly transition metals (TM), and some may contain rare-earth (RE) elements. The electrochemical high-rate performances of some of these MH alloys were compared in 2010 and the RE-based AB_5 MH alloy had the best high-rate dischargeability (HRD) performance [13]. The magnetic susceptibilities of these alloys, and a couple new ones, were compared in a 2013 article [14]. With several new MH alloys, especially the recently discovered superlattice A_2B_7-based ones with improved electrochemical properties, it is important to update the comparison results.

A few pre-activation processes, such as surface fluorination, alkaline bath, acid etch, mechanical alloying, etc., were proposed to shorten the activation process and improve the electrochemical performance of the MH alloys (see a review in [15]). Among these processes, the alkaline bath is very effective in dissolving the native oxide, so as to form a porous oxide surface with catalytic Ni clusters imbedded [13,16–22] and to increase the surface reactive area [23]. In the past, we have used this technique to prepare the alloy for the study of activated surfaces without going through electrochemical formation cycling [24].

2. Experimental Set-Up

MH alloys were prepared by either vacuum induction melting (VIM) or arc melting (AM). Details of the melting process were reported before [25,26]. The main difference between these two preparation methods is the size of ingot. While the former is usually used in large production (1–1000 kg), the latter is mostly used in laboratories (5–200 g) [15]. Some alloys went through annealing either in vacuum or Ar. The obtained ingot was hydrided first, followed by grinding and sifting into a −200 mesh size powder. To make the electrode, the powder was compacted directly onto a 1 cm × 1 cm expanded Ni substrate without any binder or conductive metal powder. The loading of powder per electrode ranged from 70 to 100 mg. The KOH etching (activation) experiment was performed on the electrode assembly in 45 wt % KOH at 110 °C for 2 h. After the etching process, the electrode was pulled out, rinsed by deionized water, dried, and tested. A Suzuki-Shokan multi-channel pressure-concentration-temperature system (PCT, Suzuki Shokan, Tokyo, Japan) was used to measure the MH-hydrogen interaction in gaseous phase. PCT measurements at various temperatures were performed after activation, which consisted of a 2 h thermal cycle between room temperature and 300 °C under 2.5 MPa H_2 pressure. A Varian Liberty 100 inductively coupled plasma optical emission spectrometer (ICP-OES, Agilent Technologies, Santa Clara, CA, USA) was used to study the chemical composition of the solution after etching. A Philips X'Pert Pro X-ray diffractometer (XRD, Philips, Amsterdam, The Netherlands) was used to perform the phase analysis, and a JEOL-JSM6320F scanning electron microscope (SEM, JEOL, Tokyo, Japan) was also used to investigate the phase distribution and composition.

A 30 wt % KOH electrolyte, a Hg/HgO reference electrode, and a sintered $Ni(OH)_2$/NiOOH counter electrode were used for the electrochemical testing. The internal resistance measurement was performed in the following steps:

(1) Charge at 0.1 C and discharge at 0.1 C (to -0.7 V vs. Hg/HgO). Record discharge capacity;
(2) Charge at 0.1 C, discharge 0.1 C to 80% state-of-charge (SOC);
(3) Conduct internal resistance test at 80% SOC (1st internal resistance test):

    a. Put in open circuit for 10 min;
    b. Discharge at 0.5 C for 10 s;
    c. Charge at 0.5 C for 10 s;
    d. Put in open circuit for 10 min;
    e. Discharge at 2 C for 10 s;
    f. Charge at 2 C for 10 s;
    g. Put in open circuit for 10 min;
    h. Calculate internal resistance.

(4) Charge at 0.1 C back from 80% SOC to 100% SOC;
(5) Conduct 4 more capacity tests at 0.2 C charge rate and 0.1 C discharge rate. Record discharge capacities;
(6) Repeat step 3 (2nd internal resistance test);
(7) Repeat step 4;
(8) Conduct rate test:

    a. Charge at 0.2 C, discharge at 0.1 C;
    b. Charge at 0.2 C, discharge at 0.2 C;
    c. Charge at 0.2 C, discharge at 0.5 C;
    d. Charge at 0.2 C, discharge at 1 C;
    e. Charge at 0.2 C, discharge at 2 C;
    f. Charge at 0.2 C, discharge at 3 C;

  g.   Charge at 0.2 C, discharge at 5 C.

(9)   Repeat step 3 (3rd internal resistance test).

A Solartron 1250 Frequency Response Analyzer (Solartron Analytical, Leicester, UK) with a sine wave amplitude of 10 mV and a frequency range of 0.5 mHz to 10 kHz was used for the alternative current (AC) impedance measurements.

3. Results and Discussions

3.1. Alloys Selection

Eight MH alloys (**A–H**) were selected for this study. The composition and preparation method of each alloy are summarized in Table 1. Details in microstructures of these alloys can be found in the cited reference at the last column in Table 1. **A** is the most commonly used RE-based AB_5 alloy in the Ni/MH battery. **B** and **C** are TM-based AB_2 alloys with dominating C14 and C15 structures, respectively. **D** is a Laves-phase related BCC solid solution MH alloy composed of a main BCC phase and C14 and TiNi secondary phases, and shows a high capacity at a moderate rate that is suitable for electric vehicle application [12]. Alloys **E–H** are RE-and Mg-containing A_2B_7-based superlattice alloys used mainly in the high-performance consumer Ni/MH batteries [7,27,28]. PCT isotherms measured at 30 °C and room temperature half-cell discharge capacities in the first 10 electrochemical cycles for eights alloys in this study are compared in Figures 1 and 2, respectively. The capacities of these alloys are summarized in Table 1. The BCC-C14 alloy (**D**) has the highest hydrogen-storage capacity whereas C14 AB_2 alloys (**B**), and superlattice alloys (**E, G,** and **H**) are next, followed by the standard AB_5 alloy (**A**). The half-cell discharge capacity of C15 AB_2 alloy (**C**) is the lowest due to its relatively high hydrogen equilibrium pressure, which makes it difficult to charge into higher state-of-charge in the open-air environment. Alloy F has the lowest discharge capacity in the superlattice alloy family (**E–H**) because of its high Sm-content. From Figure 2, the ease of activation in order is **E, F, G > C, H > A > D, B**. In general, Mg-containing superlattice alloys are easier to activate, and alloys without La (**B, D**) take more cycles to reach their maximum capacities. The addition of rare elements to a TM-based MH alloy to improve the activation behavior is a well-known recipe for the preparation of electrochemical applications [29–36].

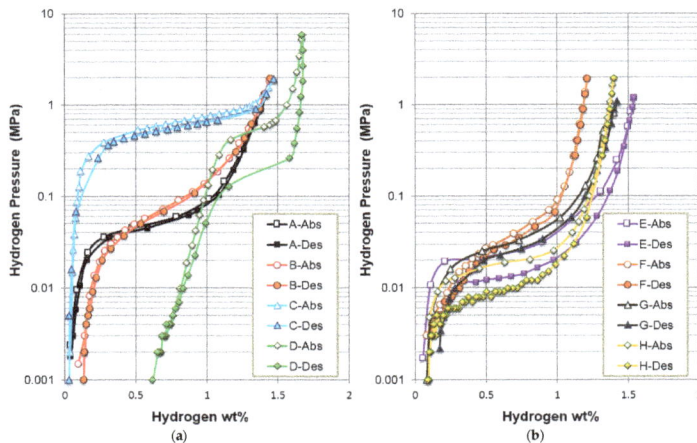

Figure 1. Pressure-concentration-temperature system (PCT) isotherms measured at 30 °C for as-is alloys (**a**) **A–D** and (**b**) **E–H**. Open and solid symbols represent the absorption and desorption curves, respectively.

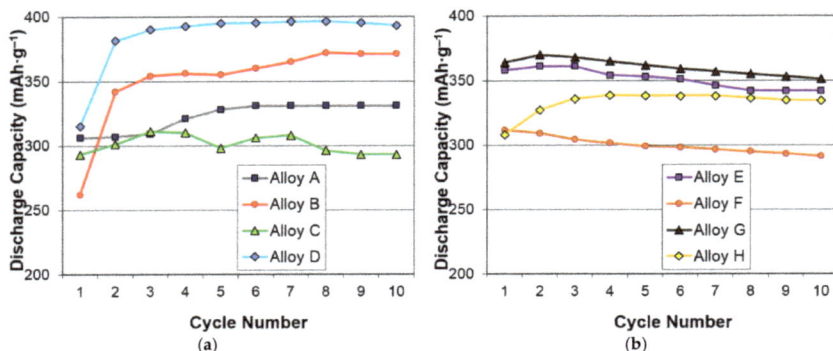

Figure 2. Electrochemical full-capacity activation behavior for as-is alloys (a) **A–D** and (b) **E–H**.

Table 1. Preparation and properties of alloys used in this study. Low-rate discharge capacity and saturated magnetic susceptibility (M_s) are in the units of mAh·g^{-1} and emu·g^{-1}, respectively. HRDs are the ratio of capacities between 50 and 4 mA·g^{-1} for **B–D** and capacities between 100 and 8 mA·g^{-1} for all other alloys.

Alloys	System	Composition	Melting	Annealing	Capacity	HRD	M_S	Reference
A	AB$_5$	La$_{10.5}$ Ce$_{4.3}$Pr$_{0.5}$Nd$_{1.4}$Ni$_{60}$ Co$_{12.7}$Mn$_{5.9}$Al$_{4.7}$	VIM	In vacuum	331	0.99	0.43	A in [13]
B	AB$_2$-C14	Zr$_{21.5}$Ti$_{12}$V$_{310}$ Cr$_{7.5}$Mn$_{8.1}$ Co$_8$Ni$_{32.2}$Sn$_{0.3}$Al$_{0.4}$	VIM	None	354	0.90	0.04	Mo0 in [37]
C	AB$_2$-C15	Zr$_{25}$Ti$_{6.5}$V$_{3.9}$Mn$_{22.2}$Fe$_{3.8}$ Ni$_{38}$Sn$_{0.3}$La$_{0.3}$	AM	None	307	0.99	0.04	C15 in [38]
D	BCC-C14	Zr$_{2.1}$Ti$_{15.6}$V$_{44}$ Cr$_{11.2}$Mn$_{6.9}$ Co$_{1.4}$Ni$_{18.5}$Al$_{0.3}$	VIM	None	397	0.95	0.39	P17 in [39]
E	A$_2$B$_7$	La$_{16.3}$Mg$_7$Ni$_{65.1}$ Co$_{11.6}$	VIM	In Ar	361	0.98	0.37	C in [13]
F	A$_2$B$_7$	La$_6$Sm$_{13.8}$Mg$_3$Ni$_{73.8}$Al$_{3.4}$	VIM	In Ar	320	0.99	0.60	A3 in [40]
G	A$_2$B$_7$	La$_{11.3}$Pr$_{1.7}$Nd$_{5.1}$Mg$_{4.5}$Ni$_{63.6}$ Co$_{13.6}$Zr$_{0.2}$	VIM	In Ar	370	0.98	1.19	B in [41]
H	A$_2$B$_7$	La$_{3.9}$Pr$_{7.7}$Nd$_{7.7}$Mg$_{3.9}$Ni$_{68.1}$ Co$_{4.7}$Al$_4$	VIM	In Ar	354	0.94	0.62	C3 in [42]

3.2. Electrochemical Results

Four electrodes from each alloy went through half-cell electrochemical testing. Examples of capacities measured at different rates and three internal resistances (R) from alloy **A** (as-is), alloy **B** (as-is and etched), and alloy E (as-is and etched) are plotted in Figures 3 and 4. The voltage dropped very quickly with the increase in the discharge current, which resulted in a significant decrease in measured capacity. The etched C14 AB$_2$ alloy (**B**) shows the highest discharge capacity, up to a 2 C rate. In the comparison of R, the etched E shows the lowest resistance, and, in general, the HRD capability is in the order of A$_2$B$_7$ (**E**) > AB$_5$ (**A**) > AB$_2$ (**B**) regardless of being etched. This finding is consistent with our previous reports [40,43].

The electrochemical testing results are summarized in Table 2. The reported capacity and internal resistance are highest when obtained with a 0.1 C discharge rate and the lowest value in the three measurements, respectively. The specific power (P) was estimated by using formula:

$$P = 2(0.45 - V_{oc})^2/9R \tag{1}$$

where 0.45 and V_{oc} are the voltage of Ni(OH)$_2$/NiOOH electrode and the open-circuit voltage of the tested alloy electrode vs. Hg/HgO reference electrode, respectively. The estimated specific power is inversely proportional to the measured internal resistance as shown in Figure 5. From Table 2, it is obvious that KOH-etching (activation) is very effective to reduce the internal resistance in C15 AB_2 (**C**, -25%) and Sm-based A_2B_7 (**F**, -19%) alloys. The reduction in the internal resistance is less significant in La-based A_2B_7 (**E**, -12%), C14 AB_2 (**B**, -9%) and Mm-based Al-free A_2B_7 (**G**, -8%) alloys. Etching in KOH (activation) even increases the internal resistance of AB_5 (**A**, +36%) and BCC-C14 (**D**, +30%) alloys by large percentages. Three A_2B_7 alloys (**E**, **F**, and **G**) show lower R than that of the standard AB_5 (**A**) after KOH-etching. The as-prepared (no etch) E and G show even lower R than that from A and have been used in the high-power design of Ni/MH batteries [41,44]. Overall, G shows the lowest R and the highest estimated specific power of 390 $W \cdot kg^{-1}$, which is considerably higher than that from the standard **A** (317 $W \cdot kg^{-1}$).

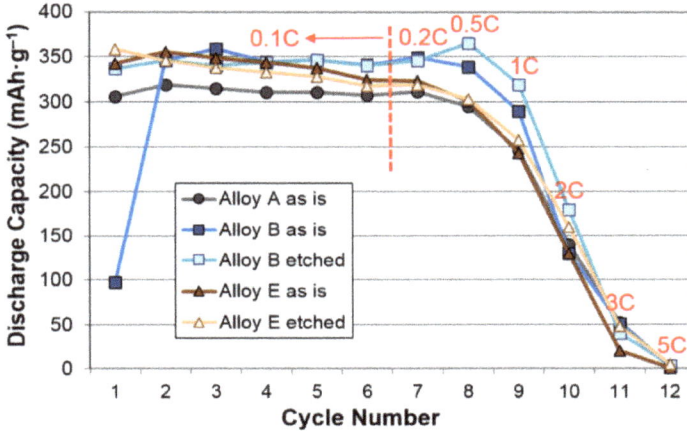

Figure 3. Examples of half-cell discharge capacities measured with 0.1 C (cycle 1–6), 0.2 C (cycle 7), 0.5 C (cycle 8), 1 C (cycle 9), 2 C (cycle 10), 3 C (cycle 11), and 5 C (cycle 12) rates from **A** (as-in only), **B** (both as-is and etched) and **E** (both as-is and etched).

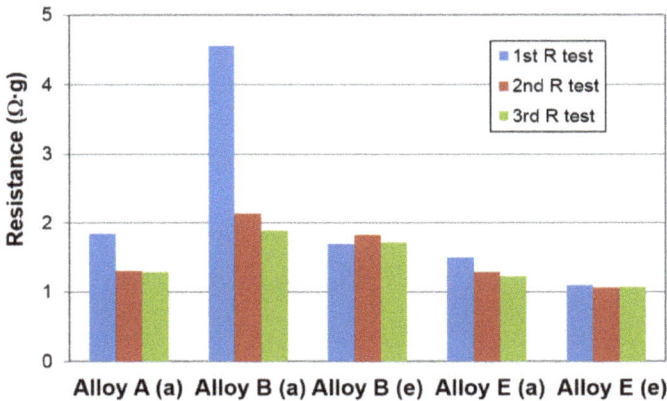

Figure 4. Examples of three internal resistance results from **A** (as-is only), **B** (both as-is and etched), and **E** (both as-is and etched). (a) and (e) in parentheses represent for as-is and etched.

Table 2. Electrochemical discharge capacity measured at 50 mA·g⁻¹, surface charge-transfer resistance (R) from AC impedance measurement, and specific power density.

Alloys	Treatment	Capacity (mAh·g⁻¹)	Resistance (Ω·g)	Specific Power (W·kg⁻¹)
A	None	321	1.28	317
	KOH-etch	321	1.74	238
B	None	360	1.88*	228*
	KOH-etch	352	1.71*	261*
C	None	290	2.25*	192*
	KOH-etch	291	1.68*	256*
D	None	390	1.80	235
	KOH-etch	349	2.34	181
E	None	356	1.22*	334*
	KOH-etch	358	1.07*	388*
F	None	300	1.42*	287*
	KOH-etch	304	1.15*	355*
G	None	353	1.13*	347*
	KOH-etch	353	1.04*	390*
H	None	345	1.86	218
	KOH-etch	320	2.04	202

Note: * The reduced R from the KOH-etch treatment.

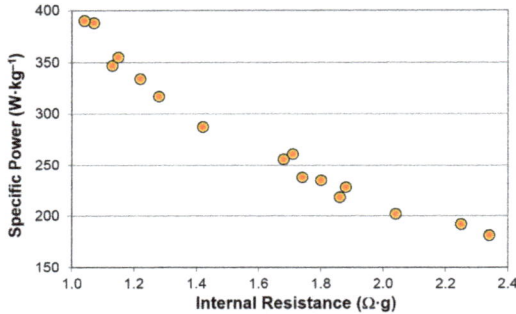

Figure 5. Plot of the calculated specific power vs. internal resistance for as-is and etched **A–H**.

The superiority of **E** vs. **A** in the high-rate performance has been verified by room-temperature AC impedance measured at different frequencies as shown in Figure 6. The radius of the semi-circle represents the surface charge-transfer resistance and shows the trend of as-is **A** > as-is **E** > etched **E**, which is in total agreement with the internal resistance measurement.

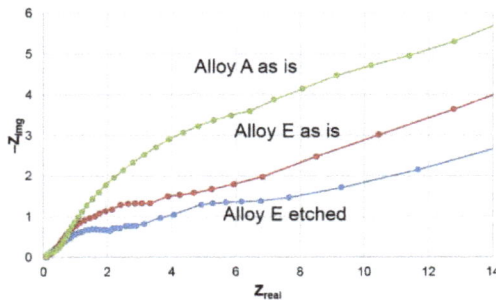

Figure 6. Cole-Cole plots obtained from room temperature alternative current (AC) impedance measurements of **A** (as-is only) and **E** (both as-is and etched).

The HRD capability of MH alloy was associated with the metallic nickel clusters embedded in its surface oxide after activation [45,46]. The saturated magnetic susceptibility (M_S) obtained from the magnetic susceptibility measurement shows clear correlation to the HRD of a series of A_2B_7 MH alloys [13]. However, we also reported that some MH alloys with HRD were more related to the structure of the metal/oxide interface [47,48]. Therefore, it is interesting to compare the internal resistance of the etched alloy with the M_S values reported in the literature (Table 1). The resulting plot is shown in Figure 7 (red dots) and does not reveal any clear correlation. The difference in R before and after etching is also plotted in the same figure (triangles) and no clear trend can be seen. We conclude that an M_S estimation of surface catalytic ability can be used only in the comparison of a series of MH alloys with a similar composition/structure and cannot be elaborated freely among various alloy systems.

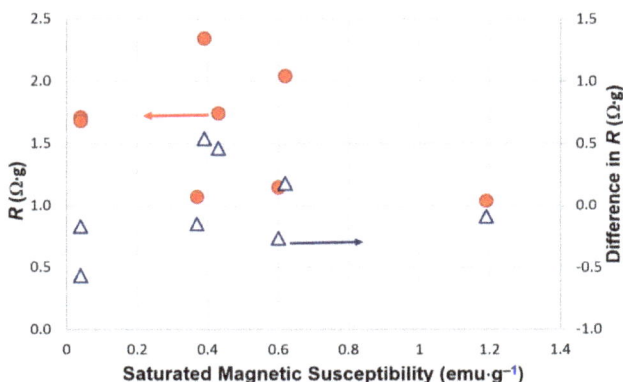

Figure 7. Plots of the internal resistance (R measured from the annealed alloys and the difference in R before and after annealing (triangle) vs. the saturated magnetic susceptibility (M_S) (red dots)).

3.3. Microsnalysis of the Activated Surface

SEM micrographs taken from the surfaces of etched alloys are shown in Figure 8. While the surfaces of RE-based AB_5 (**A**) and A_2B_7 (**E–H**) alloys are covered with $RE(OH)_3$ needles, those from TM-based AB_2 (**B** and **C**) and BCC-C14 (**D**) are modified by patches of ZrO_2. Needles from $La(OH)_3$ are smaller compared to hydroxides from other RE elements [24]. The microstructures of RE- and TM-based MH alloys were studied by transmission electron microscope before and can be summarized as a 50 nm buffer oxide (amorphous) and a 100 nm surface oxide with Ni-inclusion with $RE(OH)_3$ needles on top in the former and a 100 nm buffer oxide layer and a 200 nm surface oxide with Ni-inclusion with patches of ZrO_2 on top in the latter [24]. While the RE in AB_5 and A_2B_7 alloys formed an impeccable passive surface oxide when reacting with hot KOH, the TM just leaches out into electrolyte. The weight comparison of three **B** (AB_2) and three **E** (A_2B_7) electrodes before and after KOH-etching (dried in vacuum over for 24 h) were conducted and results are listed in Table 3 and plotted in Figure 9. The average weight loss (gain) for **B** and **E** (MH powder only) are −2.2% and +1.0%. The leached-out species were further analyzed by examining the composition of the alkaline solution after the etching experiment with ICP and results are summarized in Table 4. In the solutions with RE-based alloys, only Al (**A**, **F**, and **H**) and a very small amount of Mg (**E**, **F**, and **G**) are detected. In the TM-based alloys (**B–D**), larger concentrations of Ti, V, and Zr are found. The Ni-concentrations in the solutions from TM-based alloys are much higher than those from the RE-based one, while the former have smaller Ni-content in their compositions. This is additional evidence showing that when the TM-based alloys were attached by hot KOH they went through a preferential leach-out and the RE-based alloys formed a passive hydroxide layer under the same situation.

To confirm the results from SEM and ICP analysis, XRD was performed, and the resulting patterns obtained before and after KOH-etch for each alloy in this study are plotted in Figure 10. All RE-based MH alloys (**A**, **E**–**H**) show peaks of RE(OH)$_3$ after etching. Only etched **B** shows ZrO$_2$ phase, and both AB$_2$ alloys (**B** and **C**) contain metallic Ni as a product of corrosion after etching. **D** exhibited a small portion of partially hydride (α-phase) formed by storing the hydrogen generated from the high level of oxidation of TM (mainly V [11]). The metal-hydrogen bond of the product from the initial hydrogenation is very strong, as seen from the large portion of low-pressure irreversible hydrogen storage in **D** (PCT in Figure 1). The XRD analysis results indicate that the leaching-out (corrosion) of the alloy surface by hot KOH is in the order of **D** > **C** > **B** > the rest.

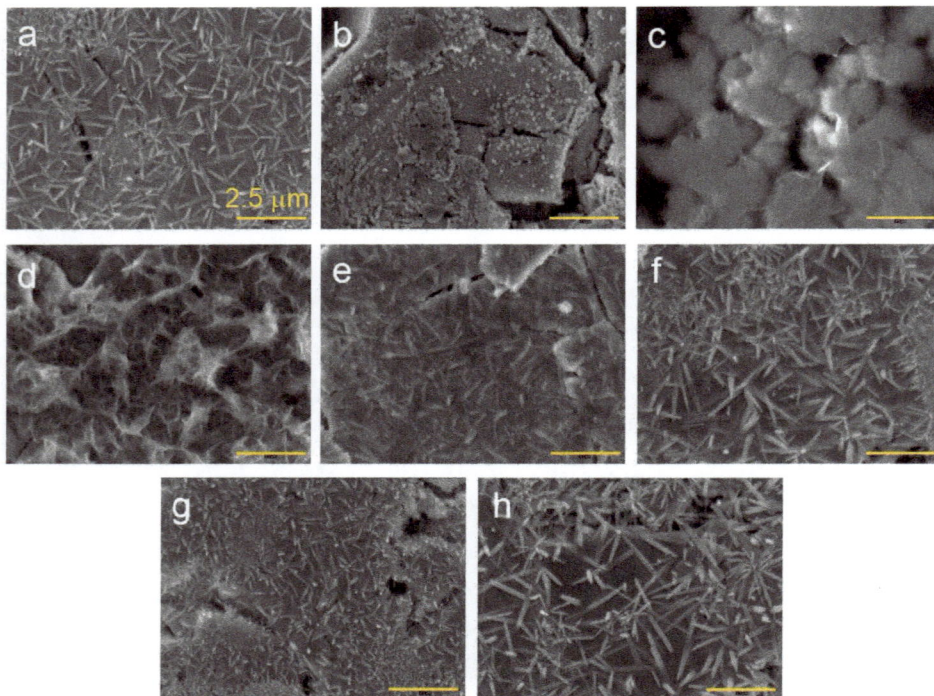

Figure 8. Scanning electron microscopy (SEM) micrographs from the etched alloys (a) **A**, (b) **B**, (c) **C**, (d) **D**, (e) **E**, (f) **F**, (g) **G** and (h) **H**. The bar at the lower right corner in each micrograph represents 2.5 μm.

Table 3. Measurement of weight change before and after KOH-etch for **A** and **B**. All numbers are in mg.

Sample #	Substrate Weight	Electrode Wright Before	Electrode Weight After	Weight Difference
B 1	192.8	262.3	260.8	−1.5
B 2	182.7	257.4	255.7	−1.7
B 3	178.4	252.0	250.5	−1.5
E 1	199.5	279.6	280.3	0.7
E 2	181.3	278.7	279.8	1.1
E 3	186.4	289.0	290.1	1.1

Batteries 2017, 3, 30

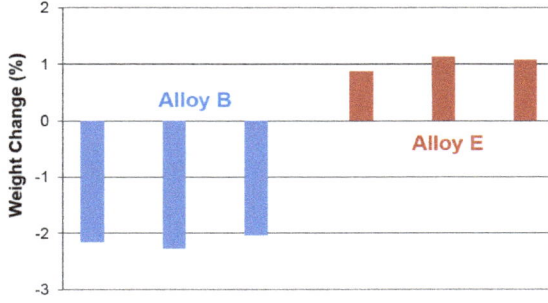

Figure 9. Weight changes after KOH-etching for three electrodes from **B** (left) and three from **E** (right).

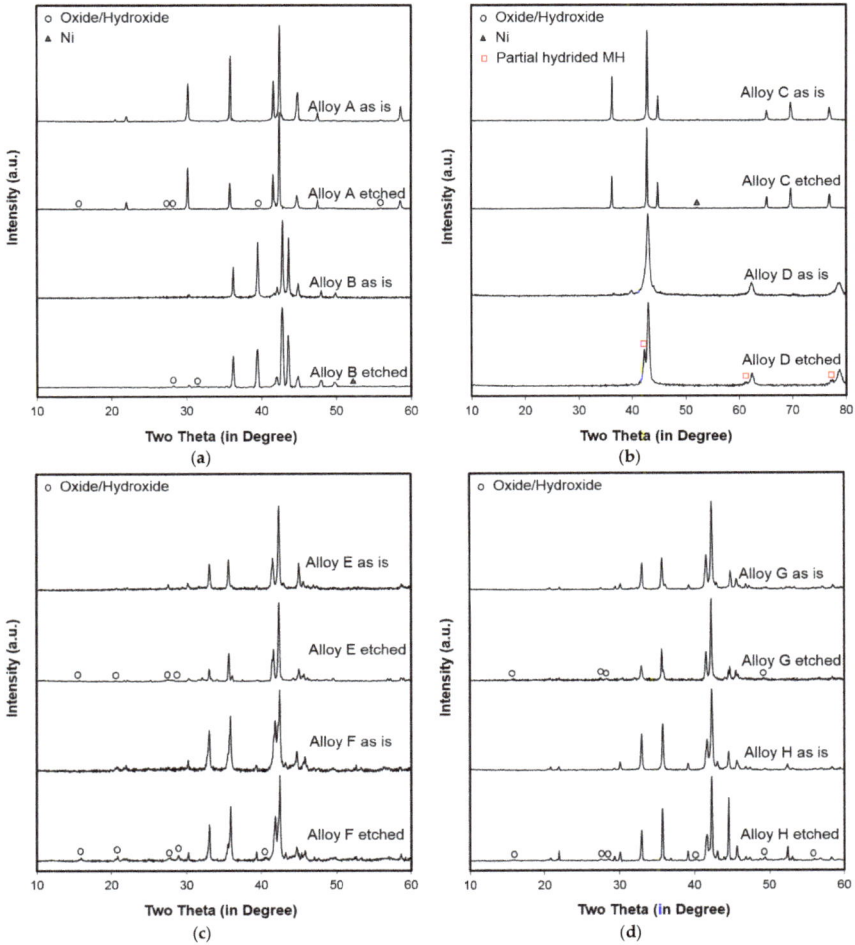

Figure 10. X-ray diffractometer (XRD) patterns using Cu-K$_\alpha$ as the radiation source for alloys (**a**) **A**, **B**, (**b**) **C**, **D**, (**c**) **E**, **F** and (**d**) **G**, **H** before and after KOH-etching.

Table 4. ICP results (in unit of ppm) from the solutions after etching experiment. N.D. and LLD denote non-detectable (below LLD) and lowest limit of detection, respectively.

Alloy	A	B	C	D	E	F	G	H	LLD
Al	58	1.6	N.D.	29	N.D.	78	N.D.	20	0.012
Co	N.D.	0.1	N.D.	1.0	N.D.	N.D.	1.2	N.D.	0.004
Cr	N.D.	4.3	N.D.	48	N.D.	N.D.	N.D.	N.D.	0.054
Fe	N.D.	N.D.	1.0	N.D.	N.D.	N.D.	N.D.	N.D.	0.002
Mg	N.D.	N.D.	N.D.	N.D.	0.05	0.3	0.5	N.D.	0.001
Mn	N.D.	2.0	0.1	8.0	N.D.	N.D.	N.D.	N.D.	0.001
Ni	N.D.	6.0	0.3	5.0	N.D.	N.D.	N.D.	N.D.	0.001
Ti	N.D.	150	7	416	N.D.	N.D.	N.D.	N.D.	0.004
V	N.D.	125	92	988	N.D.	N.D.	N.D.	N.D.	0.013
Zr	N.D.	495	518	309	N.D.	N.D.	N.D.	N.D.	0.002

4. Conclusions

Combining the electrochemical testing results and microstructure analysis, we have the following findings: The improvement in the high-rate capability (reduction in R) by KOH-etch is the most prominent in C15 AB_2 (**C**) with a small La-content, followed by two A_2B_7 alloys (**E** and **F**) with Mg, and then C14 AB_2 (**A**) and an A_2B_7 with Mg and no Al (**G**). The KOH-etch deteriorates the high-rate performance (increase in R) in the standard Mg-free AB_5 (**A**) and V-rich BCC-C14 (**D**) alloys, whereas Mg-content in A_2B_7 MH alloys responds well with the KOH-etch and La also helps the decrease in surface charge-transfer resistance by etching. BCC phase in a multiple phase alloy is more robust against the KOH corrosion [49] and is inert in the electrochemical environment [50–52]. Etching at a high temperature will increase the density of BCC phase on the surface and impede the electrochemical reaction, which explains the severe degradation in high-power performance of alloy **D** after KOH-etch.

Acknowledgments: The authors would like to thank the following individuals from BASF-Ovonic for their help: Taihei Ouchi, Jean Nei, Shuan Chang, Su Cronogue, Baoquan Huang, William Mays, Diana F. Wong, David Pawlik, Allen Chan, and Ryan J. Blankenship.

Author Contributions: Tiejun Meng performed the experiment and analyzed results. Kwo-Hsiung Young contributed to data analysis and manuscript preparation. Chaolan Hu and Benjamin Reichman provided the KOH-etching recipe and helped in AC impedance measurement and data interpretation.

Conflicts of Interest: The authors declare no conflict of interest.

Abbreviations

The following abbreviations are used in this manuscript:

MH	Metal hydride
IMC	Intermetallic compound
Ni/MH	Nickel/metal hydride
BCC	Body-centered-cubic
TM	Transition metal
RE	Rare earth
HRD	High-rate dischargeability
VIM	Vacuum induction melting
AM	Arc melting
PCT	Pressure–concentration–temperature
ICP-OES	Inductively coupled plasma-optical emission spectrometer
XRD	X-ray diffractometer
SEM	Scanning electron microscope
SOC	State of charge

M_S	Saturated magnetic susceptibility
AC	Alternative current
R	Internal resistance
P	Specific power
V_{oc}	Open-circuit voltage

References

1. Young, K. Metal Hydride. In *Elsevier Reference Module in Chemistry, Molecular Sciences and Chemical Engineering*; Reedijk, J., Ed.; Elsevier: Waltham, MA, USA, 2013.
2. Goo, N.H.; Woo, J.H.; Lee, K.S. Mechanism of Rapid Degradation of Nanostructured Mg_2Ni Hydrogen Storage Alloy Electrode Synthesized by Mechanical Alloying and the Effect of Mechanically Coating with Nickel. *J. Alloys Compd.* **1999**, *288*, 286–293. [CrossRef]
3. Zhang, Q.A.; Lei, Y.Q.; Wang, C.S.; Wang, F.S.; Wang, Q.D. Structure of the Secondary Phase and Its Effects on Hydrogen-storage Properties in a $Ti_{0.7}Zr_{0.2}V_{0.1}Ni$ Alloy. *J. Power Sour.* **1998**, *75*, 288–291. [CrossRef]
4. Nei, J.; Young, K. Gaseous Phase and Electrochemical Hydrogen Storage Properties of $Ti_{50}Zr_1Ni_{44}X_5$ (X = Ni, Cr, Mn, Fe, Co, or Cu) for Nickel Metal hydride Battery Applications. *Batteries* **2016**, *2*, 24. [CrossRef]
5. Züttel, A.; Meli, F.; Schlapbach, L. Electrochemical and Surface Properties of $Zr(V_xNi_{1-x})_2$ Alloys as Hydrogen-absorbing Electrodes in Alkaline Electrolyte. *J. Alloys Compd.* **1994**, *203*, 235–241.
6. Liao, B.; Lei, Y.Q.; Chen, L.X.; Lu, G.L.; Pan, H.G.; Wang, Q.D. A Study on the Structure and Electrochemical Properties of $La_2Mg(Ni_{0.95}M_{0.05})_9$ (M = Co, Mn, Fe, Al, Cu, Sn) Hydrogen Storage Electrode Alloys. *J. Alloys Compd.* **2004**, *376*, 186–195. [CrossRef]
7. Yasuoka, S.; Magari, Y.; Murata, T.; Tanaka, T.; Ishida, J.; Nakamura, H.; Nohma, T.; Kihara, M.; Baba, Y.; Teraoka, H. Development of High-capacity Nickel-metal Hydride Batteries Using Superlattice Hydrogen-Absorbing Alloys. *J. Power Sour.* **2006**, *156*, 662–666. [CrossRef]
8. Zhao, Y.; Zhang, L.; Ding, Y.; Cao, J.; Jia, Z.; Ma, C.; Li, Y.; Han, S. Comparative Study on the Capacity Degradation Behavior of $Pr_5 Co_{19}$-type Single-phase Pr_4MgNi_{19} and La_4MgNi_{19} Alloys. *J. Alloys Compd.* **2017**, *694*, 1089–1097. [CrossRef]
9. Willems, J.J.G.; Buschow, K.H.J. From Permanent Magnets to Rechargeable Hydride Electrodes. *J. Less Common Met.* **1987**, *129*, 13–30. [CrossRef]
10. Young, K.; Ouchi, T.; Huang, B.; Nei, J. Structure, Hydrogen Storage, and Electrochemical Properties of Body-centered-cubic $Ti_{40}V_{30} Cr_{15}Mn_{13}X_2$ alloys (X = B, Si, Mn, Ni, Zr, Nb, Mo, and La). *Batteries* **2015**, *1*, 74–90. [CrossRef]
11. Young, K.; Nei, J.; Wong, D.F.; Wang, L. Structural, Hydrogen Storage, and Electrochemical Properties of Laves Phase-related Body-centered-cubic Solid Solution Metal Hydride Alloys. *Int. J. Hydrog. Energy* **2014**, *39*, 21489–21499. [CrossRef]
12. Young, K.; Ng, K.Y.S.; Bendersky, L.A. A Technical Report of the Robust Affordable Next Generation Energy Storage System-BASF program. *Batteries* **2016**, *2*, 2. [CrossRef]
13. Young, K.; Ouchi, T.; Huang, B.; Chao, B.; Fetcenko, M.A.; Bendersky, L.A.; Wang, K.; Chiu, C. The Correlation of C14/C15 Phase Abundance and Electrochemical Properties in the AB_2 Alloys. *J. Alloys Compd.* **2010**, *506*, 841–848. [CrossRef]
14. Young, K.; Nei, J. The Current Status of Hydrogen Storage Alloy Development for Electrochemical Applications. *Materials* **2013**, *6*, 4574–4608. [CrossRef] [PubMed]
15. Young, K.; Chang, S.; Lin, X. C14 Laves Phase Metal Hydride Alloys for Ni/MH Batteries Applications. *Batteries* **2017**, *3*, 27. [CrossRef]
16. Schlapbach, L.; Seiler, A.; Siegmann, H.C.; Waldkirch, T.V.; Zücher, P.; Brundle, C.R. Self Restoring of the Active Surface in $LaNi_5$. *Int. J. Hydrog. Energy* **1979**, *4*, 21–28. [CrossRef]
17. Fetcenko, M.A.; Young, K.; Ovshinsky, S.R.; Reichman, B.; Koch, J.; Mays, W. Modified Electrochemical Hydrogen Storage Alloy Having Increased Capacity, Rate Capability and Catalytic Activity. US Patent 6,740,448, 25 May 2004.
18. Meli, F.; Schlapbach, L. Surface Analysis of AB_5-type electrodes. *J. Less Common Met.* **1991**, *172*, 1252–1259. [CrossRef]

19. Broom, D.P.; Kemali, M.; Ross, D.K. Magnetic Properties of Commercial Metal Hydride Battery Materials. *J. Alloys Compd.* **1999**, *255*, 293–295. [CrossRef]
20. Yu, B.; Chen, L.; Wen, M.F.; Tong, M.; Chen, D.M.; Tian, Y.W.; Zhai, Y.C. Surface Modification of AB_2 and AB_5 Hydrogen Storage Alloy Electrodes by the Hot-charging Treatment. *J. Mater. Sci. Technol.* **2001**, *17*, 247–251.
21. Shen, Y.; Peng, F.; Kontos, S.; Noréus, D. Improved NiMH Performance by a Surface Treatment that Creates Magnetic Ni-clusters. *Int. J. Hydrog. Energy* **2016**, *41*, 9933–9938. [CrossRef]
22. Tan, S.; Shen, Y.; Şahin, E.Q.; Noréus, D.; Öztürk, T. Activation Behavior of an AB_2 Type Metal Hydride Alloy for NiMH Batteries. *Int. J. Hydrog. Energy* **2016**, *41*, 9948–9953. [CrossRef]
23. Ye, Z.; Noréus, D. Metal hydride electrodes: The Importance of Surface Area. *J. Alloys Compd.* **2016**, *664*, 59–64. [CrossRef]
24. Young, K.; Chao, B.; Liu, Y.; Nei, J. Microstructures of the Oxides on the Activated AB_2 and AB_5 Metal Hydride Alloys Surface. *J. Alloys Compd.* **2014**, *606*, 97–104. [CrossRef]
25. Chang, S.; Young, K.; Ouchi, T.; Meng, T.; Nei, J.; Wu, X. Studies on Incorporation of Mg in Zr-based AB_2 Metal Hydride Alloys. *Batteries* **2016**, *2*, 11. [CrossRef]
26. Young, K.; Wong, D.F.; Nei, J. Effects of Vanadium/nickel Contents in Laves Phase-related Body-centered-cubic Solid Solution Metal Hydride Alloys. *Batteries* **2015**, *1*, 34–53. [CrossRef]
27. Teraoka, H. Development of Low Self-Discharge Nickel-Metal Hydride Battery. Available online: http://www.scribd.com/doc/9704685/Teraoka-Article-En (accessed on 9 April 2016).
28. Kai, T.; Ishida, J.; Yasuoka, S.; Takeno, K. The Effect of Nickel-Metal Hydride Battery's Characteristics with Structure of the Alloy. In Proceedings of the 54th Battery Symposium in Japan, Osaka, Japan, 7–9 October 2013; p. 210.
29. Liu, F.J.; Kitayama, K.; Suda, S. La and Ce-incorporation Effects on the Surface Properties of the Fluorinated $(Ti,Xr)(Mn,Cr,Ci)_2$ Hydriding Alloys. *Vacuum* **1996**, *47*, 903–906. [CrossRef]
30. Kim, S.; Lee, J. Activation behaviour of $ZrCrNiM_{0.05}$ metal hydride electrode (M = La, Mm (misch metal), Nd). *J. Alloys Compd.* **1992**, *185*, L1–L4. [CrossRef]
31. Park, H.Y.; Chang, I.; Cho, W.I.; Cho, B.W.; Jang, H.; Lee, S.R.; Yun, K.S. Electrode Characteristics of the Cr and La Doped AB_2-type Hydrogen Storage Alloys. *Int. J. Hydrog. Energy* **2001**, *26*, 949–955. [CrossRef]
32. Liu, F.J.; Sandrock, G.; Suda, S. Surface and Metallographic Microstructure of the La-added AB_2 Compound $(Ti, Zr)(Mn, Cr, Ni)_2$. *J. Alloys Compd.* **1995**, *231*, 392–396. [CrossRef]
33. Sun, D.; Latroche, M.; Percheron-Guégan, A. Effects of Lanthanum or Cerium on the Equilibrium of $ZrNi_{1.2}Mn_{0.6}V_{0.2}Cr_{0.1}$ and Its Related Hydrogenation Properties. *J. Alloys Compd.* **1997**, *248*, 215–219.
34. Yang, X.G.; Lei, Y.Q.; Shu, K.Y.; Lin, G.F.; Zhang, Q.A.; Zhang, W.K.; Zhang, X.B.; Lu, G.L.; Wang, Q.D. Contribution of Rare-earths to Activation Property of Zr-based Hydride Electrode. *J. Alloys Compd.* **1999**, *293*, 632–636. [CrossRef]
35. Sun, J.C.; Li, S.; Ji, S.J. The Effects of the Substitution of Ti and La for Zr in $ZrMn_{0.7}V_{0.2}Co_{0.1}Ni_{1.2}$ Hydrogen Storage Alloys on the Phase Structure and Electrochemical Properties. *J. Alloys Compd.* **2007**, *446*, 630–634.
36. Chen, W.X. Effects of Addition of Rare-earth Element on Electrochemical Characteristics of $ZrNi_{1.1}Mn_{0.5}V_{0.3}Cr_{0.1}$ Hydrogen Storage Alloy Electrodes. *J. Alloys Compd.* **2001**, *319*, 119–123.
37. Young, K.; Ouchi, T.; Huang, B.; Reichman, B.; Fetcenko, M.A. Effect of Molybdenum Content on Structural, Gaseous Storage, and Electrochemical Properties of C14-predominant AB_2 Metal Hydride Alloys. *J. Power Sour.* **2011**, *196*, 8815–8821. [CrossRef]
38. Young, K.; Nei, J.; Wan, C.; Denys, R.V.; Yartys, V.A. Comparison of C14- and C15-predominated AB_2 Metal Hydride Alloys for Electrochemical Applications. *Batteries* **2017**, *3*, 22. [CrossRef]
39. Young, K.; Ouchi, T.; Meng, T.; Wong, D.F. Studies on the Synergetic Effects in Multi-phase Metal Hydride Alloys. *Batteries* **2016**, *2*, 15. [CrossRef]
40. Young, K.; Ouchi, T.; Nei, J.; Koch, J.M.; Lien, Y. Comparison among constituent phases in superlattice metal hydride alloys for battery applications. *Batteries* **2017**. submitted.
41. Koch, J.M.; Young, K.; Nei, J.; Hu, C.; Reichman, B. Performance comparison between AB_5 and superlattice metal hydride alloys in sealed cells. *Batteries* **2017**. submitted.
42. Wang, L.; Young, K.; Meng, T.; Ouchi, T.; Yasuoka, S. Partial Substitution of Cobalt for Nickel in Mixed Rare Earth Metal Based Superlattice Hydrogen Absorbing Alloy—Part 1 Structural, Hydrogen Storage and Electrochemical Properties. *J. Alloys Compd.* **2016**, *660*, 407–415. [CrossRef]

43. Young, K.; Yasuoka, S. Past, present, and future of metal hydride alloys in nickel-metal hydride batteries. In Proceedings of the 14th International Symposium on Metal-Hydrogen Systems, Manchester, UK, 21–25 July 2014.

44. Zhou, X.; Young, K.; West, J.; Regalado, J.; Cherisol, K. Degradation Mechanisms of High-energy Bipolar Nickel Metal Hydride Battery with AB_5 and A_2B_7 Alloys. *J. Alloys Compd.* **2013**, *580*, S373–S377. [CrossRef]

45. Stucki, F.; Schlapbach, L. Magnetic Properties of $LaNi_5$, FeTi, Mg_2Ni and Their Hydrides. *J. Less Common Met.* **1980**, *74*, 143–151. [CrossRef]

46. Schlapbach, L.; Stucki, F.; Seiler, A.; Siegmann, H.C. Magnetism and Hydrogen Storage in $LaNi_5$, FeTi and Mg_2Ni. *J. Magn. Magn. Mater.* **1980**, *15–18*, 1271–1272. [CrossRef]

47. Young, K.; Chao, B.; Nei, J. Microstructures of the Activated Si-containing AB_2 Metal Hydride Alloy Surface by Transmission Electron Microscope. *Batteries* **2016**, *2*, 4. [CrossRef]

48. Young, K.; Chao, B.; Pawlik, D.; Shen, H.T. Transmission Electron Microscope Studies in the Surface Oxide on the La-containing AB_2 Metal Hydride Alloy. *J. Alloys Compd.* **2016**, *672*, 356–365. [CrossRef]

49. Young, K.H.; Fetcenko, M.A.; Ovshinsky, S.R.; Ouchi, T.; Reichman, B.; Mays, W.C. Improved Surface Catalysis of Zr-based Laves Phase Alloys for NiMH Batteries. In *Hydrogen at Surface and Interfaces*; Jerkiewicz, G., Feliu, J.M., Popov, B.N., Eds.; Electrochemical Society: Pennington, NJ, USA, 2000.

50. Lee, H.; Chourashiya, M.G.; Park, C.; Park, C. Hydrogen Storage and Electrochemical Properties of the $Ti_{0.32}Cr_{0.43-x-y}V_{0.25}Fe_xMn_y$ (x = 0–0.055, y = 0–0.080) Alloys and Their Composites with $MmNi_{3.99}Al_{0.29}Mn_{0.3}Co_{0.6}$ Alloy. *J. Alloys Compd.* **2013**, *566*, 37–42. [CrossRef]

51. Inoue, H.; Arai, S.; Iwakura, C. Crystallographic and Electrochemical Characterization of $TiV_{4-x}Ni_x$ Alloys for Nickel-metal Hydride Batteries. *Electrochim. Acta* **1996**, *41*, 937–939. [CrossRef]

52. Yu, X.B.; Wu, Z.; Xia, B.J.; Xu, N.X. A Ti-V-based bcc Phase Alloy for Use as Metal Hydride Electrode with High Discharge Capacity. *J. Chem. Phys.* **2004**, *121*, 987–990. [CrossRef] [PubMed]

batteries

MDPI

Article

Fabrications of High-Capacity Alpha-Ni(OH)$_2$

Kwo-Hsiung Young [1,2,*], Lixin Wang [2], Shuli Yan [1], Xingqun Liao [3,4], Tiejun Meng [2], Haoting Shen [2] and William C. Mays [2]

1   Department of Chemical Engineering and Materials Science, Wayne State University, Detroit, MI 48202, USA; shuliyan2010@gmail.com
2   BASF/Battery Materials—Ovonic, 2983 Waterview Drive, Rochester Hills, MI 48309, USA; lixinwang@a123systems.com (L.W.); tiejun.meng@partners.basf.com (T.M.); htshen@ufl.edu (H.S.); william.c.mays@basf.com (W.C.M.)
3   Shenzhen Highpower Technology Co., Luoshan Industrial Zone, Pinhu, Longgang, Shenzhen 518111, Guangdong, China; xqliao@highpowertech.com
4   College of Chemistry and Chemical Engineering, Central South University, South Lushan Road, Changsha 410083, Hunan, China
*   Correspondence: kwo.young@basf.com; Tel.: +1-248-293-7000

Academic Editor: Hua Kun Liu
Received: 10 January 2017; Accepted: 2 March 2017; Published: 7 March 2017

Abstract: Three different methods were used to produce α-Ni(OH)$_2$ with higher discharge capacities than the conventional β-Ni(OH)$_2$, specifically a batch process of co-precipitation, a continuous process of co-precipitation with a phase transformation step (initial cycling), and an overcharge at low temperature. All three methods can produce α-Ni(OH)$_2$ or α/β mixed-Ni(OH)$_2$ with capacities higher than that of conventional β-Ni(OH)$_2$ and a stable cycle performance. The second method produces a special core–shell β-Ni(OH)$_2$/α-Ni(OH)$_2$ structure with an excellent cycle stability in the flooded half-cell configuration, is innovative and also already mass-production ready. The core–shell structure has been investigated by both scanning and transmission electron microscopies. The shell portion of the particle is composed of α-Ni(OH)$_2$ nano-crystals embedded in a β-Ni(OH)$_2$ matrix, which helps to reduce the stress originating from the lattice expansion in the β-α transformation. A review on the research regarding α-Ni(OH)$_2$ is also included in the paper.

Keywords: alpha nickel hydroxide; continuous stirring single reactor process; core–shell structure; nickel metal hydride battery

1. Introduction

Nickel hydroxide has been used as the positive electrode for various rechargeable alkaline batteries for a long time. The history extends from Ni-Fe [1] and Ni-Cd [2], which were invented over 100 years ago, to Ni-H, Ni-Zn, and the current Ni/metal hydride (MH) batteries. The advantages of using nickel hydroxide include relatively low cost, adequate redox voltage (close to the oxygen gas evolution potential to maximize the operation voltage), good high-rate dischargeability (HRD) performance, and wide operation and storage temperature ranges. US Patents [3] and Japanese Patent Applications [4] for nickel hydroxide as a positive electrode active material were previously reported by our group. While the conventional nickel hydroxide used in the battery went through a β-Ni(OH)$_2$-Ni(II)/β-NiOOH-Ni(III) redox electrochemical reaction, Bode et al. demonstrated another redox reaction between α-Ni(OH)$_2$-Ni(II)/γ-NiOOH-Ni(III) [5]. The α–γ transformation may involve more than one electron transfer per Ni atom (up to 1.6 to 1.67 electrons [6,7]) due to the non-integral average oxidation states of the α and γ phases that occur because of the presence of anions (NO$_3^-$, CO$_3^{2-}$, SO$_4^{2-}$, Cl$^-$, etc.) and cations (Li$^+$, Na$^+$, K$^+$, etc.) in the water layers in α-Ni(OH)$_2$ and γ-NiOOH, respectively [8–10] (see Bode's diagram in Figure 1). The relatively large theoretic capacity of α–γ transformation (462–480 mAh·g^{-1}), compared to that in the β(II)–β(III) transformation (289 mAh·g^{-1}), makes it a very attractive research topic for electrochemical material scientists. These efforts are summarized in Table 1. The most common preparation methods involve co-precipitations using metal nitrates or sulfates. Since the beginning, pure α-Ni(OH)$_2$ was found to be unstable and will convert to β-Ni(OH)$_2$ through a slow aging process [11]. Trivalent elements, such as Al, Y, and La, have been added to stabilize the α-Ni(OH)$_2$ phase. Other elements, such as Mg, Fe, Co, Cu, Zn, Ce, Nd, and Yb, have also been previously reported. In the case of battery operation using the β–β one-electron transition, the α-Ni(OH)$_2$ phase formed during accidental overcharge will cause the early failure of the battery [12] and, thus, Zn is co-precipitated in commercial Ni(OH)$_2$ batteries to suppress the formation of α-Ni(OH)$_2$ [13,14]. The research target of α-Ni(OH)$_2$ switched from capacity and cycle stability in the early days, to HRD, then to tap density, which are all important criteria for a suitable battery material. With research funding from the U.S. Department of Energy [15], we were able to further investigate the feasibility of mass production, in particular integration of β-Ni(OH)$_2$ with the currently used fabrication method.

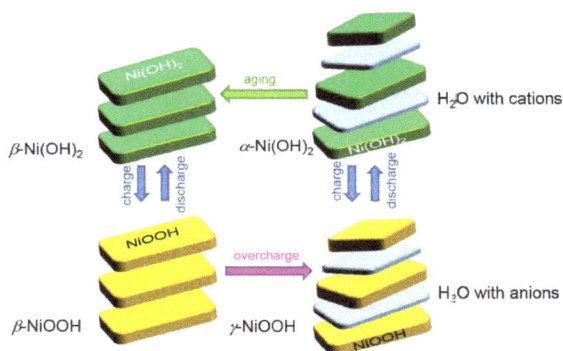

Figure 1. Bode's diagram showing the relationship among various hydroxides and oxyhydroxides of Ni. The lattice constants c in α-Ni(OH)$_2$ and γ-NiOOH are larger than those in β-Ni(OH)$_2$ and β-NiOOH, due to the insertion of an extra layer of water between the two Ni-containing sheets.

2. Experimental Setup

Three α-Ni(OH)$_2$ fabrication methods were adopted in this research: a batch process designed for small-scale laboratory implementation and benchmark sample preparation; a continuous process with the identical design principle to our mass-production equipment; and a low-temperature formation process for converting β-Ni(OH)$_2$ into α-Ni(OH)$_2$ material. The sample names, compositions, and fabrication methods are summarized in Table 2. The batch process was composed of the following steps. First, 11.6 g (0.063 mol) of Ni(NO$_3$)$_2$ and 3 g (0.014 mol) of Al(NO$_3$)$_3$ were dissolved in 500 mL distilled water and the solution was denoted as Solution 1. Next, 60 g (0.075 mol) of NaOH and 15.6 g (0.15 mol) of Na$_2$CO$_3$ were dissolved in 500 mL distilled water and called Solution 2. Finally, 6.4 g (0.12 mol) of NH$_4$Cl was dissolved in 100 mL distilled water, and NH$_3$·H$_2$O was added until the pH value reached 10, which formed Solution 3. Solutions 1 and 2 were gradually pumped into Solution 3 while maintaining a pH value in a range of 10–10.2. The color of the mixture solution was purple. After Solution 1 was exhausted, Solution 2 was constantly pumped into the mixed solution to increase the pH value to a range of 11.0–11.2. During this process, the solution color changed from purple to blue-green. This mixture was stirred for 5 h at room temperature. The final mixture was aged at 60 °C for 8 h, and then filtered and washed with deionized water. The obtained solid was dried at 60 °C in air for 5 h and used as the Ni-2 sample in this experiment. The Ni-1 sample was fabricated in the same way, except for the sole 14.7 g (0.077 mol) Ni(NO$_3$)$_2$ used in Solution 1.

Table 1. Summaries of prior works on producing α-Ni(OH)$_2$ and mixed α/β-Ni(OH)$_2$ for electrochemical applications. EP, CP, and ECI denote electrochemical precipitation, chemical precipitation, and electrochemical impregnation, respectively. In addition, α-Ni(OH)$_2$ and β-Ni(OH)$_2$ are represented by α and β, respectively. The concentration of the main dopant in molar percentage is listed in the Main Dopant column.

Main Dopant	Preparation Method	Main Findings	Reference
Co (6%, 45%, 75%)	EP	• First used with Co to stabilize α and γ	[16]
Co (20%, 55%)	CP	• Co stabilizes α up to 50 cycles with 345 mAh·g^{-1}	[17]
Fe (10%, 20%, 30%)	CP	• First used with Fe to stabilize α up to 60 cycles with 232 mAh·g^{-1}	[18]
Al, Fe, Cr, Mn (25%)	EP	• First used with other trivalent cations to promote α • Al shows the highest columbic efficiency	[19]
Al (10%–25%)	CP from nitrates	• Discharge capacity of 240 mAh·g^{-1} • A lower resistance and a high self-discharge rate found on α	[20]
Al (20%)	CP from nitrates	• α has a higher self-discharge than β	[21]
None	Sintering	• Discharge capacity of 225 mAh·g^{-1} • First use of a non-precipitation method	[22]
Mn (20%–40%)	CP	• Mn stabilizes α up to 120 cycles	[23]
Al (20%)	CP from sulfates	• Discharge capacity of 343 mAh·g^{-1} • Better reversibility of α	[24,25]
Al (30%)	Sintering	• Discharge capacity of ca. 450 mAh·g^{-1}	[26]
Al (25%)	CP from nitrates	• Discharge capacity of 381 mAh·g^{-1}	[27]
Al (5%–20%)	CP	• Discharge capacity of 260 mAh·g^{-1}	[28]
Fe (0%–40%)	CP	• 20% Fe is needed to stabilize α	[29]
None	CP from nitrates	• Capacity of 180 mAh·g^{-1} for an α/β mixture • Aging process from α to β	[30]
Y (1%–10%)	CP from nitrates	• Tap density of 1.6 g·cc^{-1} and 330 mAh·g^{-1}	[31]
Al (8.8%–25.4%)	CP from nitrates	• Al stabilizes α up to 100 cycles	[32]
Mn, Zn, Co (13%, 4.5%, 3%)	CP	• α/β mixture showing high tap density and capacity of 375 mAh·g^{-1}	[33]

Table 1. *Cont.*

Main Dopant	Preparation Method	Main Findings	Reference
Al (10%–30%)	CP from nitrates	• Discharge capacity of ca. 350 mAh·g^{-1} with 10 at % Al	[34]
Al (10%, 15%)	ECI	• Discharge capacity of 230–240 mAh·g^{-1} after 200 cycles	[35]
Al (20%)	Solid-state synthesis	• Discharge capacity of 336 mAh·g^{-1}	[36]
Al (13.2%–19.5%)	CP from nitrates	• Discharge capacity of 360 mAh·g^{-1} after 300 cycles	[37]
Al (10%, 25%), Co	CP from nitrates	• Al and Co together improve rate capability	[38]
Al (10%, 25%)	Sintering	• 25% Al is better than 10% Al for battery application	[39]
La (3.26%–9.19%)	CP from nitrates	• La increases the plateau voltage and cycle stability	[40]
Ti (11%)	CP from sulfates	• 1.22 electron transfer per Ni • Good for high temperature (80 °C)	[41]
Al (10%)	CP from nitrates	• Discharge capacity of 380–400 mAh·g^{-1} • Low tap density	[42]
Co, Zn, Mn (10%–20%)	CP	• Mn improves the conductivity and reversibility	[43]
Al (20%)	CP	• Nitrate is better than sulfate as raw material • Performance controlled by bulk diffusion	[44]
Al (10%), Y (5%), Nd (5%)	CP from sulfates	• Y and Nd improve the hydrogen diffusion	[45]
Al (10%), Y (5%)	CP from sulfates	• Discharge capacity of 388 mAh·g^{-1}	[46]
Al (20%)	CP from nitrates	• Capacity degradation due to Al-migration onto surface	[47]
None	Rapid freezing micro-emulsion	• Discharge capacity of 340 mAh·g^{-1} after 35 cycles	[48]
Al (15%), Y (4%)	CP from sulfates	• Discharge capacity of 351 mAh·g^{-1} due to high defect density	[49]
Co (20%), Cd (3%)	Ultrasound radiation	• Lower the oxidation potential to avoid oxygen evolution during charge	[50]
Al (10%), Ce (5%)	CP + rapid freezing	• Discharge capacity of 363.2 mAh·g^{-1}	[51]
Al (10%)	CP from nitrates	• Capacity, high-rate, and high temperature performance improved by addition of metallic Co or Y(OH)$_3$	[52]
None	Liquid phase deposition	• α/carbon composite was made	[53]
Al (20%)	CP from nitrates	• Nitrate shows better results than sulfate	[54]

Table 1. *Cont.*

Main Dopant	Preparation Method	Main Findings	Reference
La (4.3%), Zn (2.4%)	CP from sulfates	• La and Zr improve the high rate, cycle stability, and capacity (373 mAh·g^{-1})	[55]
Al (15%), Mg (5%)	CP from nitrates	• Mg increases defect density and capacity (359 mAh·g^{-1})	[56]
Al	Conversely migrates	• Nanowire for capacitor use	[57]
Al, Co, Zn, Y (16%)	Supersonic CP	• Capacity of 346 mAh·g^{-1} due to synergetic effect	[58]
Y (16%)	Supersonic CP	• Y increases the capacity to 358 mAh·g^{-1}	[59,60]
Co (50%)	CP from chlorides	• Hexamethylenetetramine is a better hydrolytic agent than urea	[61]
Zn (5%–20%)	ECI	• Increased surface area for supercapacitor application	[62]
Al (9%)	CP from nitrates	• α/β mixture showing high tap density and capacity of 325 mAh·g^{-1} @ 200 mA·g^{-1}	[63]
Co, Al, Mn, Ca	CP	• β formation is the main failure mode	[64]
Cu (9%–23%)	Supersonic CP	• Cu improves both capacity (310 mAh·g^{-1}) and cycle stability	[65]
Al (3%–7%)	CP from nitrates	• Al improves both capacity (324.5 mAh·g^{-1}) and cycle stability	[66]
Al (17.2%)	Coating with β	• β coating improves both capacity and cycle stability	[67]
Y (17%)	CP from chlorides	• Influence from the buffer Na$_2$CO$_3$ was studied	[68]
Al (20%)	Drying and hydrothermal	• High tap density: 1.84 g·cc^{-1}	[69]
Al (20%)	Anion exchange from chloride	• High tap density: 1.89 g·cc^{-1}	[70]
Al (20%)	Drying	• Cl$^-$ improves both high rate and cycle stability	[71]
Al (9%), Mn (4.5%), Yb (9%)	Supersonic CP	• Capacity of 309 mAh·g^{-1} due to synergetic effects	[72]
None	Hydrothermal + calcination	• Large capacitance for capacitor application	[73]
Co (11.1%)	Hydrothermal	• Large capacitance for capacitor application	[74]
Sn (2%–17%)	Supersonic CP	• Sn improves capacity, cycle stability, reaction reversibility, and bulk diffusion	[75]

Table 2. List of samples used in this study. WSU and SZHP are abbreviations for Wayne State University (Detroit, MI, USA) and Shenzhen Highpower Technology Co. (Shenzhen, China).

Name	Composition	Fabrication Method	Fabrication Location	Comment
Ni-1	$Ni(OH)_2$	Batch process	WSU	β-$Ni(OH)_2$
Ni-2	$Ni_{0.86}Al_{0.14}(OH)_2$	Batch process	WSU	α-$Ni(OH)_2$
AP50	$Ni_{0.91}Co_{0.045}Zn_{0.045}(OH)_2$	Continuous process	BASF	Commercial β-$Ni(OH)_2$
WM02	$Ni_{0.91}Co_{0.05}Al_{0.04}(OH)_2$	Continuous process	BASF	α/β-$Ni(OH)_2$
WM12	$Ni_{0.84}Co_{0.12}Al_{0.04}(OH)_2$	Continuous process	BASF	α/β-$Ni(OH)_2$
YRM3	$Ni_{0.93}Co_{0.02}Al_{0.05}(OH)_2$	Continuous process	SZHP	Commercial β-$Ni(OH)_2$

The second method used the same continuous stirring tank reactor (CSTR) technique that we developed to mass-produce NiCoZn hydroxide (for cathode active material in Ni/MH battery) and NiMnCo hydroxide (for precursor of NiMnCo (NMC) cathode material used in Li-ion battery) spherical particles [76]. The prototype of the equipment is installed in Troy, MI, USA and is capable of producing 1 ton of powder per day. A smaller version of the CSTR, capable of producing 1 kg of powder per day, was used for this experiment. In a CSTR, reactants in solution, sulfates and NaOH, and ammonia buffer solution were pumped into a reactor at different rates (three in Figure 2), while maintaining a constant reactor temperature (70 °C) and pH value. Small sized crystallites nucleated at the bottom of the reactor (1) and brought up to the top of the liquid surface by a stirring blade (2) at the bottom of the reactor. During the moving from the lower part to the higher part of the reactor, the hydroxide plates conglomerated into spherical shapes and were carried out by the overflow (4) into a storage container (5). At the end of each run, the solid product in the container was rinsed with deionized water and then dried to powder in hot air. Two controls used in BASF (AP50) and Shenzhen HighPower Co. (YRM3) were made by the commercial scale of CSTR process (Figure 2a). In preparation for WM02 (lower Co-content) and WM12 (higher Co-content), an addition Al source from $Al(NO_3)_3$ solution was added in the feed-in solution throughout the entire process. With a higher Co-level, core–shell structured spherical particles can be produced with a higher Al-content in the shell than in the core (WM12). The core–shell structure is formed by the different solubility of Al at different stages of particle growth (Figure 2b). This is very different from another core ($Ni(OH)_2$)-shell ($Ni_{0.5}Mn_{0.5}(OH)_2$) material prepared by a two-step method (core first and followed by the precipitation of shell with different composition solution) reported by Camardese et al. [77]. When the Co-content is lower, the differential solubility of Al disappears and the distribution of Al is uniform throughout the particle (Figure 2a, WM02). The full-cell testing of WM12 was done in a pouch-cell [15]. The negative electrode used here was a dry-compacted AB_5 ($La_{10.5}Ce_{4.3}Pr_{0.5}Nd_{1.4}Ni_{60.0}Co_{12.7}Mn_{5.9}Al_{4.7}$) MH alloy electrode with a 20% increase in capacity (an negative/positive (N/P) design of 1.2). The electrolyte used was the standard 30% KOH and the separator was a grafted polyethylene (PE)/polypropylene (PP). The cell construction was electrolyte-flooded.

The third method is to convert β-$Ni(OH)_2$ into α-$Ni(OH)_2$ in a sealed cell by using a low-temperature formation scheme. AA-sized Ni/MH cells with a nominal capacity of 1200 mAh were made using a standard commercially available AB_5 ($La_{10.5}Ce_{4.3}Pr_{0.5}Nd_{1.4}Ni_{60.0}Co_{12.7}Mn_{5.9}Al_{4.7}$) paste negative electrode, co-precipitated YRM3 ($Ni_{0.93}Co_{0.02}Al_{0.05}(OH)_2$) paste positive electrode, a grafted polypropylene (PP)/polyethylene (PE) separator, and 30% KOH as the electrolyte. After being filled with electrolyte and sealed, cells underwent a standard three-cycle electrical formation process performed at room temperature. For capacity measurements before the low-temperature treatment, these cells were charged at 120 mA (C/10 rate) for 16 h (160% state of charge, SOC) at 20 °C and put to rest for 4 h at the same temperature. Cells were then discharged with 240 mA (C/5 rate) at 20 °C until a cut-off voltage of 1 V was reached, following which the discharge capacities were recorded. The cells were cooled to 10 °C and then overcharged with a current of 120 mA (C/10) for 30 days at that temperature. Cells were discharged with 240 mA (C/5) at 10 °C until a cut-off voltage of 1 V was reached and capacities were recorded. Cells were then brought back to 20 °C and charged with 120 mA for 16 h, followed by a 4 h rest, and discharged with 240 mA until 1 V was reached and the capacities were recorded.

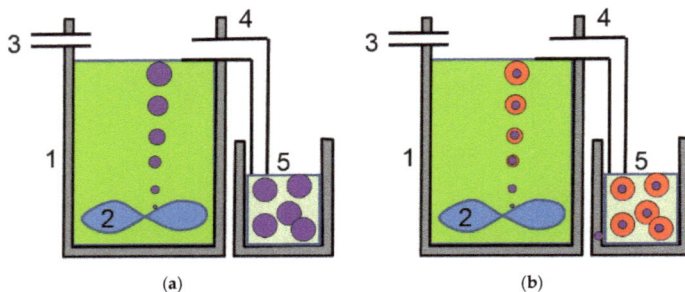

Figure 2. Schematic diagram of (**a**) a continuous stirring tank reactor (CSTR) used to co-precipitate hydroxides of metals with similar solubilities, such as Co, Ni, and Zn, and a consistent composition throughout the spherical particles; and (**b**) a CSTR to incorporate larger atoms, such as Al and Y, which have lower precipitation rates at initial nucleation stage (purple core) and higher precipitation rates at the growth stage (red shell). **1**: stainless-steel vessel; **2**: mixer blade; **3**: raw material inlet; **4**: product overflow; and **5**: product container.

For electrochemical testing of the powder, carbon black and polyvinylidene fluoride (PVDF) were used in the preparation of the positive electrode to enhance conductivity and electrode integrity. To this end, 100 mg of active material was first stir-mixed thoroughly with carbon black and PVDF in a weight ratio of 3:2:1 and applied onto a 0.5×0.5 in^2 nickel mesh with a nickel mesh tab leading out of the square substrate for the electrical connection. The electrode was then pressed under three tons of pressure for 5 s. The negative counter electrode used was a dry-compacted AB$_5$ type alloy. The positive and negative electrodes were sandwiched together with a polypropylene/polyethylene separator in a flooded half-cell configuration. The capacity of the negative electrode was significantly more than that of the positive electrode, resulting in a positive limited design. Electrochemical testing was performed with an Arbin electrochemical testing station (Arbin Instrument, College Station, TX, USA). X-ray diffraction (XRD) analysis was performed with a Philips X'Pert Pro X-ray diffractometer (Philips, Amsterdam, The Netherlands) and the generated patterns were fitted and peaks indexed by the Jade 9 software (Jade Software Corp. Ltd., Christchurch, New Zealand). A JEOL-JSM6320F scanning electron microscope (SEM, JEOL, Tokyo, Japan) with energy dispersive spectroscopy (EDS) was applied in investigating the phase distributions and compositions of the powders. An FEI Titan 80–300 (scanning) transmission electron microscope (TEM/STEM, Hillsboro, OR, USA) was employed to study the microstructure of the core–shell structure of the α/β mixture from the continuous process. For TEM characterization, mechanical polishing was used to thin samples, followed by ion milling.

3. Results

3.1. Batch Process

Two different conditions (Ni-1 and Ni-2) were employed in the batch process [15]. While Ni-1 has a composition that incorporates pure Ni as the cation source, which exhibits an exclusively β-Ni(OH)$_2$ structure, Ni-2 is doped with 14 at % Al and show a typical α-Ni(OH)$_2$ structure. The discharge capacities for the three samples from the Ni-2 process are shown in Figure 3. It took approximately 15 cycles to stabilize the capacity. The highest discharge capacity obtained with a 25 mA·g^{-1} rate was 346 mAh·g^{-1} at the 25th cycle.

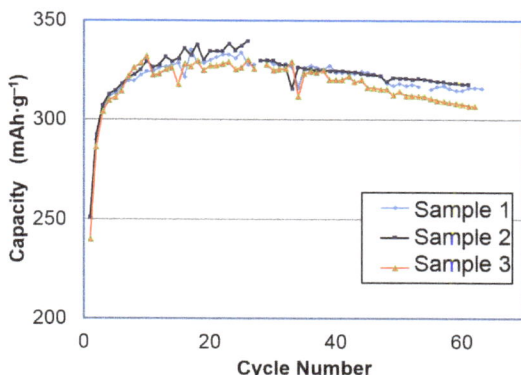

Figure 3. Half-cell capacity measurement of three α-Ni(OH)$_2$ samples prepared by the batch process Ni-2 with charge and discharge currents at 25 mA·g^{-1} (charged for 18.5 h). Capacity calculations are based on the weight of the active material.

The topologies of the particles before and after electrochemical cycling were studied by SEM and the corresponding micrographs are shown in Figure 4. The pristine material shows a granular structure and conglomerates into larger particles with cycling. The growth in size with cycling retards cracking of the particle, due to the lattice expansion from β-Ni(OH)$_2$ to α-Ni(OH)$_2$, which is a common failure mode for capacity degradation in α-Ni(OH)$_2$ [12,78]. Small amounts of capacity degradation can be seen after 25 cycles due to partial electrode disintegration in the flooded-configuration.

(a)   (b)

Figure 4. Scanning electron microscope (SEM) micrographs of α-Ni(OH)$_2$ prepared by the Ni-2 batch process (**a**) before and (**b**) after the flooded half-cell electrochemical capacity measurements.

XRD analysis results indicate a β-Ni(OH)$_2$ structure for the material before cycling, which turned into an α-Ni(OH)$_2$–predominant structure after cycling (Figure 5). The different peak widths of the XRD patterns (especially in Figure 5a) are the products of preferential growth of Ni(OH)$_2$ flakes on the *ab*-plane and various stacking faults [79]. The product may be further improved (especially the cycle stability) for battery applications, but the process itself is complicated and currently limited to a laboratory scale. Mass production of the batch process will not be cost-effective.

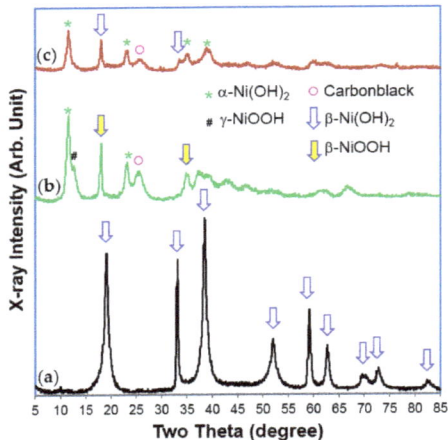

Figure 5. X-ray diffraction (XRD) patterns, using Cu-K$_\alpha$ as the radiation source, for α-Ni(OH)$_2$ samples prepared by the Ni-2 batch process in the (**a**) pristine; (**b**) charge; and (**c**) discharge state. After formation, the material changes in structure from a β-Ni(OH)$_2$ to an α-Ni(OH)$_2$–predominant phase. During charge, α-Ni(OH)$_2$ was converted into a γ-NiOOH structure.

3.2. Continuous Production

In the Robust Affordable Next Generation Energy Storage System (RANGE) program funded by the US Department of Energy, a series of research efforts were dedicated to the synthesis of a durable and high capacity nickel hydroxide using the CSTR method [15]. We first confirmed that Al is effective in promoting nucleation of the α-Ni(OH)$_2$ phase. The XRD patterns from three samples with 0, 4, and 20 at % Al are shown in Figure 6. While the pristine Al-free sample shows a pure β-Ni(OH)$_2$ structure (Figure 6a), the sample with 4% Al shows a small hint of α-Ni(OH)$_2$, and the third sample, with 20% Al-content, is dominated by the α-Ni(OH)$_2$ structure even without electrochemical cycling. For the rest of the materials, two chemistries were chosen for comparison: WM02 and WM12 (compositions listed in Table 3). Their preparation parameters and key performances are also listed in Table 3.

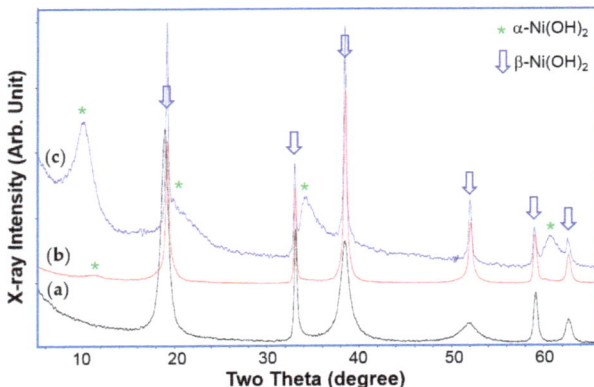

Figure 6. XRD patterns, using Cu-K$_\alpha$ as the radiation source, from Ni(OH)$_2$ samples prepared by the CSTR process with (**a**) 0; (**b**) 4; and (**c**) 20 at % Al in the pristine stage.

Table 3. Parameter differences between WM02 and WM12.

Parameter and Properties	WM02	WM12
Target composition	$Ni_{0.91}Co_{0.05}Al_{0.04}(OH)_2$	$Ni_{0.84}Co_{0.12}Al_{0.04}(OH)_2$
Tap density	$1.4\ g\cdot cc^{-1}$	$0.9\ g\cdot cc^{-1}$
Original structure	β-Ni(OH)$_2$	β-Ni(OH)$_2$
Structure after activation	Uniform α-Ni(OH)$_2$	α/β-Ni(OH)$_2$ core–shell
Half-cell cycle stability	4	100
BET surface area	$30.35\ m^2\cdot g^{-1}$	$51.97\ m^2\cdot g^{-1}$
Surface pore density	$0.016\ cc\cdot g^{-1}$	$0.027\ cc\cdot g^{-1}$
Average pore diameter	24.6 Å	24.6 Å

The results from the capacity measurement of WM02, WM12, and AP50 (a control sample of β-Ni(OH)$_2$ with a cation composition of $Ni_{0.91}Zn_{0.045}Co_{0.045}$) are presented in Figure 7. Both WM02 and WM12 show a higher discharge capacity than AP50. WM02 has a higher initial capacity, but also a more severe degradation in capacity. The XRD patterns of WM02 and WM12 in pristine, charged, and discharged states are compared in Figure 8. According to the XRD results, both WM02 and WM12 started with a pure β-Ni(OH)$_2$ structure and are converted into α-Ni(OH)$_2$-predominated and α/β mixed Ni(OH)$_2$ state, respectively. The XRD peaks in the pristine WM12 are broader than those in the pristine WM02, indicating a smaller crystallite (platelet) in WM12, which is confirmed by the TEM work showed later in this session.

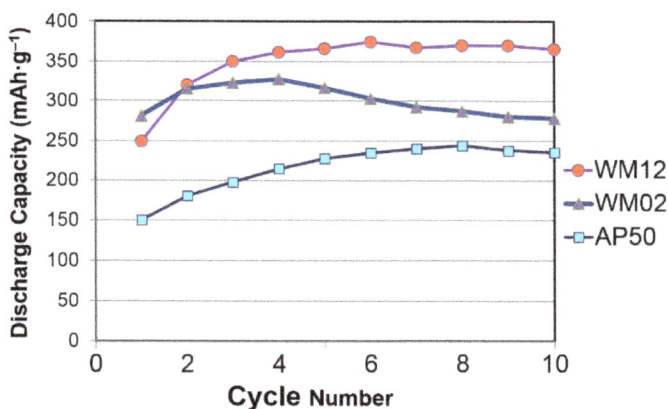

Figure 7. Half-cell capacity measurements for α/β Ni(OH)$_2$ (WM12), α-Ni(OH)$_2$ (WM02), and β-Ni(OH)$_2$ (AP50 with a cation composition of $Ni_{91}Co_{4.5}Zn_{4.5}$) prepared by the CSTR process with charge and discharge currents at $25\ mA\cdot g^{-1}$ (charge for 18.5 h). Capacity calculations are based on the weight of the active material.

SEM analysis results show that cracking in the cycled WM02 spherical particles is due to swelling caused by the β-to-α transition (Figure 9b) and that WM12 maintained the same shape after 20 cycles (Figure 10). In addition, the surface morphologies of the two materials are different. While WM02 spherical particles have a more compact surface with a granular texture, WM12 spherical particles have a less dense surface with crystallite plates aligning perpendicular to the surface. The surface area and pore density of WM02 are smaller than those of WM12 with the same average pore diameter (Table 3).

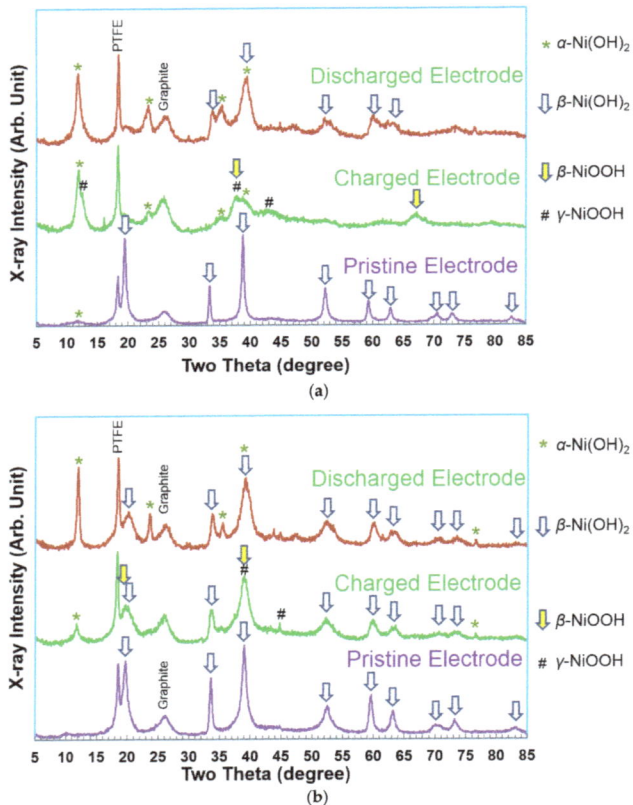

Figure 8. XRD patterns, using Cu-K$_\alpha$ as the radiation source, of (**a**) α-Ni(OH)$_2$ (WM02) and (**b**) αβ-Ni(OH)$_2$ (WM12) prepared by the CSTR process at three different stages.

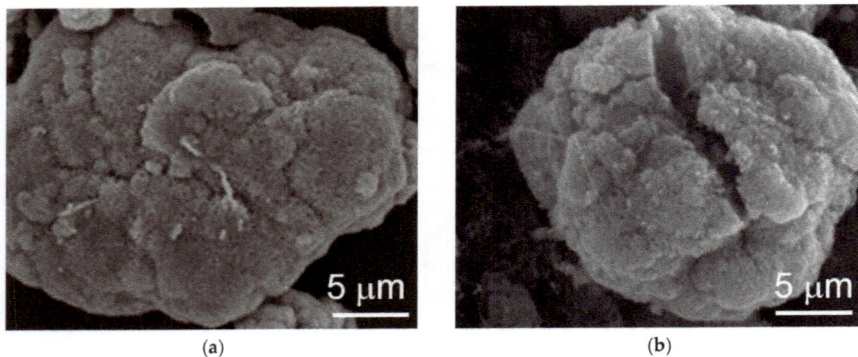

Figure 9. SEM micrographs of α-Ni(OH)$_2$ (WM02) prepared by the CSTR process (**a**) before and (**b**) after 20 cycles.

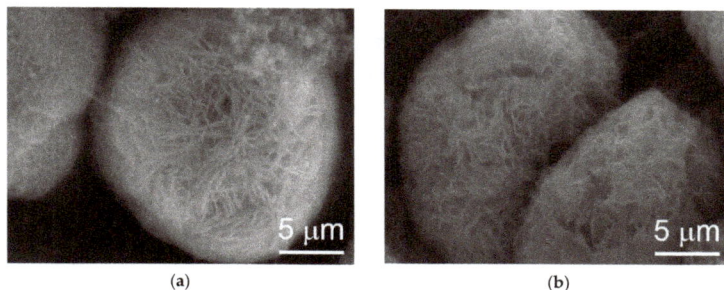

Figure 10. SEM micrographs of α/β-Ni(OH)$_2$ (WM12) prepared by the CSTR process (**a**) before and (**b**) after 20 cycles.

In another comparison test, three samples (two WM12 and one control β-Ni(OH)$_2$) (YRM3: Ni$_{0.93}$Co$_{0.02}$Al$_{0.05}$(OH)$_2$) underwent nine different charge/discharge conditions (Table 4) and the obtained capacities are plotted in Figure 11. While the increases in charge current density from 50 to 75 and 125 mA$\cdot$g^{-1} in stages III and IV boosted capacities for both WM12 and β-Ni(OH)$_2$, further increase in the charge current density (stage V) improved the capacity of WM12, but not β-Ni(OH)$_2$. It seems that WM12 benefits, but β-Ni(OH)$_2$ deteriorates with faster charge. From this comparison, the superiorities in the discharge capacity and cycle stability of WM12, compared to the regular material, are validated. The final failure mechanisms for both materials (WM12 and the β-Ni(OH)$_2$) are the same: electrode disintegration due to particle pulverization.

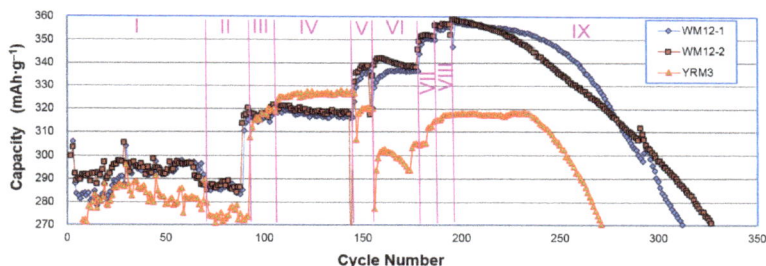

Figure 11. Half-cell capacity measurements of two α/β Ni(OH)$_2$ (WM12-1 and WM12-2) and one β-Ni(OH)$_2$ (YRM3 with a cation composition of N$_{93}$Zn$_5$Co$_2$) samples prepared by the CSTR. The charge and discharge current densities for each stage are listed in Table 4. Capacity calculations are based on the weight of the active material.

Table 4. Various test conditions for the half-cell capacity measurements of two α/β core–shell samples from the continuous process (WM12-1 and WM12-2) and a control β-Ni(OH)$_2$ sample (YRM3). Charge and discharge currents are in mA$\cdot$g^{-1} units and time is in h. Total amount of charge-in is in mAh$\cdot$g^{-1}.

Stage	Cycle Number	Charge Current	Charge Time	Amount of Charge-In	Discharge Current
I	1-65	25	18	450	150
II	66-95	50	7	350	150
III	96-105	75	6	450	150
IV	106-145	125	3	375	150
V	146-155	250	1.5	375	150
VI	155-180	300	1.5	450	150
VII	181-190	200	2	400	200
VIII	191-201	250	2	500	150
IX	201-end	200	2	400	150

In a separate cycle life experiment, we performed a 100 mA·g^{-1} rate charge for 5.5 h and discharged at the same rate. Results are shown in Figure 12. The highest capacity of 376 mAh·g^{-1} was obtained during the 61th cycle and the capacity at the 100th cycle was 371 mAh·g^{-1}. With the success in the half-cell experiment (with 50% additional binder and electrical conduction enhancer), we began the full-cell measurement of WM12 and WM12 in the newly developed pouch-type Ni/MH battery [15]. Only 10% binder and electrical conduction enhancer were added in the positive electrode and the capacity results, based on the total material weight (active plus additives), are plotted in Figure 13. Discharge capacities of 329 mAh·g^{-1} and 311 mAh·g^{-1} were obtained for the positive electrodes of the WM12 and WM02 full cells, respectively. From this point, further tests on the full cells are planned and more results will be reported in the future, especially for the phase stability under storage condition since α-Ni(OH)$_2$ is known to have an aging issue when stored in an alkaline solution [30]. It is obvious that the same WM12 material required less activation cycles and achieved a lower capacity in the sealed-cell configuration than that in the half-cell (6 versus 20). This is due to the reliance of the β–NiOOH to γ–NiOOH transition on the amount of over-charge. In the half-cell operated in the open atmosphere, the over-charging process has to compete with oxygen gas evolution and thus is less effective, compared to reactions operated in the sealed-cell configuration (more activation cycle is needed). The difference in the maximum capacities between two measurements is related to the expansion of unit cell that occurs in the β–NiOOH to γ–NiOOH transition. In the half-cell testing, the electrode was allowed to expand and, therefore, more NiOOH was converted, which differs from the limited space available in the sealed-cell.

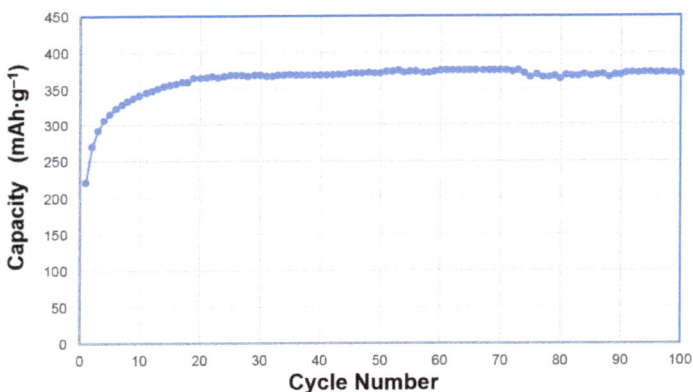

Figure 12. Half-cell capacity measurement of α/β Ni(OH)$_2$ WM12 prepared by the CSTR process with charge and discharge currents at 100 mA·g^{-1} (charge for 5.5 h). Capacity calculations are based on the weight of active material. The highest capacity of 376 mAh·g^{-1} was obtained at the 61th cycle and the capacity at 100th cycle is 371 mAh·g^{-1}.

The microstructure of WM12 was studied by both SEM and TEM. Cross-section SEM backscattering electron images of activated WM12 at different magnifications are shown in Figure 14. Different contrasts can be observed in the shell (A, darker) and core (B, lighter). EDS analysis results show that the surface region has a higher Al-content (ca. 4 at %) than the core region (ca. 1.5 at %) [15]. The core–shell structure was produced in the CSTR with a differential precipitation of Al in the tank (Figure 2b). In the nucleation stage, near the bottom of the tank, less Al becomes crystallite due to its relatively large size and, thus, an Al-lean core forms. As the particles are moved into the upper portion of the reactor by the stirring blade, they grow in size with a higher Al-content in the shell part of the particle. Materials produced at this stage are still β-Ni(OH)$_2$. Later, during electrochemical formation, the shell, with a relatively higher Al-content, develops a β-Ni(OH)$_2$ structure.

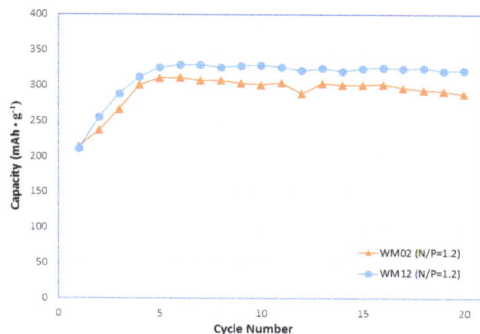

Figure 13. Full-cell capacity measurement of α Ni(OH)$_2$ WM02 and α/β Ni(OH)$_2$ WM12 prepared by the CSTR process with charge and discharge currents at 100 mA·g^{-1} (charge for 5.5 h). Capacity calculations are based on the weight of active material. The active material loading was increased from 50 wt % in the half-cell to 90 wt % in the full-cell with a 120% capacity AB$_5$ dry-compacted negative electrode (negative/positive (N/P) = 1.2).

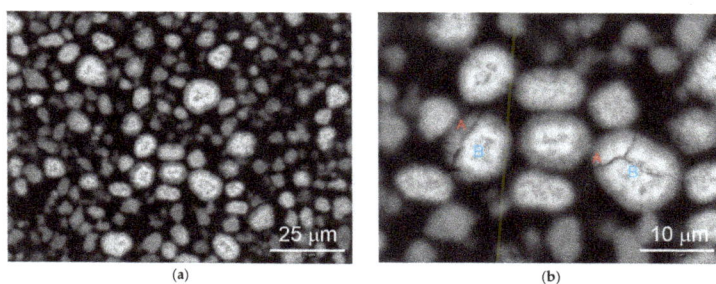

Figure 14. Cross-section SEM micrographs at different magnifications with scale bars indicating (**a**) 25 and (**b**) 10 μm. of α/β-Ni(OH)$_2$ (WM12) prepared by the CSTR process showing a core (B)/shell (A) structure. After activation, the Al-rich A region (shell) becomes α-Ni(OH)$_2$, while the Al-lean B region (core) remains β-Ni(OH)$_2$.

The microstructure of WM12 particle's shell was further investigated by TEM. A representative TEM micrograph is shown in Figure 15a. The high-resolution TEM images taken from areas ⓑ (brighter) and ⓒ (darker) are shown in Figure 15b,c, respectively. Area ⓑ has a smaller inter-planar distance is β-Ni(OH)$_2$, while area ⓒ, which has a larger inter-planar discharge, is α-Ni(OH)$_2$. The electron diffraction from a representative selective 10-μm region, which covers the most of the particle shown in Figure 15a, is shown in Figure 15d. The integrated electron diffraction intensities were collected from the diffraction pattern for areas ⓐ–ⓔ. For the convenience of comparison between the TEM and XRD results, the distance in reciprocal space obtained from TEM electron diffraction has been converted to a degree based system on the wave length of a Cu K$_\alpha$ X-ray and the results are shown in Figure 15e with the conversion to the standard XRD suing Cu K$_\alpha$ as the radiation source. Although the electron density plot is not identical to the XRD pattern, due to different scattering factors between X-rays and electron beams, the main features from a and b can still be distinguished and areas ⓐ–ⓔ have been identified as β, β, β, β, and mixed α/β structures, respectively. Therefore, we conclude that the shell region of WM12 is composed of nano-sized α-Ni(OH)$_2$ imbedded in a β-Ni(OH)$_2$ matrix, which helps to distribute the stress from the lattice expansion during the α-β transition. The broader XRD peaks in the pristine WM12 indicate a small crystallite form, even before the α-β transition occurs.

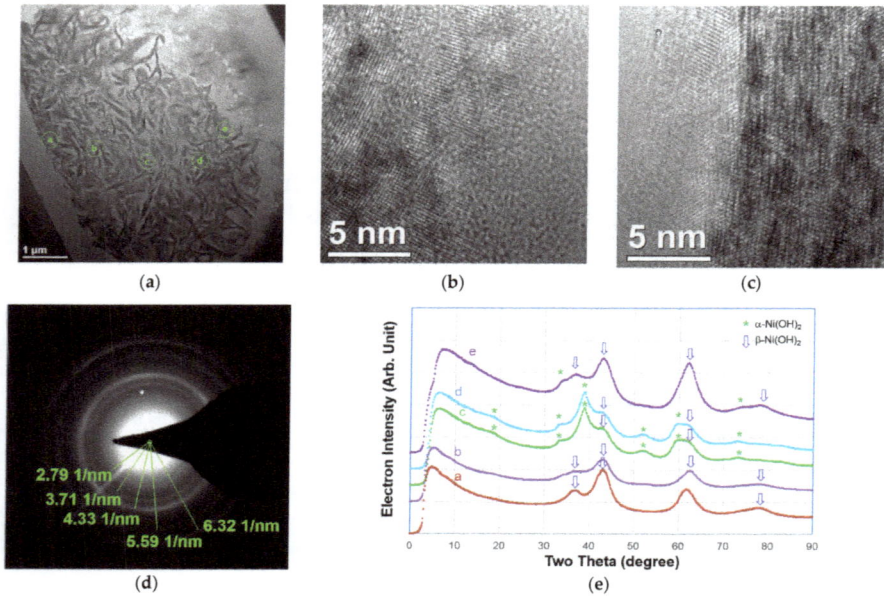

Figure 15. (**a**) TEM micrograph; (**b**) high-resolution images from area ⓑ (**c**) from area ⓒ in (**a**); (**d**) selective area diffraction pattern; and (**e**) integrated diffraction electron intensity (with distance in the reciprocal space obtained from TEM electron diffraction converted to a degree based system on the wavelength of Cu-K$_\alpha$ X-ray) of the shell-portion of a discharged electrode from WM12.

3.3. Low-Temperature Formation

It is well known that applying overcharge to a positive-limited Ni/MH battery will result in oxygen gas evolution. During normal operation of Ni/MH batteries, the produced oxygen gas will recombine with the hydrogen stored in the negative electrode, forming water and generating heat. A charging process in the $Ni(OH)_2$ based rechargeable alkaline battery is always a competition between oxidizing Ni(II) into Ni(III) and oxygen gas evolution. At room temperature, the oxygen evolution potential is raised by the co-precipitation of Co and Zn into spherical particles and, therefore, the charging process can be completed without any oxygen gas generation. At higher temperatures, where the potential of oxygen evolution is significantly reduced, the regular Ni-electrode cannot be charged fully and special high temperature Ni-electrodes have been developed to lower the oxidation potential of the Ni(II)/Ni(III) reaction [80]. When the cell is overcharged, part of its β-NiOOH will be converted into γ-NiOOH (see Bode's diagram in Figure 1), but the conversion efficiency is low because most of the overcharge is consumed during oxygen gas evolution. If the overcharge is performed at a relative low temperature, then the β–γ conversion happens at a higher rate. This concept initiated the following experiment. Ten AA-cells were made from regular AB$_5$ and YRMS ($Ni_{0.93}Co_{0.02}Al_{0.05}(OH)_2$) as the active materials in the negative and positive electrodes, respectively. These cells were exposed to a low-temperature overcharge process (C/10 rate at 10 °C for 30 days). Their discharge capacities before and after the treatment are listed in Table 5. More than a 15% increase in the discharge capacity was observed. The XRD patterns before and after the low-temperature treatment show a transition from 89% β-NiOOH/11% γ-NiOOH to 11% β-NiOOH/89% γ-NiOOH (Figure 16). This validates the effectiveness of low-temperature formation for α/γ-NiOOH. After disassembling the cell, we found swelling in the positive electrode due to the formation of a α/γ-NiOOH phase with an enlarged unit cell (about 74% [5]) from the insertion of a water layer between the NiOOH layers (Figure 1).

The swelling of the positive electrode causes an approximate 8% increase in the internal resistance (Table 5) due to the breakdown in the Co- conductive network [12]. Future work will be focused on a solution that addresses this electrode swelling. For example, swelling could be alleviated through a coating of CoOOH, Yb(OH)$_3$, or both on the spherical particles before electrode fabrication [12].

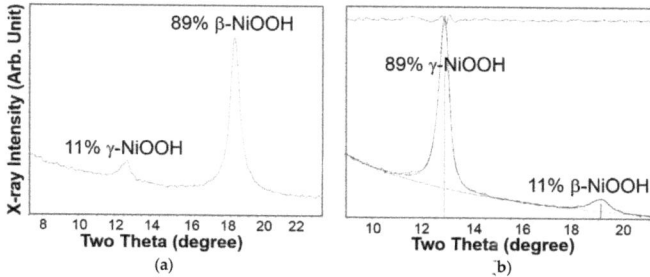

Figure 16. XRD patterns, using Cu-K$_\alpha$ as the radiation source, of the positive electrode from cells (a) before and (b) after the low-temperature overcharge treatment.

Table 5. Capacities (in mAh) and internal resistance (in mΩ) of the AA cells measured at a C/10 rate before and after the overcharge (OC) at 10 °C.

Cell Number	Capacity before OC	Capacity Immediately after OC, Measured at 10 °C	Capacity after OC, Measured at 20 °C	Internal Resistance before OC	Internal Resistance after OC
31	1252	1584	1475	16.5	19.2
32	1232	1522	1444	16.4	17.3
33	1248	1557	1466	17.1	18.2
34	1221	1553	1447	17.4	18.8
35	1217	1508	1436	17.4	18.4
36	1240	1537	1444	16.4	17.0
37	1238	1550	1450	18.4	19.8
38	1204	1513	1403	17.5	18.6
39	1230	1608	1452	17.1	18.3
40	1227	1558	1423	17.5	19.0

3.4. Comparisons

In order to better compare the three α- (α/β-mixed) Ni(OH)$_2$ fabrication methods, a summary table is present (Table 6). A few key directions for future development are also included for each fabrication method.

Table 6. Comparison of three fabrication methods for high-capacity Ni(OH)$_2$ powder. DOD denoted degree of disorder in both composition and phase structure.

Characteristics	Batch Process	Continuous Process	Over-Charge in Low-Temperature
Scale	Small, laboratory only	Large, mass-production	Large, mass-production
Cost	High	Low	Low
Loading	High	Low	High
α-phase	80%–90%	50%–80%	90%
Discharge Capacity	346 mAh·g^{-1}	376 mAh·g^{-1}	340 mAh·g^{-1}
Half-cell Stability	20	100	10
Failure Mode	Pulverization	Electrode disintegration	Pulverization
Directions for Improvement	• Addition of other elements to increase DOD • Increase in α/β grain boundary area	• Addition of other elements to increase DOD • Increase in the tap density • Increase in the high-rate capability	• Addition of other elements to improve the cycle stability—anti-cracking and aging • Electrolyte optimization

4. Conclusions

Three different fabrication methods for high-capacity α-Ni(OH)$_2$ were presented. The batch and continuous processes employ an α-Ni(OH)$_2$ promoter, such as Al, and reached stable cycle performance using a unique surface morphology and core–shell structure. While the batch process is currently only suitable for laboratory use, the continuous process with a later β-to-α formation process is much more cost-effective and already used in mass production. The low-temperature overcharge process does not require the Ni(OH)$_2$ promoter. However, the particle size and paste additives require further development to address the increase in internal resistance due to the swollenness of the positive electrode.

Acknowledgments: This work was financially supported by the U.S. Department of Energy Advanced Research Projects Agency-Energy under the RANGE Program (DE-AR0000386).

Author Contributions: Kwo-Hsiung Young was the Principle Investigator of the program, designed the experiments, and organized the test results while Shuli Yan, William Mays, and Xingqun Liao conducted the material preparation in batch, continuous, and low-temperature formation processes, respectively. Lixin Wang and Tiejun Meng performed the electrochemical measurement and Haoting Shen did the TEM analysis.

Conflicts of Interest: The authors declare no conflict of interest.

Abbreviations

CP	Co-precipitation
CSTR	Continuous stirring tank reactor
DOD	Degree of disorder
ECI	Electrochemical impregnation
EDS	X-ray energy dispersive spectroscopy
HRD	High-rate dischargeability
MH	Metal hydride
NMC	NiMnCo
N/P	Negative/positive
OC	Overcharge
PE	Polyethylene
PP	Polypropylene
PVDF	Polyvinylidene fluoride
RANGE	Robust Affordable Next Generation Energy Storage System
SEM	Scanning electron microscopy
SOC	State of charge
TEM	Transmission electron microscopy
XRD	X-ray diffraction

References

1. Edison, T.A. Reversible Galvanic Battery. U.S. Patent 678,722, 16 July 1901.
2. Edison, T.A. Reversible Galvanic Battery. U.S. Patent 692,507, 4 February 1902.
3. Chang, S.; Young, K.; Nei, J.; Fierro, C. Reviews on the U.S. Patents Regrading nickel/metal hydride batteries. *Batteries* **2016**, *2*, 10. [CrossRef]
4. Ouchi, T.; Young, K. Reviews on the Japanese patent applications Regrading nickel/metal hydride batteries. *Batteries* **2016**, *2*, 21. [CrossRef]
5. Bode, H.; Dehmelt, K.; White, J. Zur kenntnis der nickelhydroxidelektrode—I. Über das nickel (II)-hydroxidhydrat. *Electrochim. Acta* **1966**, *11*, 1079–1087. (In German) [CrossRef]
6. Corrigan, D.A.; Knight, S.L. Electrochemical and spectroscopic evidence on the participation of quadrivalent nickel in the nickel hydroxide redox reaction. *J. Electrochem. Soc.* **1989**, *136*, 613–619. [CrossRef]
7. Corrigan, D.A.; Bendert, R.M. Effect of coprecipitated metal ions on the electrochemistry of nickel hydroxide thin films: cyclic voltammetry in 1 M KOH. *J. Electrochem. Soc.* **1989**, *136*, 723–728. [CrossRef]

8. Oliva, P.; Leonardi, J.; Laurent, J.F.; Delmad, C.; Braconnier, J.J.; Figlarz, M.; Fievet, F.; Guibert, A. Review of the structure and the electrochemistry of nickel hydroxides and oxy-hydroxides. *J. Power Sources* **1982**, *8*, 229–255. [CrossRef]

9. Delmas, C.; Braconnier, J.J.; Borthomieu, Y.; Hagenmuller, P. New families of cobalt substituted nickel oxyhydroxides and hydroxides obtained by soft chemistry. *Mater. Res. Bull.* **1987**, *22*, 741–751. [CrossRef]

10. Delmas, C.; Braconnier, J.J.; Borthomieu, Y. From sodium nickelate to nickel hydroxide. *Solid State Ion.* **1988**, *28–30*, 1132–1137. [CrossRef]

11. Bernard, M.C.; Bernard, P.; Keddam, M.; Senyarich, S.; Takenouti, H. Characterization of new nickel hydroxides during the transformation of α Ni(OH)$_2$ to β Ni(OH)$_2$ by aging. *Electrochim. Acta* **1996**, *41*, 91–93. [CrossRef]

12. Young, K.; Yasuoka, S. Capacity degradation mechanisms in nickel/metal hydride batteries. *Batteries* **2016**, *2*, 3. [CrossRef]

13. Jayashree, R.S.; Kamath, P.V. Suppression of the α → β-nickel hydroxide transformation in concentrated alkali: Role of dissolved cations. *J. Appl. Electrochem.* **2001**, *31*, 1315–1320. [CrossRef]

14. Tessier, C.; Guerlou-Demourgues, L.; Faure, C.; Denage, C.; Delatouche, B.; Delmas, C. Influence of zinc on the stability of the β(II)/β(III) nickel hydroxide system during electrochemical cycling. *J. Power Sources* **2001**, *102*, 105–111. [CrossRef]

15. Young, K.; Ng, K.Y.S.; Bendersky, L.A. A technical report of the robust affordable next generation energy storage system-BASF program. *Batteries* **2016**, *2*, 2. [CrossRef]

16. Armstrong, R.D.; Charles, E.A. Some effects of cobalt hydroxide upon the electrochemical behavior of Nckel hydroxide electrodes. *J. Power Sources* **1989**, *25*, 89–97. [CrossRef]

17. Faure, C.; Delmas, C.; Willmann, P. Electrochemical behavior of α-cobalted nickel hydroxide electrodes. *J. Power Sources* **1991**, *36*, 497–506. [CrossRef]

18. Demourgues-Guerlou, L.; Delmas, C. Effect of iron on the electrochemical properties of the nickel hydroxide electrode. *J. Electrochem. Soc.* **1994**, *141*, 713–717. [CrossRef]

19. Indira, L.; Dixit, M.; Kamath, P.V. Electrosynthesis of layered double hydroxides of nickel with trivalent cations. *J. Power Sources* **1994**, *52*, 93–97. [CrossRef]

20. Kamath, P.V.; Dixit, M.; Indira, L.; Shukla, A.K.; Kumar, V.G.; Munichandraiah, N. Stabilized α-Ni(OH)$_2$ as electrode material for alkaline secondary cells. *J. Electrochem. Soc.* **1994**, *141*, 2956–2959. [CrossRef]

21. Kumar, V.G.; Munichandraiah, N.; Kamath, P.V.; Shukla, A.K. On the performance of stabilized α-nickel hydroxide as a nickel-positive electrode in alkaline storage batteries. *J. Power Sources* **1995**, *56*, 111–114. [CrossRef]

22. Dixit, M.; Kamath, P.V.; Munichandraiah, N.; Shukla, A.K. An electrochemically impregnated sintered-nickel electrode. *J. Power Sources* **1996**, *63*, 167–171. [CrossRef]

23. Guerlou-Demourgues, L.; Delmas, C. Electrochemical behavior of the manganese-substituted nickel hydroxides. *J. Electrochem. Soc.* **1996**, *143*, 561–566. [CrossRef]

24. Liu, B.; Wang, X.Y.; Yuan, H.T.; Zhang, Y.S.; Song, D.Y.; Zhou, Z.X. Physical and electrochemical characteristics of aluminum-substituted nickel hydroxide. *J. Appl. Electrochem.* **1999**, *29*, 855–860. [CrossRef]

25. Liu, B.; Yuan, H.; Zhang, Y.; Zhou, Z.; Song, D. Cyclic voltammetric studies of stabilized α-nickel hydroxide electrode. *J. Power Sources* **1999**, *79*, 277–280.

26. Dixit, M.; Jayashree, R.S.; Kamath, P.V.; Shukla, A.K.; Kumar, V.G.; Munichandraiah, N. Electrochemically impregnated aluminum-stabilized α-nickel hydroxide electrode. *Electrochem. Solid State Lett.* **1999**, *2*, 170–171. [CrossRef]

27. Sugimoto, A.; Ishida, S.; Janawa, K. Preparation and characterization of Ni/Al-layered double hydroxide. *J. Electrochem. Soc.* **1999**, *146*, 1251–1255. [CrossRef]

28. Dai, J.; Li, S.F.Y.; Xiao, T.D.; Wang, D.M.; Reisner, D.E. Structural stability of aluminum stabilized alpha nickel hydroxide as a positive electrode material for alkaline secondary batteries. *J. Power Sources* **2000**, *89*, 40–45. [CrossRef]

29. Leng, Y.; Ma, Z.; Zhang, C.; Zhang, J.; We, Y.; Cao, C. Preparation, structure and electrochemical performance of iron-substituted nickel hydroxide. *Chin. J. Power Sources* **2000**, *24*, 32–35. (In Chinese)

30. Freitas, M.B.J.G. Nickel hydroxide powder for NiO·OH/Ni(OH)$_2$ electrodes of the alkaline batteries. *J. Power Sources* **2001**, *93*, 163–174. [CrossRef]

31. Yang, W.; Chen, G.; Yin, Y.; Chen, H.; Jia, J. Study on the Y-doped α-Ni(OH)$_2$. *Chin. J. Appl. Chem.* **2001**, *18*, 689–692. (In Chinese)

32. Wang, C.Y.; Zhong, S.; Konstantinov, K.; Walter, G.; Liu, H.K. Structural study of Al-substituted nickel hydroxide. *Solid State Ion.* **2002**, *148*, 503–508.

33. Wang, X.; Luo, H.; Parkhutik, P.V.; Millan, A.; Matveeva, E. Studies of the performance of nanostructural multiphase nickel hydroxide. *J. Power Sources* **2003**, *115*, 153–160. [CrossRef]

34. Hu, W.K.; Noréus, D. Alpha nickel hydroxides as lightweight nickel electrode materials for alkaline rechargeable cells. *Chem. Mater.* **2003**, *15*, 974–978. [CrossRef]

35. Pan, T.; Wang, J.M.; Zhao, Y.L.; Chen, H.; Xiao, H.M.; Zhang, J.Q. Al-stabilized α-nickel hydroxide prepared by electrochemical impregnation. *Mater. Chem. Phys.* **2003**, *78*, 711–718. [CrossRef]

36. Xu, J.; Zhou, Y.; Tang, Y.; Wang, Q.; Lu, T. Solid state synthesis and electrochemical performance of α-Ni(OH)$_2$ including 20% Al. *Chin. J. Inorg. Chem.* **2003**, *19*, 535–538. (In Chinese)

37. Zhao, Y.L.; Wang, J.M.; Chen, H.; Pan, T.; Zhang, J.Q.; Cao, C.N. Al-substituted α-nickel hydroxide prepared by homogenous precipitation method with urea. *Int. J. Hydrog. Energy* **2004**, *29*, 889–896. [CrossRef]

38. Zhao, Y.L.; Wang, J.M.; Chen, H.; Pan, T.; Zhang, J.Q.; Cao, C.N. Different additives-substituted α-nickel hydroxide prepared by urea decomposition. *Electrochim. Acta* **2004**, *50*, 91–98. [CrossRef]

39. Liu, B.; Yuan, H.; Zhang, Y. Impedance of Al-substituted α-nickel hydroxide electrodes. *Int. J. Hydrog. Energy* **2004**, *29*, 453–458. [CrossRef]

40. Yang, Y.; Zhang, P.; Zhang, Y.; Tang, Y.; Liu, K.; Sang, S. Preparation and electrochemical performance of lanthanum doping α-Ni(OH)$_2$. *J. Cent. South Univ. (Sci. Technol.)* **2005**, *36*, 898–993. (In Chinese)

41. Liu, H.; Xiang, L.; Jin, Y. Synthesis and high-temperature performance of Ti substituted α-Ni(OH)$_2$. *Trans. Nonferrous Met. Soc. China* **2005**, *15*, 823–827.

42. Hu, W.; Gao, X.; Noréus, D.; Burchardt, T.; Nakstad, N.K. Evaluation of nano-crystal sized α-nickel hydroxides as an electrode material for alkaline rechargeable cells. *J. Power Sources* **2006**, *160*, 704–710. [CrossRef]

43. Luo, F.; Chen, Q.; Yin, Z. Electrochemical performance of multiphase nickel hydroxide. *Trans. Nonferrous Met. Soc. China* **2007**, *17*, 654–658. [CrossRef]

44. Qi, J.; Xu, P.; Lv, Z.; Liu, X.; Wen, A. Effect of crystallinity on the electrochemical performance of nanometer Al-stabilized α-nickel hydroxide. *J. Alloys Compd.* **2008**, *462*, 164–169. [CrossRef]

45. Liu, C.; Song, S.; Li, Y.; Liu, A. Investigations on structure and proton diffusion coefficient of rare earth ion (Y^{3+}/Nd^{3+}) and aluminum co-doped α-Ni(OH)$_2$. *J. Rare Earths* **2008**, *26*, 594–597. [CrossRef]

46. Zhao, W.; Liu, C.; Song, S.; Li, G. Structure and electrochemical performances of Y and Al co-doped α-Ni(OH)$_2$ electrode. *Chin. J. Rare Met.* **2009**, *33*, 366–370. (In Chinese)

47. Morishita, M.; Kakeya, T.; Ochiai, S.; Ozaki, T.; Kawabe, Y.; Watada, M.; Sakai, T. Structural analysis by synchrotron X-ray diffraction, X-ray absorption fine structure and transmission electron microscopy for aluminum-substituted α-type nickel hydroxide electrode. *J. Power Sources* **2009**, *1893*, 871–877. [CrossRef]

48. Liu, C.; Li, Y. Synthesis and characterization of amorphous α-nickel hydroxide. *J. Alloys Compd.* **2009**, *478*, 415–418. [CrossRef]

49. Liu, C.; We, H.; Li, Y. Structure and electrochemical performance of Y(III) and Al(III) co-doped amorphous nickel hydroxide. *J. Phys. Chem. Solids* **2009**, *70*, 723–726. [CrossRef]

50. Vidotti, M.; Salvador, R.P.; Torresi, S.I.C. Synthesis and characterization of stable Co and Cd doped nickel hydroxide nanoparticles for electrochemical applications. *Ultrason. Sonochem.* **2009**, *16*, 35–40. [CrossRef] [PubMed]

51. Liu, C.; Song, S.; Liu, A.; Li, G. Electrochemical performance and action mechanism of Ce/Al-codoped α-Ni(OH)$_2$ as electrode material. *Rare Met. Mater. Eng.* **2009**, *38*, 540–543. (In Chinese)

52. Wu, Q.D.; Liu, S.; Li, L.; Yan, T.Y.; Gao, X.P. High-temperature electrochemical performance of Al-α-nickel hydroxides modified by metallic cobalt or Y(OH)$_3$. *J. Power Sources* **2009**, *186*, 521–527. [CrossRef]

53. Béléké, A.B.; Hosokawa, A.; Mizuhata, M.; Deki, S. Preparation of α-nickel hydroxide/carbon composite by the liquid phase deposition method. *J. Ceram. Soc. Jpn.* **2009**, *117*, 392–394. [CrossRef]

54. Li, Y.W.; Yao, J.H.; Liu, C.J.; Zhao, W.M.; Deng, W.X.; Zhong, S.K. Effect of interlayer anions on the electrochemical performance of Al-substituted α-type nickel hydroxide electrodes. *Int. J. Hydrog. Energy* **2010**, *35*, 2539–2545. [CrossRef]

55. Liu, C.; Li, P.; Zhao, Y.; Huang, L. Electrochemical performance of α-nickel hydroxide co-doped with La and Zn. *CIESC J.* **2010**, *61*, 2743–2747.

56. Li, P.; Chen, S.; Zhao, W.; Liu, C. Electrochemical performance of Mg and Al substituted α-Ni(OH)$_2$. *Guangdong Chem. Ind.* **2010**, *37*, 142–144. (In Chinese)
57. Wang, Y.; Hu, Z.; Wu, H. Preparation and electrochemical performance of alpha-nickel hydroxide nanowire. *Mater. Chem. Phys.* **2011**, *126*, 580–583. [CrossRef]
58. Zhang, Z.J.; Zhu, Y.J.; Bao, J.; Lin, X.R.; Zheng, H.Z. Electrochemical performance of multi-element doped α-nickel hydroxide prepared by supersonic co-precipitation method. *J. Alloys Compd.* **2011**, *509*, 7034–7037. [CrossRef]
59. Ye, X.; Zhu, Y.; Wu, S.; Zhang, Z.; Zhou, Z.; Zheng, H.; Lin, X. Study on the preparation and electrochemical performance of rare earth doped nano-Ni(OH)$_2$. *J. Rare Earths* **2011**, *29*, 787–792. [CrossRef]
60. Wu, S.; Zhu, Y.; Zhang, Z.; Zhou, Z.; Ye, X.; Zheng, H.; Lin, X.; Bao, J. Effect of supersonic power and pH value on the structure and electrochemical performance of Y doped nano-Ni(OH)$_2$. *J. Mater. Eng.* **2011**, *6*, 27–31. (In Chinese)
61. Liu, J.; Wang, X.; Yao, X.; Wang, J.; Liu, Z. Homogeneous precipitation of α-phase Co-Ni hydroxides hexagonal platelets. *Particuology* **2012**, *10*, 24–28. [CrossRef]
62. You, Z.; Shen, K.; Wu, Z.; Wang, X.; Kong, X. Electrodeposition of Zn-doped α-nickel hydroxide with flower-like nanostructure for supercapacitors. *Appl. Surf. Sci.* **2012**, *258*, 8117–8123. [CrossRef]
63. Li, Y.; Yao, J.; Zhu, Y.; Zou, Z.; Wang, H. Synthesis and electrochemical performance of mixed phase α/β nickel hydroxide. *J. Power Sources* **2012**, *203*, 177–183. [CrossRef]
64. Máca, T.; Vondrák, J.; Sedlaříková, M.; Nezgoda, L. Incorporation of multielement doping into LDH structure of alpha nickel hydroxide. *ECS Trans.* **2012**, *40*, 119–131.
65. Bao, J.; Zhu, Y.; Zhang, Z.; Xu, Q.; Zhao, W.; Chen, J.; Zhang, W.; Han, Q. Structure and electrochemical properties of nanometer Cu substituted α-nickel hydroxide. *Mater. Res. Bull.* **2013**, *48*, 422–428. [CrossRef]
66. Huang, J.; Cao, P.; Lei, T.; Yang, S.; Zhou, X.; Xu, P.; Wang, G. Structural and electrochemical performance of Al-substituted β-Ni(OH)$_2$ nanosheet electrodes for nickel metal hydride battery. *Electrochim. Acta* **2013**, *111*, 713–719. [CrossRef]
67. Yao, J.; Li, Y.; Li, Y.; Zhu, Y.; Wang, H. Enhanced cycling performance of Al-substituted α-nickel hydroxide by coating with β-nickel hydroxide. *J. Power Sources* **2013**, *224*, 236–240. [CrossRef]
68. Zheng, H.Z.; Zhu, Y.J.; Lin, X.R.; Zhuang, Y.H.; Zhao, R.D.; Liu, Y.L.; Zhang, S.J. The influence of Na$_2$Co$_3$ content and Ni^{2+} concentration on the physicochemical properties of nanometer Y-substituted nickel hydroxide. *Mater. Sci. Eng. B* **2013**, *178*, 1365–1370. [CrossRef]
69. Li, J.; Shangguan, E.; Guo, D.; Tian, M.; Wang, Y.; Li, Q.; Chang, Z.; Yuan, X.; Wang, H. Synthesis, characterization and electrochemical performance of high-density aluminum substituted α-nickel hydroxide cathode material for nickel-based rechargeable batteries. *J. Power Sources* **2014**, *270*, 121–130. [CrossRef]
70. Li, J.; Shangguan, E.; Nie, M.; Jin, Q.; Zhao, K.; Chang, Z.; Yuan, X.; Wang, H. Enhanced electrochemical performance of high-density Al-substituted α-nickel hydroxide by a novel anion exchange method using NaCl solution. *Int. J. Hydrog. Energy* **2015**, *40*, 1852–1858. [CrossRef]
71. Shangguan, E.; Li, J.; Guo, D.; Guo, L.; Nie, M.; Chang, Z.; Yuan, X.; Wang, H. A comparative study of structural and electrochemical properties of high-density aluminum substituted α-nickel hydroxide containing different interlayer anions. *J. Power Sources* **2015**, *282*, 158–168. [CrossRef]
72. Miao, C.; Zhu, Y.; Huang, L.; Zhao, T. The relationship between structural stability and electrochemical performance of multi-element doped alpha nickel hydroxide. *J. Power Sources* **2015**, *274*, 186–193. [CrossRef]
73. Yao, M.; Hu, Z.; Xu, Z.; Liu, Y.; Liu, P.; Zhang, Q. High-performance electrode materials of hierarchical mesoporous nickel oxide ultrathin nanosheets derived from self-assembled scroll-like α-nickel hydroxide. *J. Power Sources* **2015**, *273*, 914–922. [CrossRef]
74. Xue, J.; Ma, W.; Zhang, F.; Wang, M.; Cui, H. Construction of cobalt substituted α-Ni(OH)$_2$ hierarchical nanostructure from nanofibers on nickel foam and its electrochemical performance. *Solid State Ion.* **2015**, *281*, 38–42. [CrossRef]
75. Miao, C.; Zhu, Y.; Huang, L.; Zhao, T. Synthesis, characterization, and electrochemical performances of alpha nickel hydroxide by coprecipitating Sn^{2+}. *Ionics* **2015**, *21*, 2295–2302. [CrossRef]
76. Fierro, C.; Fetcenko, M.A.; Young, K.; Ovshinsky, S.R.; Sommers, B.; Harrison, C. Nickel Hydroxide Positive Electrode Material Exhibiting Improved Conductivity and Engineered Activation Energy. U.S. Patent 6,228,535, 8 May 2001.

77. Camardese, J.; McCalla, E.; Abarbanel, D.W.; Dahn, J.R. Determination of shell thickness of spherical core-shell Ni$_x$Mn$_{1-x}$(OH)$_2$ particles via absorption calculations of X-ray diffraction patterns. *J. Electrochem. Soc.* **2014**, *161*, A814–A820. [CrossRef]

78. Zhou, X.; Young, K.; West, J.; Regalado, J.; Cherisol, K. Degradation mechanisms of high-energy bipolar nickel metal hydride battery with AB$_5$ and A$_2$B$_7$ alloys. *J. Alloys Compd.* **2013**, *580*, S373–S377. [CrossRef]

79. Wong, D.F.; Young, K.; Wang, L.; Nei, J.; Mg, K.Y.S. Evolution of stacking faults in substituted nickel hydroxide spherical powders. *J. Alloys Compd.* **2017**, *695*, 1763–1769. [CrossRef]

80. Fierro, C.; Zallen, A.; Koch, J.; Fetcenko, M.A. The influence of nickel-hydroxide composition and microstructure on the high-temperature performance of nickel metal hydride batteries. *J. Electrochem. Soc.* **2006**, *153*, A492–A496. [CrossRef]

Article

Effects of Cs$_2$CO$_3$ Additive in KOH Electrolyte Used in Ni/MH Batteries

Shuli Yan [1,2], Jean Nei [2], Peifeng Li [1,2], Kwo-Hsiung Young [1,2,*] and K. Y. Simon Ng [1]

[1] Department of Chemical Engineering and Materials Science, Wayne State University, Detroit, MI 48202, USA; shuli.yan@partners.basf.com (S.Y.); peifeng.li@partners.basf.com (P.L.); sng@wayne.edu (K.Y.S.N.)

[2] BASF/Battery Materials—Ovonic, 2983 Waterview Drive, Rochester Hills, MI 48309, USA; jean.nei@basf.com

* Correspondence: kwo.young@basf.com; Tel.: +1-248-293-7000

Received: 31 August 2017; Accepted: 24 November 2017; Published: 18 December 2017

Abstract: The effects of Cs$_2$CO$_3$ addition in a KOH-based electrolyte were investigated for applications in nickel/metal hydride batteries. Both MgNi-based and Laves phase-related body-centered cubic solid solution metal hydride alloys were tested as the anode active materials, and sintered β-Ni(OH)$_2$ was used as the cathode active material. Certain amounts of Cs$_2$CO$_3$ additive in the KOH-based electrolyte improved the electrochemical performances compared with a conventional pure KOH electrolyte. For example, with Laves phase-related body-centered cubic alloys, the addition of Cs$_2$CO$_3$ to the electrolyte improved cycle stability (for all three alloys) and discharge capacity (for the Al-containing alloys); moreover, in the 0.33 M Cs$_2$CO$_3$ + 6.44 M KOH electrolyte, the discharge capacity of Mg$_{52}$Ni$_{39}$Co$_3$Mn$_6$ increased to 132%, degradation decreased to 87%, and high-rate dischargeability stayed the same compared with the conventional 6.77 M KOH electrolyte. The effects of Cs$_2$CO$_3$ on the physical and chemical properties of Mg$_{52}$Ni$_{39}$Co$_3$Mn$_6$ were characterized by Fourier transform infrared spectroscopy, X-ray diffraction, transmission electron microscopy, inductively coupled plasma, and electrochemical impedance spectroscopy. The results from these analyses concluded that Cs$_2$CO$_3$ addition changed both the alloy surface and bulk composition. A fluffy layer containing carbon was found covering the metal particle surface after cycling in the Cs$_2$CO$_3$-containing electrolyte, and was considered to be the main cause of the reduction in capacity degradation during cycling. Also, the Cs$_2$CO$_3$ additive promoted the formations of the C–O and C=O bonds on the alloy surface. The C–O and C=O bonds were believed to be active sites for proton transfer during the electrochemical process, with the C–O bond being the more effective of the two. Both bonds contributed to a higher surface catalytic ability. The addition of 0.33 M Cs$_2$CO$_3$ was deemed optimal in this study.

Keywords: nickel metal hydride battery; electrochemistry; hydrogen storage alloys; nickel hydroxide; alkaline electrolyte; salt additive

---

1. Introduction

Since the commercialization of nickel/metal hydride (Ni/MH) batteries in 1980s, they have been widely used as energy storage devices in hybrid electric vehicles, vacuum cleaners, electric toys, power tools, and cordless phones, to name a few uses [1–3]. Ni/MH batteries have many superior properties over rival battery technologies, such as high specific power, long cycle life, robust abuse tolerance, and a wide temperature operation range [2]. In the past decades, Ni/MH batteries have repeatedly attracted attention from both researchers and markets.

Basic Ni/MH battery electrochemistry is shown in the following reactions:

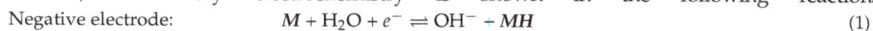

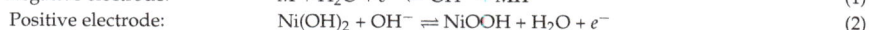

Negative electrode: $\qquad M + H_2O + e^- \rightleftharpoons OH^- + MH$ $\qquad\qquad$ (1)

Positive electrode: $\qquad Ni(OH)_2 + OH^- \rightleftharpoons NiOOH + H_2O + e^-$ $\qquad$ (2)

The reaction taking place at the negative electrode (anode) is described in Equation (1). M is a metal hydride (MH) alloy capable of storing hydrogen reversibly, and MH is the corresponding hydrided metal. During charge, the added voltage splits the water molecule into a proton and a hydroxide ion. Driven by voltage and diffusion difference, protons transfer from the electrolyte to the surface of the MH alloy particles, and then into the bulk of alloy. During discharge, protons travel in a reverse route. Equation (2) represents the reaction at the positive electrode (cathode). During charge, protons are dissociated from $Ni(OH)_2$, then move to the cathode surface, and finally recombine with the hydroxide ions in the electrolyte.

In order to promote proton transfer and improve electrochemical performance, many researchers and companies have focused on developing new anode/cathode materials, with a particular focus on anode materials. A wide set of hydrogen storage MH alloys have been studied for electrochemical applications, including AB_2, AB_5, A_2B_7, body-centered cubic (BCC) solid solution, BCC-AB_2 composite, Mg_xNi_y, TiNi and its composite, etc. [4–9]. Yu et al. reported a $Ti_{40}V_{30}Cr_{15}Mn_{15}$ alloy with an initial capacity of 814 mAh·g^{-1} at a rate of 10 mA·g^{-1} and 80 °C, which is more than twofold higher than the capacity obtained from the conventional rare earth-based AB_5 MH alloy (350 mAh·g^{-1}) [10]. However, degradation for this alloy was very high due to the pulverization caused by hydrogen evolution inside MH alloy particles. Young and Nei reported various MgNi-based amorphous/microcrystalline MH alloys with a theoretical capacity as high as 1080 mAh·g^{-1} [11]. However, most of these alloys demonstrated rapid decay during cycling. Nei reported an 80% or higher decay in capacity after 20 cycles for MgNi-based MH alloys [12]. Many methods have been attempted to improve the electrochemical performance of MH alloys. Recently, additives such as B [13], Ti [14,15], Pt [15], Pd [14], Nd [16], Cr [17], La [18], Co [19], Ni [20], Li [21], and Cu [22] were added to the bulk or surface of MH alloys to enhance the capacity, cycle stability, and high-rate dischargeability (HRD). For MgNi-based MH alloys, Ni coating [23], the addition of TiO_2 [24], and substitutions of Mn [25,26] and Nb [26] have been intensively studied with the goal of improving electrochemical performances. Many studies on the alkaline electrolytes [27–49] and salt additives [50,51] were conducted before in NiMH and other alkaline batteries; however, focuses were on the rare earth-based AB_5 and Zr-based AB_2 MH alloys. Young et al. have pointed out that electrolyte modification is one of the most economic and effective methods to alter electrode performances, since it does not affect the battery gravimetric and volumetric energy densities [11]. Nei et al. reported the conductivity and corrosion behaviors of several hydroxides [16]. Later, Yan et al. published a screen test of 32 salt additives in the KOH electrolyte, of which 12 salt additives were found to efficiently decrease the corrosion of the traditional KOH electrolyte on alloy AR3 (an MgNi-based MH alloy with a nominal composition of $Mg_{52}Ni_{39}Co_3Mn_6$) [52]. However, no detailed investigation was done on these 12 salt additives.

This study is a continuous work from Yan's previous report [52]. A systematic investigation of the effects of Cs_2CO_3 addition in a conventional KOH electrolyte on various MH alloys is performed. This additive was originally used in the KOH electrolyte for Ni/Zn batteries to extend the cycle life [53]. The total concentration of Cs_2CO_3 and KOH is fixed at 6.77 M. The influence of Cs_2CO_3 on the cell performance, electrolyte properties, and surface and bulk structure of MH alloy electrodes are examined. A possible proton transfer process for the Cs_2CO_3-containing electrolyte system is also discussed.

2. Experimental Setup

Both the sintered β-$Ni(OH)_2$ and MH alloys (AR3, P31, P32, and P37) were produced in-house. The compositions and fabrication methods of the MH alloys are summarized in Table 1. AR3 is an amorphous/microcrystalline MgNi-based MH alloy made by a melt–spin method, followed by mechanical alloying [52]. The P-series MH alloys belong to a family of Laves phase-related body-centered cubic (BCC) solid solution alloys [54], which were developed during a United States (U.S.) Department of Energy-funded research program [4]. The P-series of alloys were produced by induction melting, followed by annealing under optimized conditions (900 °C for 12 h [55]). KOH and Cs_2CO_3 were purchased from the Sigma-Aldrich Corporation (St. Louis, MO, USA).

Table 1. Compositions of the four metal hydride (MH) alloys used in this study. BCC: body-centered cubic.

Alloy	Alloy System	Composition	Preparation Method
AR3	Amorphous/microcrystalline MgNi	$Mg_{52}Ni_{39}Co_3Mn_6$	Melt-spin and mechanical alloying
P31	Laves phase-related BCC	$Ti_{15.6}Hf_{2.1}V_{44}Cr_{11.2}Mn_{6.9}Co_{1.4}Ni_{18.5}Al_{0.3}$	Induction melting and thermal annealing
P32	Laves phase-related BCC	$Ti_{15.6}Hf_{2.4}V_{44}Cr_{11.2}Mn_{6.9}Co_{1.4}Ni_{18.5}$	Induction melting and thermal annealing
P37	Laves phase-related BCC	$Ti_{14.5}Zr_{1.7}V_{46.6}Cr_{11.9}Mn_{6.5}Co_{1.5}Ni_{16.9}Al_{0.4}$	Induction melting and thermal annealing

Electrochemical charge/discharge cycling tests were performed with an Arbin BT2000 battery tester (Arbin, College Station, TX, USA) at room temperature. In the test cells, the cathode was sintered β-$Ni(OH)_2$, the anode was made from directly dry-compacting the alloy powder onto an expanded Ni substrate without using any binder, and the separator was hydrophilic nonwoven polyolefin. Charge/discharge processes were the same as reported before by Yan et al. [52] The cell was charged at 100 mA·g^{-1} for 5 h, and discharged first at 100 mA·g^{-1} to a cutoff voltage of 0.9 V. The initial discharge was followed by a 30 s rest for the voltage to recover, and then the cell was discharged at 24 mA·g^{-1} to reach a cutoff voltage of 0.9 V. The cell was put to rest for 30 s again before the final discharge at 8 mA·g^{-1} to 0.9 V. Testing for each alloy/electrolyte combination was repeated three times. When the total discharge capacity (sum of capacities at 100, 24, and 8 mA·g^{-1}) of the cell decreased by 70%, it was considered to be cell failure.

Discharge capacity degradation and HRD were calculated and compared with those of a traditional 6.77 M KOH electrolyte. Degradation was determined as Yan previously reported [52]. The percent capacity loss per cycle within the initial 10 cycles is shown by the following equation:

$$Degradation \% = \frac{Cap_{high} - Cap_{low}}{\left(n_{high} - n_{low}\right) \times Cap_{high}} \times 100\% \tag{3}$$

where Cap_{high} is the highest value of discharge capacity achieved in the initial 10 cycles, Cap_{low} is the lowest value of discharge capacity in the initial 10 cycles, n_{high} is the cycle number of the highest discharge capacity in the initial 10 cycles, and n_{low} is the cycle number of the lowest discharge capacity in the initial 10 cycles. HRD is defined as the ratio of capacities measured at 100 and 8 mA·g^{-1}.

Fourier transform infrared (FTIR) spectroscopy was performed on a Perkin Elmer Spectrum Spotlight 200™ (Perkin Elmer, Waltham, MA, USA). Powder X-ray diffraction (XRD) patterns were taken with a Rigaku RU2000 rotating anode powder diffractometer (Rigaku Americas Corporation, The Woodlands, TX, USA) equipped with Cu-K_α radiation (40 kV, 200 mA). Transmission electron microscopy (TEM) was carried out using a JEOL 2010 (JEOL, Tokyo, Japan) operated at 200 kV for microstructural and morphological studies. Inductively coupled plasma-optical emission spectroscopy (ICP-OES) was performed on a Perkin Elmer Optima TM 2100 DV ICP-OES system (Perkin Elmer, Waltham, MA, USA). Electrochemical impedance spectroscopy was measured on a Solartron S1287 potentiostat/galvanostat with a S1255 frequency response analyzer (Solartron, Hampshire, UK).

3. Results and Discussion

3.1. Electrochemical Performances for Electrolytes with Cs_2CO_3 Addition

Generally, KOH solutions with concentrations varying from 4.0 M to 8.5 M are used for Ni/MH batteries in the research field and for commercial applications [11,56–58]. In this study, the electrolyte concentration (KOH + Cs_2CO_3) is fixed at 6.77 M. Concentrations of KOH and Cs_2CO_3 in various electrolytes are shown in Table 2.

Table 2. Normalized discharge capacities, degradations, and high-rate dischargeabilities (HRDs) of AR3 cycled in five different electrolytes.

Electrolyte Composition	Normalized Capacity (%)	Normalized Degradation (%)	Normalized HRD (%)
6.77 M KOH	100.00	100.00	100.00
6.44 M KOH + 0.33 M Cs$_2$CO$_3$	132.45	86.65	99.86
6.11 M KOH + 0.66 M Cs$_2$CO$_3$	120.24	98.30	101.32
5.77 M KOH + 1.00 M Cs$_2$CO$_3$	117.62	97.70	91.41
5.44 M KOH + 1.33 M Cs$_2$CO$_3$	97.28	87.41	65.25

The effects of Cs$_2$CO$_3$ addition on the electrochemical performances of alloys P31, P32, and P37 are shown in Figure 1. All of the cells require approximately three to five cycles to be activated. For alloy P31, Cs$_2$CO$_3$ slightly increases the initial discharge capacity, and greatly decreases the degradation. The initial discharge capacities of alloy P31 in the 6.77 M KOH and 6.44 M KOH + 0.33 M Cs$_2$CO$_3$ electrolytes are 349 and 375 mAh·g^{-1}, respectively. In the 6.77 M KOH electrolyte, the capacity of alloy P31 begins to fade after the 20th cycle, while the capacity fade begins at the 55th cycle in the 6.44 M KOH + 0.33 M Cs$_2$CO$_3$ electrolyte. The same trend was observed in alloy P37, where the addition of Cs$_2$CO$_3$ increases the initial discharge capacity from 364 mAh·g^{-1} to 375 mAh·g^{-1}, and changes the beginning of the capacity fade from the 30th to the 65th cycle. For alloy P32, Cs$_2$CO$_3$ does not increase the initial discharge capacity; however, it greatly decreases the decay, as shown in Figure 1. In comparison to alloy P32, both alloys P31 and P37 contain Al, raising the possibility that the presence of Al results in the alloy surface reacting with the CO$_3^{2-}$ ions in the electrolyte and forming Al$_2$(CO$_3$)$_3$. Al$_2$(CO$_3$)$_3$, which is known to be unstable in water [59]. Such phenomenon may assist in the dissolution of Al into the highly alkaline electrolyte, and therefore increase the reactive surface area.

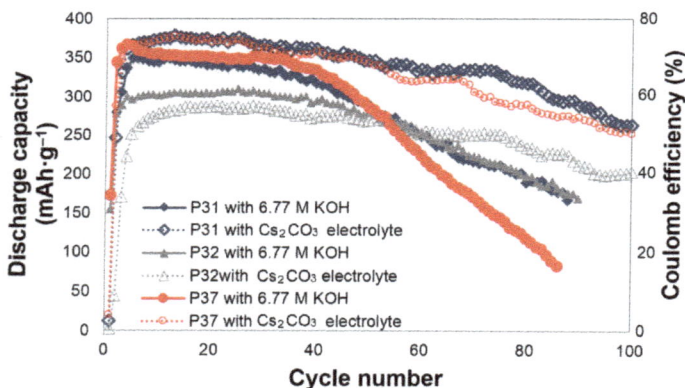

Figure 1. Discharge capacities of alloys P31, P32, and P37 in the 6.77 M KOH and 6.44 M KOH + 0.33 M Cs$_2$CO$_3$ electrolytes.

The effects of Cs$_2$CO$_3$ on the AR3 alloy electrode performances are shown in Figure 2. The AR3 MH alloy is very reactive to the KOH electrolyte due to the alloy's high content of Mg and the high porosity caused by the mechanical alloying preparation [12]. A rapid decay in the KOH electrolyte was previously reported [5,6,12]. Figure 2 shows an increase in initial discharge capacity and a decrease in capacity decay with the addition of Cs$_2$CO$_3$. The discharge capacity, degradation, and HRD of the Cs$_2$CO$_3$-containing electrolytes are normalized to those of the 6.77 M KOH electrolyte and presented in Table 2. The optimized conditions were obtained at the concentration of 6.44 M KOH + 0.33 M Cs$_2$CO$_3$, which exhibited the highest discharge capacity and lowest degradation. Table 2 also indicates that a small addition of Cs_2CO_3 has an insignificant effect on HRD, while a high concentration of Cs_2CO_3 in the electrolyte has a negative influence on HRD.

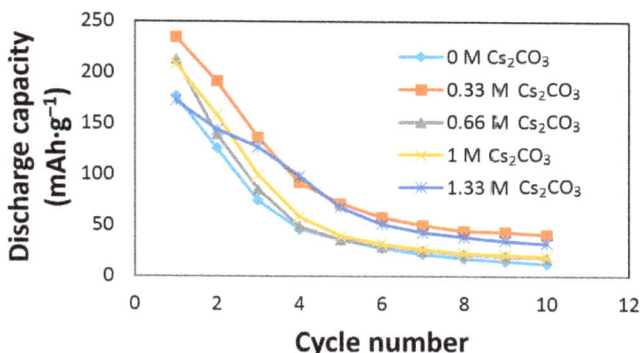

Figure 2. Discharge capacities of AR3 in 6.77 M KOH, 6.44 M KOH + 0.33 M Cs_2CO_3, 6.11 M KOH + 0.66 M Cs_2CO_3, 5.77 M KOH + 1.00 M Cs_2CO_3, and 5.44 M KOH + 1.33 M Cs_2CO_3 electrolytes.

3.2. Effects of Cs_2CO_3 Addition on MgNi Alloy

Weights of the cycled AR3 alloy electrodes were measured. After 10 cycles, weights of the alloy electrodes cycled in all of the electrolytes increased due to surface metal oxidation and the deposition of some salts. Surface metal oxidation during charge leads to the formation of surface metal hydroxide, such as $Mg(OH)_2$ [11,16]. The majority of salt depositions are carbonates and bicarbonates [52]. Weight gains of the alloy electrodes cycled in the Cs_2CO_3-containing electrolytes are normalized to that in the 6.77 M KOH electrolyte and presented in Table 3. As the Cs_2CO_3 concentration in the electrolyte increases, the weight gain decreases. The addition of Cs_2CO_3 changes the physical and chemical properties of the electrolyte. Cs_2CO_3 reacts with the MH alloy surface during electrochemical cycling, which results in a protective layer and greatly decreases the reaction rate of metal oxidation. Therefore, the weight gain due to surface oxidation (listed in Table 3) is substantially reduced by the addition of Cs_2CO_3 in the electrolyte.

Table 3. Normalized weight gains of the AR3 alloy electrodes cycled in five different electrolytes.

Electrolyte Composition	Normalized Electrode Weight Gain (%)
6.77 M KOH	100.0
6.44 M KOH + 0.33 M Cs_2CO_3	76.0
6.11 M KOH + 0.66 M Cs_2CO_3	64.8
5.77 M KOH + 1.00 M Cs_2CO_3	54.2
5.44 M KOH + 1.33 M Cs_2CO_3	46.9

TEM micrographs of the cycled AR3 indicate that a thin layer of fluffy material covers the surface of the MH alloy (Figure 3a,b). FTIR was used to characterize the surface structure, and the results are shown in Figure 4. For the fresh MH alloy, there is no clear diffraction peak, which indicates that the alloy surface is clean. For the alloy cycled in the 6.77 M KOH electrolyte, a peak at approximately 1409 cm^{-1} is observed, which is related to the vibration of surface metal–O bonds [60]. For the electrode cycled in the 6.44 M KOH + 0.33 M Cs_2CO_3 electrolyte, the same peak is seen, but shifted slightly to a lower wavelength, implying a decrease in bond strength. However, the peak intensity increases, which suggests the appearance of C–O bonds on the alloy surface, since the stretching vibration of the C–O bond occurs at approximately the same wavelength [61] as the metal–O bond. With further increases in concentration of Cs_2CO_3 in the electrolyte, the metal–O bond becomes weaker with the peak shifting even lower, to around 1375 cm^{-1}, and the amount of C–O bond decreases, as shown by a reduction in peak intensity compared with the electrode cycled in the 6.77 M KOH electrolyte. Moreover, peaks at approximately 1730 and 1210 cm^{-1} start to appear as the content of Cs$_2$CO$_3$ increases, which is related to the vibration of the C=O bond in carbonate [61]. FTIR results demonstrate the changes in the surface groups on the MH alloy surface with varying Cs$_2$CO$_3$ concentration.

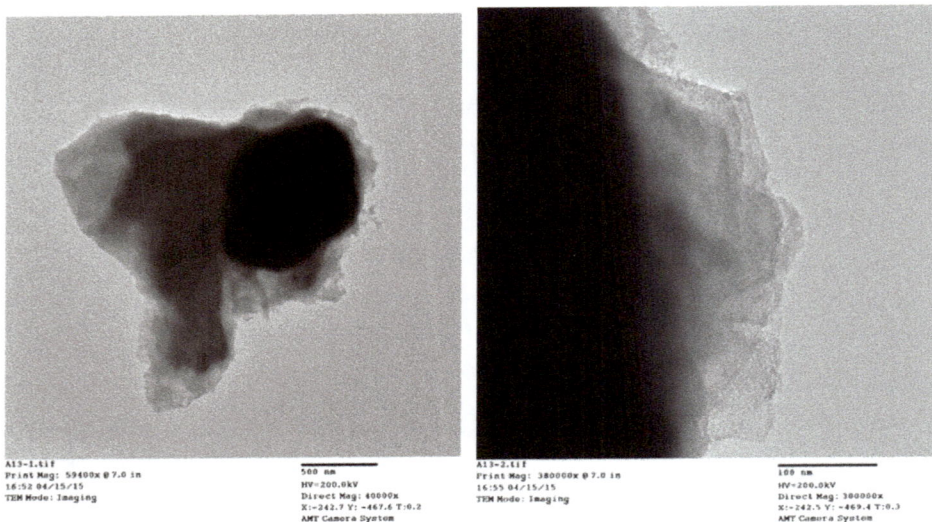

Figure 3. Transmission electron microscopy (TEM) micrographs at (**a**) ×40,400 and (**b**) ×300,000 magnification of AR3 cycled in the 6.44 M KOH + 0.33 M Cs$_2$CO$_3$ electrolyte.

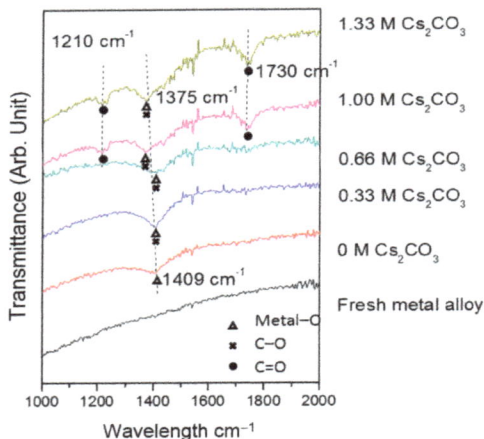

Figure 4. Fourier transform infrared (FTIR) spectra of the fresh and cycled AR3 in the 6.77 M KOH, 6.44 M KOH + 0.33 M Cs$_2$CO$_3$, 6.11 M KOH + 0.66 M Cs$_2$CO$_3$, 5.77 M KOH + 1.00 M Cs$_2$CO$_3$, and 5.44 M KOH + 1.33 M Cs$_2$CO$_3$ electrolytes.

XRD patterns before and after electrochemical cycling are shown in Figure 5. As our previous study has shown [62], the two broad peaks at approximately 23° and 42° (marked as "A") are from the broadening of the $MgNi_2$ and Mg_2Ni phases. These two broad peaks exist in all of the cycled AR3 samples, suggesting high stability for $MgNi_2$ and Mg_2Ni during cycling. Except for the peaks marked as "A", there are some strong peaks related to metallic Mn and Co (marked as "1" and "2" in Figure 5, respectively). Mn and Co fine particles are the remnants from the mechanical alloying process, and act as catalysts for hydrogen storage to enhance the electrochemical capacity [11]. $Mg(OH)_2$ is also found in all of the cycled AR3 (marked as "3" in Figure 5), and is the oxidation product from high-Mg AR3. The peaks at approximately 18° and 33° are the (001) and (100) diffraction peaks for the hexagonal $Mg(OH)_2$. Results from the phase deconvolution by Jade 9.0 software (MDI, Livermore, CA, USA) are summarized in Tables 4 and 5. While no obvious trends are found in the crystallite sizes of phases with cycling in different electrolytes, clear trends are observed in phase abundances. An increase in Cs_2CO_3 concentration in the exchange of KOH reduces both the amount of $Mg(OH)_2$ (decrease in oxidation) and the Mg_2Ni-to-$MgNi_2$ ratio (last column in Table 5). The Mg_2Ni phase is more oxidable compared with the $MgNi_2$ phase, due to its higher Mg content. Upon contacting the KOH electrolyte, the Mg_2Ni phase is oxidized into $Mg(OH)_2$, which results in a reduction in the Mg_2Ni-to-$MgNi_2$ ratio from 1.50 to 0.67. Partial replacement of the corrosive KOH with Cs_2CO_3 increases the Mg_2Ni-to-$MgNi_2$ ratio, which is another indication of decrease in alloy oxidation.

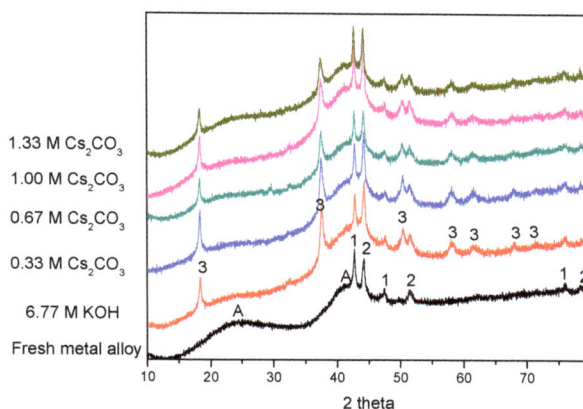

Figure 5. X-ray diffraction (XRD) patterns of the pristine and cycled AR3 in the 6.77 M KOH, 6.44 M KOH + 0.33 M Cs_2CO_3, 6.11 M KOH + 0.66 M Cs_2CO_3, 5.77 M KOH + 1.00 M Cs_2CO_3, and 5.44 M KOH + 1.33 M Cs_2CO_3 electrolytes. Note: Peaks marked as "A" represent the microcrystalline/amorphous components of $MgNi_2$ and Mg_2Ni; peaks marked as "1" and "2" represent metallic Mn and Co, respectively, and peaks marked as "3" represent $Mg(OH)_2$.

Table 4. Crystallite sizes in nm of the Mg_2Ni, $MgNi_2$, Co, Mn, and $Mg(OH)_2$ phases obtained from the XRD patterns in Figure 5.

Condition	Mg_2Ni	$MgNi_2$	Co	Mn	$Mg(OH)_2$
Pristine	0.5	0.6	28	41	-
Cycled in 6.77 M KOH	0.4	0.4	36	41	28
Cycled in 6.44 M KOH + 0.33 M Cs_2CO_3	0.5	0.6	34	62	28
Cycled in 6.11 M KOH + 0.66 M Cs_2CO_3	0.5	0.5	50	42	30
Cycled in 5.77 M KOH + 1.00 M Cs_2CO_3	0.6	0.5	34	40	22
Cycled in 5.44 M KOH + 1.33 M Cs_2CO_3	0.5	0.5	24	41	26

Table 5. Abundances in wt % of the Mg_2Ni, $MgNi_2$, Co, Mn, and $Mg(OH)_2$ phases, and the ratio between Mg_2Ni and $MgNi_2$ obtained from the XRD patterns in Figure 5.

Condition	Mg_2Ni	$MgNi_2$	Co	Mn	$Mg(OH)_2$	$Mg_2Ni/MgNi_2$
Pristine	53.5	35.7	6.5	4.3	-	1.50
Cycled in 6.77 M KOH	29.4	44.0	4.0	3.0	19.6	0.67
Cycled in 6.44 M KOH + 0.33 M Cs_2CO_3	39.2	32.1	4.3	2.6	21.8	1.22
Cycled in 6.11 M KOH + 0.66 M Cs_2CO_3	41.2	32.3	3.7	5.8	17.0	1.27
Cycled in 5.77 M KOH + 1.00 M Cs_2CO_3	42.8	35.1	6.7	4.9	10.5	1.22
Cycled in 5.44 M KOH + 1.33 M Cs_2CO_3	45.3	32.8	7.1	3.6	11.2	1.38

The chemical compositions of fresh and cycled MH alloys determined by ICP are compared in Table 6. Similar to the results in previous reports, the bulk composition changes slightly after cycling [5,6]. The loss of Mg occurs at approximately 1.5%, and results in an increase in the concentrations of other elements.

Table 6. ICP results in at % of the fresh and cycled AR3 in five different electrolytes.

Condition	Co	Ni	Mn	Mg
Pristine	2.94	38.17	5.69	53.20
Cycled in 6.77 M KOH	3.15	37.89	5.86	50.92
Cycled in 6.44 M KOH + 0.33 M Cs_2CO_3	3.10	39.16	5.59	52.25
Cycled in 6.00 M KOH + 0.77 M Cs_2CO_3	3.04	39.41	5.88	51.67
Cycled in 5.77 M KOH + 1.00 M Cs_2CO_3	3.04	39.56	6.19	51.21
Cycled in 5.44 M KOH + 1.33 M Cs_2CO_3	2.99	41.04	6.18	49.79

Cole–Cole plots obtained from the alternating current (AC) impedance measurements are shown in Figure 6. The reported equivalent circuit model for Ni-MH batteries using Mg-based alloy anodes is shown in Figure 7 [63–67]. Constant phase elements are used in this circuit model, due to the inhomogeneity properties of the electrode surface, such as porosity and roughness. Results obtained from the Cole–Cole plots are presented in Table 7. R_0 represents the resistance of ions traveling through the electrolyte and separator. The electrolyte containing 6.44 M KOH + 0.33 M Cs_2CO_3 shows the lowest R_0, which is consistent with its high discharge capacity, low degradation, and similar HRD compared with the 6.77 M KOH electrolyte. R_1 is the resistance among alloy particles and increases as the Cs_2CO_3 concentration increases. R_2 is the whole electrode resistance, and the addition of Cs_2CO_3 greatly decreases R_2 compared with the pure KOH electrolyte. C_1 represents the particle capacitance, which is closely related to the contact area among alloy particles. With increasing Cs_2CO_3 concentration, C_1 decreases. C_2 is the electrode capacitance, which is an indication of the amount of the active area in the electrode. Table 7 shows that the addition of Cs_2CO_3 decreases C_2 as well. The product of R_2 and C_2 represents the electrode activity performance [68,69], and a smaller value suggests better electrochemical performances. Table 7 shows an optimized $R_2 \cdot C_2$ value at 5.77 M KOH + 1.00 M Cs_2CO_3.

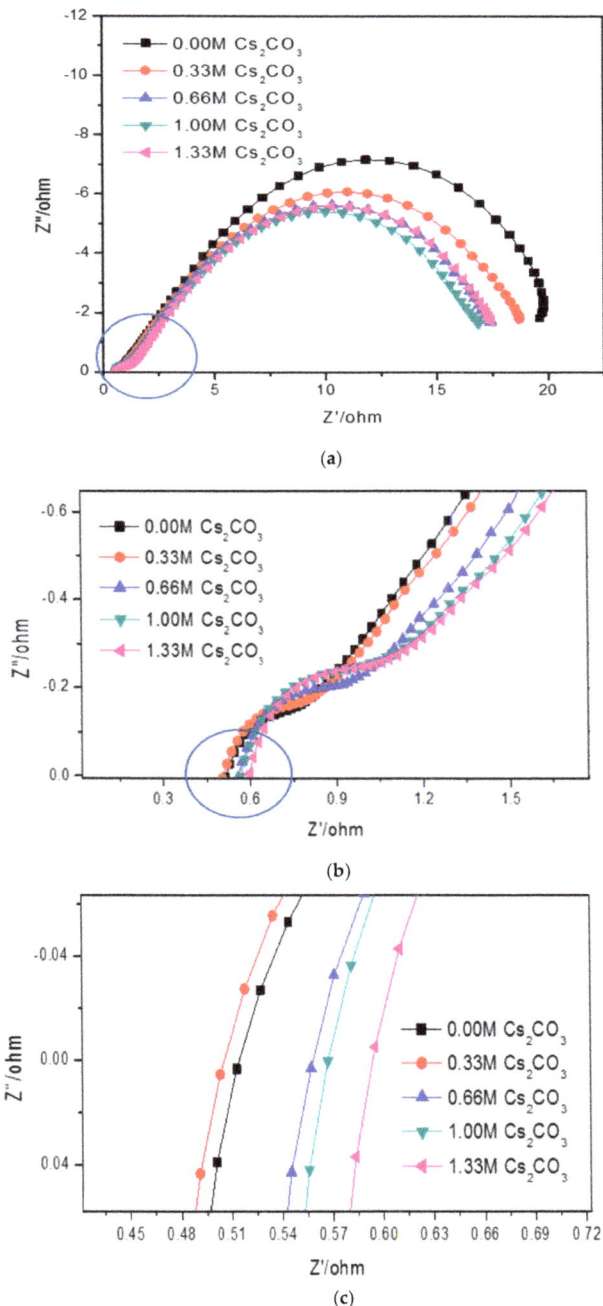

Figure 6. (**a**) Full Cole–Cole plots of AR3 in the 6.77 M KOH, 6.44 M KOH + 0.33 M Cs_2CO_3, 6.11 M KOH + 0.66 M Cs_2CO_3, 5.77 M KOH + 1.00 M Cs_2CO_3, and 5.44 M KOH + 1.33 M Cs_2CO_3 electrolytes, (**b**) magnification of the circled section in (**a**), and (**c**) magnification of the circled section in (**b**).

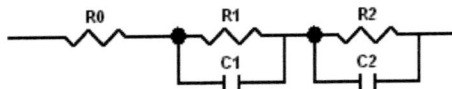

Figure 7. Proposed equivalent circuit model for NiMH batteries using Mg-based anodes [63].

Table 7. Data obtained by the AC impedance measurements of AR3 cycled in five different electrolytes.

Electrolyte Composition	R_0 (Ω·g)	R_1 (Ω·g)	C_1 (F·g^{-1})	R_2 (Ω·g)	C_2 (F·g^{-1})	$R_1 \cdot C_1$ (s)	$R_2 \cdot C_2$ (s)
6.77 M KOH	0.0369	0.0326	0.0033	1.2483	0.5929	1.079×10^{-4}	0.7401
6.44 M KOH + 0.33 M Cs$_2$CO$_3$	0.0363	0.0353	0.0022	1.1811	0.5786	7.812×10^{-5}	0.6834
6.00 M KOH + 0.77 M Cs$_2$CO$_3$	0.0400	0.0372	0.0019	1.0730	0.5357	7.115×10^{-5}	0.5748
5.77 M KOH + 1.00 M Cs$_2$CO$_3$	0.0407	0.0423	0.0018	1.0269	0.4629	7.490×10^{-5}	0.4753
5.44 M KOH + 1.33 M Cs$_2$CO$_3$	0.0417	0.0428	0.0017	1.0713	0.4643	7.099×10^{-5}	0.4974

3.3. Discussion

Results obtained from the current study indicate that the addition of Cs$_2$CO$_3$ alters the chemical and physical properties of the KOH electrolyte. As shown in Figure 6 and Table 7, an addition of 0.33 M Cs$_2$CO$_3$ decreases the solution resistance. Even if the electrolyte concentration was fixed at 6.77 M, the number of conductive ions increased with the addition of Cs$_2$CO$_3$. Therefore, the 6.44 M KOH + 0.33 M Cs$_2$CO$_3$ electrolyte has a decreased resistance and increased conductivity. While a low Cs$_2$CO$_3$ concentration promotes proton transfer in the system and improves both the discharge capacity and cycling performance, a higher Cs$_2$CO$_3$ concentration increases the solution resistivity due to the increased concentrations of larger cations (Cs$^+$) and anions (CO$_3^{2-}$) in the electrolyte, according to Stokes' Law [12].

Cs$_2$CO$_3$ also changes the alloy surface structure during cycling. The FTIR results show that a small addition of Cs$_2$CO$_3$ decreases the strength of surface metal–O bond, but generates the C–O bond on the alloy surface. By further increasing the Cs$_2$CO$_3$ concentration, the C=O bond begins to appear, in relation with decreasing C–O bond. The changes in surface groups by electrolyte additive consequently change the chemical and physical properties of the alloy particles. In our previous work [52], 32 types of salt additives in KOH electrolytes were tested, and some oxyacid salts were reported to create more surface groups that promoted proton transfer. In the current study, the addition of Cs$_2$CO$_3$ provides C–O and C=O bonds as new active sites for proton transfer. However, protons bond to the two active sites differently; protons are covalently bound to C–O, but electrostatically bound to C=O [52,66]. With the stronger attraction to C–O, more protons can be bound and later transferred (driven by voltage). Therefore, the largest number of C–O bonds, which occur at a small addition of Cs$_2$CO$_3$ (6.44 M KOH + 0.33 M Cs$_2$CO$_3$), were demonstrated to be the most effective in improving the electrochemical performances among all of the electrolytes tested in the current study.

The addition of Cs$_2$CO$_3$ also changes the bulk structure of the alloy particles. The TEM images show a layer of solid covering the MH alloy particle (Figure 3), which ranges from 20 nm to 500 nm. This surface layer decreases the contact area among alloy particles (as shown by the decrease in C_1) and increases the barrier for proton transfer among the particles (as shown by the increase in R_1). The ICP results show that a small amount of Cs$_2$CO$_3$ results in this decrease, but further increases in Cs$_2$CO$_3$ concentration increase the loss of Mg after cycling. For the pure KOH electrolyte, the alloy particles are covered by Mg(OH)$_2$. However, a small addition of Cs$_2$CO$_3$ reduces the particle size of Mg(OH)$_2$ on the surface, as indicated by the increase in full width at half maximum of the Mg(OH)$_2$ peaks (Figure 5). Smaller Mg(OH)$_2$ crystals are more strongly adsorbed on the surface of the MH alloy, which is not easily removed from the electrolyte, and also protects the bulk alloy from further oxidation. As the Cs$_2$CO$_3$ concentration further increases in the electrolyte, more MgCO$_3$ starts to form on the alloy surface. Since the solubility of MgCO$_3$ in KOH solution is greater than that of Mg(OH)$_2$, the loss of Mg occurs at a higher rate at higher Cs$_2$CO$_3$ concentrations. Therefore, high Cs$_2$CO$_3$ concentrations are not suggested for Mg–Ni alloys. On the other hand, if properly balanced with the loss of Mg, an appropriate amount of carbonate formation by adding Cs_2CO_3 can promote the dissolution of surface oxidation products and consequently reveal a clean metal surface exposed to the electrolyte, which can lead to a decrease in electrode resistance (as shown by the decrease in R_2 as the Cs_2CO_3 concentration increases).

4. Conclusions

The effects of Cs_2CO_3 addition in a KOH-based electrolyte for Ni/MH batteries were investigated. Four different MH alloys (three Laves phase-related BCC and one MgNi-based) were used as the anode materials, and β-$Ni(OH)_2$ was used as the cathode material. A proper amount of Cs_2CO_3 addition greatly improved electrochemical performances. For the Laves phase-related BCC alloys, adding Cs_2CO_3 into the electrolyte improved the cycle stability (for all three alloys tested) and the discharge capacity (for Al-containing alloys). For the MgNi-based alloy, the discharge capacity increased to 132%, while degradation decreased to 87% in the 6.44 M KOH + 0.33 M Cs_2CO_3 electrolyte (compared with those in the 6.77 M KOH electrolyte). The effects of Cs_2CO_3 addition on the electrolyte and alloy properties are summarized as follows:

(1) A small addition of Cs_2CO_3 decreases the electrolyte resistance and increases the conductivity.
(2) A newly-formed fluffy C-containing surface oxide by the addition of Cs_2CO_3 is believed to be the main cause of the decrease in capacity decay during cycling.
(3) The addition of Cs_2CO_3 in the electrolyte changes the alloy surface structure after cycling by creating more surface groups in addition to metal–O bonds, including C–O and C=O bonds, and the C–O bond is more effective than the C=O bond during proton transfer.
(4) For MgNi-based alloys, the addition of Cs_2CO_3 changes the alloy bulk structure after cycling. A small addition of Cs_2CO_3 strengthens the $Mg(OH)_2$ layer on the alloy surface and prevents loss of Mg. However, a large addition of Cs_2CO_3 causes the formation of $MgCO_3$ with higher solubility in the KOH solution, and consequently a more severe loss of Mg.

Acknowledgments: The authors would like to thank the following individuals from BASF—Ovonic for their help: Su Cronogue, Baoquan Huang, Diana F. Wong, Taihei Ouchi, Tiejun Meng, and Shiuan Chang.

Author Contributions: Shuli Yan designed and Peifeng Li performed the experiments, and analyzed the results. Jean Nei, Kwo-Hsiung Young, and Simon Ng provided guidance and helped in manuscript preparation.

Conflicts of Interest: The authors declare no conflict of interest.

Abbreviations

Ni/MH	Nickel/metal hydride
M	Metal
MH	Metal hydride alloy
MH	Hydrided metal
HRD	High-rate dischargeability
BCC	Body-centered-cubic
Cap_{high}	The highest value of discharge capacity in the initial 10 cycles
Cap_{low}	The lowest value of discharge capacity in the initial 10 cycles
n_{high}	The cycle number of the highest discharge capacity in the initial 10 cycles
n_{low}	The cycle number of the lowest discharge capacity in the initial 10 cycles
$n_{0,high}$	The cycle number of the highest discharge capacity in the initial 10 cycles for 6.77 M KOH electrolyte
$n_{0,low}$	The cycle number of the lowest discharge capacity in the initial 10 cycles for 6.77 M KOH electrolyte
FTIR	Fourier transform infrared
XRD	X-ray diffraction
TEM	Transmission electron microscopy

ICP-OES	Inductively coupled plasma-optical emission spectroscopy
AC	Alternating current
R_0	Resistance of ions traveling through the electrolyte and separator
R_1	Resistance among alloy particles
R_2	Whole electrode resistance
C_1	Particle capacitance
C_2	Electrode capacitance

References

1. Linden, D.; Reddy, T.B. *Handbook of Batteries*; McGraw-Hill: New York, NY, USA, 2002.
2. Fetcenko, M.A.; Ovshinsky, S.R.; Reichman, B.; Young, K.; Fierro, C.; Koch, J.; Zallen, A.; Mays, W.; Ouchi, T. Recent advances in NiMH battery technology. *J. Power Sources* **2007**, *165*, 544–551. [CrossRef]
3. Young, K.; Cai, X.; Chang, S. Reviews on the Chinese Patents regarding nickel/metal hydride battery. *Batteries* **2017**, *3*, 24. [CrossRef]
4. Young, K.; Ng, K.Y.S.; Bendersky, L.A. A Technical report of the robust affordable next generation energy storage system-BASF program. *Batteries* **2016**, *2*, 2. [CrossRef]
5. Chang, S.; Young, K.-H.; Nei, J.; Fierro, C. Reviews on the US Patents regarding nickel/metal hydride batteries. *Batteries* **2016**, *2*, 10. [CrossRef]
6. Ouchi, T.; Young, K.-H.; Moghe, D. Reviews on the Japanese Patent Applications regarding nickel/metal hydride batteries. *Batteries* **2016**, *2*, 21. [CrossRef]
7. Züttle, A. Materials for Hydrogen Storage. *Mater. Today* **2003**, 24–33. [CrossRef]
8. Zhao, X.; Ma, L. Recent progress in hydrogen storage alloys for nickel/metal hydride secondary batteries. *Int. J. Hydrog. Energy* **2009**, *34*, 4788–4796. [CrossRef]
9. Ouyang, L.; Huang, J.; Wang, H.; Liu, J.; Zhu, M. Progress of hydrogen storage alloys for Ni-MH rechargeable power batteries in electric vehicles: A review. *Mater. Chem. Phys.* **2017**, *200*, 164–178. [CrossRef]
10. Yu, X.B.; Wu, Z.; Xia, B.J.; Xu, N.X. A Ti-V-Based BCC phase alloy for use as metal hydride electrode with high discharge capacity. *J. Chem. Phys.* **2004**, *121*, 987–990. [CrossRef] [PubMed]
11. Young, K.; Nei, J. The current status of hydrogen storage alloy development for electrochemical applications. *Materials* **2013**, *6*, 4574–4608. [CrossRef] [PubMed]
12. Nei, J.; Young, K.; Rotarov, D. Studies on MgNi-based metal hydride electrode with aqueous electrolytes composed of various hytproxides. *Batteries* **2016**, *2*, 27. [CrossRef]
13. Redzeb, M.; Zlatanova, Z.; Spassov, T. Influence of boron on the hydriding of nanocrystalline Mg_2Ni. *Intermetallics* **2013**, *34*, 63–68. [CrossRef]
14. Nikkuni, F.R.; Santos, S.F.; Ticianelli, E.A. Microstructures and electrochemical properties of $Mg_{49}Ti_6Ni_{45-x}M_x$ (M = Pd and Pt) alloy electrodes. *Int. J. Energy Res.* **2013**, *37*, 706–712. [CrossRef]
15. Zhang, X.; Belharouak, I.; Li, L.; Lei, Y.; Elam, J.W.; Nie, A.; Chen, X.; Yassar, R.S.; Axelbaum, R.L. Structural and electrochemical study of Al_2O_3 and TiO_2 coated $Li_{1.2}Ni_{0.13}Mn_{0.54}Co_{0.13}O_2$ cathode material using ALD. *Adv. Energy Mater.* **2013**, *3*, 1299–1307. [CrossRef]
16. Zhang, Y.; Li, C.; Cai, Y.; Hu, F.; Liu, Z.; Guo, S. Highly improved electrochemical hydrogen storage performances of the Nd-Cu-added Mg_2Ni-type alloys by melt spinning. *J. Alloy. Compd.* **2014**, *584*, 81–86. [CrossRef]
17. Wang, Y.T.; Wan, C.B.; Wang, R.L.; Meng, X.H.; Huang, M.F.; Ju, X. Effect of Cr substitution by Ni on the cycling stability of Mg_2Ni alloy using EXAFS. *Int. J. Hydrog. Energy* **2014**, *39*, 14858–14867. [CrossRef]
18. Hou, X.; Hu, R.; Zhang, T.; Kou, H.; Song, W.; Li, J. Microstructure and electrochemical hydrogenation/dehydrogenation performance of melt-spun La-doped Mg_2Ni alloys. *Mater. Charact.* **2015**, *106*, 163–174. [CrossRef]
19. Verbovytskyy, Y.; Zhang, J.; Cuevas, F.; Paul-Boncour, V.; Zavaliy, I. Synthesis and properties of the $Mg_2Ni_{0.5}Co_{0.5}H_{4.4}$ hydride. *J. Alloy. Compd.* **2015**, *645*, S408–S411. [CrossRef]
20. Li, M.; Zhu, Y.; Yang, C.; Zhang, J.; Chen, W.; Li, L. Enhanced electrochemical hydrogen storage properties of Mg_2NiH_4 by coating with nano-nickel. *Int. J. Hydrog. Energy* **2015**, *40*, 13949–13956. [CrossRef]

21. Shang, J.; Ouyang, Z.; Liu, K.; Xing, C.; Liu, W.; Wang, L. Effect of Li atom infiltration by the way of electro-osmosis on electrochemical properties of amorphous $Mg_{65}Ni_{27}La_8$ alloy used as negative electrode materials for the nickel–metal hydride secondary batteries. *J. Non-Cryst. Solids* **2015**, *415*, 30–35. [CrossRef]
22. Shao, H.; Li, X. Effect of nanostructure and partial substitution on gas absorption and electrochemical properties in Mg_2Ni-based alloys. *J. Alloy. Compd.* **2016**, *667*, 191–197. [CrossRef]
23. Ohara, R.; Lan, C.-H.; Hwang, C.-S. Electrochemical and structural characterization of electroless nickel coating on Mg_2Ni hydrogen storage alloy. *J. Alloy. Compd.* **2013**, *580*, S368–S372. [CrossRef]
24. Shahcheraghi, A.; Dehghani, F.; Raeissi, K.; Saatchi, A.; Enayati, M.H. Effects of TiO_2 additive on electrochemical hydrogen storage properties of nanocrystalline/amorphous Mg_2Ni intermetallic alloy. *Iran. J. Mater. Sci. Eng.* **2013**, *10*, 1–9.
25. Haghighat-Shishavan, S.; Bozorg, F.K. Nano-crystalline $Mg_{2-x}Mn_xNi$ compounds synthesized by mechanical alloying: Microstructure and electrochemistry. *J. Ultrafine Grained Nanostruct. Mater.* **2014**, *47*, 43–49.
26. Venkateswari, A.; Nithya, C.; Kumaran, S. Electrochemical behaviour of $Mg_{67}Ni_{33-x}Nb_x$ (x = 0, 1, 2 and 4) alloy synthesized by high energy ball milling. *Proc. Mater. Sci.* **2014**, *5*, 679–687. [CrossRef]
27. Rubin, E.J.; Baboian, R. A correlation of the solution properties and the electrochemical behavior of the nickel hydroxide electrode in binary aqueous alkali hydroxides. *J. Electrochem. Soc.* **1971**, *118*, 428–433. [CrossRef]
28. Barnard, R.; Randell, C.F.; Tye, F.L. Studies concerning changes nickel hydroxide electrodes. IV. Reversible potentials in LiOH, NaOH, RbOH and CdOH. *J. Appl. Electrochem.* **1981**, *11*, 517–523. [CrossRef]
29. Oliva, P.; Leonardi, J.; Laurent, J.F.; Delmas, C.; Braconnier, J.J.; Figlarz, M.; Fievet, F.; Guibert, A. Review of the structure and the electrochemistry of nickel hydroxides and oxy-hydroxides. *J. Power Sources* **1982**, *8*, 229–255. [CrossRef]
30. Leblanc, P.; Jordy, C.; Knosp, B.; Blanchard, Ph. Mechanism of alloy corrosion and consequences on sealed nickel-metal hydride battery performance. *J. Electrochem. Soc.* **1998**, *145*, 860–863. [CrossRef]
31. Knosp, B.; Vallet, L.; Blamchard, P. Performance of an AB_2 alloy in sealed Ni-MH batteries for electric vehicles: Qualification of corrosion rate and consequences on the battery performance. *J. Alloy. Compd.* **1999**, *293–295*, 770–774. [CrossRef]
32. Jeong, Y.H.; Kim, H.G.; Jung, Y.H.; Ruhmann, H. Effect of LiOH, NaOH and KOH on Corrosion and Oxide Microstructure of Zr-Based Alloys. Available online: http://www.iaea.org/inis/collection/NCLCollectionStore/_Public/30/060/30060383.pdf (accessed on 26 February 2016).
33. Liu, J.; Wang, D.; Liu, S.; Feng, X. Improving high temperature performance of MH/Ni battery by orthogonal design. *Battery Bimon.* **2003**, *33*, 218–220.
34. Hou, X.; Nan, J.; Han, D.; Zhao, J. Preparation and performance of high-rated A-type MN-Ni batteries. *Chin. J. Appl. Chem.* **2004**, *21*, 1169–1173.
35. Lv, J.; Liu, X.; Zhang, J.; Fan, L.; Wang, L.; Zhang, Z. Studies on high-power nickel-metal hydride battery. *Chin. J. Power Sources* **2005**, *29*, 826–830.
36. Li, X.; Dong, H.; Zhang, A.; Wei, Y. Electrochemical impedance and cyclic voltammetry characterization of a metal hydride electrode in alkaline electrolytes. *J. Alloy. Compd.* **2006**, *426*, 93–96. [CrossRef]
37. Park, C.; Shim, J.; Jang, M.; Park, C.; Choi, J. Influences of various electrolytes on the low-temperature characteristics of Ni-MH secondary battery. *Trans. Korean Hydrog. New Energy Soc.* **2007**, *18*, 284–291.
38. Chen, R.; Li, L.; Wu, F.; Qiu, X.; Chen, S. Effects of low temperature on performance of hydrogen-storage alloys and electrolyte. *Min. Metall. Eng.* **2007**, *27*, 44–46.
39. Yang, D.C.; Park, C.N.; Park, C.J.; Choi, J.; Sim, J.S.; Jang, M.H. Design of additives and electrolyte for optimization of electrode characteristics of Ni-MH secondary battery at room and low temperatures. *Trans. Korean Hydrog. New Energy Soc.* **2007**, *18*, 365–373.
40. Zhang, X.; Chen, Y.; Tao, M.; Wu, C. Effect of electrolyte concentration on low-temperature electrochemical properties of $LaNi_5$ alloy electrode at 233 K. *J. Rare Earths* **2008**, *26*, 402–405. [CrossRef]
41. Zhang, X.; Chen, Y.; Tao, M.; Wu, C. Effect of electrolyte on the low-temperature electrochemical properties of $LaNi_5$ alloy electrode at 253 K. *Rare Metal Mater. Eng.* **2008**, *37*, 2012–2015.
42. Pei, L.; Yi, S.; He, Y.; Chen, Q. Effect of electrolyte formula on the self-discharge properties of nickel-metal hydride batteries. *J. Guangdong Univ. Technol.* **2008**, *25*, 10–12.
43. Khaldi, C.; Mathlouthi, H.; Lamloumi, J. A comparative study of 1 M and 8 M KOH electrolyte concentrations used in Ni-MH batteries. *J. Alloy. Compd.* **2009**, *469*, 464–471. [CrossRef]

44. Guiose, B.; Cuevas, F.; Décamps, B.; Leroy, E.; Percheron-Guégan, A. Microstructural analysis of the aging of pseudo-binary (Ti, Zr)Ni intermetallic compounds as negative electrodes of Ni-MH batteries. *Electrochim. Acta* **2009**, *54*, 2781–2789. [CrossRef]
45. Qiu, Z.; Wu, A. Study on wide temperature characteristics of Ni-MH battery. *J. South China Norm. Univ.* **2009**, *1*, 79–81.
46. Song, M.; Chen, Y.; Tao, M.; Wu, C.; Zhu, D.; Yang, H. Some factors affecting the electrochemical performances of LaCrO$_3$ as negative electrodes for Ni/MH batteries. *Electrochim. Acta* **2010**, *55*, 3103–3108. [CrossRef]
47. Ma, H.; Cheng, F.; Chen, J. Nickel-metal hydride (Ni-MH) rechargeable battery. In *Electrochemical Technologies for Energy Storage and Conversion*; Zhang, J., Zhang, L., Liu, H., Sun, A., Liu, R., Eds.; John Wiley & Sons, Inc.: New York, NY, USA, 2011; p. 204.
48. Karwowska, M.; Jaron, T.; Fijalkowski, K.J.; Leszczynski, P.J.; Rogulski, Z.; Czerwinski, A. Influence of electrolyte composition and temperature on behavior of AB$_5$ hydrogen storage alloy used as negative electrode in Ni-MH batteries. *J. Power Sources* **2014**, *263*, 304–309. [CrossRef]
49. Giza, K. Influence of electrolyte on capacity and corrosion resistance of anode material used in Ni-MH cells. *Ochr. Przed Koroz.* **2016**, *59*, 167–169. [CrossRef]
50. Shangguan, E.; Li, J.; Chang, Z.; Tang, H.; Li, B.; Yuan, X.; Wang, H. Sodium tungstate as electrolyte additive to improve high-temperature performance of nickelemetal hydride batteries. *Int. J. Hydrog. Energy* **2013**, *38*, 5133–5138. [CrossRef]
51. Vaidyanathan, H.; Robbins, K.; Rao, G.M. Effect of KOH concentration and anions on the performance of an Ni-H$_2$ battery positive plate. *J. Power Sources* **1996**, *63*, 7–13. [CrossRef]
52. Yan, S.; Ng, K.Y.S.; Young, K.-H. Effects of salt additives to the KOH electrolyte used in Ni/MH batteries. *Batteries* **2015**, *1*, 54–73. [CrossRef]
53. Li, L. Non-Toxic Alkaline Electrolyte with Additives for Rechareable Zinc Cells. U.S. Patent 2010/0062327, 11 March 2010.
54. Young, K.; Nei, J.; Wong, D.; Wang, L. Structural, hydrogen storage, and electrochemical properties of Laves-phase related body-centered-cubic solid solution metal hydride alloys. *Int. J. Hydrog. Energy* **2014**, *39*, 21489–21499. [CrossRef]
55. Young, K.; Ouchi, T.; Nei, J.; Wang, L. Annealing effects on Laves phase-related body-centered-cubic solid solution metal hydride alloys. *J. Alloy. Compd.* **2016**, *654*, 216–225. [CrossRef]
56. Ruiz, F.C.; Martínez, P.S.; Castro, E.B.; Humana, R.; Peretti, H.A.; Visintin, A. Effect of electrolyte concentration on the electrochemical properties of an AB$_5$-type alloy for Ni/MH batteries. *Int. J. Hydrog. Energy* **2013**, *38*, 240–245. [CrossRef]
57. Martínez, P.S.; Ruiz, F.C.; Visintin, A. Influence of different electrolyte concentrations on the performance of an AB$_2$-type alloy. *J. Electrochem. Soc.* **2014**, *161*, A326–A329. [CrossRef]
58. Yasuoka, S.; Magari, Y.; Murata, T.; Tanaka, T.; Ishida, J.; Nakamura, H.; Nohma, T.; Kihara, M.; Baba, Y.; Teraoka, H. Development of high-capacity nickel-metal hydride batteries using superlattice hydrogen-absorbing alloys. *J. Power Sources* **2006**, *156*, 662–666. [CrossRef]
59. Why is Aluminum Carbonate Unstable? Available online: https://chemistry.stackexchange.com/questions/6369/why-is-aluminium-carbonate-unstable (accessed on 16 August 2017).
60. Yang, C.; Wöll, C. IR spectroscopy applied to metal oxide surfaces: Adsorbate vibrations and beyond. *Adv. Phys.* **2017**, *2*, 373–408. [CrossRef]
61. Reig, F.B.; Adelantado, J.V.; Moya Moreno, M.C. FTIR quantitative analysis of calcium carbonate (calcite) and silica (quartz) mixtures using the constant ratio method. Application to geological samples. *Talanta* **2002**, *58*, 811–821. [CrossRef]
62. Young, K. Stoichiometry in inter-metallic compounds for hydrogen storage applications. In *Stoichiometry and Materials Science—When Numbers Matter*; Innocenti, A., Kamarulzaman, N., Eds.; InTech: Rijeka, Crotia, 2012.
63. Zhang, W.; Kumar, M.P.S.; Srinivasan, S. AC impedance studies on metal hydride electrodes. *J. Electrochem. Soc.* **1995**, *142*, 2935–2943. [CrossRef]
64. Chang, S.; Young, K.; Ouchi, T.; Meng, T.; Nei, J.; Wu, X. Studies on incorporation of Mg in Zr-based AB$_2$ metal hydride alloys. *Batteries* **2016**, *2*, 11. [CrossRef]
65. Trapanese, M.; Franzitta, V.; Viola, A. Description of hysteresis of nickel metal hydride battery. In Proceedings of the 38th Annual Conference on IEEE Industrial Electronics Society, Montreal, QC, Canada, 25–28 October 2012; pp. 967–970.

Batteries **2017**, *3*, 41

66. Trapanese, M.; Franzitta, V.; Viola, A. Description of hysteresis in lithium battery by classical Preisach model. *Adv. Mater. Res.* **2013**, *622*, 1099–1103.

67. Trapanese, M.; Franzitta, V.; Viola, A. The Jiles Atherton model for description on hysteresis in lithium battery. In Proceedings of the Twenty-Eighth Annual IEEE Applied Power Electronics Conference and Exposition (APEC), Long Beach, CA, USA, 17–21 March 2013; pp. 2772–2775.

68. Young, K.; Ouchi, T.; Nei, J.; Moghe, D. The importance of rare-earth additions in Zr-based AB_2 metal hydride alloys. *Batteries* **2016**, *2*, 25. [CrossRef]

69. Grabowski, J.S. What is the covalency of hydrogen bonding? *Chem. Rev.* **2011**, *111*, 2597–2625. [CrossRef] [PubMed]

![batteries logo] *batteries*

MDPI

Article

Ionic Liquid-Based Non-Aqueous Electrolytes for Nickel/Metal Hydride Batteries

Tiejun Meng [1], Kwo-Hsiung Young [1,2,*], Diana F. Wong [1] and Jean Nei [1]

1   BASF/Battery Materials-Ovonic, 2983 Waterview Drive, Rochester Hills, MI 48309, USA;
    tiejun.meng@partners.basf.com (T.M.); diana.f.wong@basf.com (D.F.W.); jean.nei@basf.com (J.N.)
2   Department of Chemical Engineering and Materials Science, Wayne State University, Detroit, MI 48202, USA
*   Correspondence: kwo.young@basf.com; Tel.: +1-248-293-7000

Academic Editor: Andreas Jossen
Received: 9 November 2016; Accepted: 23 January 2017; Published: 6 February 2017

Abstract: The voltage of an alkaline electrolyte-based battery is often limited by the narrow electrochemical stability window of water (1.23 V). As an alternative to water, ionic liquid (IL)-based electrolyte has been shown to exhibit excellent proton conducting properties and a wide electrochemical stability window, and can be used in proton conducting batteries. In this study, we used IL/acid mixtures to replace the 30 wt % KOH aqueous electrolyte in nickel/metal hydride (Ni/MH) batteries, and verified the proton conducting character of these mixtures through electrochemical charge/discharge experiments. Dilution of ILs with acetic acid was found to effectively increase proton conductivity. By using 2 M acetic acid in 1-ethyl-3-methylimidazolium acetate, stable charge/discharge characteristics were obtained, including low charge/discharge overpotentials, a discharge voltage plateau at ~1.2 V, a specific capacity of 161.9 mAh·g^{-1}, and a stable cycling performance for an AB$_5$ metal hydride anode with a (Ni,Co,Zn)(OH)$_2$ cathode.

Keywords: ionic liquid (IL); non-aqueous electrolyte; nickel/metal hydride (Ni/MH) batteries

1. Introduction

Aqueous-based electrolyte batteries possess apparent advantages over carbonate-based electrolyte Li-ion batteries in terms of safety and cost, despite their relatively lower energy densities. The open circuit voltage of an aqueous electrolyte-based battery is intrinsically limited by the narrow electrochemical stability window of water (1.23 V), which restricts the selection to electrodes with higher standard potentials (more positive or negative) and presents an obstacle in the improvement of energy density. As an alternative to water and flammable non-aqueous electrolytes (e.g., carbonate, acetonitrile), ionic liquids (ILs) exhibit unique and tunable physicochemical properties, including a wide electrochemical stability window, good ionic conductivity, a wide liquidus range, negligible vapor pressure, good thermal and chemical stability, inflammability, and non-toxicity, all of which have made ILs ideal candidates for many electrochemical applications, such as batteries [1–6], fuel cells [7–12], supercapacitors [13–16] and dye-sensitized solar cells [17–20].

The excellent proton conductivity of ILs has been demonstrated in both fuel cells and proton-conducting batteries. It was reported that diethylmethylammonium trifluoromethanesulfonate ([DEMA][TfO]) exhibits an ionic conductivity of 10 mS·cm^{-1}, and fuel cells using it as the electrolyte show an open circuit voltage of 1.03 V [9,10]. Our recent study on proton-conducting batteries showed a discharge capacity of 3635 mAh·g^{-1} for a hydrogenated amorphous silicon thin film anode using 1-ethyl-3-methylimidazolium acetate/acetic acid as the electrolyte [21]. For the non-aqueous electrolyte in proton-conducting batteries, a hydrogen-bond network consisting of proton donors and acceptors is required. Long range proton transport occurs in systems with weakly bonded hydrogen networks, where there is rapid hydrogen bond dissociation and formation [22]. Protons move through both vehicle and Grotthuss mechanisms [23]. In the latter case, protons hop through the hydrogen-bond network, as illustrated in [21] (Figure 1). It has been demonstrated that the Grotthuss mechanism (structural diffusion) plays an important role in proton transfer, and that the diffusion coefficient of protons subject to the Grotthuss (proton hopping) mechanism is greater than those subject to the vehicle mechanism in the case of imidazole and bis(trifluoromethanesulfonyl)imide acid (HTFSI) mixtures [23]. It has also been reported that proton hopping persists over a wide temperature range (up to 120 °C) and results in high proton conductivity and high proton transference numbers in imidazolium bis(trifluoromethylsulfonyl)imide ([Im][TFSI]) with excess imidazole [24]. Depending on the availability of exchangeable protons in the chemical structure, ILs can be categorized into protic ionic liquids (PILs) and aprotic ionic liquids (AILs). PILs are prepared through the neutralization reaction of Brønsted acids and Brønsted bases and have intrinsic exchangeable/active protons. In contrast, all other ILs are categorized as AILs, which do not have exchangeable/active protons. The aforementioned [DEMA][TfO] is a PIL with good proton conductivity. While AILs have generally poor ionic conductivity and are not proton-conductive, acids can be introduced to increase proton conductivity by functioning as proton donors. Therefore, by mixing non-aqueous acids with ILs, a proton-conductive electrolyte with a much larger electrochemical stability window (compared to aqueous electrolyte) can be obtained. Therefore, the energy density of the proton-conducting batteries can be dramatically improved through the utilization of high-potential electrode materials, which are unable to work in aqueous systems.

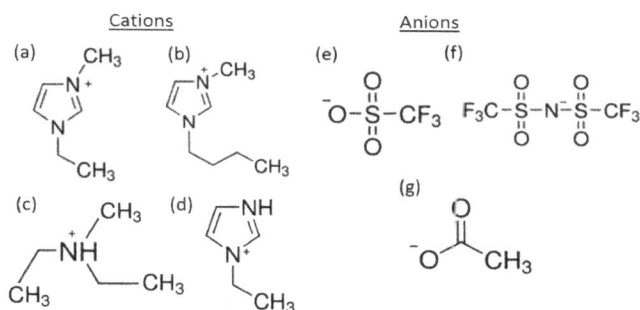

Figure 1. Structures of the cations and anions in the ionic liquids (ILs) used for this study. Cations: (**a**) 1-ethyl-3-methylimidazolium ([EMIM]$^+$); (**b**) 1-butyl-3-methylimidazolium ([BMIM]$^+$); (**c**) diethylmethylammonium ([DEMA]$^+$); and (**d**) 1-ethylimidazolium ([EIM]$^+$). Anions: (**e**) trifluoromethanesulfonate ([TfO]$^-$); (**f**) bis(trifluoromethylsulfonyl)imide ([TFSI]$^-$); and (**g**) acetate ([Ac]$^-$).

IL-based non-aqueous electrolytes combine a wide electrochemical stability window (typically at least 2–3 times larger than the water-based counterpart) and a high proton conductivity, which are promised to replace water-based electrolytes in proton conducting batteries and further improve the energy density. The nickel/metal hydride (Ni/MH) battery is one of the most commonly used secondary batteries with a high energy density and the utilization of IL-based non-aqueous electrolytes in Ni/MH batteries has not been demonstrated before. In the present work, we developed mixtures of acids and ILs to replace the conventional 30 wt % KOH aqueous electrolyte in Ni/MH batteries and studied their room temperature electrochemical performances. Unlike the alkaline electrolyte in which the hydrogen transport is in the form of the hydroxide ion OH^-, in IL-based electrolyte, the transport of protons is mainly through the Grotthuss mechanism—proton hopping through the hydrogen-bond network that consists of proton donors and acceptors. The cation and anion structures of the ILs in this study are shown in Figure 1 and the physicochemical properties of the eight developed ILs are listed in Table 1. By adding non-aqueous acid such as acetic acid into IL, the ionic conductivity can be increased to ~10^{-2} S·cm^{-1}, which is comparable to that of the carbonate-based electrolyte used in Li-ion batteries and meets the demand for battery electrolyte. The verification of the functionality for the IL-based electrolyte in a Ni/MH battery could enable the use of electrode materials with high voltage and/or capacity which leads to an energy density boost for proton conducting batteries.

Table 1. Physicochemical properties of ILs used in this study. [EMIM][TfO]: 1-ethyl-3-methylimidazolium trifluoromethanesulfonate; [BMIM][TfO]: 1-butyl-3-methylimidazolium trifluoromethanesulfonate; [EMIM][TFSI]: 1-ethyl-3-methylimidazolium bis(trifluoromethylsulfonyl)imide; [BMIM][TFSI]: 1-butyl-3-methylimidazolium bis(trifluoromethylsulfonyl)imide; [EMIM][Ac]: 1-ethyl-3-methylimidazolium acetate; [BMIM][Ac]: 1-butyl-3-methylimidazolium acetate; [EIM][TFSI]: 1-ethylimidazolium bis(trifluoromethylsulfonyl)imide; and [DEMA][TfO]: Diethyl$methylammonium trifluoromethanesulfonate.

IL's Abbreviation	T_{melt} (°C)	Viscosity (cP)	Electrochemical Window (V)	Conductivity at 25 °C (mS·cm^{-1})	References
[EMIM][TfO]	−9	45	4.1	8.6–11	[25]
[BMIM][TfO]	16	90	-	3.7	[25]
[EMIM][TFSI]	−21~−3	28–34	4.0–4.5	5.7–8.8	[25]
[BMIM][TFSI]	−4	52	4.6	3.9	[25]
[EMIM][Ac]	<−20	93	3.2	2.5	[26]
[BMIM][Ac]	<−20	554	3.1	1.1	[26]
[EIM][TFSI]	-	54	-	4	[8]
[DEMA][TfO]	−6	19.4	-	55 *	[23,27]

* Conductivity measured at 150 °C.

2. Experimental Setup

The electrochemical performance of the non-aqueous IL-based electrolytes was studied by replacing the 30 wt % KOH aqueous solution electrolyte in Ni/MH batteries with the new electrolyte. The non-aqueous IL-based electrolytes are mixtures of ILs and anhydrous acids. The ILs used in this work include 1-ethyl-3-methylimidazolium trifluoromethanesulfonate ([EMIM][TfO], 99%, Ionic Liquids Technologies (IoLiTec) Inc., Tuscaloosa, AL, USA), 1-butyl-3-methylimidazolium trifluoromethanesulfonate ([BMIM][TfO], ≥95%, Sigma-Aldrich, St. Louis, MO, USA), 1-ethyl-3-methylimidazolium bis(trifluoromethylsulfonyl)imide ([EMIM][TFSI], >97%, IoLiTec GmbH), 1-butyl-3-methylimidazolium bis(trifluoromethylsulfonyl)imide ([BMIM][TFSI], ≥98%, Aldrich), 1-ethyl-3-methylimidazolium acetate ([EMIM][Ac], >95%, IoLiTec GmbH), 1-butyl-3-methylimidazolium acetate ([BMIM][Ac], ≥95%, Aldrich), 1-ethylimidazolium bis(trifluoromethylsulfonyl)imide ([EIM][TFSI], >97%, IoLiTec GmbH), and [DEMA][TfO] (>98%, IoLiTec GmbH). Prior to use, all ILs were baked at 115 °C in a vacuum oven for 48 h to remove residual water. The water content of the IL was then determined to be <1000 ppm using a Metrohm 831 KF coulometer (Metrohm, Riverview, FL, USA). Two types of anhydrous acids, glacial acetic acid (CH$_3$COOH, 100%, anhydrous for analysis, EMD Millipore, Billerica, MA, USA) and phosphoric acid (H$_3$PO$_4$, ≥99.999%, crystalline, Sigma-Aldrich), were used to create the IL/acid mixtures. The ionic conductivity of the solutions was measured with a conductivity meter (YSI Model 3200, YSI Incorporated, Yellow Spring, OH, USA) at room temperature (25 °C).

The typical negative electrode (anode) for Ni/MH batteries uses an AB$_5$ type metal hydride alloy with a composition of La$_{10.5}$Ce$_{4.3}$Pr$_{0.5}$Nd$_{1.4}$Ni$_{60.0}$Co$_{12.7}$Mn$_{5.9}$Al$_{4.7}$ (AB5), which was supplied by Eutectix (Troy, MI, USA). In addition to AB5, two other metal hydride alloys, AR3 and BCC-B08, were also tested. AR3 is an MgNi-based AB type MH alloy with a composition of Mg$_{52}$Ni$_{39}$Co$_3$Mn$_6$. The AR3 powder was prepared by melt spinning and mechanical alloying processes using a homebuilt melt spin system and an SC-10 attritor (Union Process, Akron, OH, USA), respectively. BCC-B08 has a *bcc* crystal structure and a composition of Ti$_{40}$V$_{30}$Cr$_{15}$Mn$_{13}$Mo$_2$ and was prepared by arc melting. The electrochemical properties of AR3 and BCC-B08 have been previously reported [28,29]. The alloy powders were pressed onto a 1.25 cm (diameter) circular Ni mesh current collectors without any binder or additive. Typically, the active material weight of the negative electrode is 50 ± 10 mg. The positive electrode was sintered Ni(OH)$_2$ on a Ni mesh, which was fabricated in-house and is used as a standard positive electrode for our Ni/MH battery research. The positive electrodes were cut into 1.25 cm (diameter) circular disks and both electrodes were baked at 115 °C in a vacuum oven for 12 h before cell assembly. With the aforementioned metal hydride anode, a sintered Ni(OH)$_2$ cathode, a standard non-woven separator (0.2 mm thick), and two 1.25 cm (diameter) nickel rod current collectors, a 1/2" (diameter) Swagelok-type cell (Swagelok, Solon, OH, USA) was assembled in an Ar-filled glove box. Finally, 1 mL of IL electrolyte was injected into the cell before the cell was sealed for testing.

The electrochemical charge/discharge tests were carried out on an Arbin BT-2143 battery test station (Arbin Instruments, College Station, TX, USA) with a constant charge/discharge current. Cyclic voltammetry (CV) measurements were conducted using a Gamry's Interface 1000 (Gamry Instruments, Warminster, PA, USA) potentiostat/galvanostat in a three-electrode Swagelok cell. The working electrode was an AB5 metal hydride alloy electrode and the counter electrode was a sintered Ni(OH)$_2$ disk. A positive-to-negative capacity ratio of >10 was used to minimize the influence from the positive side. A leak-free microminiature reference electrode (1 mm diameter, Warner Instruments, Hamden, CT, USA) was inserted through the top port of the Swagelok cell as a reference, which has a standard potential of 0.242 V versus a standard hydrogen electrode in a 3.4 M KCl aqueous solution.

3. Results and Discussion

3.1. Ionic Conductivity of Ionic Liquid/Acetic Acid Electrolytes

Pure ILs exhibit much lower ionic conductivities (σ) than conventional aqueous electrolytes. The [EMIM]$^+$ cation-based ILs show high conductivity among various types of ILs, typically at ~10 Ms·cm^{-1} [25], which is comparable to the carbonate-based electrolytes used in Li-ion batteries. This is nearly two orders of magnitude lower than observed with the 30 wt % KOH aqueous solutions used in Ni/MH batteries.

The relationship between the conductivity and viscosity for dilute aqueous solutions follows the Walden rule:

$$\Lambda \eta = C \text{ (constant)} \tag{1}$$

where Λ, η, and C are the molar conductivity, viscosity, and temperature dependent constant, respectively. The Walden rule has been widely used in IL studies to explain the conductivity–viscosity relationship and to estimate the extent of ion association [30–33]. Furthermore, combining the Nernst–Einstein equation for Λ and the Stokes–Einstein equation for the diffusion of spherical particles with an effective radius r, σ can be expressed as [25]:

$$\sigma = \frac{z^2 e^2 N}{6 V \pi r \eta} \tag{2}$$

where z, e, V, and N denote the valence of the charge carrier, elementary charge, volume, and the number of charge carriers in volume V, respectively. It is well known that ILs show higher viscosity, typically 2–3 orders of magnitude higher than common molecular solvents, due to them containing large cations and/or anions. The viscosities of ILs can increase further with increasing length of the alkyl side chain, due to the longer alkyl side chain increasing the van der Waals interactions [34–36]. According to Equations (1) and (2), high viscosity results in low conductivity, which explains how the [EMIM]$^+$-based ILs with lower viscosities (as a result of shorter alkyl side chains) have higher conductivity than their [BMIM]$^+$ counterparts, as shown in Table 1. It has also been demonstrated by nuclear magnetic resonance (NMR) and molecular dynamics studies that the high viscosity of ILs is due to significant ionic association or aggregation [37–39]. Diluting ILs with solvents can separate the cations and anions in solution and reduce their aggregation, which effectively decreases the viscosity and increases the conductivity. Mixed with molecular solvents, such as acetonitrile and butanone, imidazolium-based ILs show more than an order of magnitude increase in conductivity [40]. In this study, we used anhydrous acetic acid to dilute the ILs to increase their conductivities. In addition, acetic acid can also function as a proton donor, which facilitates the proton transport across the electrolyte.

Figure 2 shows the conductivities of the IL/acetic acid mixtures as a function of acetic acid concentration. For all seven IL mixture bases, the conductivity increases with increasing acetic acid concentration in the range of 0–6 M, which is consistent with the discussion above and with previous reports regarding diluted ILs in other solvents [25,40]. The introduction of acetic acid reduces the ion association in ILs and lowers the viscosity. In correlation with the lowered viscosity, the conductivity of the binary system increases. [DEMA][TfO] was the only PIL in this group that showed high conductivity (14.3 mS·cm^{-1}) with 6 M acetic acid (9.0 mS·cm^{-1} for pure [DEMA][TfO]). The other six AILs consist of two cations, $[EMIM]^+$ and $[BMIM]^+$, and three anions, $[TfO]^-$, $[TFSI]^-$ and $[Ac]^-$. All three $[EMIM]^+$-based ILs exhibited approximately two times higher conductivities than their $[BMIM]+$ counterparts in the acetic acid concentration range of 0–6 M. $[TfO]^-$ and $[TFSI]^-$-based ILs showed higher conductivities than their $[Ac]^-$ counterparts.

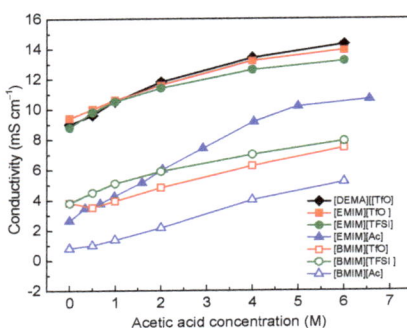

Figure 2. Ionic conductivity as a function of acetic acid concentration in ILs at room temperature.

3.2. Half-Cell Electrochemical Tests

In contrast to the electrochemical reactions occurring at the metal hydride anode in aqueous electrolyte [41], where:

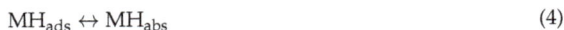

$$M + H_2O + e^- \leftrightarrow MH_{ads} + OH^- \tag{3}$$

$$MH_{ads} \leftrightarrow MH_{abs} \tag{4}$$

in non-aqueous proton conducting IL electrolytes, the following reactions occur:

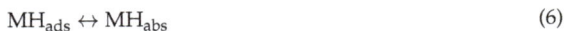

$$M + xH^+ + xe^- \leftrightarrow MH_{ads} \tag{5}$$

$$MH_{ads} \leftrightarrow MH_{abs} \tag{6}$$

where M is the hydrogen storage metal alloy; MH is metal hydride; the subscripts "ads" and "abs" denote "adsorbed" and "absorbed", respectively; forward and backward arrows represent the charge and discharge processes, respectively. The charge process can be simplified as a two-step process consisting of charge-transfer and hydrogen diffusion. First, protons diffuse in the non-aqueous electrolyte through either the vehicle or Grotthuss mechanisms towards the MH anode. They are then reduced and adsorbed at the anode–electrolyte interface. The adsorbed hydrogen atoms are further absorbed and diffused into the bulk of the MH. During discharge, protons diffuse toward the anode–electrolyte interface, where they are oxidized and desorbed.

We screened the electrolytes using 1 M acetic acid with eight different ILs in Ni/MH half-cells with an AB_5 alloy (AB5) anode and an oversized $Ni(OH)_2$ cathode. The electrochemical charge/discharge tests were performed with a constant charge/discharge current of 2 mA·g^{-1}. The anode was charged to

$20\ \text{mAh·g}^{-1}$, and the discharge terminated at a cut-off voltage of 0.2 V. Several cycles were required to activate the electrodes and stabilize the charge/discharge processes. Figure 3 shows the charge/discharge curves for eight electrolytes at the 20th cycle. Compared to 30 wt % KOH aqueous electrolytes, the conductivities of the IL/acetic acid electrolytes are more than two orders of magnitude lower. It can be observed in Figure 3 that the charge/discharge overpotentials for the IL/acetic acid electrolytes are substantial, which can be attributed to ohmic loss resulting from the high resistance of the non-aqueous electrolytes. Among the eight electrolytes, the acetate-based electrolytes [EMIM][Ac] and [BMIM][Ac] show the smallest charge/discharge overpotentials, with a charge voltage plateau at 1.5–1.7 V and a discharge voltage plateau at 1.4–1.1 V. [EMIM][Ac] has a much flatter discharge voltage plateau than [BMIM][Ac], with a voltage of 1.2 V at the middle point of the plateau. The other electrolytes, including the $[\text{TfO}]^-$- and $[\text{TFSI}]^-$-based ILs that exhibit high conductivities, showed worse charge/discharge characteristics, in general, than the $[\text{Ac}]^-$-based ILs. Although the conductivities of the $[\text{Ac}]^-$-based ILs are inferior, the shared $[\text{Ac}]^-$ anion with acetic acid may facilitate Grotthuss diffusion of protons in the electrolyte and charge-transfer at the electrode/electrolyte interface, which results in improved performance compared to other electrolytes.

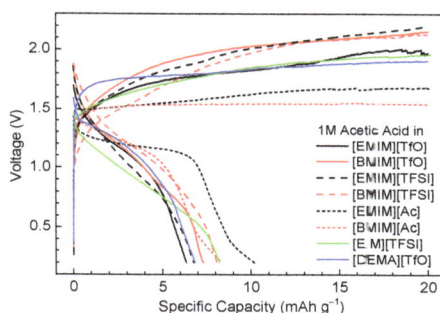

Figure 3. Electrochemical charge/discharge characteristics of nickel/metal hydride (Ni/MH) half-cells with different acetic acid/IL electrolytes at the 20th cycle. The acetic acid concentration was fixed at 1 M. Anode: AB5 MH alloy.

Based on the electrolyte screening results depicted above, a 2 M acetic acid in [EMIM][Ac] was used to conduct the cycle life testing at a higher rate. The cell was charged to a capacity of $40\ \text{mAh·g}^{-1}$ at a current density of $4\ \text{mA·g}^{-1}$ and discharged at the same current density until a cut-off voltage of 0.7 V was reached. The evolutions of specific capacity and charge/discharge curves at the 1st, 10th, 20th, 30th, and 100th cycles are shown in Figure 4a,b, respectively.

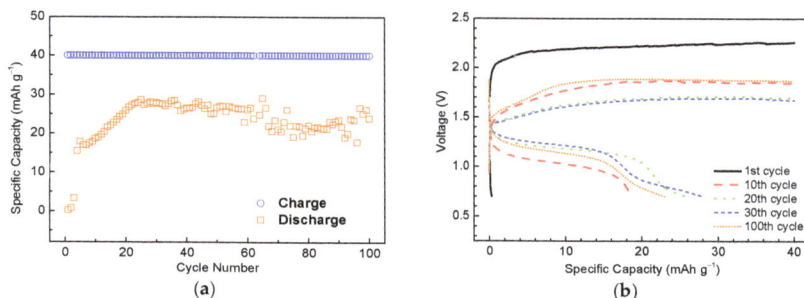

Figure 4. (a) Specific capacities as a function of cycle number and (b) charge/discharge characteristics at different cycles for a Ni/MH half-cell using 2 M acetic acid/[EMIM][Ac]. Anode: AB5 MH alloy.

The initial specific capacity was very low, and several cycles were necessary to activate the cell. It is well known that in KOH electrolyte, many types of MH anodes need activation process until their capacities reach maximum and charge/discharge overpotentials minimize. Initially, the surface of MH alloys is covered with highly resistant oxides which hinder the transport of protons. During the activation process, the MH alloy surfaces experience etching followed by re-oxidation in KOH which results in a new hydroxide/oxide surface layer (solid electrolyte interphase (SEI) for Ni/MH) embedded with metallic nickel-based nanoclusters. Such a Ni-embedded SEI layer is crucial because it not only catalyzes the electrochemical reactions but also protects the bulk alloy from corrosion. For the IL-based electrolytes, the initial cycles show similar charge/discharge characteristics with increasing specific capacities and decreasing charge/discharge overpotentials. Thus, it is speculated that a surface modification process occurs and further investigation is needed to identify the detailed composition, structure and formation mechanisms of the new SEI. After activation, the capacity increased gradually until a maximum of 28.6 mAh·g^{-1} was attained at the 25th cycle, after which the capacity slightly degraded to reach 23.9 mAh·g^{-1} at the 100th cycle. The charge/discharge curves show that there is a very high overpotential for the first charge process, and the discharge capacity is close to zero. At the 10th cycle, the charge/discharge overpotentials were dramatically decreased, with charge and discharge plateaus at 1.8 V and 1.1 V, respectively. This data demonstrates the lowest charge/discharge overpotentials and highest discharge capacities at the 20th and 30th cycle, with charge and discharge plateaus at 1.6 V and 1.2 V, respectively. At the 100th cycle, the charge/discharge overpotentials increased slightly, and the discharge capacity decreased as a consequence of cell degradation—typically as a result of MH anode pulverization/disintegration or corrosion.

Deep charge/discharge tests were carried out with AB5, in addition to AR3 and BCC-B08 alloys, using the optimized electrolyte and the results are shown in Figure 5. The cells were charged to 400 mAh·g^{-1} at a constant current of 4 mA·g^{-1} and discharged at the same current until a cut-off voltage of 0.2 V was attained.

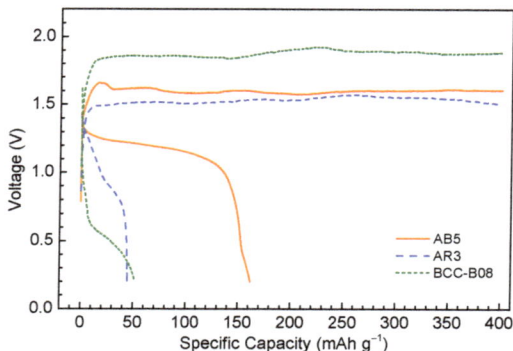

Figure 5. Deep charge/discharge characteristics of Ni/MH half-cells using three different metal hydride anodes: AB5, AR3, and BCC-B08. Electrolyte: 2 M acetic acid/[EMIM][Ac].

Among the three metal hydride alloys, AB5 exhibited the best performance, showing the highest capacity at 161.9 mAh·g^{-1} and stable charge and discharge plateaus at 1.6 V and 1.2 V, respectively. In comparison, AR3 and BCC-B08 showed much smaller capacities (~50 mAh·g^{-1}), larger charge/discharge overpotentials (except for the charge curve for AR3), and more slanted discharge curves, which may result from poor material stability in the acidic environment or a lack of stable SEI at the alloy surface. Lower specific capacities and higher charge/discharge potentials were obtained for all three MH alloys in the non-aqueous 2 M acetic acid/[EMIM][Ac] than those in the 30 wt % KOH aqueous electrolyte. For instance, AB5 exhibited a specific capacity of ~320 mAh·g^{-1} and charge/discharge voltage plateaus (mid-point) at ~1.45 V/1.25 V in 30 wt % KOH. The increase in charge/discharge overpotentials and deficit in specific capacity may result from the following causes. First, the ionic conductivity of 2 M acetic acid/[EMIM][Ac] is around two orders of magnitude lower than that in 30 wt % KOH, which leads to large charge/discharge overpotentials. Second, a stable SEI may not exist which catalyzes the electrochemical reactions and prevents the corrosion of MH alloys. Without a stable SEI layer, both slower surface reaction kinetics and more severe corrosions are expected, which lead to increased overpotentials and decreased capacities. Third, there is still a small amount of water left in the electrolyte even though the ILs experienced long-time heating/drying in a vacuum oven. With >1.5 V charge voltages, water splitting is expected to occur which leads to the low coulombic efficiency. Also, the sintered Ni(OH)$_2$ might not be stable enough in the weakly acidic environment. It could slowly react with acetic acid and produce water. The search for more stable cathodes used in the non-aqueous system is ongoing.

To improve the proton conducting properties of the IL-based electrolyte, anhydrous phosphoric acid was dissolved into ILs to act as the electrolyte. Phosphoric acid is a solid crystal at room temperature, and its melting temperature is 42 °C. Above its melting point, pure phosphoric acid is a highly viscous liquid with high conductivity (77 mS·cm^{-1} at 42 °C), which is due to the diffusion of protons [42]. There are more proton donor sites than acceptor sites in phosphoric acid and this amphoteric property makes it an ideal proton conductor [22]. Due to the fact that it occurs as a solid state at room temperature, phosphoric acid exhibits lower solubility in ILs than acetic acid. Three electrolytes, [EMIM][Ac], [EMIM][TfO], and [DEMA][TfO], in 1 M phosphoric acid were prepared and tested in Ni/MH cells with an AB5 alloy electrode. All three electrolytes were highly viscous. The cells were charged to 40 mAh·g^{-1} at a current of 4 mA/g and discharged at the same current until a cut-off voltage of 0.2 V. As shown in Figure 6, all three 1 M phosphoric acid/IL electrolytes AB5 anodes exhibit low specific capacities and large charge/discharge overpotentials. The mid-point charge/discharge voltages are >1.8 V/<1 V, compared to 1.6 V/1.2 V for 2 M acetic acid/[EMIM][Ac]. The highest specific capacity obtained is from 1 M phosphoric acid/[EMIM][TfO], ~21 mAh·g^{-1}, compared to 28 mAh·g^{-1} for 2 M acetic acid/[EMIM][Ac]. The phosphoric acid/IL electrolytes exhibit higher viscosities and lower ionic conductivities than those in the acetic acid/IL electrolytes. In addition, both the MH anode and Ni(OH)$_2$ cathode are not stable in the phosphoric acid/IL electrolytes and the battery cells exhibit short cycle life (<30 cycles). All these caused the poor performances for phosphoric acid/IL electrolytes.

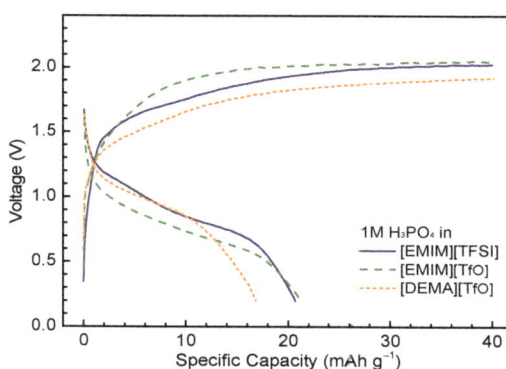

Figure 6. Electrochemical charge/discharge characteristics of Ni/MH half-cells using 1 M phosphoric acid with [EMIM][TFSI], [EMIM][TfO], and [DEMA][TfO]. Anode: AB5 MH alloy.

3.3. Cyclic Voltammetry

CV measurements were performed on a three-electrode Swagelok cell with 2 M acetic acid/[EMIM][Ac] electrolyte and the candidate that showed the best electrochemical performance in this study. The working electrode is an AB5 alloy electrode that is activated with 10 charge/discharge cycles. The counter electrode was sintered Ni(OH)$_2$, and a leak-free microminiature reference electrode (Ag/AgCl, 0.242 V versus the standard hydrogen electrode) was inserted through the top port of the Swagelok cell to act as a reference. A three-electrode cell with the 30 wt % KOH aqueous electrolyte and an Hg/HgO reference electrode was used for comparison. The cyclic voltammograms for 2 M acetic acid/[EMIM][Ac] and 30 wt % KOH are shown in Figure 7a,b, respectively. With 2 M acetic acid/[EMIM][Ac], the anodic and cathodic peaks are clearly presented. The anodic peak originates from the oxidation and desorption of absorbed hydrogen atoms at the anode–electrolyte interface, while the cathodic peak results from the reduction and adsorption of protons at the interface. The anodic peak is situated near −0.4 V versus the reference electrode with a scan rate of 50 mV·s^{-1}. As the scan rate increases, the anodic peak current increases and the peak potential shifts slightly in the positive direction. A full cathodic peak was observed with a center near −1.0 V versus the reference electrode. There is about a 0.6 V voltage gap between the hydrogen cathodic peak and the edge of the reduction of the electrolyte. Theoretically, such a voltage gap is enough to guarantee a high coulombic efficiency. However, in the cycle test, as shown in Figure 4, the coulombic efficiency is less than 70%, which is attributed to the splitting of residual water in the electrolyte during charge, as discussed in Section 3.2. Approaches to remove water completely from ILs and more stable cathodes compatible with the weakly acidic environment are under current investigation. With the 30 wt % KOH aqueous electrolyte, the CV curves exhibit wrinkle-like features starting at approximately −0.9 V versus Hg/HgO, and no obvious cathodic peaks are observed. In fact, the cathodic peak is so close to the hydrogen evolution threshold that it overlaps with the hydrogen evolution edge and cannot be differentiated, which is consistent with previous reports [41,43–45]. The acetic acid/[EMIM][Ac] electrolyte prevails over the 30 wt % KOH electrolyte in terms of the electrochemical stability window, which means that the use of redox couples with higher potentials is a possibility. A non-aqueous environment also enables the study of electrode materials that are not stable in aqueous solutions, with the goal of improving capacities. As a consequence, a marked increase in the energy density is expected for proton conducting batteries.

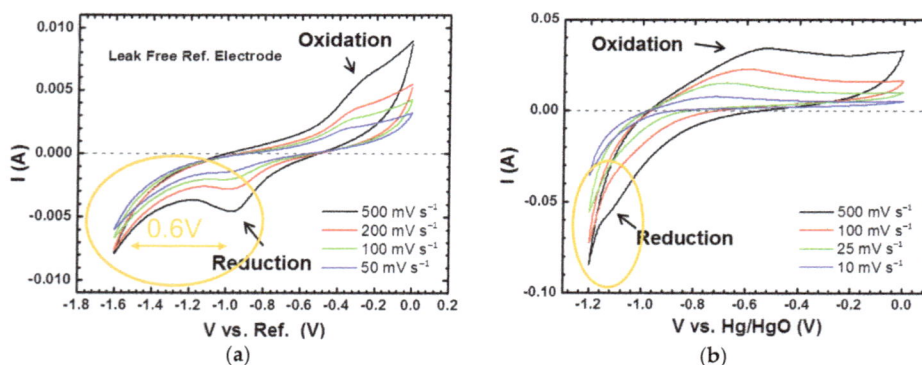

Figure 7. Cyclic voltammograms of Ni/MH half-cells using (**a**) 2 M acetic acid in [EMIM][Ac] as an electrolyte and (**b**) the 30 wt % KOH aqueous electrolyte.

4. Conclusions

IL/acid mixtures were applied in Ni/MH batteries to replace the 30 wt % KOH aqueous electrolyte, and their proton conducting character was verified through electrochemical charge/discharge tests. Pure ILs are viscous and, when diluted with acetic acid, this viscosity decreases and conductivity increases. [DEMA][TfO] with 6 M acetic acid showed a conductivity of 14.3 mS·cm^{-1}, compared to 9.0 mS·cm^{-1} for pure [DEMA][TfO]. In addition, the introduction of acetic acid introduces proton donors, which are crucial for the diffusion of protons. Screening eight different ILs, a mixture of acetic acid/[EMIM][Ac] exhibited the best electrochemical performance with a long cycle life, low charge/discharge overpotentials, and a stable discharge voltage plateau at ~1.2 V. A specific capacity of 161.9 mAh·g^{-1} was obtained with an AB5 anode in a 2 M acetic acid/[EMIM][Ac] electrolyte, which exhibited a voltage difference of approximately 0.6 V between the cathodic peak and the electrolyte reduction edge, based on CV measurements. An extended electrochemical stability window enables the use of redox couples with higher potentials than those currently in use with aqueous electrolyte, which is expected to boost the energy density of proton conducting batteries.

Acknowledgments: This work is financially supported by the U.S. Department of Energy's Advanced Research Project Agency-Energy (ARPA-E) under the Robust Affordable Next Generation EV-storage (RANGE) Program (DE-AR0000386).

Author Contributions: Tiejun Meng and Kwo-Hsiung Young designed the experiments and analyzed the results. Diana F. Wong and Jean Nei assisted in data analysis and manuscript preparation.

Conflicts of Interest: The authors declare no conflict of interest.

Abbreviations

IL	Ionic liquid
[DEMA][TfO]	Diethylmethylammonium trifluoromethanesulfonate
[Im][TFSI]	Imidazolium bis(trifluoromethylsulfonyl)imide
PIL	Protic ionic liquid
AIL	Aprotic ionic liquid
Ni/MH	Nickel/metal hydride
[EMIM][TfO]	1-ethyl-3-methylimidazolium trifluoromethanesulfonate
[BMIM][TfO]	1-butyl-3-methylimidazolium trifluoromethanesulfonate
[EMIM][TFSI]	1-ethyl-3-methylimidazolium bis(trifluoromethylsulfonyl)imide
[BMIM][TFSI]	1-butyl-3-methylimidazolium bis(trifluoromethylsulfonyl)imide
[EMIM][Ac]	1-ethyl-3-methylimidazolium acetate
[BMIM][Ac]	1-butyl-3-methylimidazolium acetate
[EIM][TFSI]	1-ethylimidazolium bis(trifluoromethylsulfonyl)imide
CV	Cyclic voltammetry
σ	Conductivity
Λ	Molar conductivity
η	Viscosity
C	Temperature dependent constant
r	Effective radius of spherical particles
z	Valence of the charge carrier
e	Elementary charge
V	Number of charge carriers
N	Volume
NMR	Nuclear magnetic resonance
M	Hydrogen storage metal alloy
MH	Metal hydride
ads	Adsorbed
abs	Absorbed
SEI	Solid electrolyte interface

References

1. Yamamoto, T.; Nohira, T.; Hagiwara, R.; Fukunaga, A.; Sakai, S.; Nitta, K.; Inazawa, S. Charge–discharge behavior of tin negative electrode for a sodium secondary battery using intermediate temperature ionic liquid sodium bis(fluorosulfonyl)amide–potassium bis(fluorosulfonyl)amide. *J. Power Sources* **2012**, *217*, 479–484. [CrossRef]
2. Nohira, T.; Ishibashi, T.; Hagiwara, R. Properties of an intermediate temperature ionic liquid NaTFSA–CsTFSA and charge–discharge properties of NaCrO$_2$ positive electrode at 423 K for a sodium secondary battery. *J. Power Sources* **2012**, *205*, 506–509. [CrossRef]
3. Khoo, T.; Somers, A.; Torriero, A.A.J.; MacFarlane, D.R.; Howlett, P.C.; Forsyth, M. Discharge behaviour and interfacial properties of a magnesium battery incorporating trihexyl(tetradecyl)phosphonium based ionic liquid electrolytes. *Electrochim. Acta* **2013**, *87*, 701–708. [CrossRef]
4. Kakibe, T.; Hishii, J.Y.; Yoshimoto, N.; Egashira, M.; Morita, M. Binary ionic liquid electrolytes containing organo-magnesium complex for rechargeable magnesium batteries. *J. Power Sources* **2012**, *203*, 195–200. [CrossRef]
5. Simons, T.J.; Howlett, P.C.; Torriero, A.A.J.; MacFarlane, D.R.; Forsyth, M. Electrochemical, transport, and spectroscopic properties of 1-ethyl-3-methylimidazolium ionic liquid electrolytes containing zinc dicyanamide. *J. Phys. Chem. C* **2013**, *117*, 2662–2669. [CrossRef]
6. Simons, T.J.; Torriero, A.A.J.; Howlett, P.C.; MacFarlane, D.R.; Forsyth, M. High current density, efficient cycling of Zn^{2+} in 1-ethyl-3-methylimidazolium dicyanamide ionic liquid: The effect of Zn^{2+} salt and water concentration. *Electrochem. Commun.* **2012**, *18*, 119–122. [CrossRef]
7. Lee, S.-Y.; Ogawa, A.; Kanno, M.; Nakamoto, H.; Yasuda, T.; Watanabe, M. Nonhumidified intermediate temperature fuel cells using protic ionic liquids. *J. Am. Chem. Soc.* **2010**, *132*, 9764–9773. [CrossRef] [PubMed]
8. Susan, M.A.B.H.; Noda, A.; Mitsushima, S.; Watanabe, M. Brønsted acid–base ionic liquids and their use as new materials for anhydrous proton conductors. *Chem. Commun.* **2003**, *8*, 938–939. [CrossRef]
9. Nakamoto, H.; Watanabe, M. Brønsted acid-base ionic liquids for fuel cell electrolytes. *Chem. Commun.* **2007**, *24*, 2539–2541. [CrossRef] [PubMed]
10. Yasuda, T.; Nakamura, S.-I.; Honda, Y.; Kinugawa, K.; Lee, S.-Y.; Watanabe, M. Effects of polymer structure on properties of sulfonated polyimide/protic ionic liquid composite membranes for nonhumidified fuel cell applications. *ACS Appl. Mater. Interfaces* **2012**, *4*, 1783–1790. [CrossRef] [PubMed]
11. Matsumoto, H.; Sakaebe, H.; Tatsumi, K.; Kikuta, M.; Ishiko, E.; Kono, M. Fast cycling of Li/LiCoO$_2$ cell with low-viscosity ionic liquids based on bis(fluorosulfonyl)imide [FSI]$^-$. *J. Power Sources* **2006**, *160*, 1308–1313. [CrossRef]
12. Ishikawa, M.; Sugimoto, T.; Kikuta, M.; Ishiko, E.; Kono, M. Pure ionic liquid electrolytes compatible with a graphitized carbon negative electrode in rechargeable lithium-ion batteries. *J. Power Sources* **2006**, *162*, 658–662. [CrossRef]
13. Tsai, W.Y.; Lin, R.Y.; Murali, S.; Zhang, L.L.; McDonough, J.K.; Ruoff, R.S.; Taberna, P.L.; Gogotsi, Y.; Simon, P. Outstanding performance of activated graphene based supercapacitors in ionic liquid electrolyte from −50 to 80 °C. *Nano Energy* **2013**, *2*, 403–411. [CrossRef]
14. Lin, R.Y.; Taberna, P.L.; Fantini, S.; Presser, V.; Perez, C.R.; Malbosc, F.; Rupesinghe, N.L.; Teo, K.B.K.; Gogotsi, Y.; Simon, P. Capacitive energy storage from −50 to 100 °C using an ionic liquid electrolyte. *J. Phys. Chem. Lett.* **2011**, *2*, 2396–2401. [CrossRef]
15. Lin, Z.; Taberna, P.-L.; Simon, P. Graphene-based supercapacitors using eutectic ionic liquid mixture electrolyte. *Electrochim. Acta* **2016**, *206*, 446–451. [CrossRef]
16. Balducci, A.; Bardi, U.; Caporali, S.; Mastragostino, M.; Soavi, F. Ionic liquids for hybrid supercapacitors. *Electrochem. Commun.* **2004**, *6*, 566–570. [CrossRef]
17. Bai, Y.; Cao, Y.; Zhang, J.; Wang, M.; Li, R.; Wang, P.; Zakeeruddin, S.M.; Gratzel, M. High-performance dye-sensitized solar cells based on solvent-free electrolytes produced from eutectic melts. *Nat. Mater.* **2008**, *7*, 626–630. [CrossRef] [PubMed]
18. Bai, Y.; Zhang, J.; Wang, Y.; Zhang, M.; Wang, P. Lithium-modulated conduction band edge shifts and charge-transfer dynamics in dye-sensitized solar cells based on a dicyanamide ionic liquid. *Langmuir* **2011**, *27*, 4749–4755. [CrossRef] [PubMed]

19. Zhang, M.; Zhang, J.; Bai, Y.; Wang, Y.; Su, M.; Wang, P. Anion-correlated conduction band edge shifts and charge transfer kinetics in dye-sensitized solar cells with ionic liquid electrolytes. *Phys. Chem. Chem. Phys.* **2011**, *13*, 3788–3794. [CrossRef] [PubMed]

20. Kawano, R.; Watanabe, M. Equilibrium potentials and charge transport of an I^-/I_3^- redox couple in an ionic liquid. *Chem. Commun.* **2003**, *3*, 330–331. [CrossRef]

21. Meng, T.; Young, K.; Beglau, D.; Yan, S.; Zeng, P.; Cheng, M.M.-C. Hydrogenated amorphous silicon thin film anode for proton conducting batteries. *J. Power Sources* **2016**, *302*, 31–38. [CrossRef]

22. Kreuer, K.-D.; Paddison, S.J.; Spohr, E.; Schuster, M. Transport in proton conductors for fuel-cell applications: Simulations, elementary reactions, and phenomenology. *Chem. Rev.* **2004**, *104*, 4637–4678. [CrossRef] [PubMed]

23. Yasuda, T.; Watanabe, M. Protic ionic liquids: Fuel cell applications. *MRS Bull.* **2013**, *38*, 560–566. [CrossRef]

24. Hoarfrost, M.L.; Tyagi, M.; Segalman, R.A.; Reimer, J.A. Proton hopping and long-range transport in the protic ionic liquid [Im][TFSI], probed by pulsed-field gradient NMR and quasi-elastic neutron scattering. *J. Phys. Chem. B* **2012**, *116*, 8201–8209. [CrossRef] [PubMed]

25. Galinski, M.; Lewandowski, A.; Stepniak, I. Ionic liquids as electrolytes. *Electrochim. Acta* **2006**, *51*, 5567–5580. [CrossRef]

26. ILCO Chemikalien GmbH. Ionic Liquid. Available online: http://www.ilco-chemie.de/downloads/Ionic%20Liquid.pdf (accessed on 24 October 2016).

27. Walsh, D.A.; Ejigu, A.; Smith, J.; Licence, P. Kinetics and mechanism of oxygen reduction in a protic ionic liquid. *Phys. Chem. Chem. Phys.* **2013**, *15*, 7548–7554. [CrossRef] [PubMed]

28. Nei, J.; Young, K.; Rotarov, D. Studies on MgNi-based metal hydride electrode with aqueous electrolytes composed of various hydroxides. *Batteries* **2016**, *2*, 27. [CrossRef]

29. Liao, X.; Yin, Z.; Young, K.; Nei, J. Studies in molybdenum/manganese content in the dual body-centered-cubic phases metal hydride alloys. *Int. J. Hydrog. Energy* **2016**, *41*, 15277–15286. [CrossRef]

30. Xu, W.; Angell, C.A. Solvent-free electrolytes with aqueous solution-like conductivities. *Science* **2003**, *302*, 422–425. [CrossRef] [PubMed]

31. Yoshizawa, M.; Xu, W.; Angell, C.A. Ionic liquids by proton transfer: Vapor pressure, conductivity, and the relevance of ΔpK_a from aqueous solutions. *J. Am. Chem. Soc.* **2003**, *50*, 15411–15419. [CrossRef] [PubMed]

32. Fraser, K.J.; Izgorodina, E.I.; Forsyth, M.; Scott, J.L.; Macfarlane, D.R. Liquids intermediate between "molecular" and "ionic" liquids: Liquid ion pairs? *Chem. Commun.* **2007**, *37*, 3817–3819. [CrossRef]

33. MacFarlane, D.R.; Forsyth, M.; Izgorodina, E.I.; Abbott, A.P.; Annat, G.; Fraser, K. On the concept of ionicity in ionic liquids. *Phys. Chem. Chem. Phys.* **2009**, *11*, 4962–4967. [CrossRef] [PubMed]

34. Bonhote, P.; Dias, A.; Papageorgiou, N.; Kalyanasundaram, K.; Gratzel, M. Hydrophobic, highly conductive ambient-temperature molten salts. *Inorg. Chem.* **1996**, *35*, 1168–1178. [CrossRef] [PubMed]

35. Yu, G.; Zhao, D.; Wen, L.; Yang, S.; Chen, X. Viscosity of ionic liquids: Database, observation, and quantitative structure-property relationship analysis. *AIChE J.* **2012**, *58*, 2885–2899. [CrossRef]

36. Rocha, M.A.A.; Neves, C.M.S.S.; Freire, M.G.; Russina, O.; Triolo, A.; Coutinho, J.A.P.; Santos, L.M.N.B.F. Alkylimidazolium based ionic liquids: Impact of cation symmetry on their nanoscale structural organization. *J. Phys. Chem. B* **2013**, *117*, 10889–10897. [CrossRef] [PubMed]

37. Wang, Y.T.; Voth, G.A. Unique spatial heterogeneity in ionic liquids. *J. Am. Chem. Soc.* **2005**, *127*, 12192–12193. [CrossRef] [PubMed]

38. Paul, A.; Kumar, P.; Samanta, A. On the optical properties of the imidazolium ionic liquids. *J. Phys. Chem. B* **2005**, *109*, 9148–9153. [CrossRef] [PubMed]

39. Lopes, J.N.A.C.; Padua, A.A.H. Nanostructural organization in ionic liquids. *J. Phys. Chem. B* **2006**, *110*, 3330–3335. [CrossRef] [PubMed]

40. Zhu, A.; Wang, J.; Han, L.; Fan, M. Measurements and correlation of viscosities and conductivities for the mixtures of imidazolium ionic liquids with molecular solutes. *Chem. Eng. J.* **2009**, *147*, 27–35. [CrossRef]

41. Yuan, X.; Xu, N. Determination of hydrogen diffusion coefficient in metal hydride electrode by cyclic voltammetry. *J. Alloy. Compd.* **2001**, *316*, 113–117. [CrossRef]

42. Dippel, T.; Kreuer, K.D.; Lassegues, J.C.; Rodriguez, D. Proton conductivity in fused phosphoric acid; A $^1H/^{31}P$ PFG-NMR and QNS study. *Solid State Ion.* **1993**, *61*, 41–46. [CrossRef]

43. Tliha, M.; Mathlouthi, H.; Khaldi, C.; Lamloumi, J.; Percheron-guegan, A. Electrochemical properties of the LaNi$_{3.55}$Mn$_{0.4}$Al$_{0.3}$Co$_{0.4}$Fe$_{0.35}$ hydrogen storage alloy. *J. Power Sources* **2006**, *160*, 1391–1394. [CrossRef]
44. Geng, M.; Feng, F.; Gamboa, S.A.; Sebastian, P.J.; Matchett, A.J.; Northwood, D.O. Electrocatalytic characteristics of the metal hydride electrode for advanced Ni/MH batteries. *J. Power Sources* **2001**, *96*, 90–93. [CrossRef]
45. Li, X.; Dong, H.; Zhang, A.; Wei, Y. Electrochemical impedance and cyclic voltammetry characterization of a metal hydride electrode in alkaline electrolytes. *J. Alloy. Compd.* **2006**, *426*, 93–96. [CrossRef]

batteries

MDPI

Article

Cell Performance Comparison between C14- and C15-Predomiated AB$_2$ Metal Hydride Alloys

Kwo-Hsiung Young [1,2,*], John M. Koch [2], Chubin Wan [3,4], Roman V. Denys [3,5] and Volodymyr A. Yartys [3,5]

[1] Department of Chemical Engineering and Materials Science, Wayne State University, Detroit, MI 48202, USA
[2] BASF/Battery Materials—Ovonic, 2983 Waterview Drive, Rochester Hills, MI 48309, USA; john.m.koch@basf.com
[3] Institute for Energy Technology, P.O. Box 40, Kjeller NO-2027, Norway; cbinwan@gmail.com (C.W.); roman.v.denys@gmail.com (R.V.D.); volodymyr.yartys@ife.no (V.A.Y.)
[4] Institute for Energy Technology, University of Science and Technology Beijing, 30 Xueyuan Rd., Haidian Dist., Beijing 100083, China
[5] Department of Chemical Engineering, Norwegian University of Science and Technology, Høgskoleringen 1, Trondheim NO-7491, Norway
* Correspondence: kwo.young@basf.com; Tel.: +1-248-293-7000

Academic Editor: Hua Kun Liu
Received: 14 July 2017; Accepted: 22 August 2017; Published: 25 September 2017

Abstract: The performance of cylindrical cells made from negative electrode active materials of two selected AB$_2$ metal hydride chemistries with different dominating Laves phases (C14 vs. C15) were compared. Cells made from Alloy C15 showed a higher high-rate performance and peak power with a corresponding sacrifice in capacity, low-temperature performance, charge retention, and cycle life when compared with the C14 counterpart (Alloy C14). Annealing of the Alloy C15 eliminated the ZrNi secondary phase and further improved the high-rate and peak power performance. This treatment on Alloy C15 showed the best low-temperature performance, but also contributed to a less-desirable high-temperature voltage stand and an inferior cycle stability. While the main failure mode for Alloy C14 in the sealed cell is the formation of a thick oxide layer that prevents gas recombination during overcharge and consequent venting of the cell, the failure mode for Alloy C15 is dominated by continuous pulverization related to the volumetric changes during hydride formation and hysteresis in the pressure-composition-temperature isotherm. The leached-out Mn from Alloy C15 formed a high density of oxide deposits in the separator, leading to a deterioration in charge retention performance. Large amounts of Zr were found in the positive electrode of the cycled cell containing Alloy C15, but did not appear to harm cell performance. Suggestions for further composition and process optimization for Alloy C15 are also provided.

Keywords: metal hydride; nickel metal hydride batteries; Laves phase alloy; electrochemistry; synergetic effects

1. Introduction

Nickel/metal hydride (Ni/MH) battery technology is important for future transportation [1] and stationary energy storage applications [2]. Using Laves phase-based AB$_2$ metal hydrides (MHs) as the active material in the negative electrode can increase the gravimetric energy of a Ni/MH battery [3]. Two different Laves phases MH alloys are available for the electrochemical applications, specifically a C14 with a hexagonal crystal structure and a C15 with a face-centered-cubic crystal structure. Both structures have the same number of tetrahedral hydrogen occupation sites per AB$_2$ formula, which are indicted by yellow (A$_2$B$_2$), blue (AB$_3$), and pink (B$_4$) tetrahedrons in Figure 1.

Comparisons in crystal structure, hydrogen-storage (H-storage) characteristics in the gaseous phase (GP), and electrochemical (EC) properties of these two state-of-art representatives were reported in a separate paper [4]. It was suggested that the C14-predominated MH alloy was more suitable for high-capacity and long-life applications, while the C15-predominated MH alloy can be used in areas requiring improved high-rate (HR) and low-temperature (LT) performances [5]. In order to verify the prediction from the previous half-cell study, complete sealed cells (C-size) were made from a C14- and a C15-predominated MH alloys and their electrochemical performances were evaluated and compared in this study.

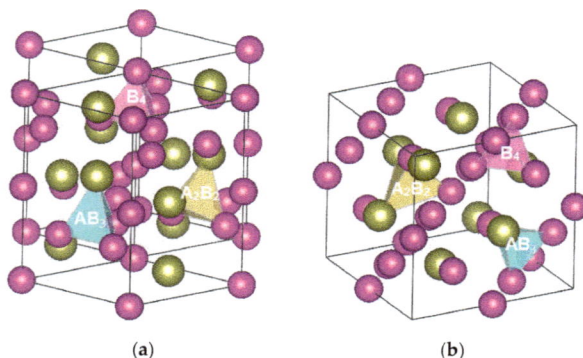

(a)   (b)

Figure 1. Schematics of (**a**) a C14 and (**b**) a C15 crystal structures. Yellow, blue, and pink tetrahedrons are the A_2B_2, AB_3, and B_4 hydrogen occupation sites, respectively.

Two compositions, $Zr_{21.5}Ti_{12.0}V_{10.0}Cr_{7.5}Mn_{8.1}Co_{8.0}Ni_{32.2}Sn_{0.3}Al_{0.4}$ (Alloy C14) and $Zr_{25.0}Ti_{6.5}V_{3.9}Mn_{22.2}Fe_{3.8}Ni_{38.0}La_{0.3}$ (Alloy C15), were selected for this comparative study. Since the annealing effects on the C14 MH alloys were well studied and previously reported, only part of the ingot from the Alloy C15 was annealed at 960 °C for 3 h (Alloy C15A). The structures, and GP and EC H-storage characteristics of these three alloys (C14, C15, and C15A) have been studied before [4] and the results are summarized in Table 1. The Alloy C14 has a C14 major phase and two secondary phases (C15 and TiNi), while Alloy C15 is predominately C15 with a ZrNi secondary phase, which was eliminated upon annealing and formation of Alloy C15A, and is very close to a single-phase alloy according to X-ray diffractometer (XRD) studies. The GP full capacities of Alloy C14 and Alloy C15 are similar, and that in C15A is much reduced, due to the lack of a beneficial secondary phase to create synergetic effects [6]. The hydrogen equilibrium plateau pressures of Alloy C15 and Alloy C15A are one order of magnitude higher than that of Alloy C14, which may cause an early venting in a sealed cell. The hysteresis in the pressure-composition-temperature (PCT) isotherms of Alloy C14 is much smaller than those in Alloy C15 and Alloy C15A, which contributes to a better mechanical integrity during pulverization [7]. The EC capacities of Alloy C15 and Alloy C15A are considerably lower than that of Alloy C14, due to the open-air configuration in the half-cell measuring apparatus, where the MH alloy cannot be charged much higher than 0.1 MPa (one atmosphere). The high-rate dischargeability (HRD) of Alloy C15 is higher than that of Alloy C14, due to a higher surface exchange current (I_0). Annealing of C15A reduced both the diffusion coefficient (D) and I_0, compared to Alloy C15. The relatively low I_0 of Alloy C15A is the result of the reduced amount of catalytic Ni-inclusions [8] on the surface, due to annealing, as judged by the relatively low saturated magnetic susceptibility (M_S).

Table 1. Summary of properties of alloys used in this study [4]. GP and EC denote gaseous phase and electrochemistry, respectively. HRD, D, I_o, and M_S represent high-rate dischargeability, diffusion constant, surface exchange current, and saturated magnetic susceptibility. XRD: X-ray diffraction.

Properties	Alloy C14	Alloy C15	Alloy C15A
Composition	$Zr_{21.5}Ti_{12.0}V_{10.0}Cr_{7.5}Mn_{8.1}Co_{8.0}$ $Ni_{32.2}Sn_{0.3}Al_{0.4}$	$Zr_{25.0}Ti_{6.5}V_{3.9}Mn_{22.2}Fe_{3.8}$ $Ni_{38.0}Sn_{0.3}La_{0.3}$	Same as C15
Preparation	Vacuum induction melting	Vacuum induction melting	Vacuum induction melting
Annealing	No	No	960 °C for 6 h
Minor phases	C15 (5.2%) + TiNi (1.2%)	ZrNi (0.7%)	Not detectable by XRD
GP full capacity	1.45%	1.46%	0.95%
GP reversible capacity	1.32%	1.44%	0.94%
GP desorption pressure (MPa)	0.078	0.87	0.90
GP hysteresis	0.04	0.13	0.31
EC capacity (mAh·g^{-1})	354	311	277
EC HRD	0.90	0.99	0.98
D (10^{-10}·cm^{2}s^{-1})	2.5	2.4	1.6
I_o (mA·g^{-1})	22.5	46.8	39.4
M_S (memu·g^{-1})	37	42	17

2. Experimental Setup

Cylindrical Ni/MH batteries (C-size) were assembled for electrochemical testing. A photograph of cells before and after closing is shown in Figure 2a. MH alloy powder was attained from an induction melting-prepared ingot (2 kg batch size) with hydrogenation and mechanical grinding to a −200 mesh size. The MH alloy powder was dry-compacted onto a nickel mesh substrate by a compaction mill and formed into negative electrodes, while the counter positive electrode was pasted with a mixture of 89% standard AP50 [9] $Ni_{0.91}Co_{0.045}Zn_{0.045}(OH)_2$ (BASF-Ovonic, Rochester Hills, MI, USA), 5 wt% Co, and 6% CoO powders into a nickel foam substrate. Photographic examples of both electrodes are shown in Figure 2b. Scimat 700/79 acrylic acid grafted polypropylene/polyethylene separators were used (Freudenberg Group, Weinheim, Germany). The negative-to-positive capacity ratio (N/P) was set at 1.3 to 1.4 to maintain a good balance between the over-charge and over-discharge reservoirs [10]. A 30% KOH solution with LiOH (1.5 wt%) additive was used as the electrolyte. After the electrodes/separator were coiled into a jelly roll bundle, the bundle was inserted into the empty can, and the cell assembly was filled with electrolyte to the top then allowed to rest for 30 min before the un-absorbed electrolyte was sucked out through a plastic pipette. The amount of electrolyte absorbed was calculated from the weight difference between the before and after electrolyte filling steps. Twenty cells were built with each alloy for use in benchmark performance testing. An electrochemical formation process was performed using a Maccor Battery Cycler (Maccor, Tulsa, OK, USA) after the batteries were sealed, and was composed of six cycles of charging at a 0.1C rate to 50%, 100%, 120%, 150%, 150%, and 150% of the calculated capacity and a 0.2 C discharge rate (except the sixth cycle using a 0.5 C discharge rate) to a cutoff voltage of 0.9 V.

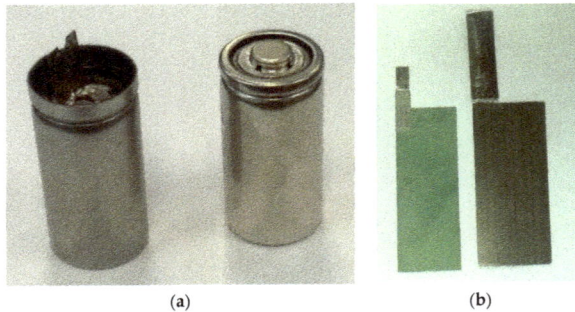

(a)   (b)

Figure 2. Photographs of (**a**) C-sized cells before (left) and after (right) closing; and (**b**) a pasted positive electrode (left) and a compacted negative electrode (right).

Cells from each alloy were cycled for 20 cycles and the peak power of the cells was measured for each cycle at a 50% depth of discharge (DOD) using a pulse-discharge method. In each measurement, the cell was discharged first at a rate of C/3 to 50% DOD, followed by a 2/3C pulse for a period of 30 s. The resistivity was calculated by dividing the difference in voltages with the different in currents. Peak power was obtained by the formula:

$$\text{Peak power} = 1/4 \, V_{oc}^2/R \tag{1}$$

where the V_{oc} and R represent open-circuit voltage (electromotive force) and internal resistance, respectively. Charge retention was measured at the end of the 7th, 14th, and 30th days at room temperature (RT) after full charging at a 0.1 C rate in the beginning of the test. The remaining capacity after each testing period was discharged at a rate of 0.2 C and was normalized to a 0.2 C discharge capacity before the charge retention experiment. The shelf life in each cell was measured by placing a 100% state-of-charge (SOC) battery (charged at 0.2 C) in an oven at 45 °C and recording the V_{oc} decay every 15 days until the V_{oc} dropped to 0 V. The cells were cooled to RT to test the capacity loss after three cycles of charge to 150% SOC at 0.1 C, followed by discharge to 0.9 V at 0.2 C. The capacity at cycle 3 was compared with the original formation capacity to determine the charge retention.

The cycle life of each cell was tested under a 0.5 C rate charge/discharge cycling at RT with a discharge cutoff voltage of 0.9 V. The charging process was terminated using a $-\Delta V$ method, specifically a 3 mV (or 5 mV) voltage drop from the maximum cell voltage. The voltage drop indicated the end of the charging process, when oxygen began to evolve at the positive electrode. The recombination of oxygen from the positive electrode with hydrogen from the negative electrode released heat and caused the voltage to drop, in accordance with the Nernst equation. End of life was reached when the measured capacity dropped below 50% of the initial capacity immediately after activation.

For the failure mechanism analysis, a JEOL-JSM6320F scanning electron microscope (SEM, JEOL, Tokyo, Japan) with energy dispersive spectroscopy (EDS) capability was used to study the morphology and composition of the electrodes after cycling. A Philips X'Pert Pro X-ray diffractometer (XRD, Philips, Amsterdam, The Netherlands) was used to study the crystal structure of the negative electrode after cycling. A JEOL JSM7100 field-emission SEM with EDS capability was used to map metal oxides that migrated into the separator.

3. Results and Discussion

Cylindrical cells (C-size) were made with Alloys C14, C15, and C15A. The thickness and weight of both electrodes of the cells used for the capacity measurement are listed in Table 2. The weight of the negative electrode in the cell with Alloy C14 (cell C14) is smaller due to consideration for its relatively higher EC capacity (Table 1) and match in N/P ratio. The amount of electrolyte in the cell containing

Alloy C15 (cell C15) is approximately 1 cc larger than the other two cells, due to the consumption of electrolyte from the oxidation of alloy surface in the electrolyte-filling process. In the following sections, a few key battery performances (HR, LT, charge retention, peak power, and cycle life) of cells C14, C15, and C15A are compared.

Table 2. Summary of cell builds and performance results. LT-Cap was obtained at $-10\ °C$ with a C/2 rate to a cutoff voltage of 0.9 V.

Properties	Cell C14	Cell C15	Cell C15A
Negative electrode thickness (mm)	0.328	0.322	0.318
Weight (g)	22.8	26.8	25.3
Positive electrode thickness (mm)	0.840	0.775	0.778
Weight (g)	27.0	26.7	26.8
Electrolyte amount (cc)	6.82	7.98	6.86
N/P	1.3	1.4	1.4
RT-capacity @C/5 (Ah)	4.88	4.74	4.77
RT-capacity @C/2 (Ah)	4.71	4.67	4.71
RT-capacity @C (Ah)	4.57	4.64	4.69
RT-capacity @2C (Ah)	3.87	3.66	4.24
LT-capacity @C/2 (Ah)	3.29	0.19	3.69
Energy density (Wh·kg^{-1})	73	73	75
Energy density (Wh·l^{-1})	241	236	239
Change retention @ 30 days	35%	54%	56%
Maximum peak power (W·kg^{-1})	157	151	176
Cycle life- C/2	350	305	255
Cycle life- C	180	100	60

3.1. Room-Tempearture Capacities with Various Discharge Rates

The discharge voltage profiles measured at RT and under four different discharge rates (C/5, C/2, C, and 2C) from cells C14, C15, and C15A are shown in Figure 3a–c, respectively. In the middle of the polarization curve (ohmic region) the cell voltage (V) is related to the discharge current (i) and V_{oc} by the formula:

$$V = V_{oc} - iR \qquad (2)$$

where R is the internal resistance. V decreases with increasing i, which explains the lower voltage profile with higher rate seen for each of the plots in Figure 3. V_{oc} in cells containing Alloy C14 is expected to be lower than those in cells containing either the C15 or C15A alloy, due to the relatively low plateau pressure of Alloy C14 and the connection through Nernst equation [11]:

$$E_{MH,\ eq}\ (\text{vs. Hg/HgO, in V}) = -0.934 - 0.029 \log P(H_2)\ (\text{in bar}) \qquad (3)$$

where $E_{MH,\ eq}$ is the equilibrium potential of H-storage electrode vs. the Hg/HgO reference electrode, and $P(H_2)$ is the equilibrium hydrogen gas pressure. An order of magnitude lower $P(H_2)$ for C14 results in a 29 mV voltage decrease. The voltage suppression with the increasing discharge current in cell C14 is worse than those from cells C15 and C15A, which correlated well with the low HRD in Alloy C14 found in the half-cell measurement [4]. The unstable voltage profile in the 2 C discharge curve of cell C15 shown in Figure 3b was from the competition between better active material utilization and lower voltage due to the increase in the cell temperature from the high-rate discharge. Capacities obtained with a cutoff voltage of 0.8 V for cells with all three alloys are listed in Table 2. Cell C14 showed the highest capacity measured at a C/5 rate, due to its relatively high loading in the positive electrode and its higher capacity. Cell C15A showed the highest capacity measured at the 2 C rate. The gravimetric and volumetric energy densities of these three cells were calculated by their capacity, weight, and volume, and these are listed in Table 2. Cell C15A has the highest gravimetric energy density due to its higher voltage and cell C14 has the highest volumetric energy density due to the largest loading in the positive electrode.

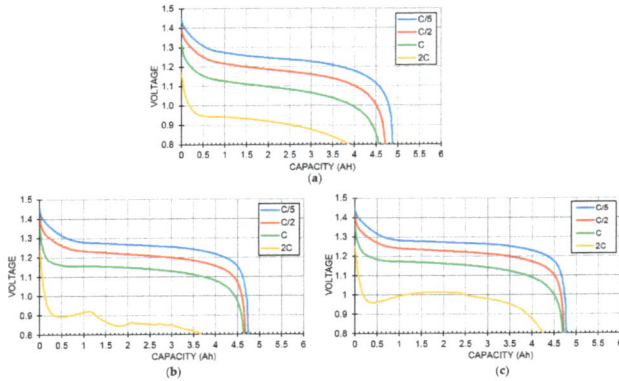

Figure 3. Room-temperature discharge voltage curves with different rates (C/5, C/2, C, and 2C) for cells (**a**) C14; (**b**) C15; and (**c**) C15A.

3.2. Low-Temperature (−10 °C) Test

The low-temperature (LT) performances were evaluated at −10 °C using a C/2 rate discharge with a cut voltage of 0.9 V and results are listed in Table 2. Cell C15A shows the highest capacity in this run, due to its relatively high voltage, and cell C15 is the worst. Both cells C15 and C15A have higher V_{oc}. However, the R in cell C15 is higher and, thus, contributed to worse HR and LT results. The comparison results in LT are the same as for HR, which is very common in Ni/MH battery studies [12]. The improvement in LT performance of Alloy C15A is not consistent with the decreased amount of embedded Ni on the activated surface and the lower surface reaction current, as previously highlighted [4]. This deviation in performance corresponds to the higher pulverization rate of Alloy C15A and the subsequent rapid increase in surface area, which contributes to the excellent HR and LT performances.

3.3. Charge Retention

Charge retention performances were compared in both RT storage and high temperature (HT, 45 °C) voltage stand measurements and the results are plotted in Figure 4. Although Alloy C14 has a relatively higher V-content (10% vs. 3.9% in Alloy C15), which is a major source of self-discharge [13], cell C14 showed the best RT charge retention result, with cell C15A showing a marginally better change retention than cell C15 did. Results from the high-temperature voltage stand are slightly different. While cell C14 showed the best HT voltage stand, which is consistent with the charge retention result, cell C15A exhibited a much worse HT voltage stand, compared to that from cell C15. The cause of this performance variation will be discussed in the failure analysis session.

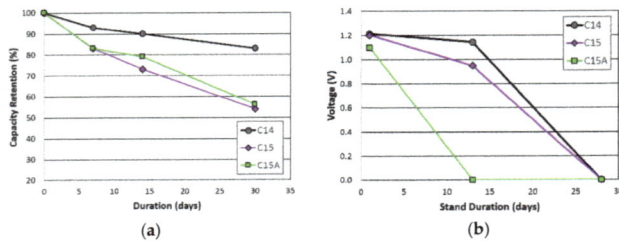

Figure 4. (**a**) Room temperature charge retentions; and (**b**) 45 °C voltage stand for cells C14, C15, and C15A.

3.4. Peak Power

Peak powers at RT for the first 20 cycles were measured and results are plotted in Figure 5. In this test, the cells did not first go through the normal six-cycle activation process. In the peak power comparison, cell C14 quickly reached a maximum while cells C15 and C15A were comparatively slower. However, at the 20th cycle, cell C15A demonstrates the highest peak power, which correlates well with its high HR performance.

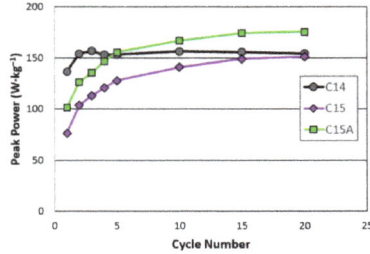

Figure 5. Room-temperature peak powers for cells C14, C15, and C15A.

3.5. Cycle Life

Cycle life performances were compared with two testing schemes: a regular one consisting of C/2 charge/C/2 discharge with a narrower $-\Delta V$ cutoff set at 0.3 mV and a fast one with C charge/C discharge with a slightly wider $-\Delta V$ at 0.5 mV. These are shown in Figure 6a,b, respectively. Both tests were conducted at RT. Cell C14 showed the best cycle stability due to the balanced designed in composition. Cell C15A showed the least cycle stability due to a lack of both a Cr and TiNi phase. The sudden drop in capacity from cell C15A originated from the venting of the cell, causing electrolyte dry-out and an increase in R. Alloy C15A has the highest hydrogen equilibrium pressure in the PCT measurement, which causes the pressure to reach the venting pressure of the safety valve (2.8 MPa).

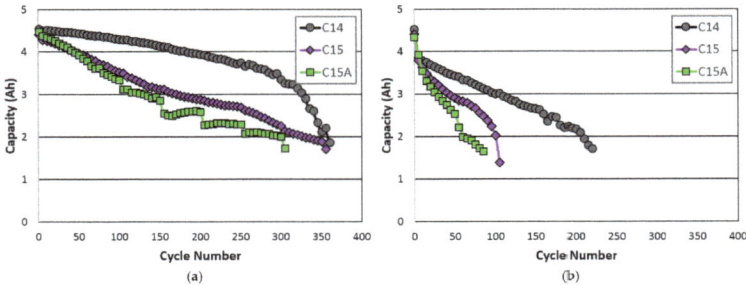

Figure 6. Room-temperature cycle life from cells C14, C15, and C15A with (**a**) a C/2 charge, a C/2 discharge, a $-\Delta V$ cutoff voltage of 3 mV and (**b**) a C charge, a C discharge, and a $-\Delta V$ cutoff voltage of 5 mV.

3.6. Failure Analysis of C15 and C15A

The failure mechanism of the V-containing C14 MH alloy in Ni/MH batteries was earlier studied in detail, and the conclusions can be summarized as follows: a relatively thick oxide impeded the O_2 gas-recombination during over-charge, causing venting and an increase in internal resistance [14]. Only failure analyses from Alloy C15 and Alloy C15A are discussed in this paper and compared to those results previously discovered for Alloy C14. Cells reaching the end of the C/2-C/2 cycling test (Figure 6a) were torn apart and the positive and negative electrodes, together with the separator, were studied. XRD patterns of the cycled positive electrodes show only a β-Ni(OH)$_2$ phase (Figure 7). Two SEM backscattering electron images (BEI) of the cycled positive electrodes are shown in Figure 8. No fractures or evidence of pulverization can be seen, which is in agreement with the XRD analysis. Severe pulverization due to lattice expansion coming from the β–α transition in the MH alloy containing Al, an α-Ni(OH)$_2$ promoter [9], was not observed here [15]. EDS analysis was taken in a few locations in each SEM micrograph and the results are summarized in Table 3. In both cases, incompletely dissolved Co (Spot 1) and CoO (Spot 2) added particles can be seen and larger amounts of Zr (leached from negative electrode) are found penetrating the spherical particles. Zr does not seem to promote the α-Ni(OH)$_2$ formation and the Co-conductive network remains intact in both cases. From the evidence collected so far, we reach the conclusion that the positive electrode is not the main course of the capacity degradation in cells with C15 and C15A MH alloys.

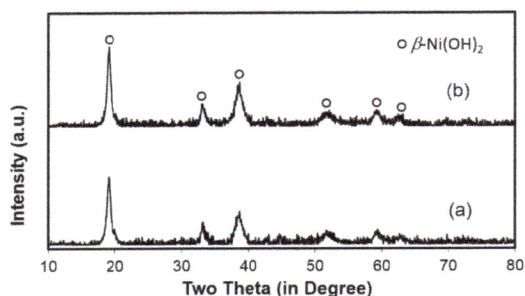

Figure 7. XRD patterns of cycled positive electrodes from (**a**) cell C15 and (**b**) cell C15A.

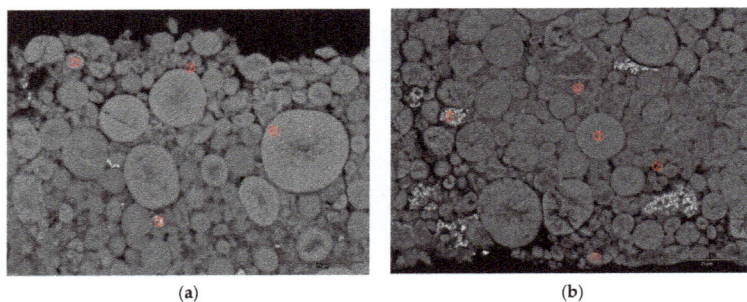

Figure 8. SEM-BEI micrographs of cycled positive electrodes from (**a**) cell C15 and (**b**) cell C15A. The scale bar at the right lower corner represents 25 microns.

Table 3. EDS results of selected areas in the cycled positive electrodes in Figure 9. Values are in at%.

Location	Ni	Co	Zn	Ti	Zr	Mn	Comment
Figure 8a-1	3.8	95.9	-	-	0.3	-	Co
Figure 8a-2	33.5	60.2	0.8	0.4	5.1	-	CoO
Figure 8a-3	86.4	9.2	2.1	0.4	1.8		Ni(OH)$_2$
Figure 8a-4	85.8	5.8	3.2	0.1	4.8	0.3	Ni(OH)$_2$
Figure 8b-1	2.9	96.9	0.2	-	-	-	Co
Figure 8b-2	43.6	52.6	1.1	-	2.6	-	CoO
Figure 8b-3	87.7	5.8	2.9	-	3.5	-	Ni(OH)$_2$
Figure 8b-4	84.2	2.4	2.5	-	10.8	-	Ni(OH)$_2$
Figure 8b-5	87.1	5.5	2.9	-	1.0	3.5	Ni(OH)$_2$

XRD patterns of the cycled negative electrodes from both cells (C15 and C15A) (Figure 9) show no evidence of oxide nor microcrystalline metallic nickel (common products of oxidizing MH alloy [16]). SEM-BEI micrographs of the cycled negative electrodes from cells C15 and C15A are shown in Figure 10. Both pictures show high degrees of pulverization. The particles in the cycled negative electrode of cell C15A are finer than those from cell C15. In addition, more electrolyte was absorbed into the negative electrode of cell C15A due to the higher surface area, which is possibly related to the pre-dry out of the separator and frequent cell venting (sudden drops in the capacity in Figure 6). The amount of oxides on the newly-formed Alloy C15 and Alloy C15A surfaces are very small compared to Alloy C14 [14]. From the comparison of the two compositions (Alloy C14 and Alloy C15), formula C15 has a high Mn content (high solubility in KOH [17]) and 0.3% La (beneficial for activation and increased surface area [18]), but no Cr (highly corrosion resistant [12,19]) and, consequently, a higher leaching rate. The chemical compositions of three locations in each SEI micrograph were analyzed by EDS and results are listed in Table 4. In the negative electrode from a cycled cell C15 (Figure 10a), the chemical composition of the large piece (Spot 1) does not vary significantly from the broken pieces (Spot 2) and a new oxide phase with very high Mn-content (Spot 3) is formed on the edge (near the separator). In the negative electrode from cycled cell C15A (Figure 10b), the chemical compositions are similar between large pieces (Spot 1) and broken ones (Spot 3). An area with a composition close to Zr_7Ni_{10} (Spot 2) retained the same shape after the cycling (no pulverization).

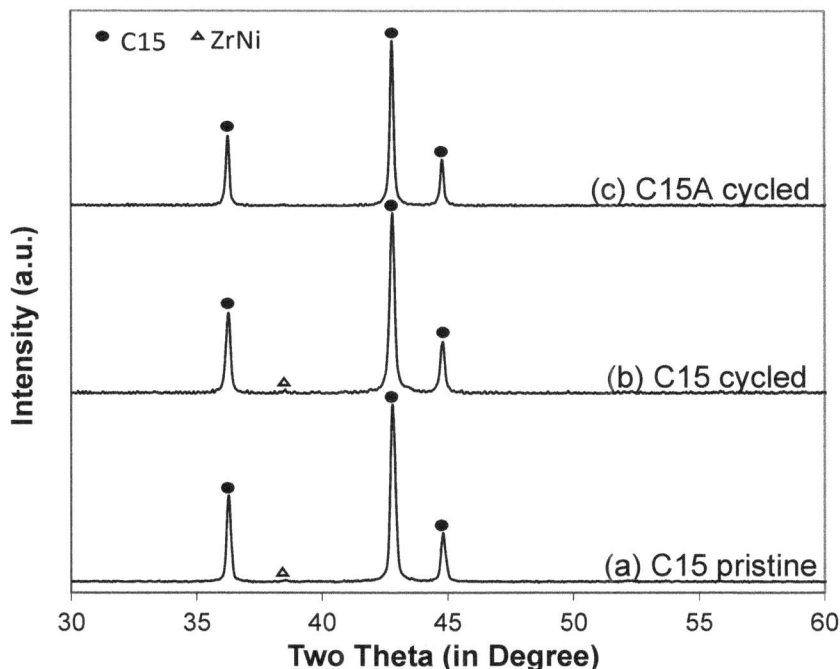

Figure 9. XRD patterns of (**a**) pristine C15 and cycled negative electrodes from (**b**) cell C15 and (**c**) cell C15A.

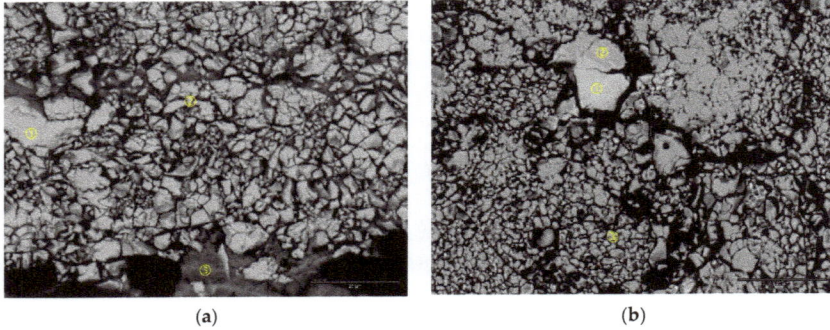

(a) (b)

Figure 10. SEM-BEI micrographs of cycled negative electrodes of (**a**) cell C15 and (**b**) cell C15A. The scale bar at the right lower corner represents 25 microns.

Table 4. EDS results of selected areas in the cycled negative electrodes shown in Figure 11. Values are in at%.

Location	Ti	Zr	V	Mn	Fe	Ni	La	B/A
Figure 10a-1	6.9	24.5	3.5	21.3	3.6	40.2	-	2.18
Figure 10a-2	5.5	24.2	4.4	24.7	5.0	36.2	-	2.36
Figure 10a-3	4.3	17.3	11.0	61.8	9.7	3.4	2.4	3.51
Figure 10b-1	7.5	25.1	4.3	17.5	4.8	40.7	0.1	2.06
Figure 10b-2	8.1	33.4	0.4	1.5	0.4	56.1	-	1.40
Figure 10b-3	6.3	25.4	3.7	22.4	4.1	38.1	-	2.16

The separator areas in the cycled cells were studied by SEM-EDS mapping and the results are present in Figure 11 (cell C15) and Figure 12 (cell C15A). A significant number of high Mn-content deposits are found in both cells. These Mn-rich deposits are the source of micro-shortening, due to their semiconductor nature [20,21], and cause a relatively inferior charge retention performance when compared to that of cell C14. The density of these Mn-rich oxides in the cycled C15 cell is higher than that in cell C15A and may explain the worse charge retention result of cell C15, when compared to cell C15A. The inferior HT voltage stand of cell C15A is more related to the higher pulverization of Alloy C15A, which expedites the degree of MH alloy oxidation in HT when compared to that of cell C15.

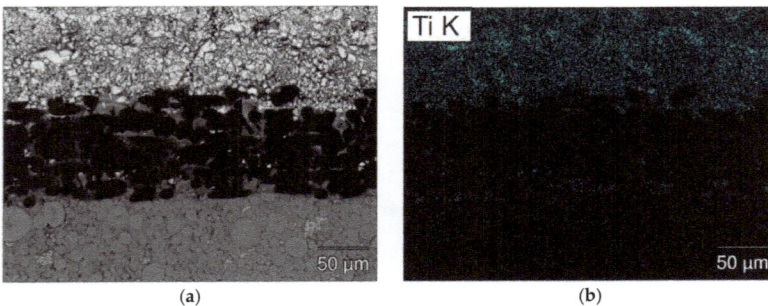

(a) (b)

Figure 11. *Cont.*

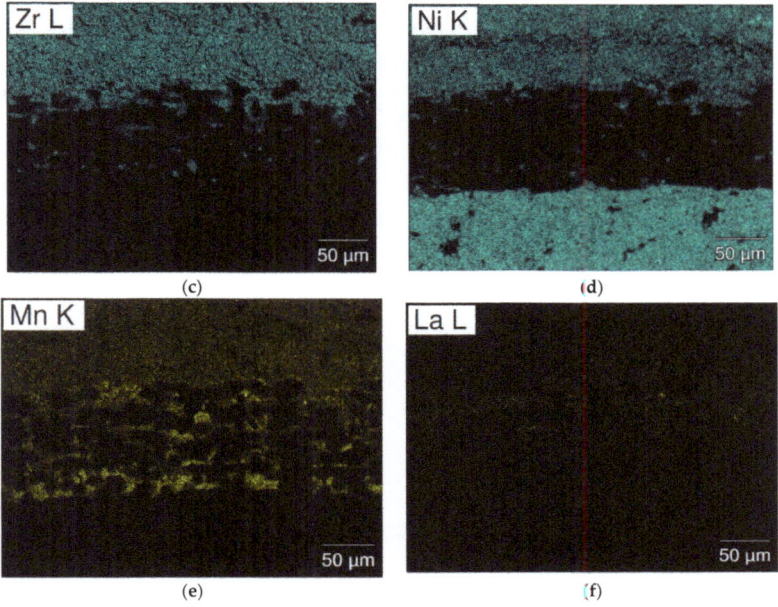

Figure 11. SEM (**a**) BEI image and EDS mappings of (**b**) Ti; (**c**) Zr; (**d**) Ni; (**e**) Mn; and (**f**) La for the separator from cycled cell C15. The upper side and lower sides are the negative and positive electrodes, respectively.

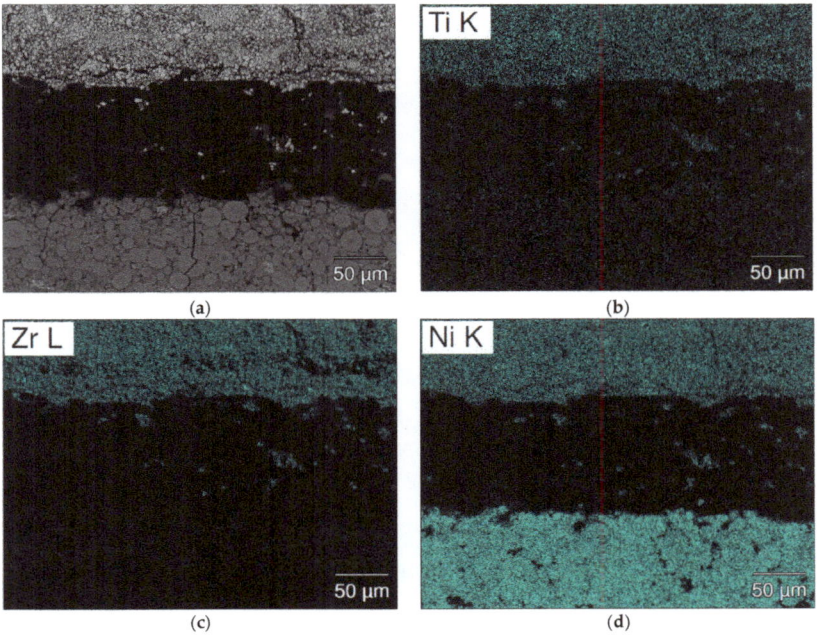

Figure 12. *Cont.*

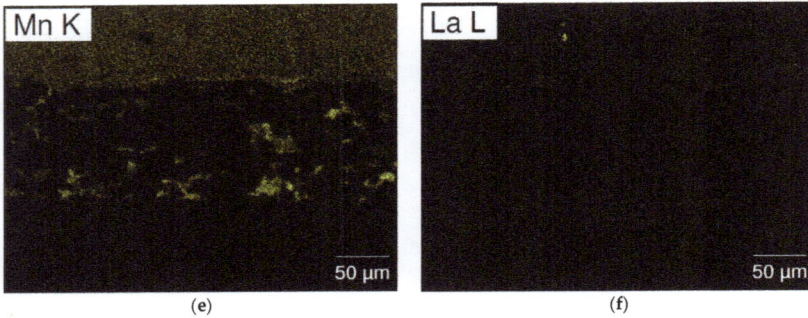

Figure 12. SEM (**a**) BEI image and EDS mappings of (**b**) Ti; (**c**) Zr; (**d**) Ni; (**e**) Mn; and (**f**) La for the separator from cycled cell C15A. The upper side and lower sides are the negative and positive electrodes, respectively.

3.7. Performance Comparison

The differences in performance of the cells made from three alloys are best represented by the radar plot shown in Figure 13. In agreement with the half-cell results presented previously [4], cell C14 showed the best cycle life and charge retention performances, while cell C15A gave the best HR and LT results. The superiority in HR performance of C15, relative to C14, has been reported many times [5,22–26]. However, the HR performance improvement in the Alloy C15 was reported here for the first time. The poor charge retention and cycle life of cell C15A could be improved by increasing the Cr and Ni-content [19], reducing the annealing condition to preserve some of the Zr_xNi_y phases, increasing the Zr/Ti ratio to lower the plateau pressure and to prevent early venting [27–29], and slowing down pulverization by reducing the PCT hysteresis through composition and annealing condition adjustments [7,19,30–33].

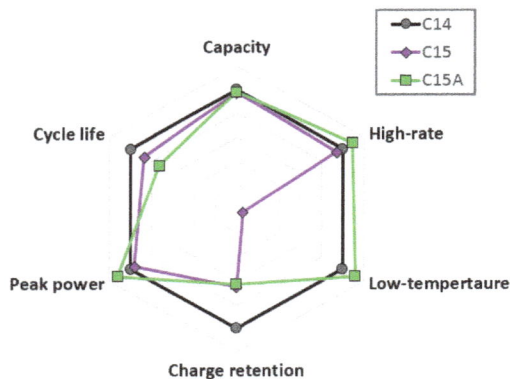

Figure 13. Key battery performance comparisons among cells C14, C15, and C15A.

4. Conclusions

The electrochemical performances of Ni/MH batteries using as-cast C14, as-cast C15, and annealed C15 AB_2 metal hydride alloys were compared. Alloys C14 and C15 with different chemical compositions were composed of mainly C14 and C15 phases, respectively. The pulverization rates with cycling exhibited the trend of Alloy C15A (annealed) > Alloy C15 (as-cast) > Alloy C14 (as-cast), which made cells with Alloy C15A render the best high-rate and low-temperature performance in the early stage of service life. However, the high pulverization rate also resulted in a steady capacity degradation, inferior cycle life, and high-temperature voltage stand. Large amounts of high-Mn oxide deposits were found in the separator of cells with Alloy C15 and Alloy C15A, and can be assumed to deteriorate the charge retention capabilities of these cells. In addition, the high plateau pressure of Alloy C15A caused ventings during the cycle life testing, which should be preventable through composition and/or process adjustments.

Acknowledgments: The authors would like to thank the following individuals from BASF-Ovonic for their assistance: Taihei Ouchi, Jean Nei, Suiling Chen, Cheryl Setterington, Nathan English, David Pawlik, Allen Chan, and Ryan J. Blankenship. The work is related to the collaboration between IFE and BASF on the project MoZEES, funded by the Norwegian Research Council.

Author Contributions: Kwo-Hsiung Young designed the experiments. John M. Koch performed the experiments. Chubin Wan, Roman V. Denys, and Volodymyr A. Yartys analyzed the results and prepared the manuscript.

Conflicts of Interest: The authors declare no conflict of interest.

Abbreviations

The following abbreviations are used in this manuscript:

Ni/MH	Nickel/metal hydride
MH	Metal hydride
H-storage	Hydrogen-storage
GP	Gaseous phase
EC	Electrochemical
HR	High rate
HT	High temperature
LT	Low temperature
PCT	Pressure-composition-temperature
HRD	High-rate dischargeability
I_o	Surface exchange current
D	Diffusion constant
M_S	Saturated magnetic susceptibility
XRD	X-ray diffraction
N/P	Negative electrode to positive electrode
DOD	Depth of discharge
V_{oc}	Open-circuit voltage
R	Internal resistance
RT	Room temperature
SOC	State-of-charge
SEM	Scanning electron microscope
EDS	Energy dispersive spectroscopy
V	Cell voltage
i	Discharge current
BEI	Backscattered electron image

References

1. Young, K.; Ng, K.Y.S.; Bendersky, L.A. A technical report of the Robust Affordable Next Generation Energy Storage System-BASF program. *Batteries* **2016**, *2*, 2. [CrossRef]
2. Zelinsky, M.; Koch, J. Batteries and Heat—A Recipe for Success? Available online: www.battcon.com/PapersFinal2013/16-Mike%20Zelinsky%20-%20Batteries%20and%20Heat.pdf (accessed on 17 May 2017).
3. Fetcenko, M.A.; Ovshinsky, S.A.; Young, K.; Reichman, B.; Fierro, C.; Koch, J.; Martin, F.; Mays, W.; Ouchi, T.; Sommers, B.; et al. High catalytic activity disordered VTiZrNiCrCoMnAlSn hydrogen storage alloys for nickel–metal hydride batteries. *J. Alloy. Compd.* **2002**, *330*, 752–759. [CrossRef]

4. Young, K.; Nei, J.; Wan, C.; Denys, R.V.; Yartys, V.A. Comparison of C14- and C15-predominated AB$_2$ metal hydride alloys for electrochemical applications. *Batteries* **2017**, *3*, 22. [CrossRef]
5. Young, K.; Nei, J.; Ouchi, T.; Fetcenko, M.A. Phase abundances in AB$_2$ metal hydride alloys and their correlations to various properties. *J. Alloy. Compd.* **2011**, *509*, 2277–2284. [CrossRef]
6. Young, K.; Ouchi, T.; Meng, T.; Wong, D.F. Studies on the synergetic effects in multi-phase metal hydride alloys. *Batteries* **2016**, *2*, 15. [CrossRef]
7. Young, K.; Ouchi, T.; Fetcenko, M.A. Pressure-composition-temperature hysteresis in C14 Laves phase alloys: Part 1. Simple ternary alloys. *J. Alloy. Compd.* **2009**, *480*, 428–433. [CrossRef]
8. Young, K.; Huang, B.; Regmi, R.K.; Lawes, G.; Liu, Y. Comparisons of metallic clusters imbedded in the surface of AB$_2$, AB$_5$, and A$_2$B$_7$ alloys. *J. Alloy. Compd.* **2010**, *506*, 831–840. [CrossRef]
9. Young, K.; Wang, L.; Yan, S.; Liao, X.; Meng, T.; Shen, H.; Mays, W.C. Fabrications of high-capacity alpha-Ni(OH)$_2$. *Batteries* **2017**, *3*, 6. [CrossRef]
10. Young, K.; Wu, A.; Qiu, Z.; Tan, J.; Mays, W. Effects of H$_2$O$_2$ addition to the cell balance and self-discharge of Ni/MH batteries with AB$_5$ and A$_2$B$_7$ alloys. *Int. J. Hydrogen Energy* **2012**, *37*, 9882–9891. [CrossRef]
11. Kleperis, J.; Wójcik, G.; Czerwinski, A.; Showronski, J.; Kopczyk, M.; Bełtowska-Brzezinska, M. Electrochemical behavior of metal hydride. *J. Solid State Electrochem.* **2001**, *5*, 229–249. [CrossRef]
12. Young, K.; Ouchi, T.; Fetcenko, M.A. Roles of Ni, Cr, Mn, Sn, Co, and Al in C14 Laves phase alloys for NiMH battery application. *J. Alloy. Compd.* **2009**, *476*, 774–781. [CrossRef]
13. Young, K.; Ouchi, T.; Koch, J.; Fetcenko, M.A. Compositional optimization of vanadium-free hypo-stoichiometric AB$_2$ metal hydride alloy for Ni/MH battery application. *J. Alloy. Compd.* **2012**, *510*, 97–106. [CrossRef]
14. Young, K.; Wong, D.F.; Yasuoka, S.; Ishida, J.; Nei, J.; Koch, J. Different failure modes for V-containing and V-free AB$_2$ metal hydride alloys. *J. Power Sources* **2014**, *251*, 170–177. [CrossRef]
15. Zhou, X.; Young, K.; West, J.; Regalado, J.; Cherisol, K. Degradation mechanisms of high-energy bipolar nickel metal hydride battery with AB$_5$ and A$_2$B$_7$ alloys. *J. Alloy. Compd.* **2013**, *580*, S373–S377. [CrossRef]
16. Wang, L.; Young, K.; Meng, T.; English, N.; Yasuoka, S. Partial substitution of cobalt for nickel in mixed rare earth metal based superlattice hydrogen absorbing alloy—Part 2 battery performance and failure mechanism. *J. Alloy. Compd.* **2016**, *664*, 417–427. [CrossRef]
17. Kong, L.; Chen, B.; Young, K.; Koch, J.; Chan, A.; Li, W. Effects of Al- and Mn-contents in the negative MH alloy on the self-discharge and long-term storage properties of Ni/MH battery. *J. Power Sources* **2012**, *213*, 128–139. [CrossRef]
18. Young, K.; Wong, D.F.; Ouchi, T.; Huang, B.; Reichman, B. Effects of La-addition to the structure, hydrogen storage, and electrochemical properties of C14 metal hydride alloys. *Electrochim. Acta* **2015**, *174*, 815–825. [CrossRef]
19. Young, K.; Chang, S.; Lin, X. C14 Laves phase metal hydride alloys for Ni/MH batteries application. *Batteries* **2017**. submitted to publication. [CrossRef]
20. Shinyama, K.; Magari, Y.; Kumagae, K.; Nakamura, H.; Nohma, T.; Takee, M.; Ishiwa, K. Deterioration mechanism of nickel metal-hydride batteries for hybrid electric vehicles. *J. Power Sources* **2005**, *141*, 193–197. [CrossRef]
21. Zhu, W.H.; Zhu, Y.; Tatarchuk, B.J. Self-discharge characteristics and performance degradation of Ni-MH batteries for storage applications. *Int. J. Hydrogen Energy* **2014**, *39*, 19789–19798. [CrossRef]
22. Nakano, H.; Wakao, S. Substitution effect of elements in Zr-based alloys with Laves phase of nickel-hydride battery. *J. Alloy. Compd.* **1995**, *231*, 587–593. [CrossRef]
23. Nakano, H.; Wakao, S.; Shimizu, T. Correlation between crystal structure and electrochemical properties of C14 Laves-phase alloys. *J. Alloy. Compd.* **1997**, *253–254*, 609–612. [CrossRef]
24. Young, K.; Ouchi, T.; Huang, B.; Chao, B.; Fetcenko, M.A.; Bendersky, L.A.; Wang, K.; Chiu, C. The correlation of C14/C15 phase abundance and electrochemical properties in the AB$_2$ alloys. *J. Alloy. Compd.* **2010**, *506*, 841–848. [CrossRef]
25. Song, X.; Zhang, X.; Leo, Y.; Wang, Q. Effect of microstructure on the properties of Zr-Mn-V-Ni AB$_2$ type hydride electrode alloys. *Int. J. Hydrogen Energy* **1999**, *24*, 455–459. [CrossRef]
26. Young, K.; Ouchi, T.; Lin, X.; Reichman, B. Effects of Zn-addition to C14 metal hydride alloys and comparisons to Si, Fe, Cu, Y, and Mo-additives. *J. Alloy. Compd.* **2016**, *655*, 50–59. [CrossRef]

27. Young, K.; Fetcenko, M.A.; Li, F.; Ouchi, T. Structural, thermodynamic, and electrochemical properties of Ti$_x$Zr$_{1-x}$(VNiCeMnCiAl)$_2$ C14 Laves phase alloys. *J. Alloy. Compd.* **2008**, *464*, 238–247. [CrossRef]

28. Jacob, I.; Stern, A.; Moran, A.; Shaltiel, D.; Davidov, D. Hydrogen absorption in (Zr$_x$Ti$_{1-x}$)B$_2$ (B ≡ Cr, Mn) and the phenomenological model for the absorption capacity in pseudo-binary Laves-phase compound. *J. Less-Common Met.* **1980**, *73*, 369–376. [CrossRef]

29. Huot, J.; Akiba, E.; Ogura, T.; Ishido, Y. Crystal structure, phase abundance and electrode performance of Laves phase compounds (Zr, A)V$_{0.5}$Ni$_{1.1}$Mn$_{0.2}$Fe$_{0.2}$ (A ≡ Ti, Nb or Hf). *J. Alloy. Compd.* **1995**, *218*, 101–109. [CrossRef]

30. Young, K.; Ouchi, T.; Fetcenko, M.A. Pressure-composition-temperature hysteresis in C14 Laves phase alloys: Part 3. Empirical formula. *J. Alloy. Compd.* **2009**, *480*, 440–448. [CrossRef]

31. Trapanese, M.; Franzitta, V.; Viola, A. Description of hysteresis of nickel metal hydride battery. In Proceedings of the 38th Annual Conference on IEEE Industrial Electronics Society, Montreal, QC, Canada, 25–28 October 2012; pp. 967–970.

32. Trapanese, M.; Franzitta, V.; Viola, A. Description of hysteresis in lithium battery by classical Preisach model. *Adv. Mater. Res.* **2013**, *622*, 1099–1103.

33. Trapanese, M.; Franzitta, V.; Viola, A. The Jiles Atherton model for description on hysteresis in lithium battery. In Proceedings of the Twenty-Eighth Annual IEEE Applied Power Electronics Conference and Exposition (APEC), Long Beach, CA, USA, 17–21 March 2013; pp. 2772–2775.

batteries

MDPI

Article

Performance Comparison between AB_5 and Superlattice Metal Hydride Alloys in Sealed Cells

John M. Koch [1], Kwo-Hsiung Young [1,2,*], Jean Nei [1], Chaolan Hu [1] and Benjamin Reichman [1]

[1] BASF/Battery Materials—Ovonic, 2983 Waterview Drive, Rochester Hills, MI 48309, USA;
     john.m.koch@basf.com (J.M.K.); jean.nei@basf.com (J.N.); sherry.hu@basf.com (C.H.);
     Benjamin.reichman@basf.com (B.R.)
[2] Department of Chemical Engineering and Materials Science, Wayne State University, Detroit, MI 48202, USA
*   Correspondence: kwo.young@basf.com; Tel.: +1-248-293-7000

Academic Editor: Sheng S. Zhang
Received: 27 September 2017; Accepted: 17 October 2017; Published: 6 November 2017

Abstract: High-power cylindrical nickel metal/hydride batteries using a misch metal-based Al-free superlattice alloy with a composition of $La_{11.3}Pr_{1.7}Nd_{5.1}Mg_{4.5}Ni_{63.6}Co_{13.6}Zr_{0.2}$ were fabricated and evaluated against those using a standard AB_5 metal hydride alloy. At room temperature, cells made with the superlattice alloy showed a 40% lower internal resistance and a 59% lower surface charge-transfer resistance compared to cells made with the AB_5 alloy. At a low temperature ($-10\,°C$), cells made with the superlattice alloy demonstrated an 18% lower internal resistance and a 60% lower surface charge-transfer resistance compared to cells made with the AB_5 alloy. Cells made with the superlattice alloy exhibited a better charge retention at $-10\,°C$. A cycle life comparison in a regular cell configuration indicated that the Al-free superlattice alloy contributes to a shorter cycle life as a result of the pulverization from the lattice expansion of the main phase.

Keywords: metal hydride (MH); nickel/metal hydride (Ni/MH) battery; hydrogen-absorbing alloy; electrochemistry; superlattice alloy

1. Introduction

Nickel/metal hydride (Ni/MH) batteries have been serving consumer portable electronics, hybrid electric vehicles, and stationary applications for more than 30 years [1–6]. Until now, the misch metal (Mm)-based AB_5 metal hydride (MH) alloy was the mainstream negative electrode active material [7]. In the last decade, the Mm-based superlattice MH alloy began to take over the market share because of its higher capacities; better high-rate dischargeability; and superior low-temperature, high-temperature, and charge retention performances compared to the conventional AB_5 MH alloy [2–6,8]. The superlattice MH alloy is composed of more than one phase with alternating A_2B_4 and AB_5 building slabs along the c-direction of the unit cell [2]. There can be one (AB_3), two (A_2B_7), three (A_5B_{19}), or more AB_5 units between two A_2B_4 slabs. Depending on the stacking sequence, either the hexagonal or rhombohedral structures are possible. The A-site of the superlattice MH alloy contains both rare-earth (RE) and alkaline earth (usually Mg) elements. While almost all academic research has focused on the single RE element (La or Nd)-based superlattice MH alloys (for reviews, see [9–11]), commercial applications have adopted the Mm composition for a higher cycle stability [2,12]. In the past, a few papers about the substitution works performed in the Mm-based superlattice alloy family with Al [13], Mn [14,15], Fe [16,17], Co [18–20], and Ce [21] were published, but a systematic performance comparison between a Mm-based superlattice MH alloy and a standard AB_5 MH alloy is absent. Therefore, we conducted a series of battery performance evaluations in the sealed cells made with both materials and report the results here.

2. Experimental Setup

Both the AB_5 and superlattice alloys were prepared by Eutectix (Troy, Michigan, USA) with a conventional 250 kg induction melting furnace [22]. The ingot was placed in a retort and annealed at 960 °C in vacuum (AB_5) or 1 atm atmosphere of Ar (superlattice) for 8 h. The annealed ingots were crushed and ground into the size of −200 mesh. A Philips X'Pert Pro X-ray diffractometer (XRD; Amsterdam, the Netherlands) and a JEOL-JSM6320F scanning electron microscope (SEM; Tokyo, Japan) were used to study the alloys' microstructures. A Suzuki Shokan multi-channel pressure–concentration–temperature system (PCT; Tokyo, Japan) was used to study the gaseous-phase hydrogen storage characteristics. Electrochemical properties were evaluated with negative electrodes made by dry compacting the annealed alloy powder onto an expanded nickel substrate. A CTE MCL2 Mini cell testing system (Chen Tech Electric MFG. Co., Ltd., New Taipei, Taiwan) was used to study the alloys' half-cell characteristics.

For the sealed-cell performance evaluation, a C-size cylindrical high-power design was chosen. Negative (0.193 mm thick) and positive electrodes (two thicknesses: 0.300 and 0.361 mm) were made using the dry-compaction and wet-paste methods [23], respectively. Positive electrode paste was composed of 89% standard AP50 [24] with a composition of $Ni_{0.91}Co_{0.045}Zn_{0.045}(OH)_2$ (BASF—Ovonic, Rochester Hills, Michigan, USA), and 5 wt % Co and 6% $Co(OH)_2$ powders on a nickel foam substrate. A Freudenberg FS2225 fluorinated acrylic acid grafted polyethylene/polypropylene non-woven fabric (Freudenberg Group, Weinheim, Germany) was used as the separator. In this high-power cell design, a high negative-to-positive capacity ratio cell design (about 1.7) was used to ensure a large amount of overcharge reservoir [25]. A 30 wt % KOH solution with LiOH (1.5 wt %) additive was used as the electrolyte. A six-cycle activation process using a Maccor battery cycler (Maccor, Tulsa, Oklahoma, USA) was conducted for each cell [26]. The battery performance testing procedures can be found in an earlier publication [27].

3. Results and Discussion

3.1. Alloy Properties Comparison

Alloy A, the most popular AB_5 alloy used in the industry with a composition of $La_{10.5}Ce_{4.3}Pr_{0.5}Nd_{1.4}Ni_{60}Co_{12.7}Mn_{5.9}Al_{4.7}$, was used as the control in this comparison work. Alloy B with a composition of $La_{11.3}Pr_{1.7}Nd_{5.1}Mg_{4.5}Ni_{63.6}Co_{13.6}Zr_{0.2}$, which shows the lowest charge-transfer resistance in a comparative study [28], was the superlattice alloy under the current study. In this composition, Pr and Nd were added to reduce the corrosion nature of the alloy, Ce and Mn were not included in the consideration of cycle stability and self-discharge [2,21], Co was added for low-temperature performance enhancement [19], and a very small amount of Zr was added for scavenging residual oxygen in the chamber. The B/A stoichiometry of 3.42 was chosen through an optimization study judging the electrochemical performance. While annealed alloy A has only one $CaCu_5$ phase, as seen from its XRD pattern (Figure 10a in [28]), alloy B shows a multi-phase structure in both pristine and annealed conditions (Figure 1). Phase abundances calculated from the XRD data are listed in Table 1. After annealing, the abundance of the desirable Nd_2Ni_7 phase [29] increased from 0 to 56.7 wt %; the unwanted $CaCu_5$ phase [30] decreased from 32.7 to 1.6 wt %; and $LaMgNi_4$ and other superlattice phases, such as $CeNi_3$, $NdNi_3$, Sm_5Ni_{19} and Ni_5Co_{19}, still existed. SEM analysis was used to confirm the XRD findings, and two representative backscattering electron micrographs for pristine and annealed alloy B are shown in Figure 2. X-ray energy-dispersive spectroscopy (EDS) was used to study the chemical compositions of a few spots in Figure 2, and the results are summarized in Table 2. In the pristine sample, the AB_5 phase (spots 2 and 3) can be identified by its relatively bright contrast due to the higher content of low-atomic weight nickel. Later, the AB_5 phase was removed by annealing. The superlattice phases are difficult to separate by contrast in the micrographs because of their similar chemical composition and stoichiometry. The microstructural analyses conclude that while annealed alloy A has only one CaCu$_5$ structure, alloy B (before or after annealing) is a superlattice-based (>95 wt %) multi-phase alloy.

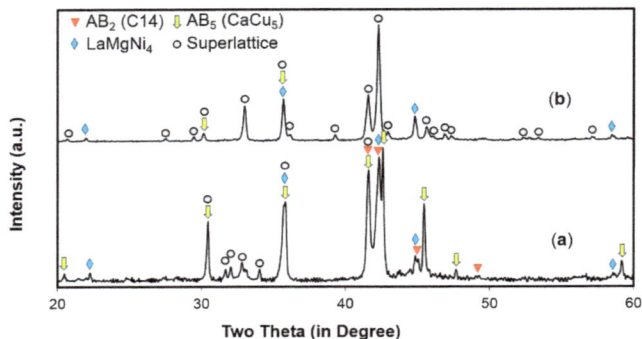

Figure 1. X-ray diffractometer (XRD) patterns using Cu-K$_\alpha$ as the radiation source for (**a**) pristine, and (**b**) annealed alloy B.

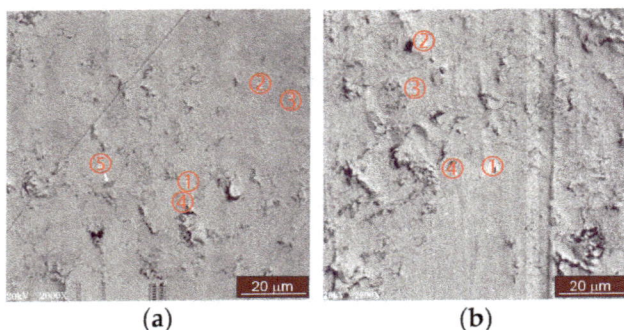

Figure 2. Scanning electron microscope (SEM) backscattering electron micrographs of (**a**) pristine, and (**b**) annealed alloy B.

Table 1. Phase abundances (in wt %) of alloy B before and after annealing determined by the X-ray diffractometer (XRD) analysis. HEX, CUB, and RHO are hexagonal, cubic, and rhombohedral, respectively.

Stoichiometry	AB$_2$		AB$_3$		A$_2$B$_7$		A$_5$B$_{19}$		AB$_5$
Structure	HEX	CUB	HEX	RHO	HEX	RHO	HEX	RHO	HEX
Phase	MgZn$_2$	LaMgNi$_4$	CeNi$_3$	NdNi$_3$	Nd$_2$Ni$_7$	Pr$_2$Ni$_7$	Sm$_5$Ni$_{19}$	Nd$_5$Co$_{19}$	CaCu$_5$
Pristine alloy B	7.9	5.7	2.0	19.3	0.0	13.1	1.6	17.7	32.7
Annealed alloy B	0.0	3.2	7.5	7.4	56.9	0.0	6.4	17.0	1.6

Both the gaseous phase and electrochemical hydrogen storage characteristics of alloys A (AB$_5$) and B (superlattice) were studied. In the gaseous phase, PCT isotherms measured at 30 °C for both alloys are plotted in Figure 3. Annealed alloy A has a higher plateau pressure and a higher reversible capacity than pristine and annealed alloy B. Annealing in alloy B flattens the isotherm, increases the storage capacity, and reduces the hysteresis, as reported previously [31]. Electrochemical testing results from the first 20 cycles of annealed alloys A and B are compared in Figure 4. Annealed alloy B exhibits a higher initial capacity, but it degrades quickly in the flooded KOH solution compared to annealed

Alloy A. The higher oxidation rate in the Mg-containing superlattice alloys is well known, and many electrode fabrication methods have been proposed to overcome this shortcoming [32]. As a result, a commercial cell capable of 6000 cycles with a superlattice alloy has been demonstrated [8]. Gaseous phase and electrochemical hydrogen storage properties of annealed alloys A and B are summarized in Table 3. Plateau pressure is defined as the equilibrium pressure corresponding to a 0.75 wt % storage capacity in the desorption isotherm, and the PCT hysteresis is defined as *ln* (absorption pressure/desorption pressure) at the same storage capacity. Although the gaseous phase capacities of the two alloys are similar, the superlattice alloy shows a higher electrochemical discharge capacity, which is close to the theoretical limit converted from the gaseous phase capacity (381 mAh·g^{-1} using the conversion of 1 wt % = 268 mAh·g^{-1}) because of the synergetic effect among the constituent phases [33]. The higher PCT hysteresis in annealed alloy B predicts a higher pulverization rate during repetitive cycling [34].

Table 2. Energy-dispersive spectroscopy (EDS) results from selected spots in Figure 2. All numbers are percentages.

Location	La	Pr	Nd	Mg	Ni	Co	Zr	B/A	Phase
Figure 2a-1	13.6	0.7	4.5	6.6	62.6	11.9	0.1	2.93	Superlattice
Figure 2a-2	11.1	1.2	5.3	0.9	67.2	14.1	0.2	4.41	AB$_5$
Figure 2a-3	9.5	0.9	5.6	0.4	69.1	14.4	0.1	5.10	AB$_5$
Figure 2a-4	11.1	0.8	5.2	18.7	57.5	6.5	0.2	1.79	LaMgNi$_4$
Figure 2a-5	5.5	0.4	3.5	0.0	17.8	3.2	69.6	9.64	ZrO$_2$
Figure 2b-1	86.0	0.0	4.5	0.8	6.4	0.8	1.5	0.1	La metal
Figure 2b-2	16.3	0.7	4.1	4.0	61.5	13.2	0.2	2.98	Superlattice
Figure 2b-3	14.4	0.9	4.4	12.4	60.3	7.5	0.1	2.11	LaMgNi$_4$
Figure 2b-4	13.2	0.7	4.7	14.2	60.4	6.7	0.1	2.04	LaMgNi$_4$

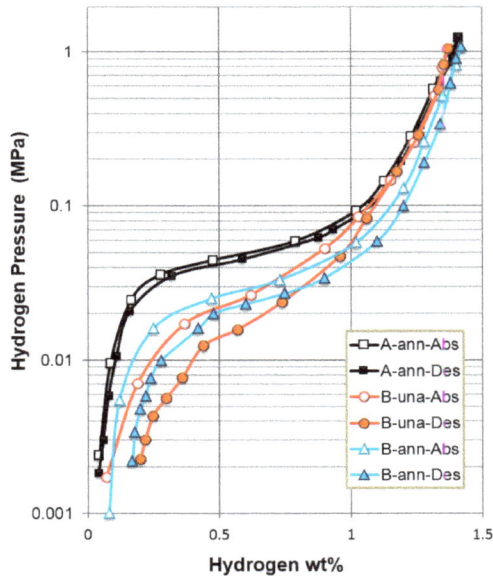

Figure 3. Pressure–concentration–temperature (PCT) isotherms measured at 30 °C. A-ann, B-una, and B-ann are annealed alloy A, pristine alloy B, and annealed alloy B, respectively. Abs and Des denote absorption and desorption, respectively.

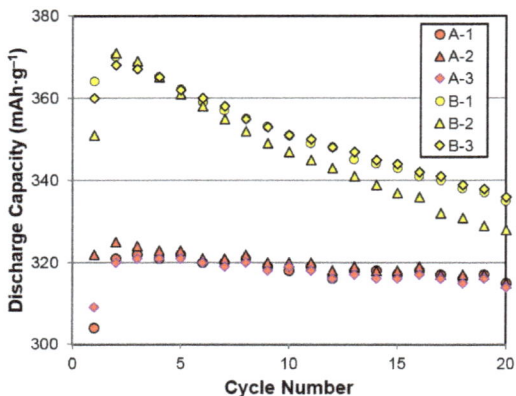

Figure 4. Half-cell capacities measured using the three-electrode cell setup for annealed alloys A and B.

Table 3. Gaseous phase and electrochemical hydrogen storage properties of annealed alloys A and B.

Alloy	Full H-Storage	Reversible H-Storage	Plateau Pressure	PCT Hysteresis	Discharge Capacity
Annealed alloy A	1.41%	1.37%	0.058 MPa	0.10	310 mAh·g^{-1}
Annealed alloy B	1.42%	1.25%	0.025 MPa	0.23	370 mAh·g^{-1}

3.2. Sealed-Cell Performance

Fifty C-size cylindrical cells in a high-power design were made with annealed alloys A (cell A) and B (cell B). After the formation process, the discharge capacities were 2.7 and 3.1 Ah from cells A and B, respectively, with a 0.6 A discharge current. Cell B shows a higher energy density (50.4 vs 41.0 Wh·kg^{-1}) than cell A because of its higher active material capacity, which allows for the matching with a thicker positive electrode (0.361 vs 0.300 mm).

3.2.1. High-Rate

Room temperature (RT) discharge voltage profiles with four different rates (C, 2C, 5C, and 10C) for cells A and B are shown in Figure 5. The cell voltage (V) decreases with the increase in the discharge current (i) following the formula:

$$V = V_{oc} - iR_{int} \qquad (1)$$

where V_{oc} and R_{int} are the open-circuit voltage (when $i = 0$) and internal resistance, respectively. Voltage suppression due to the increase in the discharge current is less severe in cell B compared to cell A, which indicates a lower R_{int} in cell B. Normalized discharge capacities (to those obtained with a 0.2C discharge rate) of cells A and B (set of four each) are listed in Table S1 and indicate a slightly lower high-rate dischargeability of cell B (average value of 84.1% in cell B vs 87.6% in cell A at a 10C rate). However, the capacities of cell B are higher than those of cell A at all discharge rates.

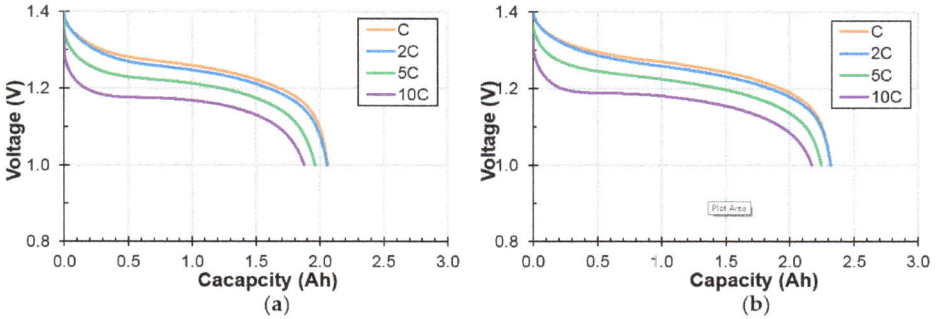

Figure 5. Discharge voltage profiles at different discharge rates of (**a**) cell A, and (**b**) cell B measured at room temperature.

3.2.2. Low Temperature

Low-temperature performances of cells A and B were evaluated by measuring the capacities at −10 °C with different discharge rates (C, 2C, 5C, and 10C). The resulting discharge voltage profiles are plotted in Figure 6. Voltage suppression due to the increase in the discharge current is more severe at a lower temperature. Only about 50% of the capacity is obtained at −10 °C with C and 2C discharge rates. The cells deliver almost no capacity with further increases in the discharge rate. Normalized −10 °C discharge capacities (to those obtained at RT with a 0.2C discharge rate) of cells A and B (set of four each) are listed and indicate a slightly better low-temperature performance of cell B (average value of 51.6% in cell B vs 49.0% in cell A at a 1C rate).

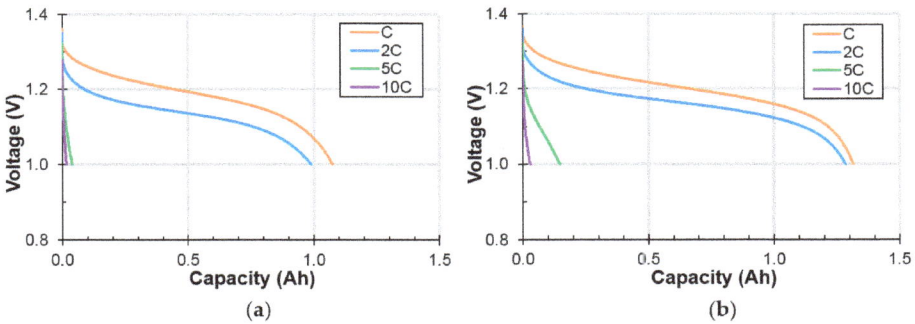

Figure 6. Discharge voltage profiles at different discharge rates of (**a**) cell A, and (**b**) cell B measured at −10 °C.

3.2.3. Charge Retention

Charge-retention behaviors of cells A and B were evaluated by both the RT and −10 °C standing voltage stabilities at an 80% state-of-charge, and the results are plotted in Figures 7 and 8, respectively. On average, cell A demonstrates a marginally better charge-retention performance at RT but a worse performance at −10 °C compared to cell B.

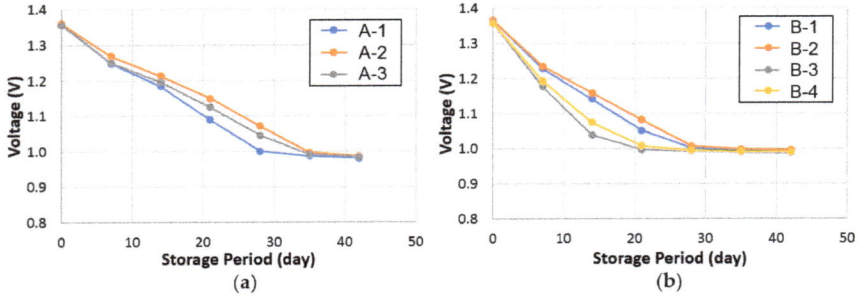

Figure 7. Open-circuit voltages of (a) cell A (set of three), and (b) cell B (set of four) stored at room temperature.

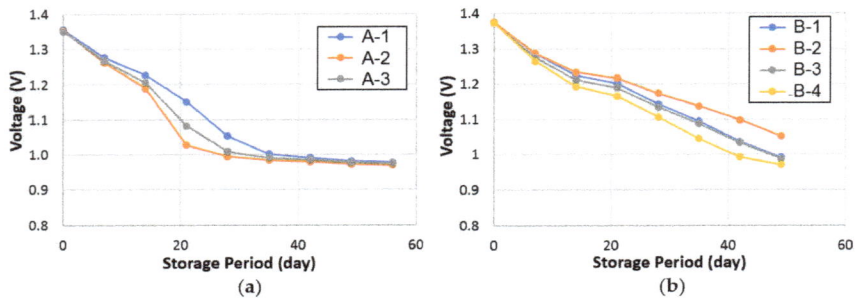

Figure 8. Open-circuit voltages of (a) cell A (set of three), and (b) cell B (set of four) stored at −10 °C.

3.2.4. Internal Resistance

Internal resistance (R_{int}) was measured by a pulse method using the formula:

$$R_{int} = \Delta V / \Delta i \qquad (2)$$

Both 1 and 10 s pulses were used to measure R_{int}s from cells A and B, and data obtained at both RT and −10 °C are listed in Table 4. RT R_{int}s decreases slightly with the increase in the discharge rate. Cell B shows a lower R_{int}s in all measurements.

Table 4. Internal resistances (R_{int}, in mΩ·m^2) measured with 1 and 10 s pulsed discharges with different discharge rates (1, 2, 5, and 10C) at both room temperature (RT) and −10 °C.

Condition	1C	2C	5C	10C
Cell A RT 1 s	0.144	0.142	0.138	0.130
Cell A RT 10 s	0.194	0.192	0.182	0.168
Cell A −10 °C 1 s	0.405	0.369	0.289	0.244
Cell A −10 °C 10 s	0.516	0.492	0.499	—
Cell B RT 1 s	0.087	0.087	0.087	0.085
Cell B RT 10 s	0.127	0.127	0.124	0.118
Cell B −10 °C 1 s	0.334	0.312	0.262	0.219
Cell B −10 °C 10 s	0.468	0.444	0.453	—

3.2.5. Surface Charge-Transfer Resistance

Out of many factors contributing to R_{int}, ohmic resistance (R_0) and surface charge-transfer resistance (R_{ct}) can be deduced from the Cole-Cole plot obtained by the alternating current (AC) impedance measurement [35]. Cole-Cole plots of cells A and B measured at RT and −10 °C are shown in Figures 9 and 10, respectively. Calculated R_0 and R_{ct} values, and double-layer capacitances (C) of cells A and B are listed in Table S2. While R_0 values in both sets are similar, R_{ct} values in cell B are lower than those in cell A at both RT and −10 °C. Part of the reason for the lower R_{ct} values in cell B is due to the larger surface reactive area (A) of the superlattice alloy from the connection:

$$C = \varepsilon A/d \tag{3}$$

where ε and d are the dielectric constant of electrolyte and the alloy surface dipole thickness. Another reason for the lower R_{ct} values in cell B is the higher surface catalytic ability of the superlattice alloy [28].

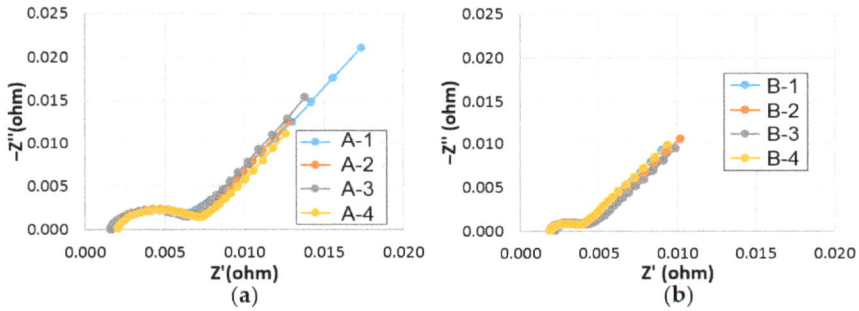

Figure 9. Cole-Cole plots of (**a**) cell A (set of four), and (**b**) cell B (set of four) measured at room temperature.

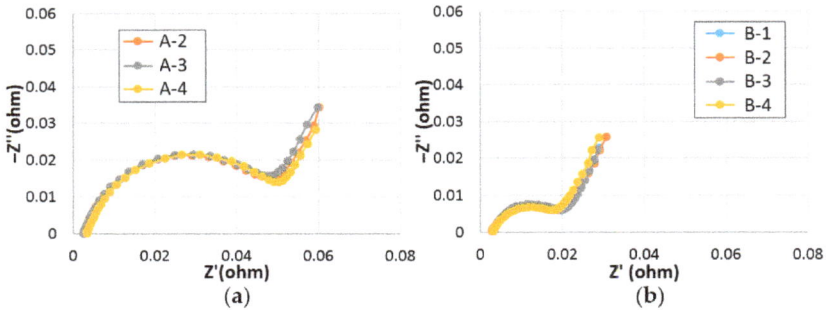

Figure 10. Cole-Cole plots of (**a**) cell A (set of three), and (**b**) cell B (set of four) measured at −10 °C.

3.2.6. Cycle Life

Because the high-power design is usually associated with shallow charge/discharge cycling, a regular C-size configuration with a nominal capacity of 4.5 Ah was used to study the cycle life performance. Cells were built with annealed alloys A and B and tested under a C/2 charge to a −ΔV of 3 mV and a C/2 discharge to a cutoff voltage of 0.9 V at RT, and the results are plotted in Figure 11. Without the protective binder commonly used in the commercial cells made with the superlattice alloys [32], the cell made with the superlattice MH alloy (alloy B) only shows half of the cycle life of a cell made with the conventional AB_5 MH alloy (alloy A). Earlier studies on the failure mode of a Mm-based Al-free superlattice MH alloy indicated that the pulverization of the main phase is the main cause of capacity degradation [30].

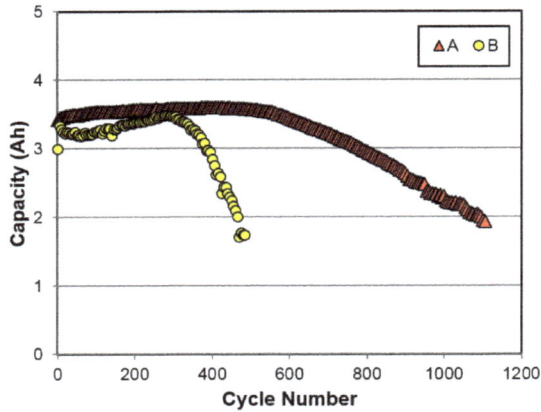

Figure 11. Room temperature cycle life (C/2-C/2) comparison between regular C-size cells (not high-power) made with annealed alloys A and B.

3.2.7. Comparison

A battery performance comparison between cells made with the AB_5 (cell A) and Al-free A_2B_7-based superlattice (cell B) MH alloys is summarized in Figure 12. Cell B has a higher capacity and a better low-temperature performance; however, it demonstrates slightly worse high-rate dischargeability and charge retention, and its cycle life is only half that of cell A. Despite the lower R_{int} and R_{ct} in cell B, it still shows a lower normalized capacity at a higher rate, which may be associated with the relatively low V_{oc} at RT caused by alloy B's relatively low equilibrium plateau pressure (Figure 2). In another article, cells made with an Al-containing superlattice MH alloy showed a comparable cycle life and better peak power and charge-retention performance compared to those made with the AB_5 alloy [19]. Therefore, the inferior cycle life observed in the superlattice alloy is only limited to the Al-free composition used in this study. Combined with the use of a hydrophobic binder in the negative-electrode paste [32], the Al-containing superlattice alloy showed even better cycle stability [8].

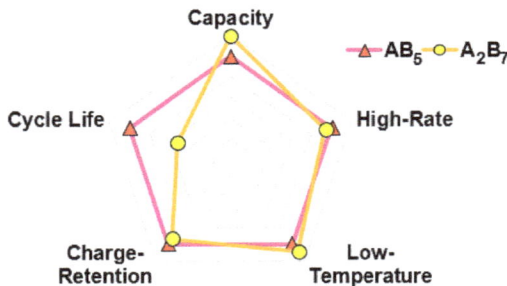

Figure 12. Sealed-cell performance comparison between cells made with the conventional AB_5 and new A_2B_7-based superlattice MH alloys.

4. Conclusions

Electrochemical performances of a misch metal-based Al-free superlattice metal hydride alloy were compared to those of a standard AB_5 metal hydride alloy in a high-power C-size cell configuration. In the sealed cell, the superlattice alloy showed higher energy densities, lower internal resistances, lower surface charge-transfer resistances at both RT and $-10\,^{\circ}C$ compared to the AB_5 alloy. For the charge-retention performance, the superlattice alloy was slightly worse at RT but outperformed the AB_5 alloy at $-10\,^{\circ}C$. The cycle stability of the superlattice alloy tested in a regular cell configuration is inferior to that of the AB_5 alloy mainly because of alloy pulverization.

Supplementary Materials: The following are available online at http://www.mdpi.com/2313-0105/3/4/35/s1, Table S1: Discharge capacities normalized to those obtained at a 0.2C rate from four Cells A and four Cells B measured at both room temperature and $-10\,^{\circ}C$. All numbers are in %. SD denotes standard deviation, Table S2: Ohmic resistances (R0 in W), surface charge-transfer resistances (Rct in W), and double-layer capacitances (C in Farad) from Cells A and B measured at both room temperature (RT) and $-10\,^{\circ}C$. SD denotes standard deviation.

Acknowledgments: The authors would like to thank the following individuals from BASF—Ovonic for their help: Su Cronogue, Taihei Ouchi, Jean Nei, Shiuan Chang, Nathan English, Sui-ling Chen, Cheryl Setterington, David Pawlik, Allen Chan, and Ryan J. Blankenship.

Author Contributions: John M. Koch designed and performed the experiments. Kwo-Hsiung Young prepared the alloys and the manuscript. Chaolan Hu and Benjamin Reichman conducted the AC impedance measurements and interpreted the data.

Conflicts of Interest: The authors declare no conflict of interest.

Abbreviations

Ni/MH	Nickel/metal hydride
Mm	Misch metal
MH	Metal hydride
RE	Rare-earth
XRD	X-ray diffractometer
SEM	Scanning electron microscope
PCT	Pressure–concentration–temperature
EDS	Energy-dispersive spectroscopy
HEX	Hexagonal
CUB	Cubic
RHO	Rhombohedral
Abs	Absorption
Des	Desorption
RT	Room temperature
V	Cell voltage
i	Current
V_{oc}	Open-circuit voltage
R_{int}	Internal resistance
R_0	Ohmic resistance
R_{ct}	Charge-transfer resistance
C	Double-layer capacitance
A	Surface reactive area
ε	Dielectric constant of electrolyte
d	Alloy surface dipole thickness

References

1. Zelinsky, M.A.; Koch, J.M.; Young, K. Performance comparison of rechargeable batteries for stationary applications (Ni/MH vs. Ni-Cd and VRLA). *Tatteries* **2017**, submitted.
2. Young, K.; Cai, X.; Chang, S. Reviews on Chinese Patents regarding the nickel/metal hydride battery. *Batteries* **2017**, *3*, 24. [CrossRef]

3. Yasuoka, S.; Magari, Y.; Murata, T.; Tanaka, T.; Ishida, J.; Nakamura, H.; Nohma, T.; Kihara, M.; Baba, Y.; Teraoka, H. Development of high-capacity nickel-metal hydride batteries using superlattice hydrogen-absorbing alloys. *J. Power Sour.* **2006**, *156*, 662–666. [CrossRef]

4. Teraoka, H. Development of Low Self-Discharge Nickel-Metal Hydride Battery. Available online: http://www.scribd.com/doc/9704685/Teraoka-Article-En (accessed on 9 April 2016).

5. Kai, T.; Ishida, J.; Yasuoka, S.; Takeno, K. The effect of nickel-metal hydride battery's characteristics with structure of the alloy. In Proceedings of the 54th Battery Symposium in Japan, Osaka, Japan, 6–9 October 2013; p. 210.

6. Teraoka, H. Development of Ni-MH EThSS with Lifetime and Performance Estimation Technology. In Proceedings of the 34th International Battery Seminar & Exhibit, Fort Lauderdale, FL, USA, 20–23 March 2017.

7. Teraoka, H. Ni-MH Stationary Energy Storage: Extreme Temperature & Long Life Developments. In Proceedings of the 33th International Battery Seminar & Exhibit, Fort Lauderdale, FL, USA, 21–24 March 2016.

8. Teraoka, H. Development of Highly Durable and Long Life Ni-MH Batteries for Energy Storage Systems. In Proceedings of the 32th International Battery Seminar & Exhibit, Fort Lauderdale, FL, USA, 9–12 March 2015.

9. Liu, Y.; Cao, Y.; Huang, L.; Gao, M.; Pan, H. Rare earth-Mg-Ni-based hydrogen storage alloys as negative electrode materials for Ni/MH batteries. *J. Alloy. Compd.* **2011**, *509*, 675–686. [CrossRef]

10. Young, K.; Nei, J. The current status of hydrogen storage alloy development for electrochemical applications. *Materials* **2013**, *6*, 4574–4608. [CrossRef] [PubMed]

11. Liu, J.; Han, S.; Li, Y.; Zhang, L.; Zhao, Y.; Yang, S.; Liu, B. Phase structures and electrochemical properties of La–Mg–Ni-based hydrogen storage alloys with superlattice structure. *Int. J. Hydrogen Energy* **2016**, *41*, 20261–20275. [CrossRef]

12. Takasaki, T.; Nishimura, K.; Saito, M.; Fukunaga, H.; Iwaki, T.; Sakai, T. Cobalt-free nickel-metal hydride battery for industrial applications. *J. Alloy. Compd.* **2013**, *580*, S378–S381. [CrossRef]

13. Yasuoka, S.; Ishida, J.; Kai, T.; Kajiwara, T.; Doi, S.; Yamazaki, T.; Kishida, K.; Inui, H. Function of aluminum in crystal structure of rare earth-Mg-Ni hydrogen-absorbing alloy and deterioration mechanism of $Nd_{0.9}Mg_{0.1}Ni_{3.4}$ and $Nd_{0.9}Mg_{0.1}Ni_{3.3}Al_{0.2}$ alloys. *Int. J. Hydrogen Energy* **2017**, *42*, 11574–11583. [CrossRef]

14. Young, K.; Wong, D.F.; Wang, L.; Nei, J.; Ouchi, T.; Yasuoka, S. Mn in misch-metal based superlattice metal hydride alloy—Part 1 Structural, hydrogen storage and electrochemical properties. *J. Power Sour.* **2015**, *277*, 426–432. [CrossRef]

15. Young, K.; Wong, D.F.; Wang, L.; Nei, J.; Ouchi, T.; Yasuoka, S. Mn in misch-metal based superlattice metal hydride alloy—Part 2 Ni/MH battery performance and failure mechanism. *J. Power Sour.* **2015**, *277*, 433–442. [CrossRef]

16. Young, K.; Ouchi, T.; Nei, J.; Yasuoka, S. Fe-substitution for Ni in misch metal-based superlattice hydrogen absorbing alloys—Part 1. Structural, hydrogen storage, and electrochemical properties. *Batteries* **2016**, *2*, 34. [CrossRef]

17. Meng, T.; Young, K.; Nei, J.; Koch, J.M.; Yasuoka, S. Fe-substitution for Ni in misch metal-based superlattice hydrogen absorbing alloys—Part 2. Ni/MH battery performance and failure mechanisms. *Batteries* **2017**, *3*, 28. [CrossRef]

18. Wang, L.; Young, K.; Meng, T.; Ouchi, T.; Yasuoka, S. Partial substitution of cobalt for nickel in mixed rare earth metal based superlattice hydrogen absorbing alloy—Part 1 structural, hydrogen storage and electrochemical properties. *J. Alloy. Compd.* **2016**, *660*, 407–415. [CrossRef]

19. Wang, L.; Young, K.; Meng, T.; English, N.; Yasuoka, S. Partial substitution of cobalt for nickel in mixed rare earth metal based superlattice hydrogen absorbing alloy—Part 2 battery performance and failure mechanism. *J. Alloy. Compd.* **2016**, *664*, 417–427. [CrossRef]

20. Meng, T.; Young, K.; Koch, J.; Ouchi, T.; Yasuoka, S. Failure mechanisms of nickel/metal hydride batteries with cobalt-substituted superlattice hydrogen-absorbing alloy anodes at 50 °C. *Batteries* **2016**, *2*, 20. [CrossRef]

21. Yasuoka, S.; Ishida, J.; Kishida, K.; Inui, H. Effects of cerium on the hydrogen absorption-desorption properties of rare earth-Mg-Ni hydrogen-absorbing alloys. *J. Power Sour.* **2017**, *346*, 56–62. [CrossRef]

22. Young, K.; Chang, S.; Lin, X. C14 Laves phase metal hydride alloys for Ni/MH batteries applications. *Batteries* **2017**, *3*, 27. [CrossRef]

23. Chang, S.; Young, K.; Nei, J.; Fierro, C. Reviews on the U.S. Patents regarding nickel/metal hydride batteries. *Batteries* **2016**, *2*, 10. [CrossRef]
24. Young, K.; Wang, L.; Yan, S.; Liao, X.; Meng, T.; Shen, H.; May, W.C. Fabrications of high-capacity alpha-Ni(OH)$_2$. *Batteries* **2017**, *3*, 6. [CrossRef]
25. Young, K.; Wu, A.; Qiu, Z.; Tan, J.; Mays, W. Effects of H$_2$O$_2$ addition to the cell balance and self-discharge of Ni/MH batteries with AB$_5$ and A$_2$B$_7$ alloys. *Int. J. Hydrogen Energy* **2012**, *37*, 9882–9891. [CrossRef]
26. Young, K.; Koch, J.M.; Wan, C.; Denys, R.V.; Yartys, V.A. Cell performance comparison between C14- and C15-predominated AB$_2$ metal hydride alloys. *Batteries* **2017**, *3*, 29. [CrossRef]
27. Yan, S.; Meng, T.; Young, K.; Nei, J. A Ni/MH pouch cell with high-capacity Ni(OH)$_2$. *Tatteries* **2017**, submitted.
28. Meng, T.; Young, K.; Hu, C.; Reichman, B. Effects of alkaline pre-etching to metal hydride alloys. *Batteries* **2017**, *3*, 30. [CrossRef]
29. Young, K.; Ouchi, T.; Koch, J.M.; Lien, Y. Comparison among constituent phases in superlattice metal hydride alloys for battery applications. *Tatteries* **2017**, submitted. [CrossRef]
30. Zhou, X.; Young, K.; West, J.; Regalado, J.; Cherisol, K. Degradation mechanisms of high-energy bipolar nickel metal hydride battery with AB$_5$ and A$_2$B$_7$ alloys. *J. Alloy. Compd.* **2013**, *580*, S373–S377. [CrossRef]
31. Young, K.; Ouchi, T.; Huang, B. Effects of various annealing conditions on (Nd, Mg, Zr)(Ni, Al, Co)$_{3.74}$ metal hydride alloys. *J. Power Sour.* **2014**, *248*, 147–153. [CrossRef]
32. Ouchi, T.; Young, K.; Moghe, D. Reviews on the Japanese Patent Applications regarding nickel/metal hydride batteries. *Batteries* **2016**, *2*, 21. [CrossRef]
33. Young, K.; Ouchi, T.; Meng, T.; Wong, D.F. Studies on the synergetic effects in multi-phase metal hydride alloys. *Batteries* **2016**, *2*, 15. [CrossRef]
34. Osumi, Y. *Suiso Kyuzou Goukin*; Agune Technology Center: Tokyo, Japan, 1999; p. 218. (In Japanese)
35. Zhang, L. AC impedance studies on sealed nickel metal hydride batteries over cycle life in analog and digital operations. *Electrochim. Acta* **1998**, *43*, 3333–3342. [CrossRef]

Table S1. Discharge capacities normalized to those obtained at a 0.2C rate from four Cells A and four Cells B measured at both room temperature and –10 °C. All numbers are in %. SD denotes standard deviation.

Cell #	Room Temperature				–10 °C			
	1C	2C	5C	10C	1C	2C	5C	10C
Cell A1	95.0	94.6	91.6	87.5	47.6	44.1	1.8	0.8
Cell A2	97.3	96.5	92.1	88.1	49.0	45.4	2.0	0.9
Cell A3	96.6	95.9	91.3	87.3	49.3	46.4	2.0	0.8
Cell A4	95.9	95.6	91.3	87.3	50.0	46.0	1.8	0.8
Average A	96.2	95.7	91.6	87.6	49.0	45.5	1.9	0.8
SD A	0.9	0.8	0.4	0.4	1.0	1.0	0.1	0.05
Cell B1	88.2	88.5	85.5	82.0	51.9	50.5	5.8	1.1
Cell B2	88.7	89.1	86.5	84.4	51.7	50.5	5.7	1.1
Cell B3	88.8	89.0	85.9	83.7	50.2	48.6	5.7	1.3
Cell B4	91.9	92.0	89.0	86.2	52.5	50.4	5.2	1.2
Average B	89.4	89.7	86.7	84.1	51.6	50.0	5.6	1.2
SD B	1.7	1.6	1.6	1.7	1.0	0.9	0.3	0.1

Table S2. Ohmic resistances (R_0 in Ω), surface charge-transfer resistances (R_{ct} in Ω), and double-layer capacitances (C in Farad) from Cells A and B measured at both room temperature (RT) and –10 °C. SD denotes standard deviation.

Cell #	R_0 at RT	R_{ct} at RT	C at RT	R_0 at –10 °C	R_{ct} at –10 °C	C at –10 °C
Cell A1	0.00164	0.00546	18.55	-	-	-
Cell A2	0.00170	0.00561	18.88	0.00333	0.04973	18.83
Cell A3	0.00159	0.00551	17.91	0.00335	0.05010	18.80
Cell A4	0.00210	0.00560	17.49	0.00351	0.05027	18.00
Average A	0.00176	0.00554	18.21	0.00340	0.05003	18.54
SD A	0.00023	0.00007	0.62	0.00010	0.00028	0.47
Cell B1	0.00187	0.00226	23.34	0.00287	0.02025	21.81
Cell B2	0.00192	0.00246	21.78	0.00297	0.01904	21.25
Cell B3	0.00227	0.00242	21.24	0.00350	0.02094	21.76
Cell B4	0.00191	0.00202	25.62	0.00313	0.01918	22.85
Average B	0.00199	0.00229	23.00	0.00311	0.01985	22.92
SD B	0.00019	0.00020	1.96	0.00027	0.00090	0.67

Article

Fe-Substitution for Ni in Misch Metal-Based Superlattice Hydrogen Absorbing Alloys—Part 2. Ni/MH Battery Performance and Failure Mechanisms

Tiejun Meng [1], Kwo-Hsiung Young [1,2,*], Jean Nei [1], John M. Koch [1] and Shigekazu Yasuoka [3]

[1] BASF/Battery Materials-Ovonic, 2983 Waterview Drive, Rochester Hills, MI 48309, USA;
    pkumeng@hotmail.com (T.M.); jean.nei@basf.com (J.N.); john.m.koch@basf.com (J.M.K.)
[2] Department of Chemical Engineering and Materials Science, Wayne State University, Detroit, MI 48202, USA
[3] Engineering Division, Ni-MH Group, FDK Corporation, 307-2, Koyagi-Machi, Takasaki, Gunma 370-0042,
    Japan; shigekazu.yasuoka@fdk.co.jp
*   Correspondence: kwo.young@basf.com; Tel.: +1-248-293-7000

Received: 19 July 2017; Accepted: 6 September 2017; Published: 18 September 2017

Abstract: The electrochemical performance and failure mechanisms of Ni/MH batteries made with a series of the Fe-substituted A_2B_7 superlattice alloys as the negative electrodes were investigated. The incorporation of Fe does not lead to improved cell capacity or cycle life at either room or low temperature, although Fe promotes the formation of a favorable Ce_2Ni_7 phase. Fe-substitution was found to inhibit leaching of Al from the metal hydride negative electrode and promote leaching of Co, which could potentially extend the cycle life of the positive electrode. The failure mechanisms of the cycled cells with the Fe-substituted superlattice hydrogen absorbing alloys were analyzed by scanning electron microscopy, energy dispersive spectroscopy and inductively coupled plasma analysis. The failure of cells with Fe-free and low Fe-content alloys is mainly attributed to the pulverization of the metal hydride alloy. Meanwhile, severe oxidation/corrosion of the negative electrode is observed for cells with high Fe-content alloys, resulting in increased internal cell resistance, formation of micro-shortages in the separator and eventual cell failure

Keywords: metal hydride (MH); nickel/metal hydride (Ni/MH) battery; hydrogen absorbing alloy (HAA); superlattice alloy; failure mechanism

1. Introduction

Misch metal (Mm)-based superlattice hydrogen absorbing alloys (HAAs) exhibit higher capacity, improved high-rate capability, lower self-discharge and wider operating temperature range than the commonly used AB_5 HAA [1,2] in nickel/metal hydride (Ni/MH) batteries. The use of Mm instead of pure La improves the cycle stability [3] and makes superlattice HAAs competitive in the consumer battery market [4]. To further improve the performances of Mm-based superlattice HAAs, the effects of adding Mn [5,6], Co [7,8] and Ce [9] on the structural, gaseous phase and electrochemical hydrogen storage, and full-cell electrochemical performances were investigated. Mn was previously reported to improve high-rate performance, but creates micro-shorts in the separator which results in severe self-discharge. Co improves the low-temperature performance in exchange for self-discharge and high-temperature performance. The use of Ce promotes the AB_2 phase formation and deteriorates the battery performance.

Previously, in the preceding paper (Part 1), the structural, gaseous phase hydrogen storage and electrochemical (in half-cell configuration) hydrogen storage properties of Fe-substituted Mm-based superlattice HAAs have been reported [10]. Fe-substitution promotes the favorable Ce_2Ni_7 (hexagonal) phase and decreases the relatively unfavorable $NdNi_3$ phase. However, the surface catalytic capability—as reflected by the total volume of surface metallic Ni clusters—was found to decrease with increasing Fe-content. No improvement in discharge capacity was observed and the high rate dischargeability (HRD) of the alloy deteriorated with increasing Fe-content. In this paper (Part 2), the performance and failure mechanisms of Ni/MH batteries made using Fe-substituted HAAs are discussed.

2. Experimental Setup

One hundred C-size Ni/MH batteries using five superlattice HAAs (Fe1 to Fe5) as negative electrode materials were assembled for electrochemical testing. For convenience, Fe1 to Fe5 refer to not only the HAAs, but also the battery cells in this manuscript. The five superlattice HAA ($Mm_{0.83}Mg_{0.17}Ni_{2.94-x}Al_{0.17}Co_{0.2}Fe_x$, $x = 0, 0.05, 0.1, 0.15, 0.2$) powder samples in this study were supplied by Japan Metals and Chemicals Co. (Tokyo, Japan) and their compositions are listed in Table 1. The design formula for all five HAAs is $AB_{3.31}$. Herein, Mm refers to the mixed rare earth metal alloy with a composition of 19.6 wt % La, 40.2 wt % Pr and 40.2 wt % Nd. Structural, gaseous phase hydrogen storage, and the electrochemical properties of the five HAAs were studied and reported in a companion paper (Part 1) [10]. The HAA powder was dry-compacted onto the nickel mesh substrate to form the negative electrode. The positive electrode consists of 94.1 wt % CoOOH-coated $Ni_{0.91}Co_{0.045}Zn_{0.045}(OH)_2$ (~2 wt % CoOOH), 4.9 wt % Co powder and 1 wt % Y_2O_3 additives, and the mixture was wet-pasted onto the nickel foam substrate, dried and then compacted. The Y_2O_3 additive was added due to the ability to increase the open-circuit voltage, decrease the impedance and extend the cycle life [11]. The separator used is Scimat 700/79 acrylic acid grafted polypropylene/polyethylene from the Freudenberg Group (Weinheim, Germany). An aqueous solution consisting of 26.8 wt % NaOH and 1.5 wt % LiOH was used as the electrolyte. The negative to positive capacity ratio was set at 2.0 to maintain a good balance between the overcharge and overdischarge reservoirs [12]. After the cells were sealed, a six-cycle electrochemical formation process was performed using a Maccor Battery Cycler (Tulsa, OK, USA). During the formation process, the cells were charged at a rate of C/10 to 50%, 100% and 120% in the first three cycles and to 150% in the next three cycles. The discharge rate was C/5 for the first five cycles and C/2 for the sixth cycle. The discharge cutoff voltage was 0.9 V.

Table 1. Summary of the compositions and properties of the superlattice hydrogen absorbing alloys (HAAs) used in this study. The gaseous phase maximum hydrogen storage capacity was obtained at 6 MPa hydrogen pressure. The electrochemical capacity was taken from the 2nd cycle at a discharge current of 8 mA g^{-1} with a cutoff voltage of 0.9 V against a standard Ni(OH)$_2$ positive electrode. HRD represents the ratio of the capacity at a discharge current of 200 mA g^{-1} to that at 8 mA g^{-1}. M_S is the saturated magnetic susceptibility, which reflects the total volume of the metallic nickel clusters imbedded in the alloy surface. C is the surface double-layer capacitance obtained from alternative current (AC) impedance measurement.

Alloy	Fe-Content (at%)	Max H Storage Capacity at 30 °C (wt %)	Discharge Capacity at 8 mA g^{-1} (mAh g^{-1})	Electrochemical to Gaseous Phase Capacity Ratio (%)	HRD (%)	M_S (Memu g^{-1})	C (Farad g^{-1})
Fe1	0.0	1.43	351	92	93	1016	0.30
Fe2	1.2	1.39	347	93	93	835	0.42
Fe3	2.4	1.37	346	94	94	718	0.37
Fe4	3.6	1.39	335	90	92	481	0.71
Fe5	4.7	1.41	345	91	87	341	0.82

Charge retention was measured using the following procedure. First, the cells were fully charged at a rate of C/10 and discharged at C/5, and the initial discharge capacity was obtained. The cells were then fully charged at C/10 and stored at room temperature (RT) for 7, 14 and 30 days before being discharged at C/5. The remaining capacities after 7, 14 and 30 days were normalized by the initial discharge capacity.

Peak power was measured using the pulse-discharge method. The cells were discharged at C/3 to a 50% depth-of-discharge (DOD) and then discharged using a 30 s pulse at a 2C/3 rate. The voltages at the end of each pulse were recorded and the peak power was calculated [6,8]. The peak power measurements were performed every 50 cycles at 50% DOD until the cells reached the end of their cycle life.

The sealed battery cells were kept in a Blue M Oven (TPS Thermal Power Solutions, White Deer, PA, USA) and tested using a Maccor Battery Cycler. The cell capacity tests were performed at different rates (C/5, C/2, 1C and 2C) and at different temperatures (RT, -10 °C and -20 °C). The cycle life tests were performed at RT and 50 °C through repeated charge/discharge cycling at a rate of C/2. For the cycling tests at RT and 50 °C, each charge process was considered complete when the cells reached 105% state-of-charge (calculated based on the initial discharge capacity), and each discharge process achieved a cutoff voltage of 0.9 V. Cycle life ended when the cell capacity dropped below 70% of the initial capacity after the formation process. After cycle life testing at RT, the cells were disassembled for failure mechanism analysis. The remaining electrolyte was removed with a Soxhlet extractor (Thermo Fisher Scientific, Walthan, MA, USA). A JEOL-JSM6320F scanning electron microscope (SEM) with energy dispersive spectroscopy (EDS) capabilities was used to study the morphology and composition of the electrodes after cycling. The solution after etching was analyzed by a Varian Liberty 100 inductively coupled plasma-optical emission spectrometer (ICP, Agilent Technologies, Santa Clara, California, USA).

3. Results and Discussion

3.1. Alloy Properties

Structural, hydrogen storage and electrochemical properties (tested in a flooded half-cell configuration) of the five superlattice HAAs (Fe1 to Fe5) were previously introduced in Part 1 [10] and their capacities (both gaseous phase and electrochemical), HRD, saturated magnetic susceptibility (M_S) and surface double-layer capacitance (C) results are shown in Table 1. Partial substitution of Ni with Fe in Mm-based superlattice HAAs has been shown to increase the abundance of the Ce_2Ni_7 (hexagonal) phase, which is favorable in terms of electrochemical performances [13–15] and decreases the relatively unfavorable $NdNi_3$ phase. However, the total volume of surface metallic Ni clusters, which reflects the surface catalytic capability [16], decreases with increasing Fe-content (shown in Table 1, M_S decreases dramatically from 1016 memu g^{-1} for Fe1 to 341 memu g^{-1} for Fe5). Consequently, the discharge capacity does not increase and HRD deteriorates with increasing Fe-content in the alloy (shown in Table 1, HRD decreases from 93% for Fe1 to 87% for Fe5 in the half-cell tests). Alloy surface treatment may be a feasible way to overcome this issue [17].

3.2. Cell Capacities at Room Temperature and Low Temperature

Full-cell test results for Fe1 to Fe5 measured at four different rates (C/5, C/2, 1C and 2C) and RT are shown in Figures 1 and 2a shows the cell capacities as a function of the Fe-content of the alloy. At lower rates the five cells exhibit similar capacities; at a discharge rate of C/5, the capacities vary between 4.34 and 4.44 Ah; at a discharge rate of C/2, the capacities decrease to 4.16 to 4.27 Ah. Further increases in discharge rate result in lower capacities and all the samples demonstrate similar capacities at the same rate, except for Fe3 which has the lowest capacities among all alloys. As seen in Table 1, M_S decreases with increasing Fe-content. Consequently, the surface catalytic capability weakens and high-rate performance deteriorates. The low capacities obtained for Fe3, especially at higher rates (1C and 2C), are due to decreased M_S. Although further increases in the Fe-content of the alloy leads to an even smaller M_S, the double-layer capacitance C increases dramatically, indicating the existence of a much larger surface area for the higher Fe-content samples. The large surface area may compensate the loss of surface Ni clusters and lead to comparable capacities obtained at high Fe-content (Fe4 and Fe5).

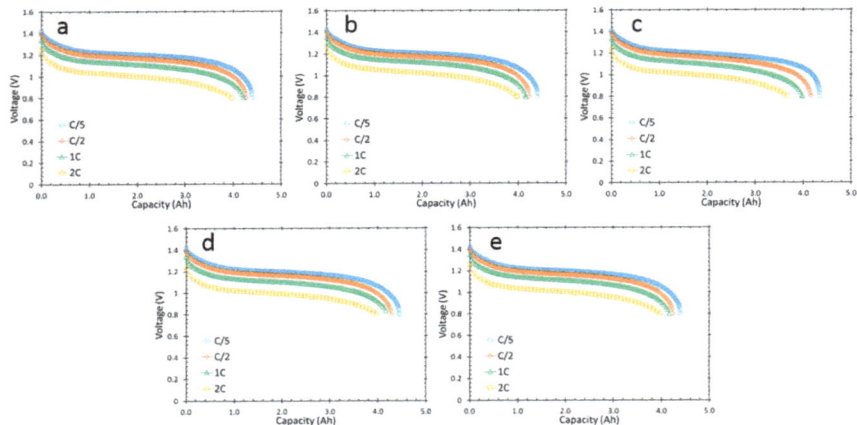

Figure 1. RT discharge voltage curves at four different rates (C/5, C/2, 1C and 2C) for cells (**a**) Fe1, (**b**) Fe2, (**c**) Fe3, (**d**) Fe4 and (**e**) Fe5.

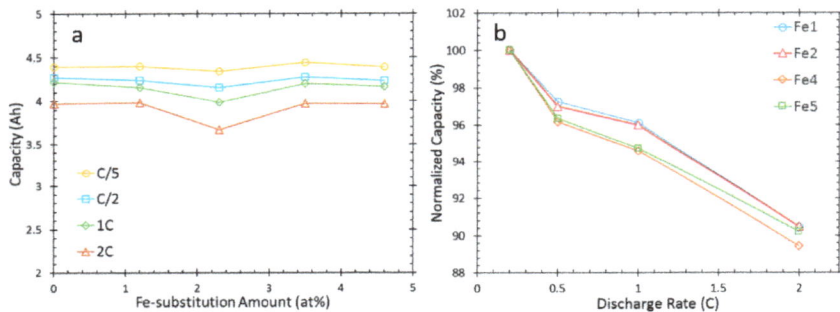

Figure 2. (**a**) Cell capacities at different rates as functions of the amount of Fe-substitution in the alloy and (**b**) normalized capacities as functions of discharge rate for cells Fe1, Fe2, Fe4 and Fe5.

It is difficult to establish a correlation between the cell capacity and Fe-content in the alloy from the RT capacity tests. Thus, the normalized capacities are compared in Figure 2b with the focus on the Fe-free (Fe1), low-Fe (Fe2) and high Fe-content alloys (Fe4 and Fe5). The capacities at various rates are normalized for each sample by the capacity at C/5. At higher rates above C/5, Fe1 and Fe2 demonstrate better rate capability than Fe4 and Fe5.

To further study the effect of the Fe-substitution on rate capability, the cells were tested at low temperatures (LT), specifically −10 °C and −20 °C. Discharge voltage curves for Fe1, Fe2, Fe4 and Fe5 at −10 °C and −20 °C are shown in Figure 3a and b, respectively, and the details of discharge capacity and mid-point voltage are listed in Table 2. A discharge rate of C/2 was used for the LT charge/discharge tests. Discharge capacities and mid-point voltages at −10 °C and −20 °C for Fe1, Fe2, Fe4 and Fe5 are summarized in Figure 3c. The discharge capacity and mid-point voltage exhibit similar trends, mainly a slight decrease as Fe-content increases to 3.6% (Fe4) followed by a sharp decrease as Fe-content increases to 4.7% (Fe5). Fe5 shows inferior LT electrochemical performance when compared to the Fe-free cell (Fe1), with a 12% lower discharge capacity and a 9% lower mid-point voltage at −20 °C. Higher Fe-content leads to a lower discharge capacity and a lower mid-point discharge voltage, which is consistent with the general trends in AC impedance and magnetic susceptibility measurements reported in Part 1 [10], where *RC* (the product of charge-transfer

313

resistance and double-layer capacitance, reflecting the surface catalytic capability) increases and Ms decreases with increasing Fe-content in the alloy. In previous studies, Fe rendered positive and negative influences on the LT performance of AB_2 [18] and AB_5 [19] HAAs, respectively. The impact of the addition of Fe on the LT performance of A_2B_7 HAA is closer to that of AB_5 HAA.

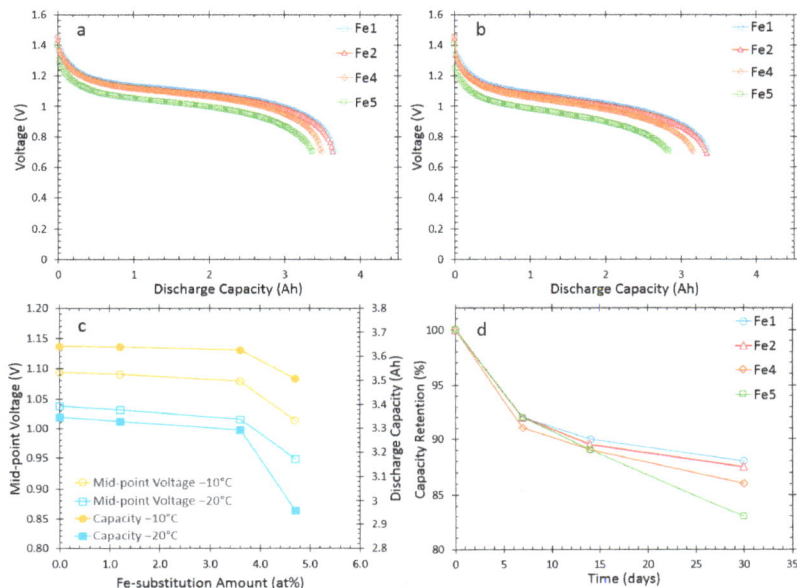

Figure 3. The discharge voltage curves obtained using a rate of C/2 at (**a**) −10 °C and (**b**) −20 °C; (**c**) a summary of the discharge capacity and mid-point voltage at −10 °C and −20 °C; and (**d**) 30-day capacity retention at RT for cells Fe1, Fe2, Fe4 and Fe5.

Table 2. Summary of discharge capacities at different rates and temperatures, mid-point voltages at different temperatures and 30-day charge retention. $Q_{C/2}$ and Q_{2C} are the discharge capacities measured at C/2 and 2C, respectively.

Alloy	$Q_{C/2}$ at RT (Ah)	Q_{2C} at 20 °C (Ah)	$Q_{C/2}$ at −10 °C (Ah)	$Q_{C/2}$ at −20 °C (Ah)	30-day Charge Retention (%)	Mid-Point Voltage at −10 °C (V)	Mid-Point Voltage at −20 °C (V)
Fe1	4.27	3.97	3.64	3.35	88	1.093	1.038
Fe2	4.27	3.97	3.63	3.34	87	1.090	1.032
Fe4	4.27	3.97	3.62	3.29	86	1.079	1.015
Fe5	4.23	3.96	3.51	2.96	83	1.013	0.949

Charge retention tests for Fe1, Fe2, Fe4 and Fe5 were performed at RT at the end of days 7, 14 and 30, and the results are shown in Figure 3d. The charge/discharge rate used for the tests was C/5 and the capacities have been normalized by the original capacity. Cells Fe4 and Fe5, with higher Fe-content, exhibited lower capacity retention than Fe1 and Fe2 after 14 days. At the end of 30 days, capacity retention followed the trend of Fe1 (88%) > Fe2 (87%) > Fe4 (86%) > Fe5 (83%). Capacity retention properties of the Mn- or Co-substituted superlattice HAAs were studied previously. The capacity loss observed in the Mn-substituted HAAs was attributed to micro-shortages between the positive and negative electrodes caused by oxides of Mn and Zn [20]. For the Co-substituted HAAs, the high corrosion rate of Co in the alkaline electrolyte resulted in capacity loss [21]. The alloys in this study were designed based on a Co-containing superlattice HAA, and Fe1 has the same composition as C3 in [21]. The lower charge retention for cells with high Fe-content HAAs indicates that, other than the corrosion of Co, there are other factors causing further reduction in capacity after storage. More severe oxidation in the Fe containing alloys may have a role in capacity loss as well, which was observed by SEM/EDS and will be presented in the failure analysis section. Fe improved the charge retention properties of AB_2 [18] HAAs and deteriorated those of AB_5 [19] HAAs. Similar to LT performance, the effects of Fe for A_2B_7 HAA are similar to those for AB_5 HAA with regard to charge retention performance.

3.3. Cycle Life and Peak Power

Cycle life performances of cells Fe1 to Fe5 were measured at RT and 50 °C. Results are shown in Figure 4 and summarized in Table 3. The charge processes were terminated when the cells reached 105% state-of-charge (calculated based on the initial discharge capacity) and the discharge process was finished at a cutoff voltage of 0.9 V. A rate of C/2 was used for both the charge and discharge processes. At RT, the Fe-free cell (Fe1) showed the highest cycle life—1055 cycles. Cycle life decreases dramatically with increasing Fe-content. Fe2 contained 1.2 at% Fe and demonstrated a cycle life of 720. When the Fe-content increased to 2.4 at% (Fe3) and above (Fe4 and Fe5), cycle life dropped to approximately 400 cycles. At 50 °C, the cycle life performances of all cells were greatly reduced, but exhibited a similar trend increasing Fe-content in the alloy. Fe1 showed a cycle life of 345, while cells with high Fe-content HAAs (Fe3, Fe4 and Fe5) demonstrated cycle numbers of less than 200. The accelerated capacity degradation at a temperature above 50 °C is common for HAA and can be attributed to higher degrees of oxidation, leaching and poisoning of the positive electrode [22,23]. The rapid deterioration in cycle life caused by increasing Fe-content in A_2B_7 HAAs was seen previously in AB_5 HAAs [19].

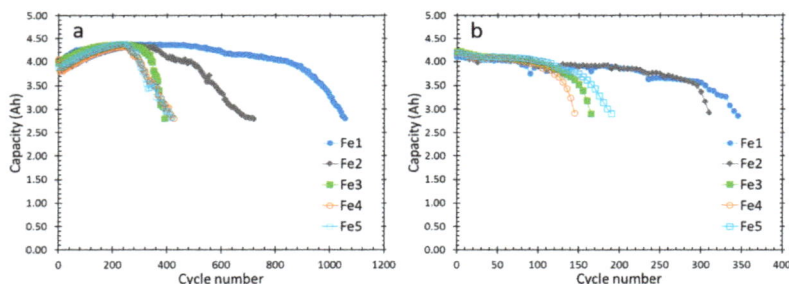

Figure 4. Cycle life performances of cells Fe1 to Fe5 measured at (**a**) RT and (**b**) 50 °C at a charge/discharge rate of C/2.

Table 3. Cycle life and peak power performances for cells Fe1 to Fe5. The cycle life tests were performed at RT and 50 °C, at a charge/discharge rate of C/2. Peak power was measured at 50% depth-of-discharge (DOD).

Alloy	Cycle life at C/2, RT	Cycle life at C/2, 50 °C	Initial Peak Power at 50% DOD (W kg^{-1})	Cycle Life (Peak Power Reached 100 W kg^{-1})
Fe1	1055	345	192	550
Fe2	720	310	195	450
Fe3	390	165	190	450
Fe4	425	145	179	450
Fe5	420	190	181	450

The peak power of cells Fe1 to Fe5 was measured every 50 cycles at 50% DOD and RT. The results are shown in Figure 5 and the initial peak power data is listed in Table 3. Details of the testing method can be found in our previous studies on Mn- and Co-substitutions in A_2B_7 [6,8]. During early cycling, cells with high Fe-content HAAs (Fe4 and Fe5) show lower peak power values than those with the Fe-free

(Fe1) and low Fe-content HAAs (Fe2 and Fe3). After 200 cycles, there was a clear trend as the peak power first decreases and then increases as Fe-content increases, with Fe3 showing the lowest peak power. Fe1 demonstrates the best power stability among all the tested cells, which was consistent with its superior cycle stability (Figure 4). The trend of peak power change with the Fe-content was consistent with the findings regarding RT rate capability (Figure 2a), since the peak power also depends on surface catalytic capability and the surface area of the HAA. Fe3 has a much lower amount of catalytic Ni clusters compared to Fe1 and a much smaller surface area than Fe4 and Fe5, which results in its low peak power.

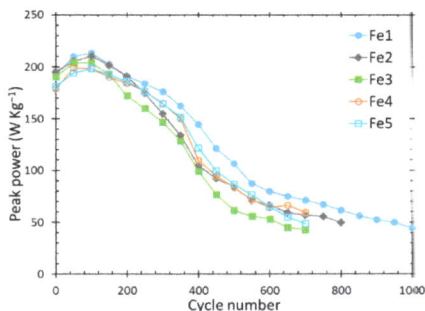

Figure 5. Peak power for cells Fe1 to Fe5 measured at 50% DOD and RT.

3.4. Failure Analysis

After cycle life tests, the cells were taken apart and their failure modes were studied by SEM, EDS and ICP. SEM backscattered electron images (BEIs) of the positive electrodes from cells Fe1 to Fe5 cycled at RT are shown in Figure 6. EDS was performed to measure the average chemical composition across a clean region without Ni foam, at 1000× magnification. The chemical compositions of the cycled positive electrodes were measured by EDS and are listed in Table 4.

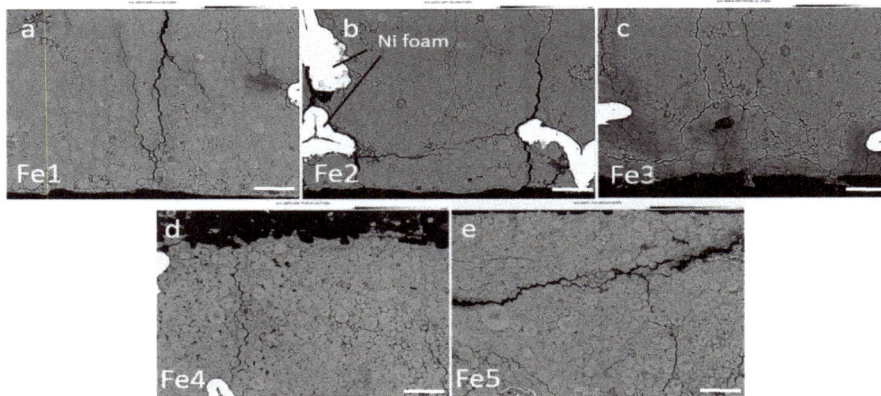

Figure 6. SEM micrographs of the positive electrodes of cells (**a**) Fe1, (**b**) Fe2, (**c**) Fe3, (**d**) Fe4 and (**e**) Fe5 after cycle life testing at RT. Magnification: 300×. The scale bar (white) represents 50 μm. The bright white regions in (**b–d**) are the Ni-foam current collectors.

Table 4. Chemical compositions (at%) measured by EDS of the positive electrodes after cycle life testing at RT.

Cell	Ni	Co	Zn	Al
Fe1	76.5	18.0	2.3	3.2
Fe2	75.7	18.5	2.6	3.2
Fe3	76.6	18.0	2.8	2.6
Fe4	74.9	20.2	2.7	2.2
Fe5	75.1	20.2	2.6	2.1

The most noticeable composition changes in the positive electrodes include the decrease in Al content and increase of Co content as the Fe-content in the alloy increases. The original positive electrode is Al free and the only Al source in the cell is the negative electrode. It has been reported that Al can leach from Co substituted superlattice HAAs and higher Co contents result in a higher degree of Al leaching [8]. However, Al is essential to superlattice HAA due to the ability to stabilize the structure against amorphization [24]. The alloys in this study contain 4.7 to 4.8 at% Co, which contributes to a high Al content (3.2 at%) in the cycled positive electrode of the cell with the Fe-free alloy (Fe1). As the Fe-content of the alloy in the negative electrode increases, the Al content in the heavily cycled positive electrode decreases to 2.1 at% for Fe5. Meanwhile, the Co content in the positive electrode increases from 18.0% to 20.2%. Therefore, Fe-substitution inhibits the leaching of Al from the superlattice HAA, but promotes the leaching of Co from the same place. Since the evolution of the Al and Co leaching are inversely correlated with the Fe-content, they eliminate the possibility of linking higher Al content to higher cycle number in cells with low Fe-content HAAs. Moreover, it is well known that the addition of Al in the positive electrode promotes the formation of α-Ni(OH)$_2$, which has a higher specific capacity than the initial β-Ni(OH)$_2$ [25]. However, the volume expansion caused by the β to α phase change leads to the swelling of the positive electrode [26]. Upon cycling, the developed stress may cause cracking and pulverization, which will eventually terminate cell activity. This is the reason α-Ni(OH)$_2$ inhibitors, such as Zn or Cd, are added to ensure a long cycle life [27–30]. Since Fe-substitution can inhibit leaching of Al from the negative electrode and, in turn, the formation of α-Ni(OH)$_2$, higher Fe-content is ideally beneficial to cycle life performance. However, the cycle life test presents the opposite result, specifically higher Fe-content leads to shorter cycle life. Furthermore, the SEM micrographs in Figure 6 show that the spherical Ni(OH)$_2$ particles are in good shape for all cycled samples and no obvious swelling or pulverization is observed. The effects of Al leaching from the negative electrodes is not observed. Therefore, it is unlikely that the shortened cycle life of Fe3 to Fe5 is caused by the degradation of the positive electrode.

The chemical compositions of the cycled negative electrodes were measured using ICP and the data is listed in Table 5. After cycle testing at RT, the composition of each negative electrode does not change significantly compared to that of the original HAA powder (data shown in Table 1 in Part 1) except that the Al content in all the samples decreases by nearly 1%, which results in a corresponding increase in Al in the positive electrodes. The increase of the Fe-content and decrease of the Ni content from Fe1 to Fe5 occur by alloy design. The SEM BEI micrographs of the cycled negative electrodes from cells Fe1 to Fe5 are shown in Figure 7. The cycled Fe1 (Figure 7a) exhibits severe pulverization and the HAA particles break into smaller pieces and gradually lose electrical connections to the current collector, which is a common failure mode for Ni/MH batteries [22,31,32]. Fe2 also suffers from severe pulverization, as shown in Figure 7b. In addition, the large darker grey colored area indicates that oxidation occurs around the pulverized HAA particles, which was confirmed by spot EDS measurements. The oxygen content in the grey area varies between 34 to 54 at%, as measured by EDS. Fe3 to Fe5 exhibit much shorter cycle life performances than Fe1, and their SEM BEI micrographs (Figure 7c–e) do not show signs of severe pulverization. The large HAA chunks do not appear to break into pieces and still occupy the largest portion of the surfaces. Instead, the large grey areas observed in Fe3 to Fe5 suggest severe oxidation, which was also confirmed by EDS measurements.

For Fe5, the grey area occupies the majority of the image, indicating that severe oxidation occurred in all the samples. Pressure-composition-temperature (PCT) hysteresis, a strong indicator for the inclination for HAA pulverization [33], for this series of alloys [10] does not support the direct link of degree-of-pulverization to the Fe-content. The heavy pulverization found in cells with Fe-free or low-Fe-content HAAs is due to the large number of cycles.

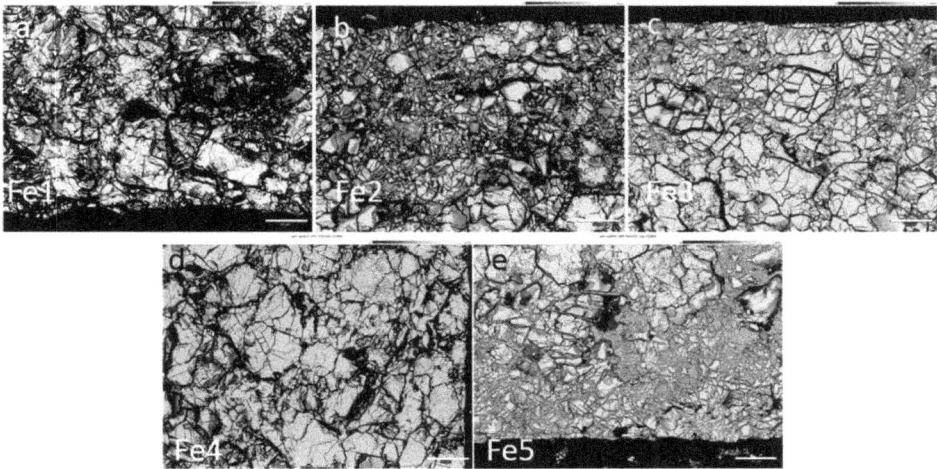

Figure 7. SEM micrographs of the negative electrodes of cells (**a**) Fe1, (**b**) Fe2, (**c**) Fe3, (**d**) Fe4 and (**e**) Fe5 after cycle life testing at RT. The scale bar represents 50 μm.

Table 5. Chemical compositions (at%), determined by ICP, of the negative electrodes after cycle life testing.

Cell	Mm	Ni	Co	Mg	Al	Fe
Fe1	19.1	69.5	4.6	3.7	3.1	0.0
Fe2	18.5	68.3	4.9	3.7	3.2	1.4
Fe3	19.2	66.7	4.6	3.8	3.3	2.4
Fe4	18.6	66.0	5.1	3.7	3.2	3.5
Fe5	19.2	64.3	4.6	3.8	3.0	5.1

SEM/EDS analyses were used to study the cross-section of the positive electrode/separator/negative electrode sandwich structure of the heavily oxidized Fe3 to Fe5 after cycle life testing at RT. The SEM BEI micrographs are shown in Figure 8a–c and their corresponding oxygen-EDS (O-EDS) micrographs are presented in Figure 8d–f. Each figure, from top to bottom of the sandwich, consists of the negative electrode, separator and positive electrode. The O-EDS mappings for Fe3 and Fe4 indicate that the oxidation of the negative electrode occurs mainly around the particle/grain boundaries, while Fe5—which has the highest Fe-content—exhibits a heavily oxidized surface across the negative electrode. This finding is consistent with the SEM BEI study above, confirming that the cell failure is caused by severe oxidation instead of pulverization with increasing Fe-content in HAAs. A heavily oxidized surface layer increases the internal resistance of the cell and leads to a shorter cycle life.

The cells after cycle life testing at 50 °C were also taken apart and investigated by SEM/EDS. Cycle life performance decreases dramatically with increasing temperature. In addition, the failure modes are similar. As shown in Figure 9, cells with the Fe-free (Fe1) and low-Fe-content (Fe2) HAAs show significant pulverization at the negative electrodes during the end of cycle life, which is the main cause of cell failure. When the Fe-content increases above 1.2 at% (Fe3 to Fe5), oxidation at the

HAA surfaces becomes more significant (Figure 9c–e). Fe5 exhibits the most severe oxidation among all alloys. Figure 10 shows the SEM BEI micrographs of the cross-sections and their corresponding O-EDS mappings for Fe2 to Fe5. O-EDS mapping does not show a significant oxygen level in the cycled negative electrode in cell Fe2, which indicates that oxidation should be excluded as the main cause of cell failure. For Fe3 to Fe5, the negative electrodes are severely oxidized, whereas no obvious pulverization is observed. In addition to negative electrode oxidation, the separators in Fe3 to Fe5 are squeezed by the expansion in the positive electrode and debris/deposits are found in the separators, which is attributed to electrolyte dry-out. Electrolyte dry-out is another typical failure mode and caused cell failure of the Co-substituted HAA cells tested at 50 °C. EDS measurements show that the debris/deposits contain Ni, Co, Mg and Al (mainly Ni and Co), which originated from the corroded negative electrode and formed a micro-shortage network that is detrimental to cycle life performance. Such debris/deposits were also observed in the high Fe-content cells Fe4 and Fe5 after cycling at RT.

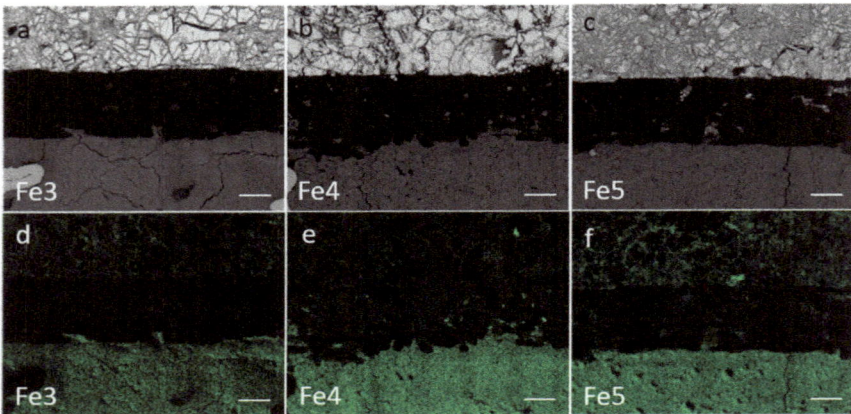

Figure 8. SEM micrographs of the cross-sections of cells (**a**) Fe3, (**b**) Fe4 and (**c**) Fe5 and their corresponding O-EDS elemental mappings (**d**) Fe3, (**e**) Fe4 and (**f**) Fe5 after cycle life testing at RT. The scale bar represents 50 μm. In each figure, the sandwich structure from top to bottom consists of the negative electrode, separator and positive electrode.

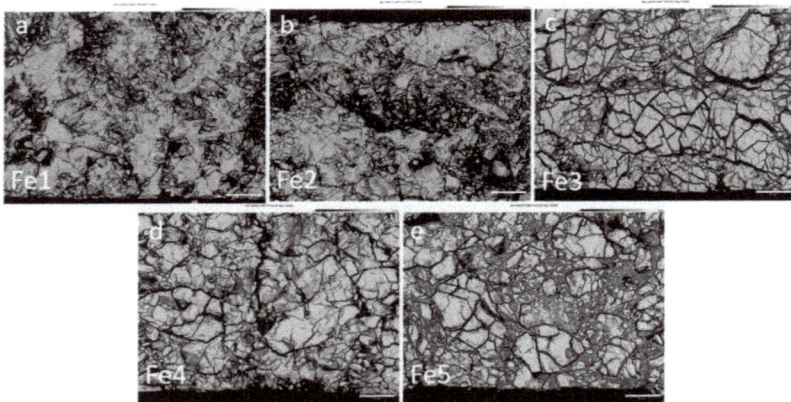

Figure 9. SEM micrographs of the negative electrodes of cells (**a**) Fe1, (**b**) Fe2, (**c**) Fe3, (**d**) Fe4 and (**e**) Fe5 after cycle life testing at 50 °C. The scale bar represents 50 μm.

The cells cycled at RT and 50 °C show similar failure mechanisms. Cells with Fe-free and low-Fe-content HAAs demonstrate long cycle life and eventually fail due to pulverization/disintegration of the negative electrode. With increasing Fe-content in HAAs, oxidation/corrosion of the negative electrode dominates and has two negative influences. First, surface oxidation may lead to increased internal resistance and deterioration of the cell capacity and cycle life. Second, during the oxidation/corrosion of HAAs, Ni, Co or other elements may leach from the negative electrode and deposit on the separator. Such deposits/debris may eventually develop into a micro-shortage network and therefore reduce cell capacity, cycle life and shelf life.

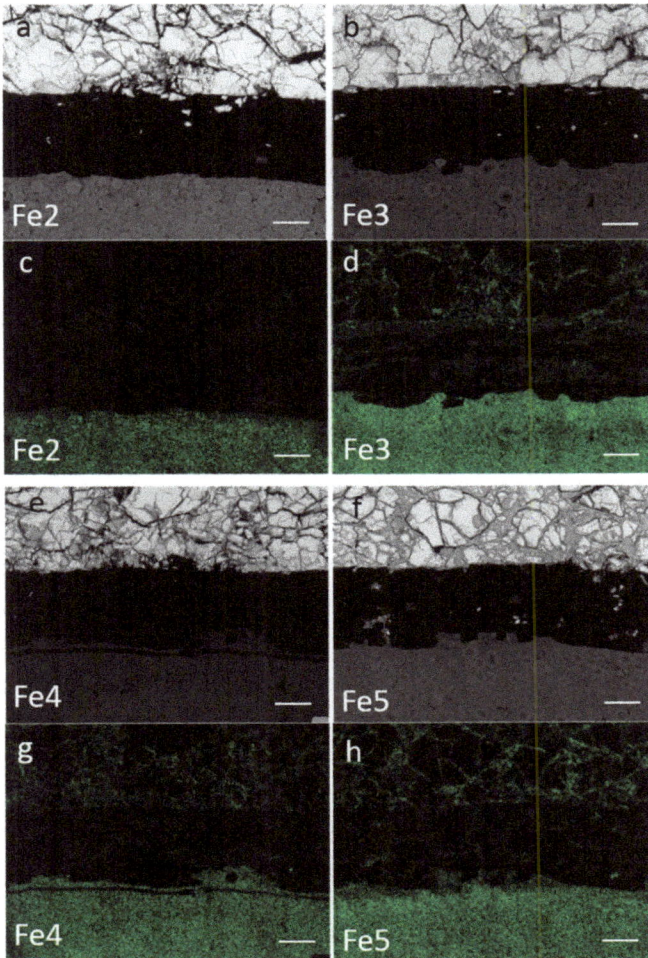

Figure 10. SEM micrographs of the cross-sections of cells (**a**) Fe2, (**b**) Fe3, (**e**) Fe4, and (**f**) Fe5 and their corresponding O-EDS elemental mappings of cells (**c**) Fe2, (**d**) Fe3, (**g**) Fe4 and (**h**) Fe5 after cycle life testing at 50 °C. The scale bar represents 50 μm. In each figure, the sandwich structure from top to bottom consists of the negative electrode, separator and positive electrode.

4. Conclusions

The effects of the Fe-substitution in Mm-based superlattice alloys on the performances of nickel/metal hydride batteries and their failure mechanisms were studied in Part 2 of this paper.

Due to the complexity involved with the Fe-substitution and limited number of compositions selected, the variations in some properties are not very smooth; however, general trends are still observable and reported. Although the incorporation of Fe promotes the formation of a favorable Ce_2Ni_7 phase, the full-cell capacities at room temperature, $-10\ ^{\circ}C$ and $-20\ ^{\circ}C$, cycle life performances at room temperature and $50\ ^{\circ}C$, capacity retention, mid-point discharge voltage and peak power all deteriorate with increasing Fe-content, especially in samples with high Fe-content. Fe in the alloy also inhibits leaching and migration of Al from the negative electrode to the positive electrode, while facilitating the leaching of Co from the negative electrode. However, the cycle life performance was dominated by the negative electrode in this study. Cells with high Fe-content alloys show much lower room temperature and $50\ ^{\circ}C$ cycle life performances and peak power than those with Fe-free (Fe1) and low Fe-content (Fe2) alloys. The failure of cells Fe1 and Fe2 is mainly attributed to the pulverization of the negative electrode, while cells with high Fe-content HAAs (Fe3 to Fe5) suffer from severe oxidation/corrosion of the negative electrode, which greatly limits cycle life performance at room temperature and $50\ ^{\circ}C$.

Acknowledgments: The authors would like to thank their coworkers at BASF–Ovonic: David A. Pawlik, Alan Chan and Ryan J. Blankenship for their help on material characterizations, and Sui-ling Chen, Cheryl Setterington and Nathan English for battery measurements.

Author Contributions: Tiejun Meng designed the experiments and analyzed the results. Kwo-Hsiung Young interpreted the data. Jean Nei prepared the electrode samples and conducted the magnetic measurements. John M. Koch prepared samples for failure analysis. Shigekazu Yasuoka designed and obtained the test samples.

Conflicts of Interest: The authors declare no conflict of interest.

Abbreviations

Mm	Misch metal
HAA	Hydrogen absorbing alloy
Ni/MH	Nickel/metal hydride
HRD	High-rate dischargeability
RT	Room temperature
DOD	Depth-of-discharge
SEM	Scanning electron microscope
EDS	Energy dispersive spectroscopy
M_S	Saturated magnetic susceptibility
C	Double-layer capacitance
LT	Low temperature
R	Charge-transfer resistance
BEI	Backscattered electron images
PCT	Pressure-composition-temperature
O-EDS	Oxygen-energy dispersive spectroscopy

References

1. Yasuoka, S.; Magari, Y.; Murata, T.; Tanaka, T.; Ishida, J.; Nakamura, H.; Nohma, T.; Kihara, M.; Baba, Y.; Teraoka, H. Development of high-capacity nickel-metal hydride batteries using superlattice hydrogen-absorbing alloys. *J. Power Sources* **2006**, *156*, 662–666.
2. Takasaki, T.; Nishimura, K.; Saito, M.; Fukunaga, H.; Iwaki, T.; Sakai, T. Cobalt-free nickel-metal hydride battery for industrial applications. *J. Alloy. Compd.* **2013**, *580*, S378–S381.

3. Liu, J.; Han, S.; Li, Y.; Yang, S.; Chen, X.; We, C.; Ma, C. Effect of Pr on phase structure and cycling stability of La-Mg-Ni-based alloys with A_2B_7- and A_5B_{19}-type superlattice structure. *Electrochim. Acta* **2015**, *184*, 257–263.
4. Kai, T.; Ishida, J.; Yasuoka, S.; Takeno, K. The effect of nickel-metal hydride battery's characteristics with structure of the alloy. In Proceedings of the 54th Battery Symposium in Japan, Osaka, Japan, 7–9 Octorber 2013.
5. Young, K.; Wong, D.F.; Wang, L.; Nei, J.; Ouchi, T.; Yasuoka, S. Mn in misch-metal based superlattice metal hydride alloy—Part 1 Structural, hydrogen storage and electrochemical properties. *J. Power Sources* **2015**, *277*, 426–432.
6. Young, K.; Wong, D.F.; Wang, L.; Nei, J.; Ouchi, T.; Yasuoka, S. Mn in misch-metal based superlattice metal hydride alloy—Part 2 Ni/MH battery performance and failure mechanism. *J. Power Sources* **2015**, *277*, 433–442.
7. Wang, L.; Young, K.; Meng, T.; Ouchi, T.; Yasuoka, S. Partial substitution of cobalt for nickel in mixed rare earth metal based superlattice hydrogen absorbing alloy—Part 1 Structural, hydrogen storage and electrochemical properties. *J. Alloy. Compd.* **2016**, *660*, 407–415.
8. Wang, L.; Young, K.; Meng, T.; English, N.; Yasuoka, S. Partial substitution of cobalt for nickel in mixed rare earth metal based superlattice hydrogen absorbing alloy—Part 2 Battery performance and failure mechanism. *J. Alloy. Compd.* **2016**, *664*, 417–427.
9. Yasuoka, S.; Ishida, J.; Kichida, K.; Iniu, H. Effects of cerium on the hydrogen absorption-desorption properties of rare Earth-Mg-Ni hydrogen-absorbing alloys. *J. Power Sources* **2017**, *346*, 56–62.
10. Young, K.; Ouchi, T.; Nei, J.; Yasuoka, S. Fe-substitution for Ni in misch metal-based superlattice hydrogen absorbing alloys—Part 1 Structural, hydrogen storage, and electrochemical properties. *Batteries* **2016**, *2*, 34.
11. Kong, L.; Chen, B.; Young, K.; Koch, J.; Chan, A.; Li, W. Effects of Al- and Mn-contents in the negative MH alloy on the self-discharge and long-term storage properties of Ni/MH battery. *J. Power Sources* **2013**, *213*, 128–139.
12. Young, K.; Wu, A.; Qiu, Z.; Tan, J.; Mays, W. Effects of H_2O_2 addition to the cell balance and self-discharge of Ni/MH batteries with AB_5 and A_2B_7 alloys. *Int. J. Hydrog. Energy* **2012**, *37*, 9882–9891.
13. Young, K.; Yasuoka, S. Past, Present, and Future of Metal Hydride Alloys in Nickel-Metal Hydride Batteries. In Proceedings of the 14th International Symposium on Metal-Hydrogen Systems, Manchester, UK, 21–25 July 2014.
14. Yan, H.; Xiong, W.; Wang, L.; Li, B.; Li, J.; Zhao, X. Investigations on AB_3-, A_2B_7- and A_5B_{19}-type La-Y-Ni system hydrogen storage alloys. *Int. J. Hydrog. Energy* **2017**, *42*, 2257–2264.
15. Young, K.; Ouchi, T.; Nei, J.; Koch, M.J.; Lien, Y. Comparison among constituent phases in superlattice metal hydride alloys for battery applications. *Batteries* **2017**, submitted.
16. Young, K.; Huang, B.; Regmi, R.K.; Lawes, G.; Liu, Y. Comparisons of metallic clusters imbedded in the surface of AB_2, AB_5, and A_2B_7 alloys. *J. Alloy. Compd.* **2010**, *506*, 831–840.
17. Young, K.; Chang, S.; Lin, X. C14 Laves phase metal hydride alloys for Ni/MH batteries applications. *Batteries* **2017**, accepted.
18. Young, K.; Ouchi, T.; Huang, B.; Reichman, B.; Fetcenko, M.A. The structure, hydrogen storage, and electrochemical properties of Fe-doped C14-predominating AB_2 metal hydride alloys. *Int. J. Hydrog. Energy* **2011**, *36*, 12296–12304.
19. Young, K.; Ouchi, T.; Reichman, B.; Koch, J.; Fetcenko, M.A. Improvement in the low-temperature performance of AB_5 metal hydride alloys by Fe-addition. *J. Alloy. Compd.* **2011**, *509*, 7611–7617.
20. Shinyama, K.; Magari, Y.; Akita, H.; Kumagae, K.; Nakamura, H.; Matsuta, S.; Nohma, T.; Takee, M.; Ishiwa, K. Investigation into the deterioration in storage characteristics of nickel-metal hydride batteries during cycling. *J. Power Sources* **2005**, *143*, 265–269.
21. Teraoka, H. Development of Highly Durable and Long Life Ni-MH Batteries for Energy Storage Systems. In Proceedings of the 32th International Battery Seminar & Exhibit, Fort Lauderdale, FL, USA, 9–12 March 2015.
22. Young, K.; Yasuoka, S. Capacity degradation mechanisms in nickel/metal hydride batteries. *Batteries* **2016**, *2*, 3. [CrossRef]
23. Meng, T.; Young, K.; Koch, J.; Ouchi, T.; Yasuoka, S. Failure mechanisms of nickel/metal hydride batteries with cobalt-substituted superlattice hydrogen-absorbing alloy anodes at 50 °C. *Batteries* **2016**, *2*, 20. [CrossRef]

24. Yasuoka, S.; Ishida, J.; Kai, T.; Kajiwara, T.; Doi, S.; Yamazaki, T.; Kishida, K.; Inui, H. Function of aluminum in crystal structure of rare Earth-Mg-Ni hydrogen-absorbing alloy and deterioration mechanism of $Nd_{0.9}Mg_{0.1}Ni_{3.5}$ and $Nd_{0.9}Mg_{0.1}Ni_{3.3}Al_{0.2}$ alloys. *Int. J. Hydrog. Energy* **2017**, *42*, 11574–11583.

25. Young, K.; Wang, L.; Yan, S.; Liao, X.; Meng, T.; Shen, H.; Mays, W.C. Fabrications of high-capacity alpha-Ni(OH)$_2$. *Batteries* **2017**, *3*, 6. [CrossRef]

26. Singh, D. Characteristics and effects of γ-NiOOH on cell performance and a method to quantify it in nickel electrode. *J. Electrochem. Soc.* **1998**, *145*, 116–120.

27. Yuan, A.; Cheng, S.; Zhang, J.; Cao, C. Effects of metallic cobalt addition on the performance of pasted nickel electrodes. *J. Power Sources* **1999**, *77*, 178–182.

28. Jayashree, R.S.; Kamath, P.V. Suppression of the α-nickel hydroxide transformation in concentrated alkali: Role of dissolved cations. *J. Appl. Electrochem.* **2001**, *31*, 1315–1320.

29. Tessier, C.; Guerlou-Demourgues, L.; Faure, C.; Denage, C.; Delatouche, B.; Delmas, C. Influence of zinc on the stability of the β(II)/β(III) nickel hydroxide system during electrochemical cycling. *J. Power Sources* **2001**, *102*, 105–111.

30. Ravikumar, C.R.; Kotteeswaran, P.; Bheema Raju, V.; Murugan, A.; Santosh, M.S.; Nagaswarupa, H.P.; Prashantha, S.C.; Anil Kumar, M.R.; Shivakumar, M.S. Influence of zinc additive and pH on the electrochemical activities of β-nickel hydroxide materials and its applications in secondary batteries. *J. Energy Storage* **2017**, *9*, 12–24.

31. Zhou, X.; Young, K.; West, J.; Regalado, J.; Cherisol, K. Degradation mechanisms of high-energy bipolar nickel metal hydride battery with AB_5 and A_2B_7 alloys. *J. Alloy. Compd.* **2013**, *560*, S373–S377.

32. Young, K.; Ouchi, T.; Koch, J.; Fetcenko, M.A. Compositional optimization of vanadium-free hypo-stoichiometric AB_2 metal hydride alloy for Ni/MH battery application. *J. Alloy. Compd.* **2012**, *510*, 97–106.

33. Young, K.; Ouchi, T.; Fetcenko, M.A. Pressure-composition-temperature hysteresis in C14 Laves phase alloys—Part 1 Simple ternary alloys. *J. Alloy. Compd.* **2009**, *480*, 428–433.

![batteries logo] **batteries**

MDPI

Article

A Ni/MH Pouch Cell with High-Capacity Ni(OH)$_2$

Shuli Yan [1], Tiejun Meng [1], Kwo-Hsiung Young [1,2,]* and Jean Nei [1]

[1] BASF/Battery Materials—Ovonic, 2983 Waterview Drive, Rochester Hills, MI 48309, USA;
    shuli.yan@partners.basf.com (S.Y.); pkumeng@hotmail.com (T.M.); jean.nei@BASF.com (J.N.)
[2] Department of Chemical Engineering and Materials Science, Wayne State University, Detroit, MI 48202, USA
* Correspondence: kwo.young@basf.com; Tel.: +1-248-293-7000

Received: 26 September 2017; Accepted: 21 November 2017; Published: 4 December 2017

Abstract: Electrochemical performances of a high-capacity and long life β-α core-shell structured $Ni_{0.84}Co_{0.12}Al_{0.04}(OH)_2$ as the positive electrode active material were tested in a pouch design and compared to those of a standard β-$Ni_{0.91}Co_{0.045}Zn_{0.045}(OH)_2$. The core-shell materials were fabricated with a continuous co-precipitation process, which created an Al-poor core and an Al-rich shell during the nucleation and particle growth stages, respectively. The Al-rich shell became α-$Ni(OH)_2$ after electrical activation and remained intact through the cycling. Pouch cells with the high-capacity β-α core-shell positive electrode material show higher charge acceptances and discharge capacities at 0.1C, 0.2C, 0.5C, and 1C, improved self-discharge performances, and reduced internal and surface charge-transfer resistances, at both room temperature and $-10\,^{\circ}C$ when compared to those with the standard positive electrode material. While the high capacity of the core-shell material can be attributed to the α phase with a multi-electron transfer capability, the improvement in high-rate capability (lower resistance) is caused by the unique surface morphology and abundant interface sites at the β-α grain boundaries. Gravimetric energy densities of pouch made with the high-capacity and standard positive materials are 127 and 110 Wh·kg^{-1}, respectively. A further improvement in capacity is expected via the continued optimization of pouch design and the use of high-capacity metal hydride alloy.

Keywords: metal hydride alloy; nickel metal hydride battery; pouch cell; electrochemistry; alpha nickel hydroxide; core shell

1. Introduction

Transportation electrification is essential for controlling the greenhouse effect by reducing CO_2 emissions from burning fossil energy. Li-ion battery technology is the mainstream energy/power source for electric vehicle (EV) applications because of its relatively high gravimetric energy density. In a comparison of commercially available Li-ion batteries (first nine rows in Table 1), cylindrical cells with the size of 18650 made by Panasonic (Tokyo, Japan) show the highest volumetric and gravimetric energy densities at the cell level. However, their gravimetric energy density drops from 233 to 140 Wh·kg^{-1} in the transition from a single cell to a complete battery pack [1]. Other vendors offer either pouch- (aluminum laminated with plastics) or prismatic- (metal case) types of Li-ion batteries that increase the package density because no space is needed between adjacent cells. In comparison, pouch cells have both higher volumetric and gravimetric energy densities than prismatic cells.

Table 1. Gravimetric and volumetric energy densities of several batteries used in electric vehicle (EVs). G and LTO are graphite and Li-titanate anodes for Li-ion batteries, respectively. LMO, NCA, NMC, and LFP are Li-spinel, LiNiCoAl oxide, LiNiMnCo oxide, and LiFePO$_4$ cathodes for Li-ion batteries, respectively.

Cell Maker	Chemistry	Configuration	Energy Density (Wh·L−1)	Energy Density (Wh·kg−1)	Used in
AESC	G/LMO-NCA	Pouch	309	155	Leaf by Nissan
LG-Chem	G/NMC-LMO	Pouch	275	157	Zoe by Renault
Li-Tec	G/NMC	Pouch	316	152	Smart by Daimler
LG-Chem	G/Ni-rich	Prismatic	208	136	Bolt by General Motor
Li-Energy	G/LMO-NMC	Prismatic	218	109	i-MiEV by Mitsubishi
Samsung	G/NMC-LMO	Prismatic	243	132	500 by Fiat
Lishen	G/LFP	Prismatic	226	116	EV by Coda
Toshiba	LTO/NMC	Prismatic	200	89	Fit by Honda
Panasonic	G/NMC	Cylindrical	630	233	Model S by Tesla
BASF–Ovonic	Ni/MH	Pouch	427	145	Not yet

There are five types of commonly used nickel/metal hydride (Ni/MH) batteries (Figure 1). Their capacities, constructions, and pros and cons are compared in Table 2. Coin cells are used in personal computer memory backup and real-time clock applications. The cylindrical design is the most popular type for various portable consumer electronics. In the past, a stick design was put in personal digital assistants and cellphones. Hybrid electric vehicles (HEV), manufactured by Toyota Motor (Tokyo, Japan), use the HEV-type prismatic Ni/MH cells because of their excellent high-rate discharge capability. Stainless steel-cased prismatic Ni/MH batteries were used to power the first commercially available EVs (EV-1 by General Motor, Inc., Detroit, MI, USA) back in 1999 [2], but they were later replaced by Li-ion batteries by all of the EV makers because of the limitation in gravimetric energy density. In 2015, a program that was funded by the United States Department of Energy, ARPA-E RANGE, investigated the possibility of using a pouch design for Ni/MH batteries [3]. Inspired by the success of pouch design for Li-ion batteries, researchers designed Ni/MH pouch cells (Figure 23 in [4]) to achieve the following advantages: higher gravimetric and volumetric energies, a higher packaging density, a lower fabrication cost, an increase in safety in case of battery puncture (with electrodes in a flooded mode), and an increase in pack design flexibility (where bending is possible) [3] over the conventional hard-cased Ni/MH cells. At the end of the program, a 100-Ah pouch design with the advanced positive and negative active materials was proposed with an expected energy density of 145 Wh·kg^{-1} discharged at C/3 [4].

Ni/MH pouch cells in the ARPA-E RANGE program achieved three major accomplishments: the discovery of high-capacity Si-anode with ionic liquid electrolyte [5], use of ionic liquid as electrolyte [6], and a high-capacity β-α core-shell Ni(OH)$_2$ positive electrode [7]. This paper elaborates on the continuing efforts to validate the third item—high-capacity positive electrode material in the pouch design tested with various regimens.

Figure 1. Nickel/metal hydride (Ni/MH) batteries in (**a**) a coin cell, (**b**) a cylindrical cell, (**c**) a stick cell (small prismatic), (**d**) a hybrid electric vehicle (HEV) module, and (**e**) an EV module (large Prismatic).

Table 2. Comparisons among various cell packaging types for Ni/MH batteries.

Type	Capacity (Ah)	Case Material	Vent Cap	Pros	Cons
Coin	0.02–0.4	Stainless steel	No	High volume mass-production	Only for low-rate application
Cylindrical	0.3–10	Stainless steel	Yes	High volume mass-production	Limited capacity
Stick	1–2	Stainless steel	Yes	High packing density	Higher cost and lower energy density
HEV-prismatic	6.5	Plastic or metal	Yes	High power density, easy packing	Lower pressure rating, poor heat transfer
EV-prismatic	20–100	Stainless steel	Yes	Large format (>100 Ah)	High manufacture cost
Pouch	0.2–100	Aluminum laminated with plastics	Yes/No	High gravimetric energy density	Low pressure rating

2. Experimental Setup

Negative electrode active material is a commercially available misch metal-based metal hydride (MH) alloy (AB$_5$) with a nominal composition of La$_{10.5}$Ce$_{4.3}$Pr$_{0.5}$Nd$_{1.4}$Ni$_{60}$Co$_{12.7}$Mn$_{5.9}$Al$_{4.7}$ and a plateau pressure of about 0.06 MPa [8]; it was supplied by Eutectix (Troy, MI, USA). Positive electrode active materials (AP50 and WM12) were fabricated by a continuous stirred-tank reactor process [7,9,10] in BASF—Ovonic (Rochester Hills, MI, USA). Electrochemical charge/discharge tests were performed on an Arbin BT-2143 battery test station (Arbin, College Station, TX, USA). AC impedance measurements were performed on a Solartron S1287 potentiostat/galvanostat, with an S1255 frequency response analyzer (Solatron, Farnborough, Hampshire, UK).

2.1. Cell Assembely

Around 1.2 g mixture of positive electrode active material (hydroxide of nickel and other transition metals, 90 wt %) and polytetrafluorethylene-acetylene black composite (10 wt %) was spread evenly and compacted onto a 1″ × 1″ nickel foam substrate by applying a pressure of 250 MPa with a hydraulic press to form the positive electrode. The negative electrode was fabricated by a continuous dry compaction process, which directly pressed the AB$_5$ MH alloy powder onto an expanded nickel substrate. A negative-to-positive (N/P) capacity ratio of 1.4 was adopted to prevent sudden gassing in the pouch cell. Future development using a MH alloy with a lower plateau pressure and a flatter pressure-concentration-temperature isotherm can reduce this N/P ratio. Individual Ni tab strips were welded onto both the negative and positive electrodes. A piece of grafted polypropylene/polyethylene was used as a separator. 30 wt % aqueous KOH solution was used as electrolyte. EQ-alf-400-7.5M (MTI Corporation, Richmond, CA, USA), an aluminum foil that was laminated with polyamide (inside) and polypropylene (outside), was used for the construction of the pouch cell. The laminated aluminum foil was cut into 4″ × 2.2″ pieces, and each piece was heat sealed at 200 °C and 0.3 MPa for 1.5 s with a MTI MSK-140 heating sealer without vacuum to form a 2″ × 2.2″ pouch for cell housing.

2.2. Activation Process

After completing the pouch assembly, each cell was set aside for 3 h. Next, each cell was charged at 0.1C (calculated based on the positive electrode active material weight) at room temperature (RT) for 10 h and relaxed for 10 min before being discharged at 0.1C to 0.9 V This process was repeated until a stablized discharge capacity was obtained.

2.3. Charge Rate Capability

After activation, each cell was set aside for 3 h at RT. Next, each cell was charged at 0.1C to 90% state-of-charge (SOC) to a cutoff voltage of 1.6 V, relaxed for 10 min, before being discharged at 0.1C to 0.9 V, and finally relaxed for 10 min. This charge/discharge process was repeated, but with charge rates of 0.2C, 0.5C, and 1C in the next three cycles.

After completing the RT charge rate capability test, each cell was charged/discharged at 0.1C for five cycles to reactivate. Each cell was then placed in a temperature-controlled chamber at −10 °C for 3 h. Next, −10 °C charge rate capability test was conducted in the same manner as the experiment at RT described above.

2.4. Discharge Rate Capability

Test procedure for discharge rate capaibility is similar to that for charge rate capability, but with the charge rate fixed at 0.1C and discharge rates of 0.1C, 0.2C, 0.5C, and 1C.

2.5. Self-Discharge

Each cell was charged at 0.1C to 60% or 80% of SOC after activation and then relaxed for 3 h. More relaxation time may be required to ensure that stablized open-circuit potential was obtained. Open-circuit potential was again obtained every seven days. Self-discharge measurements were performed at both RT and −10 °C.

2.6. Internal Resistance Measurement

After activation, each cell was charged at 0.1C to 60% SOC and relaxed for 1 h at RT. Next, each cell was discharged at 0.1C for 15 s, relaxed for 10 min, before being charged at 0.1C for 15 s, and finally relaxed for 10 min. This pulse discharge/charge process continued on, but with discharge/charge rates of 0.2C, 0.5C, and 1C. Internal resistance (R_{int}) at RT was calculated based on the equation,

$$R_{int} = \Delta V / \Delta I \tag{1}$$

where V and I are the voltage at the end of each charge or discharge pulse and corresponding rate, respectively.

After completing the RT R_{int} measurement, each cell was charged/discharged at 0.1C for two cycles to reactivate. Each cell was then placed in a temperature-controlled chamber at −10 °C for 3 h, charged at 0.1C to 60% SOC, and relaxed for 1 h. Next, −10 °C R_{int} measurement was conducted in the same manner as the experiment at RT described above. Equation 1 was also used to calculate R_{int} at −10 °C.

2.7. Charge-Transfer Resistance Measurement

Each cell was charged at 0.1C to 60% SOC and relaxed for 30 min. Next, AC impedance measurements were performed with an amplitude of 10 mV and a frequency range of 0.005 to 10 KHz at RT and 0.002 to 100 KHz at −10 °C. Charge-transfer resistances (R_{ct}) at RT and −10 °C of each cell were obtained from the resulted Cole-Cole plots.

3. Results and Discussion

3.1. Ni(OH)$_2$ Selection

Two types of positive electrode materials are compared in this study—AP50 and WM12. The important parameters of these two materials are compared in Table 3. AP50 with a nominal cation composition of $Ni_{91}Co_{4.5}Zn_{4.5}$, a standard positive electrode material for commercial Ni/MH batteries, was chosen as control. During charge, AP50 goes through a transformation from β-Ni(OH)$_2$ to β-NiOOH (and vice versa during discharge). WM12 is a newly developed high-capacity Ni(OH)$_2$-based positive electrode material [4]. In WM12, zinc—the γ-phase inhibitor in AP50—is replaced by aluminum, the γ-phase promoter, which results in a nominal cation composition of $Ni_{84}Co_{12}Al_4$. After activation, WM12 has a structure consisted of a β-rich core and a α-rich shell that was confirmed by both scanning and transmission electron microscopes [7]. Although being similar in shape (both spherical), WM12 has a highly-decorated surface while AP50's surface is relatively smooth

(Figure 2). WM12's high surface area may contribute positively in the electrochemical environment. Half-cell capacity measurement results for WM12 and AP50 (using a counter electrode made with the standard AB$_5$ MH alloy) are shown in Figure 3. After activation, WM12 demonstrates a 40% increase in discharge capacity over AP50 (350 vs. 250 mAh·g^{-1}) at 30 mA·g^{-1}. The extra capacity of WM12 comes from the α-shell, which is capable of an electron transfer of up to 1.67 electrons per Ni atom during the redox reaction [11,12]. Other than the usual transformation from β-Ni(OH)$_2$ to β-NiOOH during charge (and vice versa during discharge), the evolution in half-cell voltage profile of WM12 shows that once the activation (Figure 4a–c, where only a single voltage plateau is observed during charge or discharge) is complete, a transformation of α-Ni(OH)$_2$ to γ-NiOOH during charge (and vice versa during discharge) appears (Figure 4d,e, where two voltage plateaus are observed during charge or discharge). Although WM12 delivers a higher gravimetric energy density, its lower tap density (0.9 g·cc^{-1}) [7] when compared to that of AP50 (2.3 g·cc^{-1}) decreases the volumetric energy density of the battery. Details about the microstructures of WM12 at different states and after cycling were reported previously [7]. In this study, pouch cells made with AP50 and WM12 are identified as Cell AP50 (control) and Cell WM12 (experimental).

Table 3. Various properties of AP50 and WM12.

Materials	AP50	WM12
Composition	Ni$_{0.91}$Co$_{0.045}$Zn$_{0.045}$(OH)$_2$	Ni$_{0.84}$Co$_{0.12}$Al$_{0.04}$(OH)$_2$
Original structure	β-Ni(OH)$_2$	β-Ni(OH)$_2$
Structure after activation	β-Ni(OH)$_2$	α-β core-shell Ni(OH)$_2$
Tap density	2.3 g·cc^{-1}	0.9 g·cc^{-1}
Discharge capacity	250 mAh·g^{-1}	350 mAh·g^{-1}
BET surface area	13.7 m^2·g^{-1}	51.98 m^2·g^{-1}
Pore density	0.022 cc·g^{-1}	0.027 cc·g^{-1}
Average pore size	19.7 Å	24.6 Å

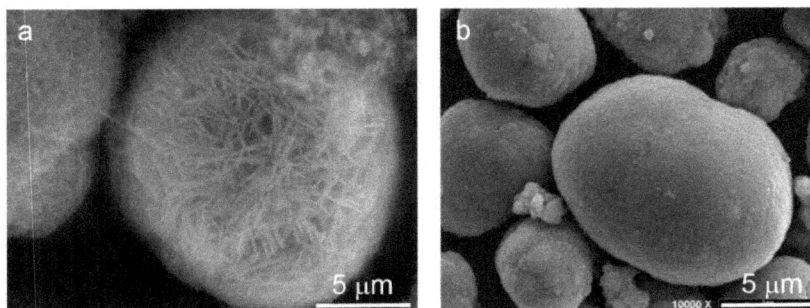

Figure 2. Scanning electron micrographs of (**a**) a heavily decorated WM12 surface and (**b**) a relatively smooth AP50 surface.

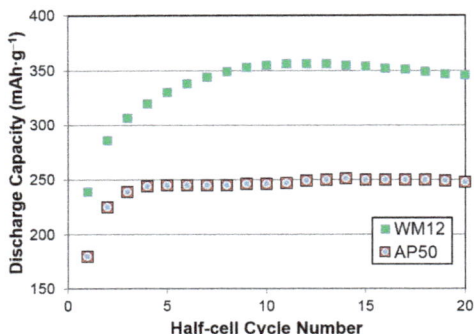

Figure 3. Half-cell capacities obtained at a discharge current of 30 mA·g^{-1} for WM12 and AP50.

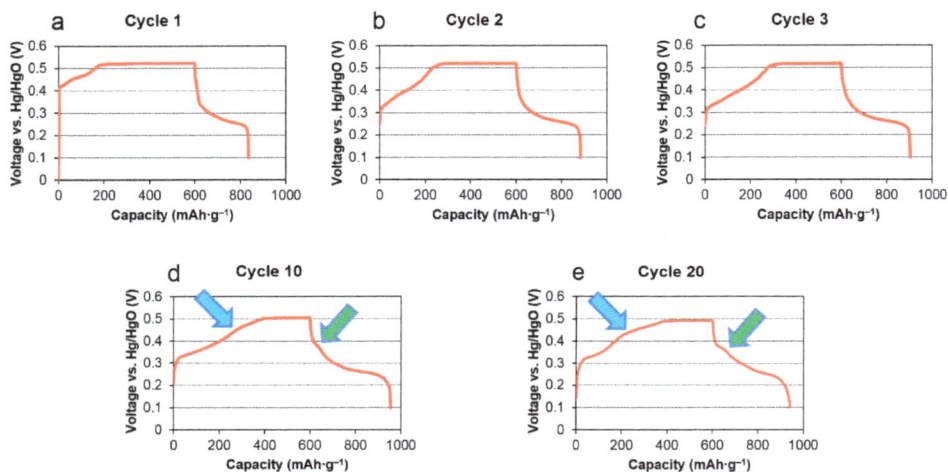

Figure 4. Half-cell voltage profiles of WM12 at (**a**) cycle 1, (**b**) cycle 2, (**c**) cycle 3, (**d**) cycle 10, and (**e**) cycle 20. The blue and greens arrows point to the α-to-γ and γ-to-α transitions, respectively.

3.2. Charge Rate Capability

Effects of varying the charge rate on pouch cell charge and discharge capacities are shown in Figure 5. Similar to earlier reports [9,10], a smaller charge rate results in a higher charge and discharge capacities, and a lower operation temperature yields lower charge and discharge capacities. Charge and discharge capacities that are obtained with different charge rates at both RT and −10 °C are listed in Table 4. Cell WM12 has higher charge and discharge capacities than Cell AP50 at both RT and −10 °C for all of the rates. Moreover, Cell WM12 shows superior high-rate charge capabilities, especially at a lower temperature, as demonstrated by the more-than-double charge/discharge capacities of Cell WM12 as compared to the corresponding charge/discharge capacities of Cell AP50 at 0.5C and 1C at −10 °C. RT energy densities obtained from Cell WM12 and Cell AP15 are 127 and 110 Wh·kg^{-1}, respectively. As the negative electrode material used is the conventional AB$_5$ MH alloy (320 mAh·g^{-1}), and the pouch design has not been optimized, further increase in energy density can be accomplished by using the new Laves phase-related body-centered-cubic MH alloy (400 mAh·g^{-1}) developed in the ARPA-E RANGE program [4] in an improved pouch design.

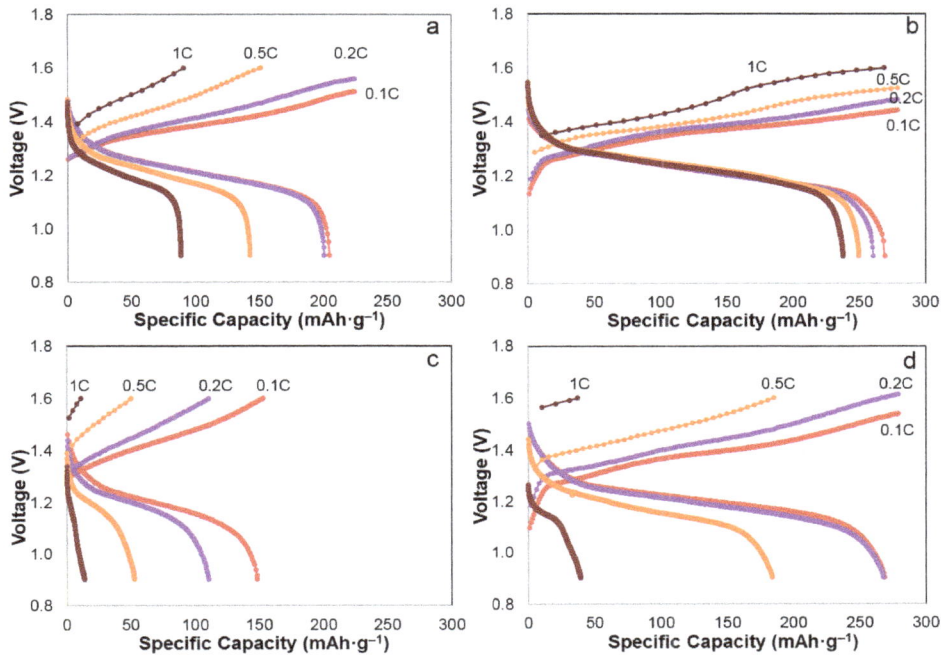

Figure 5. Charge rate evaluation by measuring both the charge and discharge capacities of (**a**) Cell AP50 and (**b**) Cell WM12 at room temperature and (**c**) Cell AP50 and (**d**) Cell WM12 at −10 °C. All measurements were discharged at 0.1C.

Table 4. Charge and discharge capacities obtained with different charge rates for Cells AP50 and WM12 at both room temperature (RT) and −10 °C.

Test #	Temperature (°C)	Rate	Step	Cell AP50 Capacity (mAh·g−1)	Cell WM12 Capacity (mAh·g−1)
1	RT	0.1C	Charge	225	279
		0.1C	Discharge	205	269
2	RT	0.2C	Charge	225	279
		0.1C	Discharge	201	260
3	RT	0.5C	Charge	151	279
		0.1C	Discharge	143	250
4	RT	1C	Charge	90	269
		0.1C	Discharge	89	238
5	−10	0.1C	Charge	153	279
		0.1C	Discharge	149	269
6	−10	0.2C	Charge	111	279
		0.1C	Discharge	110	268
7	−10	0.5C	Charge	53	186
		0.1C	Discharge	53	184
8	−10	1C	Charge	14	37
		0.1C	Discharge	14	39

3.3. Discharge Rate Capability

The effects of varying the discharge rate on pouch cell discharge capacity are shown in Figure 6. Since the charge capacity curves are very similar (all of the measurements were charged at 0.1C), only the discharged curves are presented. Discharge capacities that are obtained at different rates are summarized in Table 5. Cell WM12 has much higher discharge capacities than Cell AP50 at both RT and −10 °C for all of the rates. In addition, Cell WM12 shows superior high-rate discharge capabilities, especially at a lower temperature. For instance, at −10 °C and 1C, discharge capacity of Cell WM12 is more than six times higher than that of Cell AP50. The abnormal lowering in cell voltage during the initial high-rate discharges (0.5C and 1C) at RT in Figure 6a is related to the sudden decrease in MH alloy volume and insufficient time for the electrolyte to flow in, causing an increase in cell impedance. At −10 °C, the cell impedance is large, and the depth-of-discharge is small, which overshadow the electrolyte refilling phenomena observed in Figure 6a.

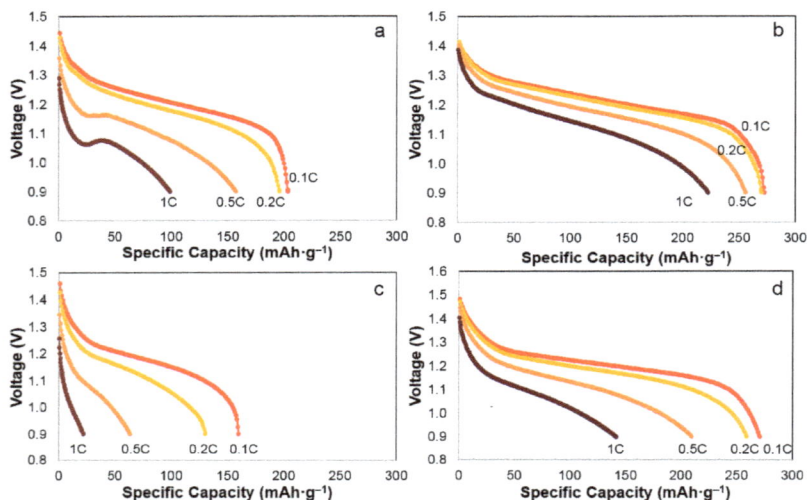

Figure 6. Discharge rate evaluation by measuring the discharge capacities of (**a**) Cell AP50 and (**b**) Cell WM12 at room temperature and (**c**) Cell AP50 and (**d**) Cell WM12 at −10 °C. All o the measurements were charged at 0.1C.

Table 5. Discharge capacities at different rates for Cells AP50 and WM12 at both room temperature (RT) and −10 °C.

Test #	Temperature (°C)	Rate	Step	Cell AP50 Capacity (mAh·g−1)	Cell WM12 Capacity (mAh·g−1)
1	RT	0.1C	Charge	225	279
		0.1C	Discharge	203	272
2	RT	0.1C	Charge	225	279
		0.2C	Discharge	196	270
3	RT	0.1C	Charge	225	279
		0.5C	Discharge	157	256
4	RT	0.1C	Charge	225	279
		1C	Discharge	99	223
5	−10	0.1C	Charge	225	279
		0.1C	Discharge	160	271
6	−10	0.1C	Charge	225	279
		0.2C	Discharge	130	259
7	−10	0.1C	Charge	225	279
		0.5C	Discharge	63	210
8	−10	0.1C	Charge	225	279
		1C	Discharge	22	142

3.4. Self-Discharge

Effects of SOC and temperature on self-discharge are shown in Figure 7a,b. A higher SOC leads to a higher open-circuit voltage for both Cell AP50 and Cell WM12. At −10 °C, both Cell AP50 and Cell WM12 have a better self-discharge performances than at RT. For the 32-day storage test, Cell WM12 shows higher voltages during storage and better self-discharge performances (higher voltages at the end of storage) than Cell AP50 at RT and −10 °C. Charge retentions that were measured at RT and 80% SOC for Cells AP50 and WM12 are plotted in Figure 7c, which also shows the superiority of WM12 over AP50 in self-discharge performance.

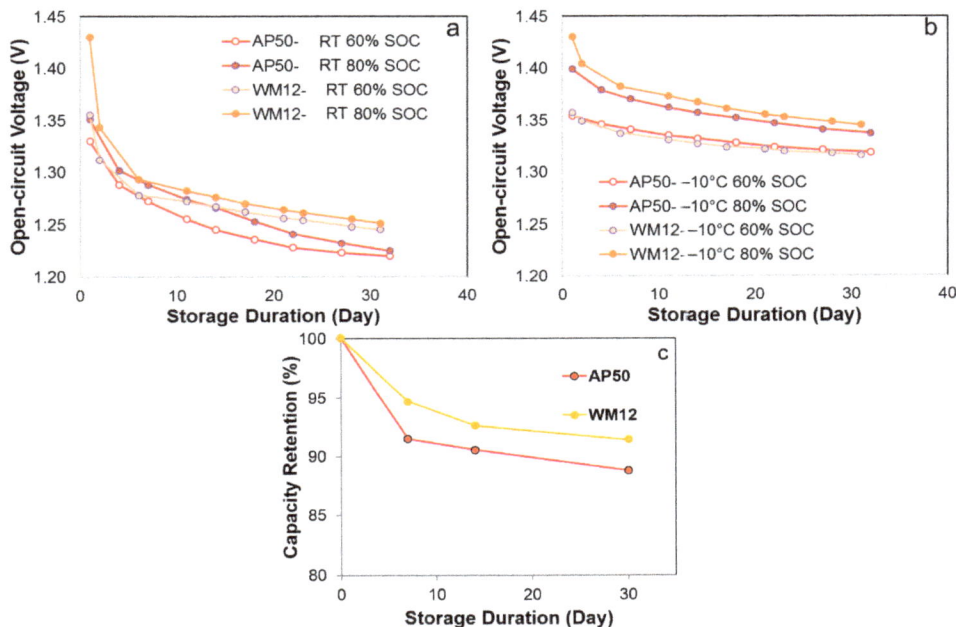

Figure 7. Self-discharge evaluations of Cell AP50 and Cell WM12 by the open-circuit voltages at both (**a**) room temperature (RT) and (**b**) −10 °C at both 60% and 80% state-of-charge (SOC), and (**c**) by the capacity retention at RT and 80% SOC.

3.5. Internal Resistance

R_{int}s of Cell AP50 and Cell WM12 are listed in Table 6. R_{int} of Cell WM12 is dramatically lower than that of Cell AP50. More specifically, a 47% decrease in R_{int} at RT and a 56% decrease in R_{int} at −10 °C are observed in Cell WM12. According to our previous works [7,13,14], the lower R_{int} of WM12 can be attributed to its higher surface area (Table 3), higher surface pore density (Table 3), large interface region between the α and β phases, and special core-shell structure.

Table 6. Charge and discharge internal resistances (R_{int}) at 60% SOC of Cells AP50 and WM12 measured at room temperature (RT) and −10 °C.

Temperature (°C)	Cell	R_{int}-Charge (Ω)	R_{int}-Charge (Ω)	Average R_{int} (Ω)
RT	AP50	0.36	0.40	0.38
RT	WM12	0.17	0.23	0.20
−10	AP50	1.06	1.07	1.07
−10	WM12	0.45	0.49	0.47

3.6. Charge-Transfer Resistance

AC impedance measurement was employed to study the surface electrochemical reaction [15]. Figure 8 (RT) and Figure 9 (−10 °C) illustrate the resulting Cole-Cole plots for Cell AP50 and Cell WM12 at the 0th, 20th, and 50th cycles. A reported equivalent circuit model [16], as shown in Figure 10, was used for the simulation of Figures 8 and 9. Table 7 lists the solution resistances (R_0) and R_{ct}s for Cell AP50 and Cell WM12 at different temperatures and cycle numbers. R_0 is the resistance of ions traveling through the electrolyte and separator and the start value of Z′ in the Cole-Cole plot (x-axis intercept). R_{ct} is closely related to the semicircle in the Cole-Cole plot. The linear part after the semicircle in the Cole-Cole plot is linked to the Warburg impedance, a parameter that is associated with the diffusion of hydrogen atoms into the electrodes. We will conduct more detailed research on Warburg impedance in our future work. From Table 7, R_{ct}s of Cell WM12 are 0.10 Ω at RT and 0.52 Ω at −10 °C, which are only 31 to 48% of those observed in Cell AP50.

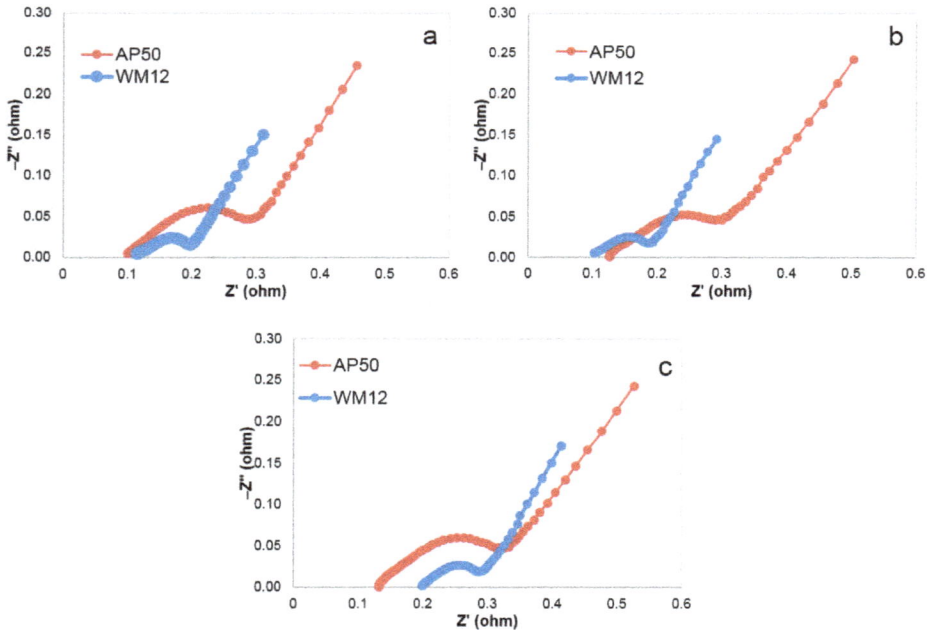

Figure 8. Room temperature Cole-Cole plots for Cell AP50 and Cell WM12 at the (**a**) 0th, (**b**) 20th, and (**c**) 50th cycles obtained at 60% SOC.

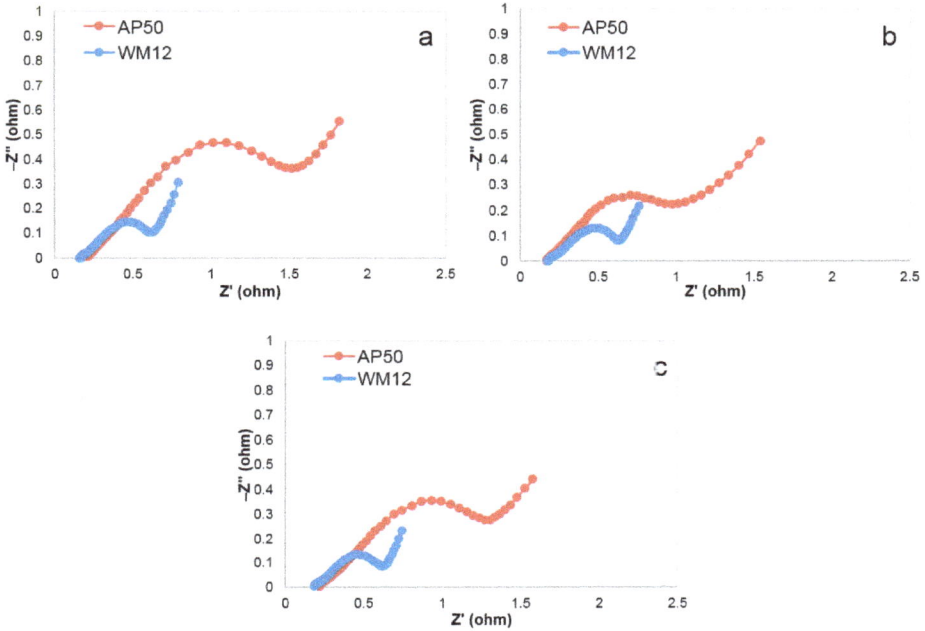

Figure 9. −10 °C Cole-Cole plots for Cell AP50 and Cell WM12 at the (**a**) 0th, (**b**) 20th, and (**c**) 50th cycles obtained at 60% SOC.

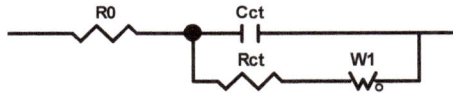

Figure 10. Proposed equivalent circuit model for pouch cells AP50 and WM12.

Table 7. Summary of solution resistances (R_0) and charge-transfer resistances (R_{ct}) for Cell AP50 and Cell WM12 measured at room temperature (RT) and −10 °C at different cycles.

Temperature (°C)	Cycle #	Cell	R_0 (Ω)	R_{ct} (Ω)
RT	0	AP50	0.10	0.23
RT	0	WM12	0.11	0.10
RT	20	AP50	0.13	0.21
RT	20	WM12	0.10	0.10
RT	50	AP50	0.13	0.23
RT	50	WM12	0.20	0.10
−10	0	AP50	0.21	1.64
−10	0	WM12	0.16	0.51
−10	20	AP50	0.17	1.18
−10	20	WM12	0.17	0.53
−10	50	AP50	0.20	1.30
−10	50	WM12	0.18	0.52

4. Summary

A systematic study on pouch cells made with a high-capacity β-α core-shell Ni(OH)$_2$ (experimental) was performed, and the charge and discharge rate capabilities, self-discharge,

Batteries **2017**, *3*, 38

and internal and charge-transfer resistances were obtained and compared with pouch cells that were made with a standard β-Ni(OH)$_2$ as control. The experimental cells exhibit superior performances in all of the tests. When compared to the control cells, the experimental cells show a higher energy density (127 vs. 110 Wh·kg^{-1}) and better charge/discharge rate capabilities, especially at a higher rate and lower temperature. For example, at 0.5 C and -10 °C, charge capacity is 3.51 times higher than the control; and at 1 C and -10 °C, discharge capacity is 6.45 times higher than the control. Furthermore, the experimental cells also demonstrate better self-discharge performances and reduced internal and charge-transfer resistances, measured at both room temperature and -10 °C.

Acknowledgments: The authors would like to thank the following individuals from BASF—Ovonic for their help: Taihei Ouchi, Shiuan Chang, Su Cronogue, Baoquan Huang, Chaolan Hu, Benjamin Reichman, William Mays, Diana F. Wong, David Pawlik, Allen Chan, and Ryan J. Blankenship.

Author Contributions: Shuli Yan composed the manuscript. Tiejun Meng performed the experiment and analyzed results. Kwo-Hsiung Young and Jean Nei helped in data analysis and manuscript preparation.

Conflicts of Interest: The authors declare no conflict of interest.

Abbreviations

The following abbreviations are used in this manuscript:

EV	Electric vehicle
G	Graphite
LTO	Li-titanate
LMO	Li-spinel
NCA	LiNiCoAl oxide
NMC	LiNiMnCo oxide
LFP	LiFePO$_4$
Ni/MH	Nickel/metal hydride
HEV	Hybrid electric vehicle
MH	Metal hydride
N/P ratio	Negative-to-positive capacity ratio
RT	Room temperature
SOC	State of charge
R_{int}	Internal resistance
R_{ct}	Charge transfer resistance
R_0	Solution resistance

References

1. Tesla Model S Battery. Available online: http://enipedia.tudelft.nl/wiki/Tesla_Model_S_Battery (accessed on 11 July 2017).
2. General Motors EV1. Available online: https://en.wikipedia.org/wiki/General_Motors_EV1 (accessed on 11 July 2017).
3. Young, K.; Nei, J.; Meng, T. Alkaline and Non-Aqueous Proton-Conducting Pouch-Cell Batteries. U.S. Patent Application 20160233461, 11 August 2016.
4. Young, K.; Ng, K.Y.S.; Bendersky, L.A. A technical report of the Robust Affordable Next Generation Energy Storage System-BASF program. *Batteries* **2016**, *2*, 2. [CrossRef]
5. Meng, T.; Young, K.; Beglau, D.; Yan, S.; Zeng, P.; Cheng, M. Hydrogenated amorphous silicon thin film anode for proton conducting batteries. *J. Power Sources* **2016**, *302*, 31–38. [CrossRef]
6. Meng, T.; Young, K.; Wong, D.F.; Nei, J. Ionic liquid-based non-aqueous electrolytes for nickel/metal hydride batteries. *Batteries* **2017**, *3*, 4. [CrossRef]
7. Young, K.; Wang, L.; Yan, S.; Liao, X.; Meng, T.; Shen, H.; Mays, W.C. Fabrications of high-capacity alpha-Ni(OH)$_2$. *Batteries* **2017**, *3*, 6. [CrossRef]
8. Meng, T.; Young, K.; Hu, C.; Reichman, B. Effects of alkaline pre-etching to metal hydride alloys. *Batteries* **2017**, *3*, 30. [CrossRef]

9. Young, K.; Ng, K.Y.S. Reviews on the Chinese Patents regarding nickel/metal hydride battery. *Batteries* **2017**, *3*, 24. [CrossRef]

10. Chang, S.; Young, K.; Nei, J.; Fierro, C. Reviews on the US Patents regarding nickel/metal hydride battery. *Batteries* **2016**, *1*, 10. [CrossRef]

11. Corrigan, D.A.; Knight, S.L. Electrochemical and spectroscopic evidence on the participation of quadrivalent nickel in the nickel hydroxide redox reaction. *J. Electrochem. Soc.* **1989**, *135*, 613–619. [CrossRef]

12. Corrigan, D.A.; Bendert, R.M. Effect of coprecipitated metal ions on the electrochemistry of nickel hydroxide thin films: cyclic voltammetry in 1M KOH. *J. Electrochem. Soc.* **1989**, *136*, 723–728. [CrossRef]

13. Fierro, C.; Zallen, A.; Koch, J.; Fetcenko, M.A. The influence of nickel-hydroxide composition and microstructure on the high-temperature performance of nickel metal hydride batteries. *J. Electrochem. Soc.* **2006**, *153*, A492–A496. [CrossRef]

14. Fierro, C.; Fetcenko, M.A.; Young, K.; Ovshinsky, S.R.; Sommers, B.; Harrison, C. Nickel Hydroxide Positive Electrode Material Exhibiting Improved Conductivity and Engineered Activation Energy. U.S. Patent 6,228,535, 8 May 2001.

15. Zhang, L. AC impedance studies on sealed nickel metal hydride batteries over cycle life in analog and digital operations. *Electrochim. Acta* **1998**, *43*, 3333–3342. [CrossRef]

16. Zhang, W.; Kumar, M.P.S.; Srinivasan, S. AC impedance studies on metal hydride electrodes. *J. Electrochem. Soc.* **1995**, *142*, 2935–2943. [CrossRef]

batteries

MDPI

Article

Performance Comparison of Rechargeable Batteries for Stationary Applications (Ni/MH vs. Ni–Cd and VRLA)

Michael A. Zelinsky *, John M. Koch and Kwo-Hsiung Young

BASF/Battery Materials–Ovonic, 2983 Waterview Drive, Rochester Hills, MI 48309, USA;
john.m.koch@basf.com (J.M.K.); kwo.young@basf.com (K.-H.Y.)
* Correspondence: michael.a.zelinsky@basf.com; Tel.: +1-248-293-7234

Received: 22 November 2017; Accepted: 22 December 2017; Published: 25 December 2017

Abstract: The stationary power market, particularly telecommunications back-up (telecom) applications, is dominated by lead-acid batteries. A large percentage of telecom powerplants are housed in outdoor enclosures where valve-regulated lead-acid (VRLA) batteries are commonly used because of their low-maintenance design. Batteries in these enclosures can be exposed to temperatures which can exceed 70 °C, significantly reducing battery life. Nickel–cadmium (Ni–Cd) batteries have traditionally been deployed in hotter locations as a high-temperature alternative to VRLA. This paper compares the performances of nickel/metal hydride (Ni/MH), Ni–Cd, and VRLA batteries in a simulated telecom environment according to published testing standards. Among these three choices, Ni/MH batteries showed the best overall performance, suggesting substantially longer operating life in high temperature stationary use.

Keywords: nickel metal hydride battery; nickel–cadmium battery; valve-regulated lead-acid battery; stationary application; telecom

1. Introduction

Nickel/metal hydride (Ni/MH) battery technology is very well suited for stationary energy storage applications because of its high power, long cycle life, compact size, unsurpassed safety, and wide operating temperature range [1–3]. These merits have been validated in laboratory testing and field evaluations alongside nickel–cadmium (Ni–Cd) and valve-regulated lead-acid (VRLA) batteries. Supported by these results, Ni/MH batteries of varying designs ranging from small cell telecom [4] and data storage [5] back-up power to substation-scale energy storage systems [6,7] have begun to appear on the stationary market (see the Supplemental Material for specific examples).

Although specialized versions had been used earlier for select applications, Ni/MH batteries first appeared on the mainstream commercial market in the late 1980s. These small rechargeable cylindrical cells created a new generation of consumer electronics by enabling widespread deployment of digital cameras, cellphones, laptop computers, and personal digital assistant. A decade later, prismatic Ni/MH batteries became the dominating technology for powering hybrid electric vehicles (HEV). Since 1997, more than 12 million HEV equipped with Ni/MH batteries have been introduced to the world's roadways [2]. The annual production of consumer-type Ni/MH battery is now over one billion cells [8]. Today's development focus on large-format Ni/MH batteries is setting the stage for stationary power, energy storage on the electrical grid, and a variety of other industrial battery applications.

A basic Ni/MH cell consists of a metal hydride (MH) negative electrode and a nickel hydroxide positive electrode in a highly conductive aqueous potassium hydroxide-based electrolyte (typically 30 wt %) in a sealed structure. This chemistry provides a nominal voltage of 1.2 volts per cell, below the electrolysis voltage of water. Although its voltage is lower than that of a VRLA, Ni/MH is characterized by higher gravimetric and volumetric energy density, higher heat tolerance, and better cycle stability at deeper discharge depths, which makes it an attractive alternative for stationary applications [1]. In addition, unlike its Ni–Cd rival, Ni/MH contains no toxic materials and is commercially recyclable, although not yet to the same extent as lead-acid batteries.

The ability of Ni/MH to tolerate high heat conditions for extended durations [3] is particularly important for stationary back-up power applications, including remote, outdoor installations. In addition to ambient environments, high heat can result from a number of other conditions including rapid cycling, high rate discharge, or fast charging. While this paper focuses on high temperature telecom environments, improved heat tolerance is equally important for grid energy storage or UPS/data center applications, where thermal management concerns add excessive cost and system complexity.

2. Experimental Section

In 2012, Telcordia (Piscataway, NJ, USA) published GR-3168-CORE, "Generic Requirements for Nickel Metal Hydride (Ni/MH) Battery Systems in Telecommunications Use" [9]. This document outlines various testing protocols and requirements for stationary Ni/MH batteries, grouped by level of deployment, covering areas such as general design, safety, performance, and service life. One of the most interesting tests specified by GR-3168-CORE is the Varying-55 °C aging test.

The Varying-55 °C Test is intended to determine the service life of an Ni/MH battery. Batteries are aged using a continuous 24 h environmental cycle with an average temperature of 55 °C. The algorithm is composed of four stages designed to simulate the temperature environment inside an outdoor telecom power cabinet from nighttime low temperatures to the highs of the day, including solar loading:

Stage 1   12 a.m.–9 a.m.: 45 °C
Stage 2   9 a.m.–12 p.m.: 45–65 °C
Stage 3   12 p.m.–9 p.m.: 65 °C
Stage 4   9 p.m.–12 a.m.: 65–45 °C

Daily exposure to 65 °C represents an extreme case and is not common in the field, especially when considering seasonal changes over the course of an entire year. In an alternative Varying-45 °C aging test, the temperature profile is reduced by 10 °C, i.e., 9 h at 55 °C, 9 h at 35 °C, and 6 h in transition between 35 and 55 °C. This temperature profile is far more commonly experienced in outdoor telecom cabinets. Nonetheless, both temperature profiles were tested to evaluate the effects of long-term heat exposure on Ni/MH and incumbent battery chemistries.

Thermotron model S-32-8200 Environmental Test Chambers (Thermotron Industries, Plano, TX, USA) were used to control the daily thermal profiles for simultaneous testing of various 12 V battery modules typically used in stationary applications, including a 100 Ah FT-type thin plate pure lead Monobloc VRLA [10], an 80 Ah (10 cells) maintenance free Ni–Cd telecom module [11], a 4 Ah (10 cell) Ni/MH module type A (standard long life industrial type), and a 4 Ah (10 cells) Ni/MH module type B (equipped with an advanced high temperature cathode).

At first glance, it may appear that this wide range of battery sizes is an inappropriate mix of test subjects. However, the objective of the investigation was to determine the effects of heat exposure on battery chemistry, not battery construction or configuration. As such, the smallest commonly deployed unit of each chemistry was selected. In actual use, multiple modules are routinely connected in parallel to achieve batteries of higher capacity. Discharge rates were scaled proportionally to the nominal capacity of each individual module (C-rate) to ensure similar loading conditions. Modules were float charged at a constant voltage of 13.5 V as they would be if installed in an actual telecom cabinet [9]. Charging and discharging were controlled using a Bitrode battery cycler (Bitrode Corporation, St. Louis, MO, USA).

Ni/MH batteries in this test were supplied by FDK Corp (Tokyo, Japan) using the new type of superlattice metal hydride alloys [12–17]. This newly developed alloy family has a higher storage capacity, an improved high-rate dischargeability, and a better cycle life when comparing to the conventional AB_5-based metal hydride alloys [18–23]. These cylindrical cells (in long-A size) were assembled with pasted negative electrode, pasted $Ni(OH)_2$-based positive electrode, polypropylene/polyethylene grafted separator, and a 30% KOH electrolyte. The cell design is targeted at a negative-to-positive ratio of 1.4 with a capacity of 3700 mAh.

3. Results and Discussion

3.1. The Varying-55 °C Test

After 30 days' float (13.5 V per 12 V module) on the Varying-55 °C profile, batteries were cooled to room temperature (RT) and stabilized for 24 h, before being discharged at a C/8 rate. Figure 1 presents results from the monthly capacity check in a unit of percentage of the battery's nominal capacity.

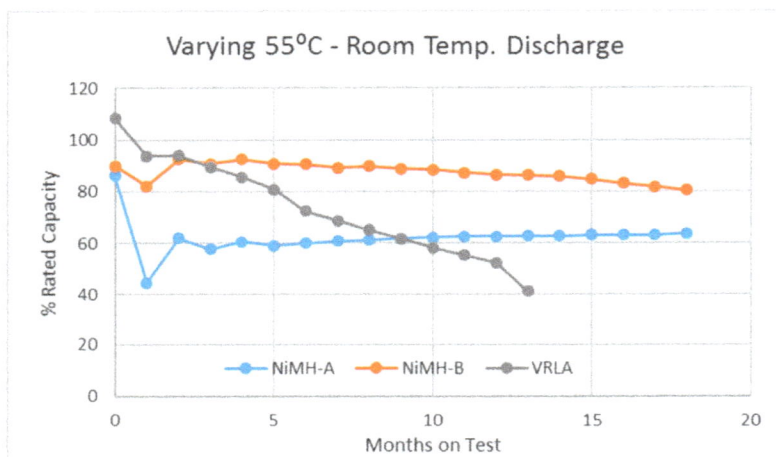

Figure 1. Eight-hour discharge at room temperature (RT) following 30 days' float on the Varying-55 °C profile.

Specifications for the state-of-the-art Ni–Cd battery state operation in temperatures ranging from −20 to +50 °C and the capability of tolerating −50 to +70 °C for short durations [11]. Daily exposure of 9 h per day at 65 °C must have exceeded such "short duration" conditions as electrolyte leakage was observed on multiple Ni–Cd cells and the module was removed from the test.

In the monthly test, the VRLA module showed a steadily decreasing capacity due to the prolonged exposure to elevated temperature. The two Ni/MH modules returned relatively stable, but with vastly different capacities. This apparent capacity inconsistency of Ni/MH modules will be explained later. It is interesting to note that the Ni–Cd battery did not achieve any recordable discharge capacity in this test.

At the end of each discharge, battery modules were recharged at RT for 24 h and discharged again at a C/8 rate. Although this procedure is not part of the GR-3168 test protocol, it still provides a valuable capacity verification. The results of the monthly capacity check are shown in Figure 2.

After recharging at RT, the VRLA results were unaffected. The module reached 50% of its original capacity and was removed from the test after 13 months. In contrast, a significant change was observed for the Ni/MH modules—both types showed almost identical results of nearly full capacity after recharge at RT after more than one year on test. The very different result comparing to Figure 1 was not surprising as it has been documented that the standard Ni/MH (like module Type A) does not charge well at temperatures higher than 50 °C [3]. The Ni/MH module Type B used an advanced cathode material [24] with a higher charging efficiency at an elevated temperature and thus did not display the same behavior as the standard Type A Ni/MH module. We reported before the failure mode of superlattice MH alloy at elevated temperature to be the continuous pulverization and surface oxidation [25]. Testing of both Ni/MH modules continues. It is interesting to note that, unlike the results in Figure 1, some discharge capacity was observed for the Ni–Cd module when charged at RT.

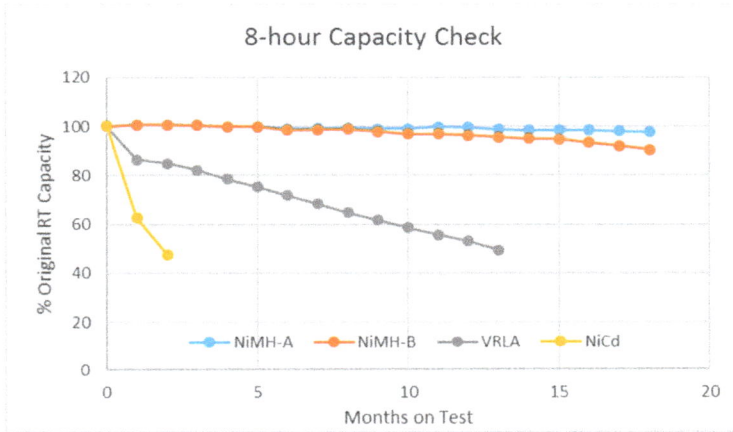

Figure 2. Varying-55 °C test; capacity check following recharge at room temperature (RT).

3.2. The Varying-45 °C Test

The Varying-45 °C aging test was conducted in a similar manner using the same types of batteries as in the Varying-55 °C test plus one additional VRLA module; a 12 V, 8 Ah monobloc module of the type commonly used in a fiber-to-the-home (FTTH) equipment [26].

Results of this ongoing testing are summarized in Figure 3. Following a significant initial jump in capacity, the 8 Ah FTTH battery began losing capacity at an increasing rate, falling below 60% of its nominal capacity after 9 months. The larger VRLA module displayed a slower, but steady capacity loss of about 3–4% with each monthly discharge. The capacity of the Ni–Cd battery showed similar behavior, losing about 2–3% per month. Capacities for both Ni/MH modules remained fairly steady, showing less than 5% reduction from their initial discharge capacity after 14 months at this elevated temperature profile.

Following each discharge, all batteries were recharged at room temperature (RT) and subsequently discharged at the C/8 rate. The results of this capacity verification are summarized in Figure 4. Similar to the results as shown in Figure 2, the standard Ni/MH battery (NiMH-A) recovered to a full capacity following recharge at RT. Since 80% of rated capacity indicates the end of life condition under which telecom batteries are replaced in the field, both VRLA modules and the Ni–Cd battery were removed from the test. Meanwhile, both Ni/MH batteries continue to return capacities above 90% and will remain on test for the foreseeable future.

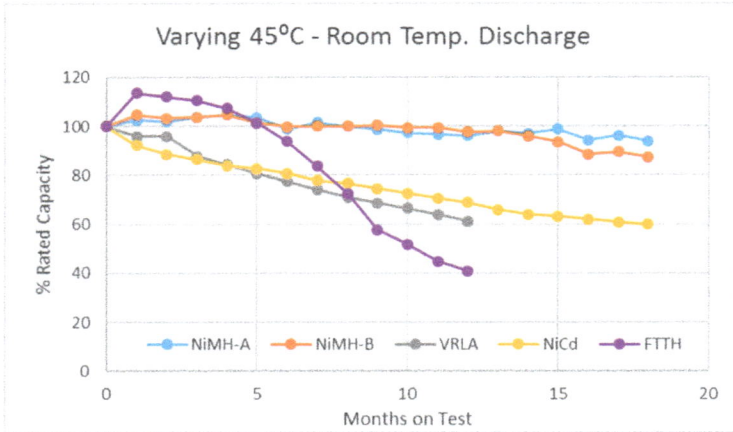

Figure 3. Eight-hour discharge at room temperature (RT) following 30 days' float on the Varying-45 °C profile.

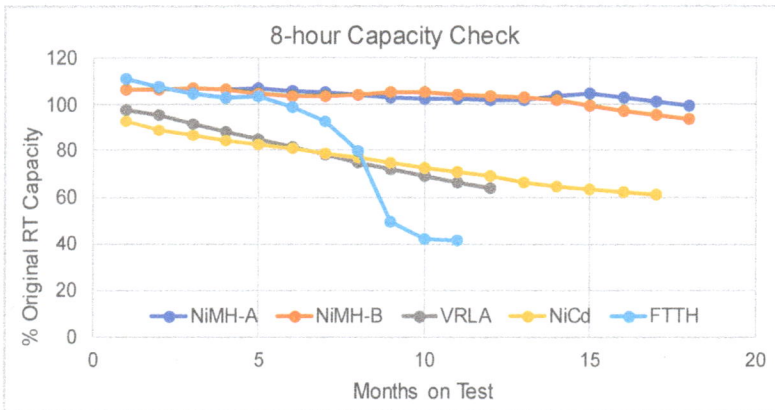

Figure 4. Varying-45 °C test; capacity check following recharge at room temperature (RT).

4. Conclusions

Battery modules of various chemistries were tested according to simulated telecom outdoor power cabinet temperature profiles. Among the three chemistries tested, nickel/metal hydride showed the most durable results in both Varying-45 and -55 °C aging tests. Longer operating life and superior capacity retention at elevated temperatures are important considerations for stationary battery users since many applications require operation at high temperature. While this paper focuses on high ambient temperature environments, high heat can result from a number of other conditions including rapid cycling, high discharge rates, and fast charging. Nickel/metal hydride, therefore, can be considered a superior alternative to valve-regulated lead-acid and nickel–cadmium batteries for a wide range of applications.

Nickel/metal hydride battery technology offers a great number of additional benefits as an advanced alternative to conventional valve-regulated lead-acid and nickel–cadmium batteries, including proven safety and reliability in more than 12 million hybrid electric vehicles, high power and energy density, and long cycle life. While small stationary Ni/MH battery products have begun to appear on the market, the development of large-format Ni/MH batteries is creating new business opportunities for substation-scale energy storage and other industrial battery applications. Examples of stationary Ni/MH batteries for specific applications are discussed in the Supplemental Materials accompanying this paper.

Supplementary Materials: The following are available online at www.mdpi.com/2313-0105/4/1/1/s1, Figure S1: A 600 W Small Cell Power System (left) and its integrated battery module with Ni/MH cells and batter management system (right). Photos are courtesy from Alpha Technologies Ltd. (Burnaby, British Columbia, Canada) and FDK Corp. (Tokyo, Japan), Figure S2: Bi-polar cell design/construction in KHI GigaCell, Figure S3: High-rate charge and discharge capability of KHI GigaCell. 5.0C equals 750 amps, Table S1: Specific energy and power for commercial large format battery modules of four different chemistries. (Ni/MH and Li-ion include integrated electronic components), Figure S4: Illustration of a wayside railroad BPS installation, Figure S5: Single day charge/discharge profile for a wayside railroad BPS installation, Figure S6 Solar farm and Ni/MH BPS near Osaka, Japan.

Acknowledgments: The authors would like to thank the following individuals: Sui-ling Chen, Cheryl Setterington, Anthony Wilde, and Nathan English from BASF.

Author Contributions: Michael A. Zelinsky and Kwo-Hsiung Young prepared the manuscript. John M. Koch performed the experiment and analyzed the data.

Conflicts of Interest: The authors declare no conflict of interest.

Abbreviations

The following abbreviations are used in this manuscript:

Telecom	telecommunications
VRLA	valve-regulated lead acid
Ni–Cd	nickel cadmium
Ni/MH	nickel/metal hydride
HEV	hybrid electric vehicle
MH	metal hydride
RT	room temperature
FTTH	fiber-to-the-home
BPS	battery power system

References

1. Zelinsky, M. Heat Tolerant Ni/MH Batteries for Stationary Power. In Proceedings of the Battcon 2010 International Stationary Battery Conference, Hollywood, FL, USA, 17–19 May 2017.
2. Fetcenko, M. Battery Materials for E-Mobility. In Proceedings of the 33rd International Battery Seminar & Exhibit, Fort Lauderdale, FL, USA, 21–24 March 2016.
3. Zelinsky, M. Batteries and Heat—A Recipe for Success? In Proceedings of the Battcon 2013 International Stationary Battery Conference, Orlando, FL, USA, 6–8 May 2013.
4. Alpha Technologies Ltd. Cellect™ 600 Product Data Sheet. Available online: https://atl.app.box.com/v/cellect-600-48v-dc (accessed on 31 July 2017).
5. Palu, J. Design considerations for data-storage memory back-up. Available online: https://www.electronicproducts.com/Power_Products/Batteries_and_Fuel_Cells/Design_considerations_for_data-storage_memory_back-up.aspx (accessed on 22 December 2017).
6. WMATA Energy Storage Demonstration Project, Federal Transit Administration Final Report. June 2015. Available online: https://www.transit.dot.gov/sites/fta.dot.gov/files/docs/FTA_Report_No._0086.pdf (accessed on 31 July 2017).
7. Nishimura, K.; Takasaki, T.; Sakai, T. Introduction of large-sized nickel-metal hydride battery GIGACELL for industrial applications. *J. Alloy. Compd.* **2013**, *580*, S353–S358. [CrossRef]
8. Bai, W. The Current Status and Future Trends of Domestic and Foreign NiMH Battery Market. Available online: http://cbea.com/u/cms/www/201406/06163842rc0l.pdf (accessed on 31 August 2017).

9. Available online: http://telecom-info.telcordia.com/site-cgi/ido/docs.cgi?ID=SEARCH&DOCUMENT= GR-3168& (accessed on 31 August 2017).
10. EnerSys PowerSafe SBS 100 Technical Specifications. Available online: http://www.enersys.com/Asia/ PowerSafe_SBS_Batteries.aspx?langType=1033 (accessed on 31 July 2017).
11. Saft Tel.X Ni–Cd Batteries for Telecom Networks Technical Manual. Available online: http://www.npstelecom. com/resources/products/fcm/_10q11wyx4nw9122mr4nbh87x1066kfv8z6smqm9djyh8pn584.pdf (accessed on 31 July 2017).
12. Yasuoka, S.; Magari, Y.; Murata, T.; Tanaka, T.; Ishida, J.; Nakamura, H.; Nohma, T.; Kihara, M.; Baba, Y.; Teraoka, H. Development of high-capacity nickel-metal hydride batteries using superlattice hydrogen-absorbing alloys. *J. Power Sources* **2006**, *156*, 662–666. [CrossRef]
13. Teraoka, H. Development of Low Self-Discharge Nickel-Metal Hydride Battery. Available online: http://www.scribd.com/doc/9704685/Teraoka-Article-En (accessed on 9 April 2016).
14. Kai, T.; Ishida, J.; Yasuoka, S.; Takeno, K. The effect of nickel-metal hydride battery's characteristics with structure of the alloy. In Proceedings of the 54th Battery Symposium, Osaka, Japan, 7–9 October 2013; p. 210.
15. Teraoka, H. Development of Ni-MH EThSS with Lifetime and Performance Estimation Technology. In Proceedings of the 34th International Battery Seminar & Exhibit, Fort Lauderdale, FL, USA, 20–23 March 2017.
16. Teraoka, H. Ni-MH Stationary Energy Storage: Extreme Temperature & Long Life Developments. In Proceedings of the 33th International Battery Seminar & Exhibit, Fort Lauderdale, FL, USA, 21–24 March 2016.
17. Teraoka, H. Development of Highly Durable and Long Life Ni-MH Batteries for Energy Storage Systems. In Proceedings of the 32th International Battery Seminar & Exhibit, Fort Lauderdale, FL, USA, 9–12 March 2015.
18. Young, K.; Yasuoka, S. Past, present, and future of metal hydride alloys in nickel-metal hydride batteries. In Proceedings of the 14th International Symposium on Metal-Hydrogen Systems, Manchester, UK, 21–25 July 2014.
19. Young, K.; Chang, S.; Lin, X. C14 Laves phase metal hydride alloys for Ni/MH batteries applications. *Batteries* **2017**, *3*, 27. [CrossRef]
20. Young, K.; Wong, D.F.; Wang, L.; Nei, J.; Ouchi, T.; Yasuoka, S. Mn in misch-metal based superlattice metal hydride alloy—Part 1 Structural, hydrogen storage and electrochemical properties. *J. Power Sources* **2015**, *277*, 426–432. [CrossRef]
21. Young, K.; Wong, D.F.; Wang, L.; Nei, J.; Ouchi, T.; Yasuoka, S. Mn in misch-metal based superlattice metal hydride alloy—Part 2 Ni/MH battery performance and failure mechanism. *J. Power Sources* **2015**, *277*, 433–442. [CrossRef]
22. Wang, L.; Young, K.; Meng, T.; Ouchi, T.; Yasuoka, S. Partial substitution of cobalt for nickel in mixed rare earth metal based superlattice hydrogen absorbing alloy—Part 1 Structural, hydrogen storage and electrochemical properties. *J. Alloy. Compd.* **2016**, *660*, 407–415. [CrossRef]
23. Wang, L.; Young, K.; Meng, T.; English, N.; Yasuoka, S. Partial substitution of cobalt for nickel in mixed rare earth metal based superlattice hydrogen absorbing alloy—Part 2 Battery performance and failure mechanism. *J. Alloy. Compd.* **2016**, *664*, 417–427. [CrossRef]
24. Fierro, C.; Zallen, A.; Koch, J.; Fetcenko, M.A. The influence of nickel-hydroxide composition and microstructure on the high-temperature performance of nickel metal hydride batteries. *J. Electrochem. Soc.* **2006**, *153*, A492–A496. [CrossRef]
25. Meng, T.; Young, K.; Koch, J.; Ouchi, T.; Yasuoka, S. Batteries with cobalt-substituted superlattice hydrogen-absorbing alloy anodes at 50 °C. *Batteries* **2016**, *2*, 20. [CrossRef]
26. GS Yuasa GoldTop Technical Specifications. Available online: http://www.gsbattery.com/fiber-home-batteries-ftth (accessed on 31 July 2017).

Performance Comparison of Rechargeable Batteries for Stationary Applications (Ni/MH vs. Ni–Cd and VRLA)

S1. Ni/MH Battery in Telecom Market

When a major telecom power systems supplier began developing an innovative solution for powering outdoor small cells, it conceptualized a low-cost system with a small battery reserve, 15 to 20 minutes, enough to power through almost 90 percent of grid power interruptions. Traditional telecom backup systems are designed to provide 4 to 8h of runtime, but these large systems are not feasible here since outdoor small cells are normally deployed on utility poles, streetlights or sides of buildings. In addition to physical size, other key requirements must also be considered, including aesthetics, weight, operating temperature range, environment, safety, mounting flexibility, and maintenance. Utilizing the high power-density of Ni/MH cells, a low profile, compact battery system with integrated battery management electronics was developed. Shown in Figure S1, the system weighs less than 12 kg and occupies less than 0.013 m³ in volume. It can be used to power loads up to 600W at ambient temperatures ranging from −40 to +55 °C [S1].

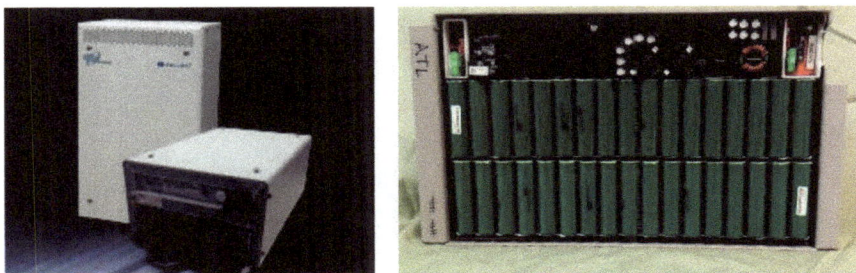

Figure S1. A 600 W Small Cell Power System (left) and its integrated battery module with Ni/MH cells and batter management system (right). Photos are courtesy from Alpha Technologies Ltd. (Burnaby, British Columbia, Canada) and FDK Corp. (Tokyo, Japan).

S2. Large Format Ni/MH Battery in Power System

In contrast to this small cylindrical cell-based battery, Kawasaki Heavy Industries (KHI, Tokyo, Japan) has introduced a large-format prismatic Ni/MH cell product [S2,S3]. Their GigaCell uses a Co-free $RE_{0.9}Mg_{0.1}Ni_{3.9}Al_{0.2}$ (RE: rare earth) metal hydride alloy and a carbon-coated $Ni(OH)_2$ as the active materials in negative and positive electrodes, respectively, and a 4.8 M KOH + 1.2 M NaOH electrolyte [S4]. The negative-to-positive ratio was set to 2.5 to guarantee superior high-rate discharge performance [S5]. At a total stored energy of 5400 Wh (36 V, 150 Ah) this product dwarfs all other commercially available Ni/MH batteries. The specific energy and power of these high-power KHI-Ni/MH batteries are compared to commonly-used stationary battery modules of other chemistries in Table S1. From the information, it is obvious that KHI trades energy density in favor of power density targeting high-power applications. A unique bi-polar battery construction designed for efficient cooling and higher delivered power (Figure S2) is employed for rapid charge/discharge applications (Figure S3). More details about the cell structures can be found in related Japanese Patent Applications reviewed before [S6].

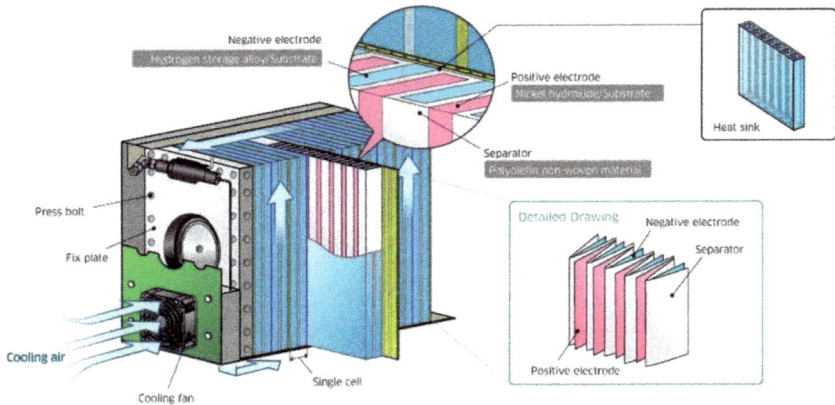

Figure S2. Bi-polar cell design/construction in KHI GigaCell [S7].

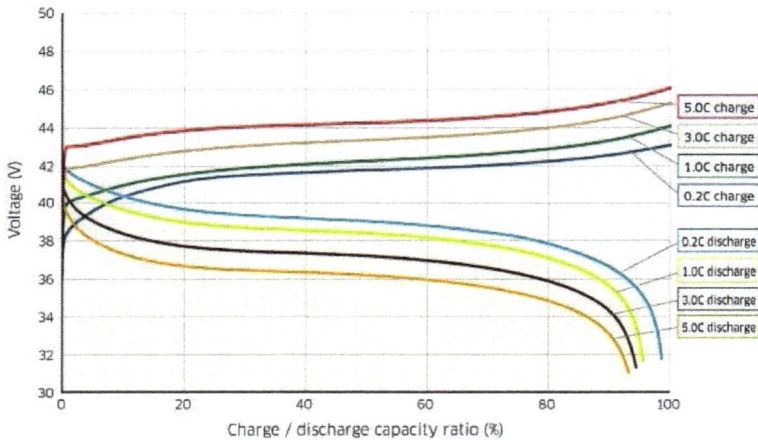

Figure S3. High-rate charge and discharge capability of KHI GigaCell. 5.0C equals 750 amps [S7].

Table S1. Specific energy and power for commercial large format battery modules of four different chemistries. (Ni/MH and Li-ion include integrated electronic components).

Chemistry	Specific Energy (in Wh·kg⁻¹)	Specific Energy (in Wh·l⁻¹)	Power Density (in W·kg⁻¹)	Power Density (in W·l⁻¹)
KHI-Ni/MH	21	52	508	1286
Ni-Cd	37	98	123	328
VRLA	45	88	111	219
Li-ion	123	163	182	215

Since 2010, numerous Ni/MH Battery Power Systems (BPS) using the KHI GigaCell product have been installed throughout Japan. The primarily use is in way-side railway storage to capture and reuse a train's regenerative braking energy (Figure S4). Reduced energy usage, lower peak power consumption, improved line voltage stabilization, and overall energy cost savings are commonly observed where these batteries are installed in subways, monorails and regional rail lines in Japan.

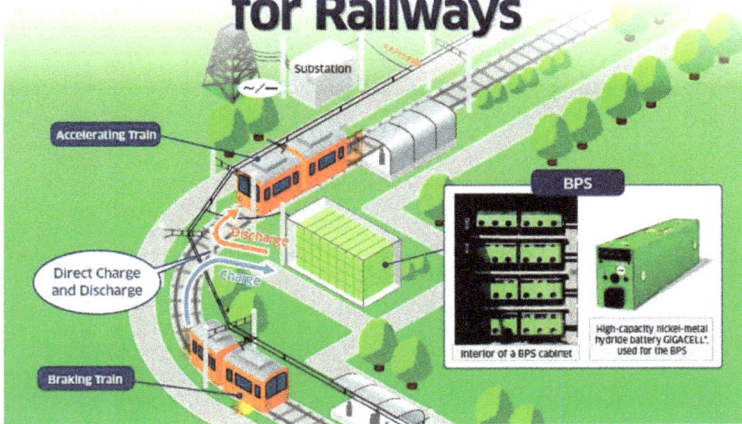

Figure S4. Illustration of a wayside railroad BPS installation [S7].

S3. WMATA Energy Storage Demonstration Project

A Ni/MH BPS was utilized for the Washington Metropolitan Area Transit Authority (WMATA) Energy Storage Demonstration Project installation of a 2 MW / 400 kWh Ni/MH BPS in 2013. The Washington-DC Metrorail system carries 720,000 passengers daily with 91 stations on six lines covering more than 117 miles of track. Trains are powered by 100 substations through a 700 Vdc third-rail distribution system. The BPS installed at the West Falls Church substation consists of four parallel units. Each of them is composed of 19 battery modules (5 kWh each). The outcomes of the two-year study, as reported by the US Department of Transportation Federal Transit Administration last year [S2], were overwhelmingly favorable with up to 15.4% energy savings, up to 12.5% peak power reduction, and a 75.5V average line stabilization.

S4. Ni/MH in Other Fields

A typical railway storage battery experiences around 4000 charge/discharge cycles daily (similar to the output of a large solar or wind farm) as shown in Figure S5. Use of Ni/MH BPS technology in grid-tied renewable energy installations is also being evaluated in Japan. An example is the 102 kWh Ni/MH BPS used to balance the electricity output of a 10 MW solar plant near Osaka as shown in Figure S6.

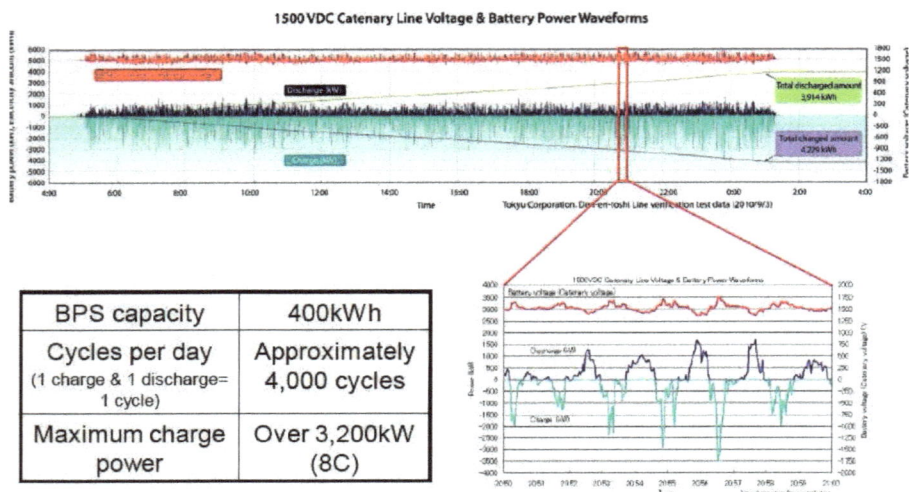

Figure S5. Single day charge/discharge profile for a wayside railroad BPS installation [S7].

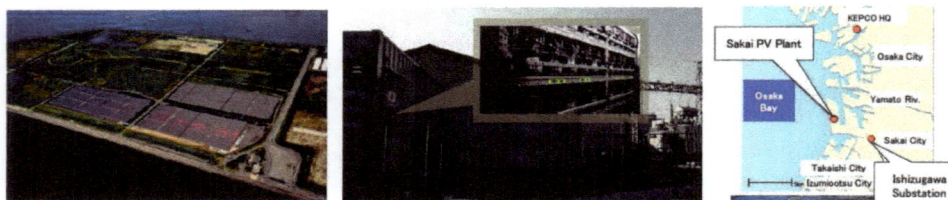

Figure S6. Solar farm and Ni/MH BPS near Osaka, Japan [S7].

References

S1. Alpha Technologies Ltd. Cellect™ 600 product data sheet. Available online: https://atl.app.box.com/v/cellect-600-48v-dc (accessed on 31 July 20017)

S2. WMATA Energy Storage Demonstration Project, Federal Transit Administration Final Report, June 2015. Available online: https://www.transit.dot.gov/sites/fta.dot.gov/files/docs/FTA_Report_No._0086.pdf (accessed on 31 July 20017).

S3. Nishimura, K.; Takasaki, T.; Sakai, T. Introduction of large-sized nickel-metal hydride battery GIGACELL for industrial applications. *J. Alloy. Compd.* **2013**, *580*, S353–S358.

S4. Takasaki, T.; Nishimura K.; Saito, M.; Fukunaga, H.; Iwaki, T.; Sakai, T. Cobalt-free nickel-metal hydride battery for industrial applications. *J. Alloys Compd.* **2013**, *580*, S378–S381.

S5. Young, K.; Wu, A.; Qiu, Z.; Tan, J.; Mays, W. Effects of H_2O_2 addition on the cell balance and self-discharge of Ni/MH batteries with AB_5 and A_2B_7 alloys. *Int. J. Hydrogen Energy* **2012**, *37*, 9882–9891.

S6. Ouchi, T.; Young, K.; Moghe, D. Reviews on the Japanese Patent Applications regarding nickel/metal hydride batteries. *Batteries* **2016**, 2, 21; doi:10.3390/batteries2030021

S7. About GIGACELL. Available online: http://global.kawasaki.com/en/energy/solutions/battery_energy/about_gigacell/index.html (accessed on 31 July 20017).

![batteries logo] *batteries*

MDPI

Article

Electron Backscatter Diffraction Studies on the Formation of Superlattice Metal Hydride Alloys

Shuli Yan [1,2], Kwo-Hsiung Young [1,2,*], Xin Zhao [3], Zhi Mei [4] and K. Y. Simon Ng [1]

[1] Department of Chemical Engineering and Materials Science, Wayne State University, Detroit, MI 48202, USA; shuli.yan@partners.basf.com (S.Y.); sng@wayne.edu (K.Y.S.N.)

[2] BASF/Battery Materials—Ovonic, 2983 Waterview Drive, Rochester Hills, MI 48309, USA

[3] School of Materials and Metallurgy, Inner Mongolia University of Science and Technology, Baotou 014010, Inner Mongolia, China; fatcatzx@163.com

[4] Department of Chemistry, Wayne State University, Detroit, MI 48202, USA; zmei@chem.wayne.edu

* Correspondence: kwo.young@basf.com; Tel.: +1-248-293-7000

Received: 2 October 2017; Accepted: 5 December 2017; Published: 13 December 2017

Abstract: Microstructures of a series of La-Mg-Ni-based superlattice metal hydride alloys produced by a novel method of interaction of a $LaNi_5$ alloy and Mg vapor were studied using a combination of X-ray energy dispersive spectroscopy and electron backscatter diffraction. The conversion rate of $LaNi_5$ increased from 86.8% into 98.2%, and the A_2B_7 phase abundance increased from 42.5 to 45.8 wt % and reduced to 39.2 wt % with the increase in process time from four to 32 h. During the first stage of reaction, Mg formed discrete grains with the same orientation, which was closely related to the orientation of the host $LaNi_5$ alloy. Mg then diffused through the *ab*-phase of $LaNi_5$ and formed the AB_2, AB_3, and A_2B_7 phases. Diffusion of Mg stalled at the grain boundary of the host $LaNi_5$ alloy. Good alignments in the *c*-axis between the newly formed superlattice phases and $LaNi_5$ were observed. The density of high-angle grain boundary decreased with the increase in process time and was an indication of lattice cracking.

Keywords: metal hydride; superlattice alloy; electron backscatter diffraction; crystallographic orientation; gaseous-state diffusion; superlattice alloy

1. Introduction

Rare earth (RE)/Mg-based superlattice metal hydride (MH) alloys are employed extensively in the consumer nickel/metal hydride (Ni/MH) batteries because of the following improvements over the conventional AB_5 MH alloys: higher hydrogen storage capacities, better high-rate dischargeability (HRD), superior low-temperature and charge-retention performances, and improved cycle stability [1–8]. Out of the six available superlattice phases (three hexagonal and three rhombohedral), Ce_2Ni_7 was found to be the most desirable phase considering general battery performance [9], and the A_2B_7 stoichiometry shows the best HRD, charge retention, and cycle life [10]. Although the superlattice MH alloys are very attractive to battery engineers, their fabrication is difficult because of the high vapor pressure of Mg [11]—an indispensable ingredient to maximize the capacity and stabilize the structure [12,13]. In the conventional melt-and-cast method, Mg was added as a late addition in the form of $MgNi_2$ [14]. Extra Mg needs to be added to compensate for the loss to vapor, which is a difficult factor to control precisely. A new method of making the Mg-containing superlattice MH alloys [15] was proposed using a gaseous-state Mg-diffusion into the AB_5 MH alloys, which can be easily produced by vacuum induction melting with a furnace size as large as one ton [16]. Early electron microscope studies indicated the feasibility of transporting Mg into the $La_{0.8}Ni_3$ alloy and forming the Mg-containing superlattice phases, but the constituent phases have not yet been confirmed [15].

Electron backscatter diffraction (EBSD) is a microstructural-crystallographic technique that allows the user to examine the crystallographic orientations of constituent phases in very localized areas (one square micron or less) of a polycrystalline material in a scanning electron microscope (SEM). Capability of EBSD can be further enhanced by including the chemical composition information gathered by X-ray energy dispersive spectroscopy (EDS) [17]. In the past, we employed EBSD in the studies of a Zr_7Ni_{10} [18], a C14-based AB_2 [19], and a C14/body-centered-cubic MH alloys [20]. In the last two cases, EBSD was used to confirm the cleanliness of the grain boundary from the strong alignment of crystallographic orientations of neighboring phases. In the current study, EBSD was used to identify the new phases formed by the Mg-diffusion into the $LaNi_5$ alloy and study the nature of grain boundary and alignment of crystallographic orientations of neighboring phases.

2. Experimental Setup

The $LaNi_5$ alloy was synthesized by induction melting La and Ni (both with purity higher than 99.5%) under an argon atmosphere. Solidification of the $LaNi_5$ alloy was operated by a rapid quenching equipment to ensure the slice thickness to be between 0.2 to 0.4 mm. The Mg-absorption alloying process was operated in a sealed internal isolation stainless-steel retort. Slices of Mg and $LaNi_5$ were placed into each side of the retort in a weight ratio of $Mg:LaNi_5 = 1:30$ and separated by foraminiferous septa. The retort was placed in an annealing furnace under an inert atmosphere (argon). Annealing (reaction) temperature was increased from the ambient temperature to 1273 K with a heating rate of $10 \ K \cdot min^{-1}$. Afterward, the vessel was cooled to room temperature in the furnace.

4-h, 8-h, 16-h, and 32-h annealed alloy samples were mounted in resin holders. The samples were polished using metallographic silicon carbide sandpapers in the sequence of 400-, 800- and, finally, 1200-grit (Buehler, Lake Bluff, IL, USA), and they were then finely polished with Buehler MicroPolish II Suspension 1-μm alumina suspension and PACE Technologies SIAMAT2 0.02-μm colloidal silica (PACE Technologies, Tucson, AZ, USA) to obtain a mirror finish. The prepared samples were kept in a sealed tank with high vacuum and low O_2^- and moisture content to avoid surface oxidization and physical and chemical absorptions.

To investigate the phase distribution, samples were studied by a JEOL JSM-7600 field emission SEM (JEOL USA, Inc., Peabody, MA, USA) equipped with an EDAX Pegasus Apex 2 Integrated EDS and EBSD System (EDAX Inc., Mahwah, NJ, USA). The EBSD data was collected and analyzed with TSL OIM data Collection 7 and TSL OIM Analysis 7 program (EDAX Inc., Mahwah, NJ, USA)., respectively. - All the measured points have confidence indices greater than 0.6, which corresponds to an accuracy higher than 95%. Fit parameter is the averaged angular difference between the detected and recalculated bands. In this case, the fit parameter is less than 0.8, showing a high degree of matching.

3. Results and Discussion

3.1. Phase Identification

As stated in the Introduction, the La-Mg-Ni-type alloys were based on the La-Ni binary intermetallic alloys. The presence of Mg destabilizes the hydride, making it suitable for room-temperature battery applications [13]. Phases commonly reported in the La-Mg-Ni superlattice alloys are $(La,Mg)Ni_5$, $(La,Mg)_5Ni_{19}$, $(La,Mg)_2Ni_7$, $(La,Mg)Ni_3$, and $(La,Mg)Ni_2$. [21]. Basic subunits for these phases are the AB_5 and A_2B_4 slabs, which alternatively stack along the *c*-axis in different patterns to form the structures [21,22]. Six types of the La-Mg-Ni superlattice phases are Pr_5Co_{19} (PDF: 04-004-1477), Nd_5Co_{19} (PDF: 04-004-1478), Ce_2Ni_7 (PDF: 04-007-1092), Pr_2Ni_7 (PDF: 01-081-8491), $CeNi_3$ (PDF: 04-007-1090), and $PuNi_3$ (PDF: 04-007-1091), and their structures are either hexagonal or rhombohedral and are generally difficult to distinguish in the X-ray diffraction patterns [21–25]. Furthermore, the AB_5 (PDF: 00-055-0277) and AB_2 (PDF: 04-001-2137) phases are also hexagonal, which adds to the difficulty in identifying the phases. EBSD provides a powerful approach to study the lattice structures of individual phases and their alignments in certain orientations in the multi-phase

La-Mg-Ni superlattice alloys. Figure 1 shows the crystal structures and computer-generated EBSD diffraction patterns of the (11$\bar{2}$0) planes of LaNi$_5$, Pr$_5$Co$_{19}$, Nd$_5$Co$_{19}$, Ce$_2$Ni$_7$, Pr$_7$Ni$_2$, CeNi$_3$, PrNi$_3$, and NbCr$_2$ (hexagonal), which are used as the base patterns in this study. Lattice parameters of the seven types of structures are different and, therefore, their EBSD patterns are different, especially away from the (0001) plane.

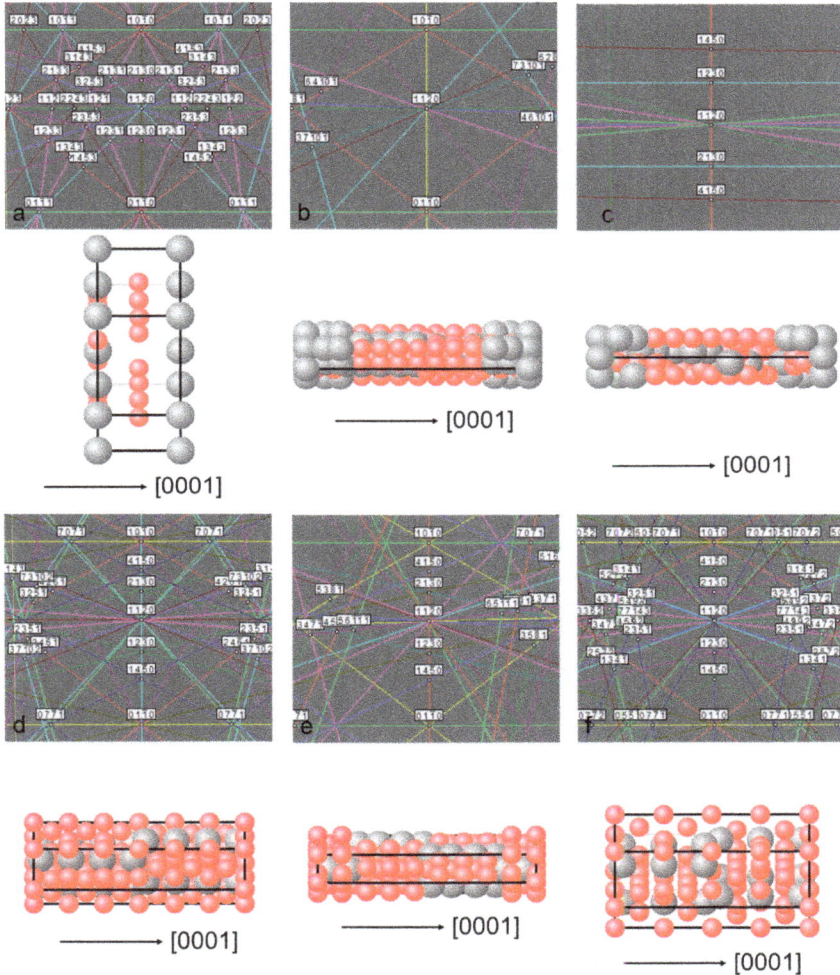

Figure 1. *Cont.*

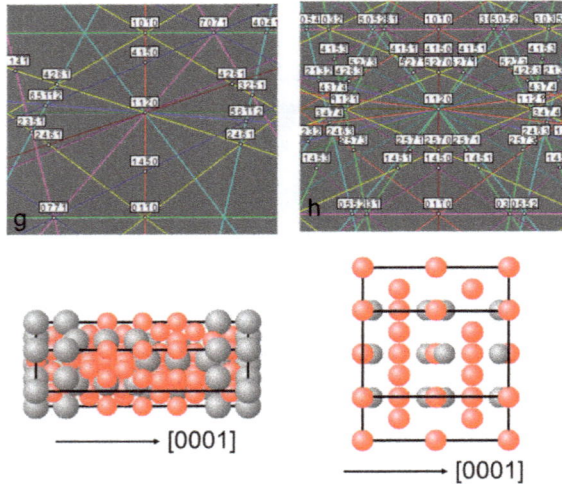

Figure 1. Computer-generated EBSD patterns for the (11$\bar{2}$0) planes of (**a**) LaNi$_5$; (**b**) Pr$_5$Co$_{19}$; (**c**) Nd$_5$Co$_{19}$; (**d**) Ce$_2$Ni$_7$; (**e**) Pr$_2$Ni$_7$; (**f**) CeNi$_3$; (**g**) PrNi$_3$; and (**h**) NbCr$_2$ (hexagonal).

SEM-backscattered electron image (BEI) of the 8-h annealed sample is shown in Figure 2, and several spots were studied in detail (spots Z1 to Z6, A to C, and Y1 to Y13). For spots Z1 to Z6, their original EBSD patterns, fitted patterns, and simulations of grain orientation are shown in Figure 3. Other than the base structure (LaNi$_5$, spot Z1), both the A$_2$B$_7$ (spots Z2 and Z3) and AB$_3$ (spot Z4) superlattice structures, Mg metal (spot Z5), and AB$_2$ (spot Z6) phases are found as the new phases. All the patterns are blurry, which can be caused by multiple factors: besides the issues of imperfection in the sample polish and limited camera resolution, strains in the alloy can also influence the band contrasts.

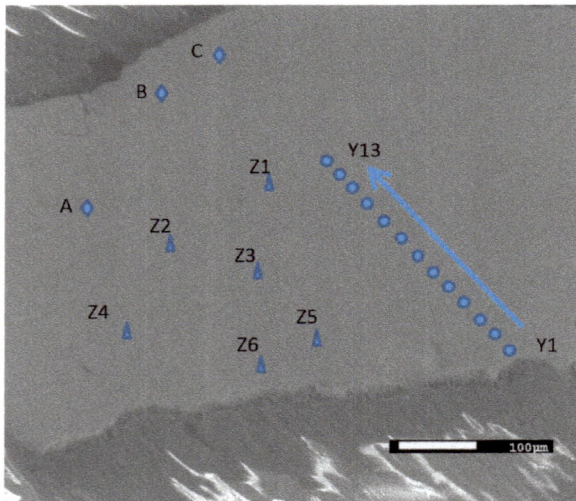

Figure 2. SEM-BEI of the 8-h annealed sample. Structures of spots Z1 to Z6 and A to C were studied by EBSD. Compositions of spots Y1 to Y13 were measured by EDS.

Figure 3. *Cont.*

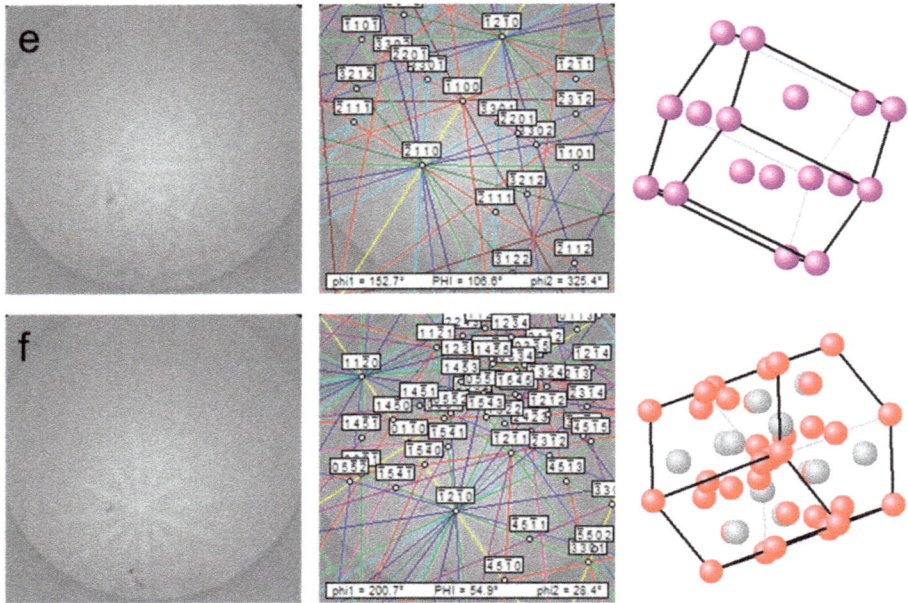

Figure 3. Original EBSD patterns, fitted patterns, and simulations of grain orientation from spots (**a**) Z1: LaNi$_5$; (**b**) Z2: Ce$_2$Ni$_7$; (**c**) Z3: Pr$_2$Ni$_7$; (**d**) Z4: CeNi$_3$; (**e**) Z5: Mg; and (**f**) Z6: NbCr$_2$ of the 8-h annealed sample (Figure 2).

Elastic and plastic strains have been reported to cause other changes in the EBSD patterns [26]. Figure 4 shows the EBSD patterns from spots A to C, and they are identified as LiNi$_5$, Ce$_2$Ni$_7$, and Pr$_2$Ni$_7$, respectively. Except for the blurriness, a new band (B2 in Figure 4b) and bands with a slight rotation (A1 in Figure 4a), a shift (A2 in Figure 4a), a narrower diffraction width (B1 in Figure 4b), and a wider diffraction width (C1 in Figure 4c) are observed. Figure 5 shows the schematic diagrams of a few lattice distortions that may cause the changes in the EBSD patterns. Elastic strains distort the crystal lattice. Winkelmann reported that if the elastic strains uniformly dilate the lattice, changes in the EBSD patterns only occur in bandwidth [27]. If other lattice distortions exist, such as partially-lengthened bonds in the lattice (Figure 5b), shifts in some zone axes in the EBSD patterns can occur [26]. Keller et al. [28] reported that a "bent" crystal, as shown in Figure 5c, leads to a slight degradation in pattern quality and a minor band rotation. Since the planes within the diffraction volume are no longer exactly parallel to each other, blurring of diffraction bands and band rotation occur and are caused by the slight changes in Bragg angles. Plastic strains lead to dislocations in the crystal lattice, and Figure 5d,e demonstrate two types of dislocations. The region in the material with a high dislocation density with a net Burgers vector of zero is considered to have statistically stored dislocations (SSD) (Figure 5d). The resulting pattern in the area containing SSDs is degraded because of the local perturbations of diffracting lattice planes that result in incoherent scattering [26]. The area with dislocations with a net-nonzero Burgers vector have geometrically-necessary dislocations (GND). Arrays of GNDs can form subgrain boundaries (Figure 5e) and degrade the EBSD pattern quality by the superposition of patterns from neighboring subgrains with a small rotation in between [26,28]. The influences of misorientation on the quality of EBSD pattern are further elaborated in Supplemental 1 with a SEM micrograph showing three locations with small misorientations (Figure S1) and the corresponding EBSD patterns (Figure S2).

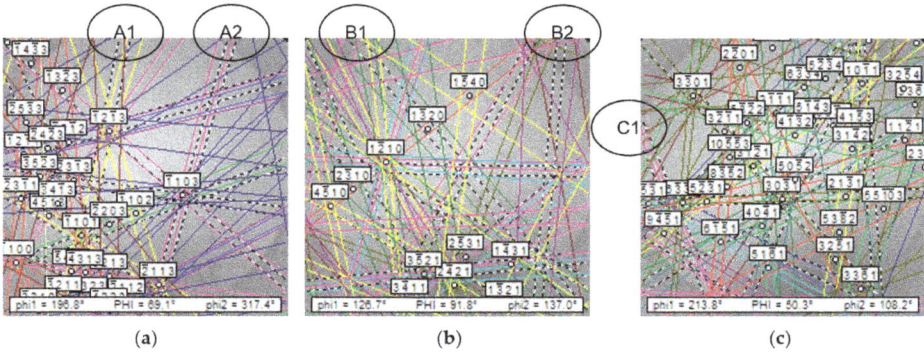

Figure 4. Comparison of original and computer-generated EBSD patterns from spots (**a**) A: LaNi$_5$; (**b**) B: Ce$_2$Ni$_7$; and (**c**) C: Pr$_2$Ni$_7$ of the 8-h annealed sample (Figure 2). Solid-color lines indicate the width of computer-generated band. Dashed-black lines show the width of actual band. A new band (B2) and bands with a slight rotation (A1), a shift (A2), a narrower width (B1), and a wider width (C1) are observed.

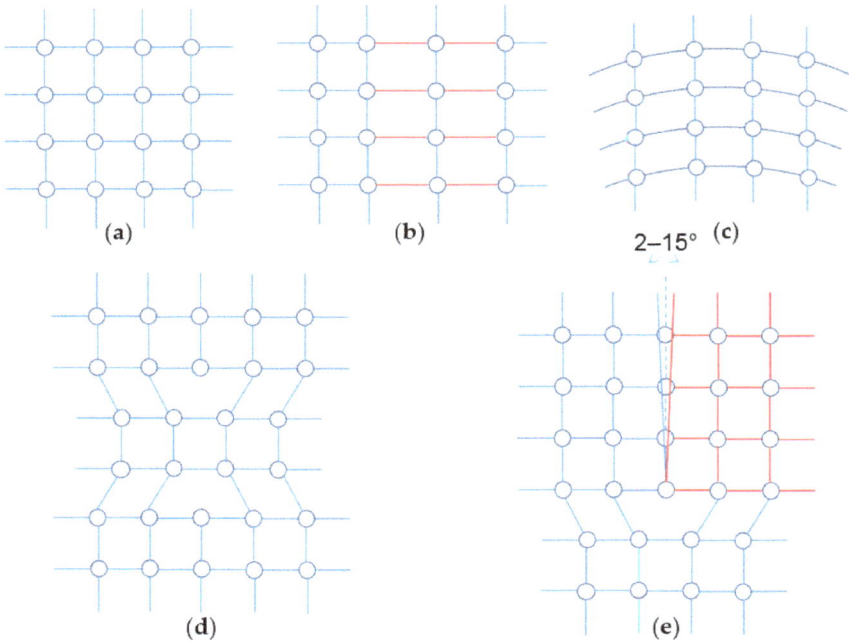

Figure 5. Schematic diagrams of (**a**) a regular crystal lattice; (**b**) a strained lattice with uniaxial lengthened bonds; (**c**) a bent lattice; (**d**) a distorted lattice with symmetric vacancy defects; and (**e**) a lattice with a subgrain boundary.

In this study, the La-Mg-Ni alloys were prepared by a solid-state method with Mg diffusing into the LaNi$_5$ alloy. Therefore, both elastic and plastic strains were formed during the Mg-diffusion process and have strong influences on the EBDS pattern clarity. During the Mg-diffusion process, defects in the raw LaNi$_5$ alloys were generated while new phases were formed. Physical and chemical properties of the La-Mg-Ni alloys prepared by this method are affected by the distribution of defects, and compositions and abundances of constituent phases. Therefore, investigating the alloys' microstructures by the EBSD technique becomes important.

3.2. Phase Distribution

3.2.1. Element Distribution

To fully characterize the alloys' microstructures, EDS elemental mappings were conducted on all four alloys (4 h-, 8 h-, 16 h-, and 32-h annealed samples), and the results are shown in Figure 6. While La and Ni distribute uniformly in all alloys, Mg shows a high concentration at the edges of all alloys. The EDS results of spots Y1 to Y13 in the SEM micrograph of the 8-h annealed sample (Figure 2) are listed in Table 1, which demonstrate the uneven distribution of Mg from the edge to center of alloy. Penetration of Mg into the 8-h annealed sample is about 100 microns. The value of Ni/(La + Mg) varies from 1.49 to 5.32, and the varying trend indicates that the phase changes along the longitudinal section (from one surface to another) of alloy. However, it must be stated that EDS is a semi-quantitative analysis method, so other technologies have to be combined with EDS to validate the phase distribution.

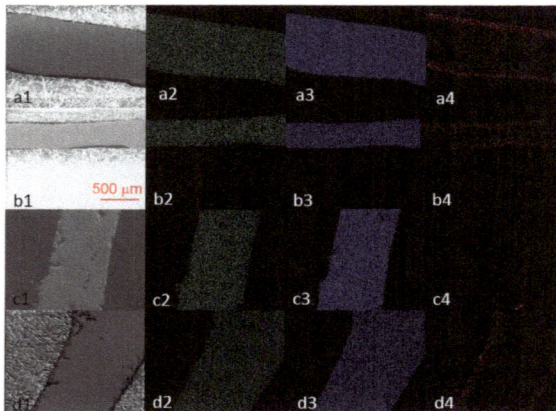

Figure 6. (a1,b1,c1,d1) SEM-BEIs and elemental mappings of (a2,b2,c2,d2) La, (a3,b3,c3,d3) Ni, and (a4,b4,c4,d4) Mg of the 4 h-, 8 h-, 16 h-, and 32-h annealed samples, respectively.

Table 1. Chemical compositions (in at%) from spots Y1 to Y13 of the 8-h annealed sample (Figure 2).

Spot	Mg	La	Ni	Ni/(La + Mg)
Y1	24.01	16.19	59.79	1.49
Y2	16.42	12.94	70.64	2.41
Y3	13.12	14.09	72.78	2.67
Y4	1.29	16.11	82.6	4.75
Y5	0.00	16.45	83.55	5.08
Y6	0.28	16.27	83.45	5.04
Y7	0.29	16.10	83.61	5.10
Y8	0.00	16.71	83.29	4.98
Y9	0.00	15.82	84.18	5.32
Y10	0.00	16.72	83.28	4.98
Y11	0.00	16.24	83.76	5.16
Y12	1.00	16.26	82.74	4.79
Y13	8.32	15.16	76.54	3.26

3.2.2. Phase Distribution

As an example, phase identification mapping by EBSD of the 16-h annealed sample is shown in Figure 7. Six phases, including $LaNi_5$, Ce_2Ni_7, Pr_2Ni_7, $CeNi_3$, $NbCr_2$ (hexagonal), and Mg, can be identified. In the investigated area of the 16-h annealed sample, the Pr_5Co_{19}, Nd_5Co_{19}, and $PuNi_3$ structures are not found. A gradient of phase abundance can be observed for most phases. $LaNi_5$, the unreacted material, is concentrated in the center. While the diffuse-in Mg is concentrated at the edge, Ce_2Ni_7, one of the important target products [9], is also concentrated at the edge. Existence of pure Mg phase is validated in Supplement 2 with SEM micrographs (Figure S3) and EDS results showing high-Mg contents (Table S1). Pr_2Ni_7, another one of the target products, is located mainly between the $LaNi_5$ and Mg phases. Abundances of the six phases in the investigated area are summarized in the table included in Figure 7. The $LaNi_5$ abundance is 27.2%, and the combined abundance of Ce_2Ni_7 and Pr_2Ni_7 is 15.8%. The $CeNi_3$ and $NbCr_2$ phases are products from the overreaction of Ce_2Ni_7 and Pr_2Ni_7 with Mg. Future work will focus on increasing the Ce_2Ni_7 and Pr_2Ni_7 phase abundances.

Edge ← Middle

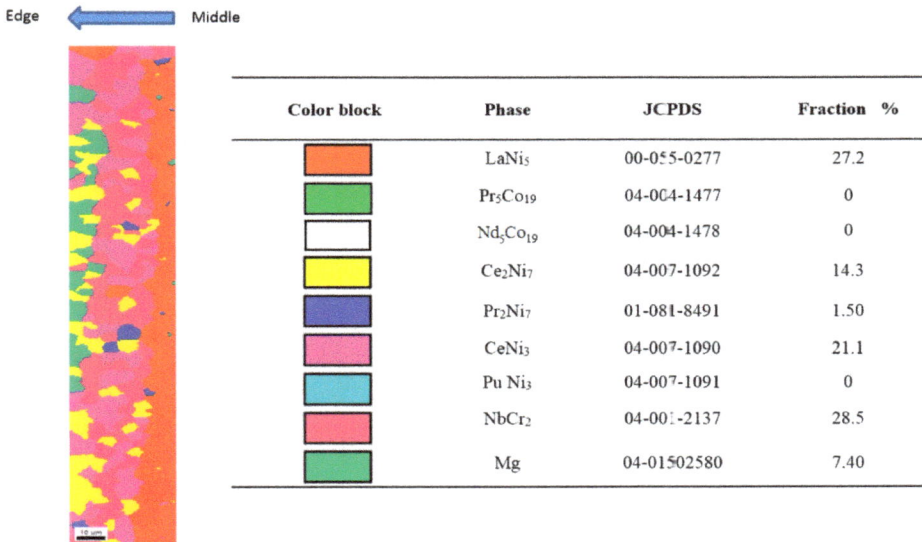

Color block	Phase	JCPDS	Fraction %
	$LaNi_5$	00-055-0277	27.2
	Pr_5Co_{19}	04-004-1477	0
	Nd_5Co_{19}	04-004-1478	0
	Ce_2Ni_7	04-007-1092	14.3
	Pr_2Ni_7	01-081-8491	1.50
	$CeNi_3$	04-007-1090	21.1
	$Pu Ni_3$	04-007-1091	0
	$NbCr_2$	04-001-2137	28.5
	Mg	04-01502580	7.40

Figure 7. EBSD phase identification mapping and quantification of the 16-h annealed sample. The grain tolerance angle is 15°, and the minimum grain size is 3 μm.

3.2.3. $LaNi_5$ and Mg Grain Distributions and Orientations

Inverse pole figure (IPF) and EBSD mapping, EBSD diffraction patterns, and crystal simulations of $LaNi_5$ of the 16-h annealed sample are shown in Figure 8. In Figure 8a,b, color channels in red, green, and blue represent, [0001], [10$\bar{1}$0], and [2$\bar{1}\bar{1}$0] of $LaNi_5$. Figure 8a shows only two grain orientations for $LaNi_5$. Grain orientations are described in the form of the Euler angle in Figure 8c,e. Figure 8a shows that grains with orientation 2 are isolated and much smaller in abundance than those with orientation 1. The right side of Figure 8a (the center of alloy) is a single $LaNi_5$ grain without reacting with Mg. The interaction with Mg stopped at the grain boundary between the $LaNi_5$ phase with orientation 1 and the $LaNi_5$ phase with orientation 2. This is the proof that the inter-diffusion of Mg into the host $LaNi_5$ alloy is through a specific direction (presumable along the *ab*-plane) and stops at the grain boundary. Therefore, increasing the grain size of the host LaNi$_5$ alloy can enhance the diffusion of Mg into the bulk.

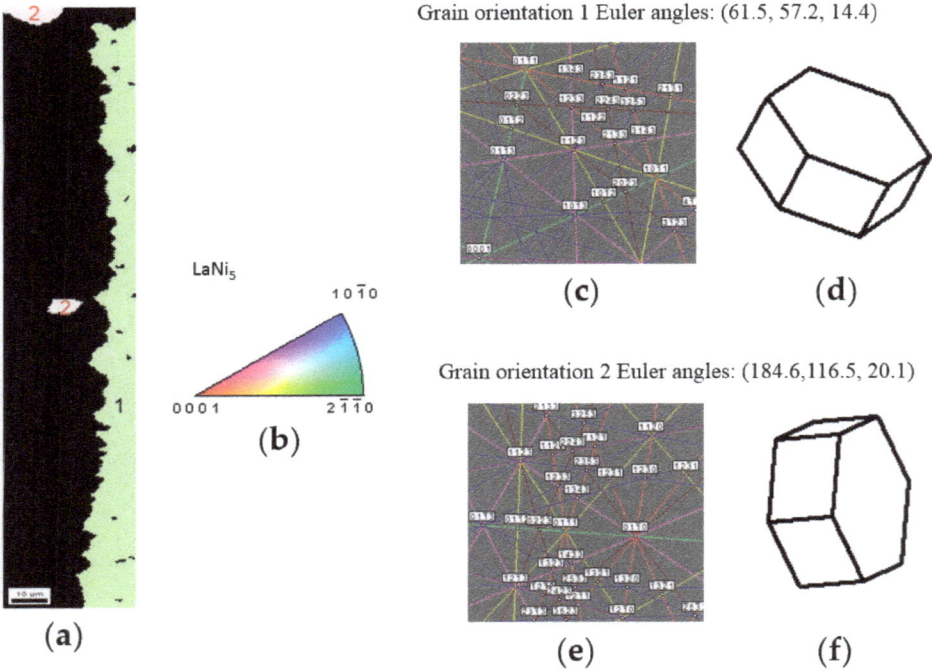

Figure 8. (a) IPF and EBSD mapping; (b) color assignment; (c) EBSD pattern for grain orientation 1 and (d) its corresponding crystal simulation; and (e) EBSD pattern for grain orientation 2 and (f) its corresponding crystal simulation of the LaNi$_5$ phase of the 16-h annealed sample.

IPF and EBSD mapping, EBSD diffraction patterns, and crystal simulations of Mg of the 16-h annealed sample are shown in Figure 9. Three different crystallographic orientations are found for Mg. Grains with orientation 1 distribute at the edge of alloy and are the largest in size. Grains with orientation 3 distribute close to the center of alloy and are the smallest. After diffusing into the host, Mg first agglomerates into individual grains with the same crystallographic orientation (related to the host orientation) before reacting with the host to form the superlattice phases. It is interesting to find that the superlattice phases are formed by the reaction of LaNi$_5$ with the Mg crystal but not the Mg vapor.

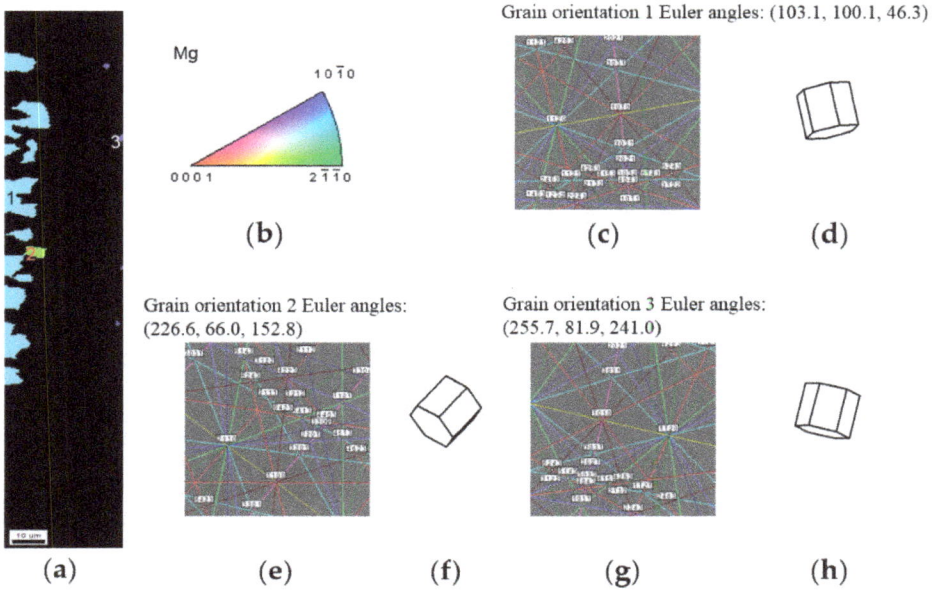

Figure 9. (a) IPF and EBSD mapping; (b) color assignment; (c) EBSD pattern for grain orientation 1 and (d) its corresponding crystal simulation; (e) EBSD pattern for grain orientation 2 and (f) its corresponding crystal simulation; and (g) EBSD pattern for grain orientation 3 and (h) its corresponding crystal simulation of the Mg phase of the 16-h annealed sample.

3.2.4. Ce_2Ni_7 and Pr_2Ni_7 Grain Distributions and Orientations

IPF and EBSD mapping, image quality (IQ) diffraction pattern and grain boundary map, and grain size distribution of Ce_2Ni_7 of the 16-h annealed sample are shown in Figure 10. Unlike $LaNi_5$ and Mg, the Ce_2Ni_7 phase shows more grain orientations (Figure 10a). IQ patterns can be used to characterize the defect distribution in grains and is especially useful for the strain mapping [20]. Figure 10c shows that some grains are darker than the others, which indicates that concentrated defects and residual strains exist in these darker grains. A grain boundary is formed by the accumulation of edge dislocations. In this study, two types of boundaries are characterized: low-angle grain boundary (LAGB) and high-angle grain boundary (HAGB). LAGB subdivides a grain into two equiaxial cells and forms subgrains, which may increase the plastic deformation. In this study, we define the lattice misorientation of LAGB-separated grain zones to be from 2° to 15°. A grain boundary with a misorientation $\geq 15°$ is denoted as HAGB, which differentiates a grain from its initial microstructure and creates a new grain. Figure 10c indicates the grain boundary distribution of the Ce_2Ni_7 phase in the 16-h annealed sample. Amount of LAGBs in the Ce_2Ni_7 phase is 40.3%, which is much higher than that in the $LaNi_5$ or Mg phase and suggests a high density of defects in the Ce_2Ni_7 grains. Grain size distribution is shown in Figure 10d. Average grain diameter of the Ce_2Ni_7 phase is 7 μm. IPF and EBSD mapping, EBSD diffraction patterns, and crystal simulations of Pr_2Ni_7 in the 16-h annealed sample are shown in Figure 11. Unlike Ce_2Ni_7, the Pr_2Ni_7 phase has only two different orientations and a smaller grain size of about 4 μm.

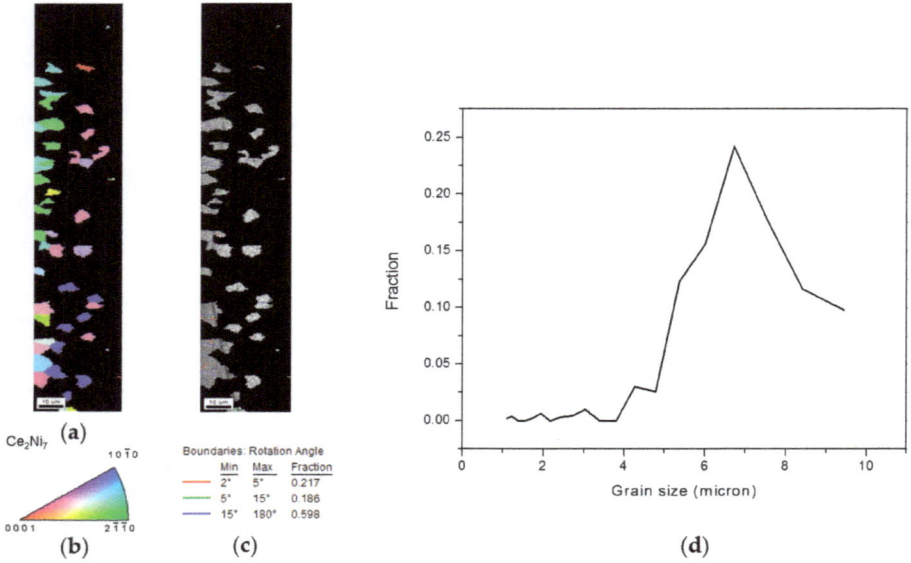

Figure 10. (**a**) IPF and EBSD mapping; (**b**) color assignment; (**c**) IQ diffraction pattern and grain boundary map; and (**d**) grain size distribution of the Ce$_2$Ni$_7$ phase of the 16-h annealed sample.

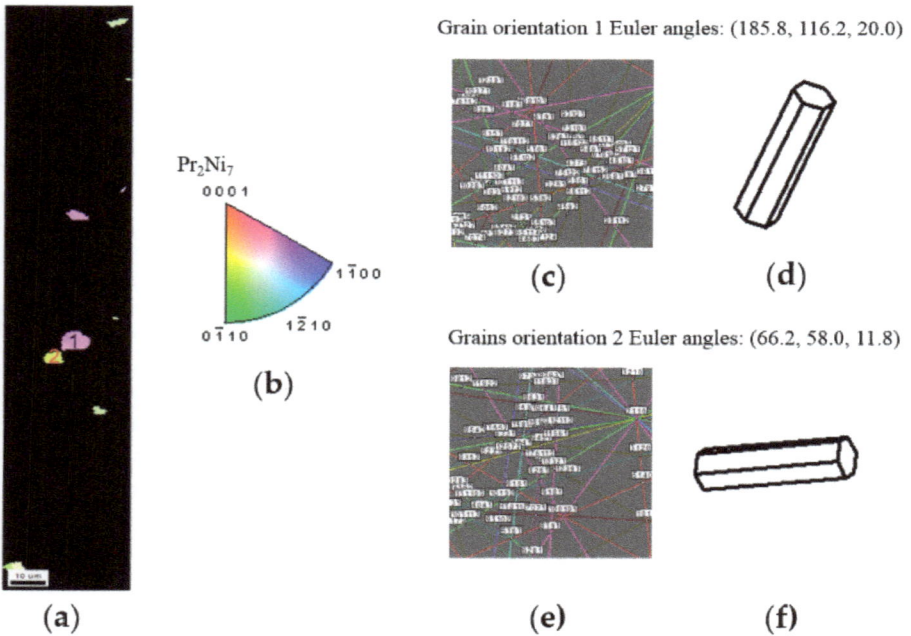

Figure 11. (**a**) IPF and EBSD mapping; (**b**) color assignment; (**c**) EBSD pattern for grain orientation 1 and (**d**) its corresponding crystal simulation; and (**e**) EBSD pattern for grain orientation 2 and (**f**) its corresponding crystal simulation of the Pr$_2$Ni$_7$ phase of the 16-h annealed sample.

3.3. Alignment in Crystallographic Orientations

Figure 12a shows the EBSD phase identification mapping of the 8-h annealed sample. [0001]s of LaNi5 and Pr2Ni7 in the green circle are found to be parallel to each other, and [0001]s of LaNi5 and Ce_2Ni_7 in the black circle are also parallel (Figure 12b–e). This information gives a hint about the grain growth mechanism of the A_2B_7-type phases during the diffusion of Mg into the AB_5 alloy. Structures of the La-Mg-Ni superlattice phases are composed of the A_2B_4 and the AB_5 slabs [23]. The alignments of LaNi5 and Pr2Ni7 in the *c*-axis and LaNi5 and Ce_2Ni_7 in the *c*-axis imply that Mg diffuses into LaNi5 and forms the A_2B_4 slab through the *ab*-plane and, therefore, the *c*-axis orientation remains unchanged after the superlattice phase formation.

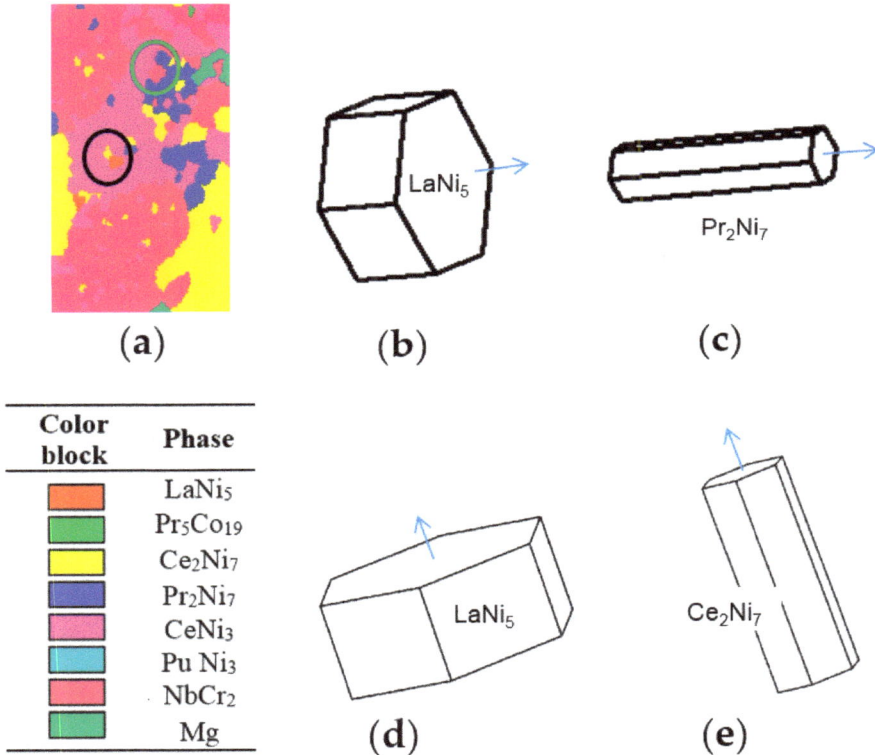

Color block	Phase
🟧	$LaNi_5$
🟩	Pr_5Co_{19}
🟨	Ce_2Ni_7
🟦	Pr_2Ni_7
🟪	$CeNi_3$
🟦	$Pu\,Ni_3$
🟥	$NbCr_2$
🟩	Mg

Figure 12. Crystallographic orientation alignments in [0001] demonstrated by (**a**) an IPF-EBSD map; orientations of (**b**) $LaNi_5$ and (**c**) neighboring Pr_2Ni_7 in the green circle; and orientations of (**d**) $LaNi_5$ and (**e**) neighboring Ce_2Ni_7 in the black circle of the 8-h annealed sample.

3.4. Effect of Process Temperature on Phase Development

EBSD mapping was performed in an area of 30×100 square microns close to the edge of each sample. The results may not be very accurate because of the limited sampling areas and large variations in distribution of the superlattice phases. Nevertheless, the calculated phase abundances are compared in Table 2. The conversion from LaNi5 to other phases is more complete with the increase in process time and reaches 98.2 wt % at a processing time of 32 h. Abundance of the most desirable Ce_2Ni_7 phase is about 25 wt % and not very sensitive to the processing time. Abundance of the A_2B_7 phases (both Ce_2Ni_7 and Pr_2Ni_7) increases from 42.5 to 45.8 wt % and reduces to 39.2 wt % as the processing time increases from 4 to 32 h. The unwanted AB_2 and AB_3 phases (with excessive Mg-content) cannot be eliminated with the increase in processing time. Future development work will focus on the reduction of the Mg-supply and/or addition of a second annealing treatment without Mg.

Table 2. Phase abundances (in wt %) obtained by EBSD mapping of the 4-h, 8-h, 16-h, and 32-h annealed samples with a grain tolerance angle of 5° and a minimum grain size of 2 μm.

Phase	4 h	8 h	16 h	32 h
$LaNi_5$	13.2	5.5	4.2	1.8
Ce_2Ni_7	23.5	27.3	27.6	25.5
Pr_2Ni_7	19.0	18.5	16.3	13.7
$CeNi_3$	22.6	19.5	26.4	21.0
$PuNi_3$	0.5	0.2	0.5	0.5
$NbCr_2$	16.2	25.5	22.3	30.3
Mg	4.9	3.4	2.8	7.2

Grain size distributions of all four samples are also compared and listed in Table 3. The 4-h annealed sample has a heavy proportion of medium-size grains (5–12 μm) while the 8-h annealed sample has a much higher percentage of small grains (1–5 μm). Longer processing time (16- and 32-h) recovers the percentage of medium-size grains. The reason for the grain size evolution is not clear and requires further studies. The last comparison is on the distribution of misorientation angle in the grain boundary, which is shown in Table 4. Amount of HAGBs (15 to 180°) decreases with the increase in process time, which suggests the occurrence of lattice cracking.

Table 3. Grain size distributions of the 4-h, 8-h, 16-h, and 32-h annealed samples.

Grain Size (μm)	4 h	8 h	16 h	32 h
1 to 5	24%	79%	35%	37%
5 to 12	61%	16%	48%	49%
>12	15%	5%	16%	14%

Table 4. Grain boundary distributions of the 4-h, 8-h, 16-h, and 32-h annealed samples.

Grain Boundary (°)	4 h	8 h	16 h	32 h
2 to 5	11.3%	14.4%	22.3%	22.1%
5 to 15	4.3%	2.5%	0.7%	11.4%
15 to 180	84.4%	83.1%	77.0%	55.5%

4. Conclusions

A gaseous-state Mg-diffusion into an AB_5 metal hydride alloy is demonstrated as an effective method to convert a $LaNi_5$ alloy to a multi-phase superlattice alloy. Four La-Mg-Ni superlattice phases are identified by EBSD in the products: Ce_2Ni_7, Pr_2Ni_7, $CeNi_3$, and $PuNi_3$. Additionally, Mg and AB_2 (with the hexagonal $NbCr_2$ structure) phases are also found. Longer process time increases the $LaNi_5$ conversion rate but shows no significant effect on increasing the abundance of the most desirable Ce_2Ni_7 phase. Defects are found abundantly in the Ce_2Ni_7 and Pr_2Ni_7 phases by EBSD, and distributions of the superlattice phases are not uniform. Future activities in reducing the Mg-loading, addition of a second annealing treatment, and increasing the initial grain size of the host alloy are suggested based on the findings in this study.

Supplementary Materials: The following are available online at http://www.mdpi.com/2313-0105/3/4/40/s1.

Acknowledgments: The authors would like to thank the following individuals for their help: Alan Chan, Jean Nei, and Diana Wong from BASF—Ovonic.

Author Contributions: Xin Zhao prepared the sample and Shuli Yan performed the EBSD study. Kwo-Hsiung and Simon Ng provided guidance and helped in manuscript preparation.

Conflicts of Interest: The authors declare no conflict of interest.

Abbreviations

RE	Rare earth
MH	Metal hydride
Ni/MH	Nickel/metal hydride
HRD	High-rate dischargeability
EBSD	Electron backscatter diffraction
SEM	Scanning electron microscope
EDS	Energy dispersive spectroscopy
BEI	Backscattered electron image
SSD	Statistically stored dislocations
GND	Geometrically necessary dislocations
IPF	Inverse pole figure
IQ	Image quality
LAGB	Low-angle grain boundary
HAGB	High-angle grain boundary

References

1. Yasuoka, S.; Magari, Y.; Murata, T.; Tanaka, T.; Ishida, J.; Nakamura, H.; Nohma, T.; Kihara, M.; Baba, Y.; Teraoka, H. Development of high-capacity nickel-metal hydride batteries using superlattice hydrogen-absorbing alloys. *J. Power Sources* **2006**, *156*, 662–666. [CrossRef]
2. Teraoka, H. Development of Low Self-Discharge Nickel-Metal Hydride Battery. Available online: http://www.scribd.com/doc/9704685/Teraoka-Article-En (accessed on 9 April 2016).
3. Kai, T.; Ishida, J.; Yasuoka, S.; Takeno, K. The effect of nickel-metal hydride battery's characteristics with structure of the alloy. In Proceedings of the 54th Battery Symposium in Japan, Osaka, Japan, 7–9 October 2013; p. 210.
4. Takasaki, T.; Nishimura, K.; Saito, M.; Fukunaga, H.; Iwaki, T.; Sakai, T. Cobalt-free nickel-metal hydride battery for industrial applications. *J. Alloys Compd.* **2013**, *580*, S378–S381. [CrossRef]
5. Teraoka, H. Development of Ni-MH EThSS with Lifetime and Performance Estimation Technology. In Proceedings of the 34th International Battery Seminar & Exhibit, Fort Lauderdale, FL, USA, 20–23 March 2017.
6. Teraoka, H. Ni-MH Stationary Energy Storage: Extreme Temperature & Long Life Developments. In Proceedings of the 33th International Battery Seminar & Exhibit, Fort Lauderdale, FL, USA, 21–24 March 2016.
7. Teraoka, H. Development of Highly Durable and Long Life Ni-MH Batteries for Energy Storage Systems. In Proceedings of the 32th International Battery Seminar & Exhibit, Fort Lauderdale, FL, USA, 9–12 March 2015.
8. Ouchi, T.; Young, K.; Moghe, D. Reviews on the Japanese Patent Applications regarding nickel/metal hydride batteries. *Batteries* **2016**, *2*, 21. [CrossRef]
9. Young, K.; Ouchi, T.; Nei, J.; Koch, J.M.; Lien, Y. Comparison among constituent phases in superlattice metal hydride alloys for batter applications. *Batteries* **2017**, *3*, 34. [CrossRef]
10. Young, K.; Yasuoka, S. Past, present, and future of metal hydride alloys in nickel-metal hydride batteries. In Proceedings of the 14th International Symposium on Metal-Hydrogen Systems, Manchester, UK, 21–25 July 2014.

11. Hayakawa, H.; Enoki, H.; Akiba, E. Annealing conditions with Mg vapor-pressure control and hydrogen storage characteristic of La$_4$MgNi$_{19}$ hydrogen storage alloy. *Jpn. Inst. Met.* **2006**, *70*, 158–161. (In Japanese) [CrossRef]

12. Crivello, J.-C.; Zhang, J.; Latroche, M. Structural stability of AB$_y$ phases in the (La,Mg)–Ni system obtained by density functional theory calculations. *J. Phys. Chem.* **2011**, *115*, 25470–25478.

13. Crivello, J.-C.; Gupta, M.; Latroche, M. First principles calculations of (La,Mg)$_2$Ni$_7$ hydrides. *J. Alloys Compd.* **2015**, *645*, S5–S8. [CrossRef]

14. Young, K.; Ouchi, T.; Huang, B. Effects of annealing and stoichiometry to (Nd, Mg)(Ni, Al)$_{3.5}$ metal hydride alloys. *J. Power Sources* **2012**, *215*, 152–159. [CrossRef]

15. Zhao, X.; Li, B.; Zhu, X.; Han, S.; Yan, H.; Ji, L.; Wang, L.; Li, J.; Xiong, W.; Jia, T. Preparation Method of Low-Melting Point Metal Alloy. Chinese Patent Application 201,511,015,059, 31 December 2015.

16. Young, K.; Chang, S.; Lin, X. C14 Laves phase metal hydride alloys for Ni/MH batteries applications. *Batteries* **2017**, *3*, 27. [CrossRef]

17. Matiland, T.; Sitzman, S. Electron backscatter diffraction (EBSD) technique and materials characterizations examples. In *Scanning Microscopy for Nanotechnology Techniques and Applications*; Zhou, W., Wang, Z.L., Eds.; Springer: New York, NY, USA, 2007.

18. Young, K.; Ouchi, T.; Liu, Y.; Reichman, B.; Mays, W.; Fetcenko, M.A. Structural and electrochemical properties of Ti$_x$Zr$_{7-x}$Ni$_{10}$. *J. Alloy. Compd.* **2009**, *480*, 521–528. [CrossRef]

19. Liu, Y.; Young, K. Microstructure investigation on metal hydride alloys by electron backscatter diffraction technique. *Batteries* **2016**, *2*, 26. [CrossRef]

20. Shen, H.; Young, K.; Meng, T.; Bendersky, L.A. Clean grain boundary found in C14/body-center-cubic multi-phase metal hydride alloys. *Batteries* **2016**, *2*, 22. [CrossRef]

21. Liu, J.; Han, S.; Li, Y.; Zhang, L.; Zhao, Y.; Yang, S. Phase structures and electrochemical properties of La–Mg–Ni-based hydrogen storage alloys with superlattice structure. *Int. J. Hydrogen Energy* **2016**, *41*, 20261–20275. [CrossRef]

22. Buschow, K.H.; Van Mal, H.H. Phase relations and hydrogen absorption in the lanthanum-nickel system. *J. Less Common Met.* **1972**, *29*, 203–210. [CrossRef]

23. Young, K.; Nei, J. The current status of hydrogen storage alloy development for electrochemical applications. *Materials* **2013**, *6*, 4574–4608. [CrossRef] [PubMed]

24. Chang, S.; Young, K.-H.; Nei, J.; Fierro, C. Reviews on the US Patents regarding nickel/metal hydride batteries. *Batteries* **2016**, *2*, 10. [CrossRef]

25. Young, K.; Ouchi, T.; Nei, J.; Yasuoka, S. Fe-substitution for Ni in misch metal-based superlattice hydrogen absorbing alloys—Part 1. Structural, hydrogen storage, and electrochemical properties. *Batteries* **2016**, *2*, 34. [CrossRef]

26. Wright, S.I.; Nowell, M.M.; Field, D.P. A review of strain analysis using electron backscatter diffraction. *Microsc. Microanal.* **2011**, *17*, 316–329. [CrossRef] [PubMed]

27. Winkelmann, A. Dynamical effects of anisotropic inelastic scattering in electron backscatter diffraction. *Ultramicroscopy* **2008**, *108*, 1546–1550. [CrossRef] [PubMed]

28. Keller, R.R.; Roshko, A.; Geiss, R.H.; Bertness, K.A.; Quinn, T.P. EBSD measurement of strains in GaAs due to oxidation of buried AlGaAs layers. *Microelecron. Eng.* **2004**, *75*, 96–102. [CrossRef]

Electron Backscatter Diffraction Studies on the Formation of Superlattice Metal Hydride Alloys

Shuli Yan, Kwo-Hsiung Young, Xin Zhao, Zhi Mei and K. Y. Simon Ng

1. Influence of Misorientation on the Quality of EBSD Pattern

A SEM image of the 32-h annealed sample is shown in Figure S1a. EBSD mapping was performed in the area enclosed by the green rectangle in Figure S1a. Figure S1b shows the distribution of the $NbCr_2$ grains, and Figure S1c provides the color assignment. Only three large grains are discovered in the area enclosed by the green rectangle. We collected the EBSD diffraction data in spots A, B, and C in the green grain.

EBSD pattern of spot A has the most diffraction lines among those of all three spots. Intensities of the diffraction lines are different. Generally speaking, band intensity could be influenced by camera resolution, background subtraction, sample surface, and some other parameters, leading to the difference in numbers of diffraction lines distinguished by naked eye and computer.

Spots B and C represent the crystals with misorientation less than 2° compared to spot A (A2, B2, and C2 in Figure S2). B1 and C1 of Figure S2 show that intensities of the diffraction lines are different compared to those in A1 of Figure S2. Moreover, other variations are observed: some lines disappear; new lines appear; some lines move parallelly; some lines rotate slightly; and widths of some lines change. Therefore, patterns may appear to be different by naked eye, especially C1 of Figure S2.

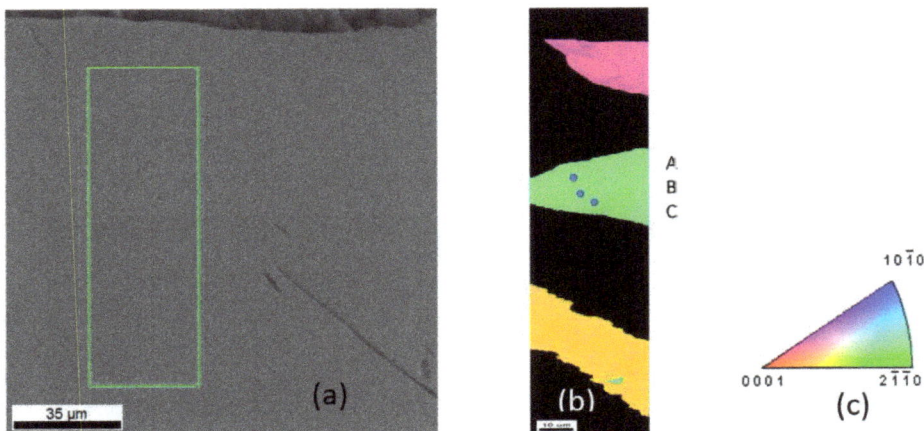

Figure S1. (a) SEM image of the 32-h annealed sample, (b) EBSD mapping of $NbCr_2$, and (c) color assignment.

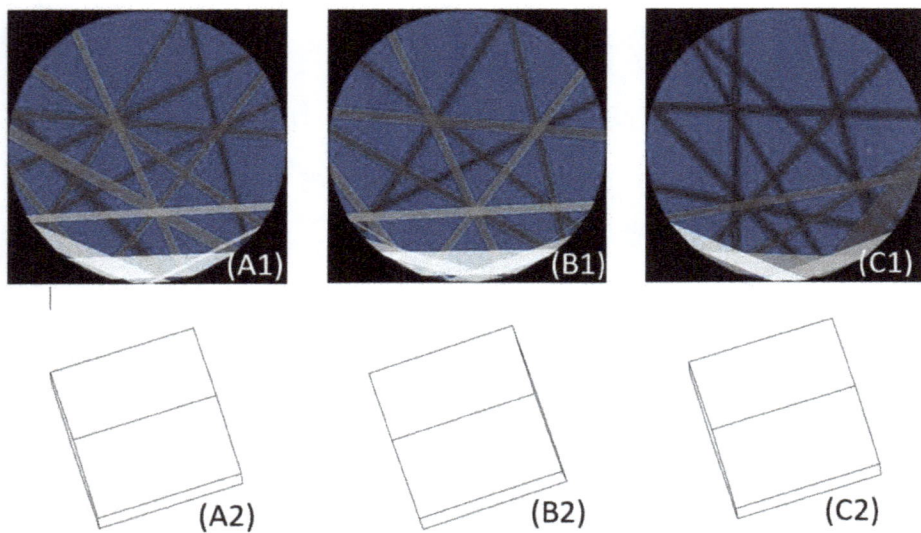

Figure S2. EBSD diffraction patterns (A1, B1, C1) and corresponding crystals (A2, B2, C2) for spots A, B, and C in Figure S1b.

2. Existence of the Mg Phase

Formation of an isolated Mg phase within the alloy bulk by the inter-diffusion of Mg is unlikely. However, EBSD patterns of some spots with perfect fitting to that of the crystalline Mg in several treated alloys are observed, and EDS reveals very high Mg-content in those spots. For example, in two areas of the 4-h annealed sample, EDS measurements were performed (Figure S3). The EDS results are summarized in Table S1 and show high Mg-content in some spots. Since the solubilities of La and Ni in Mg are negligible, these areas with high Mg-content (> 75 at%) are considered to be the pure Mg phase with the La and Ni signals from the neighboring phases.

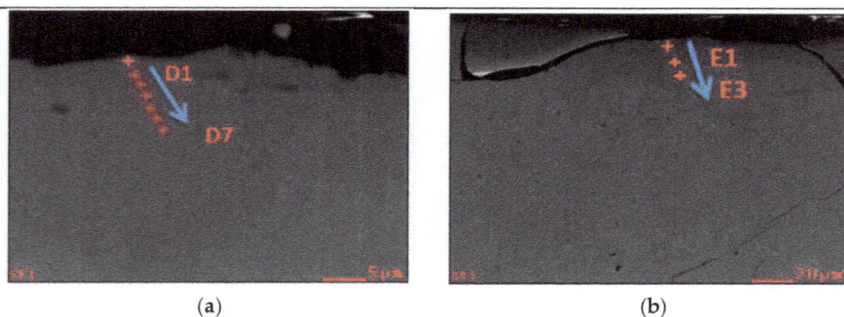

Figure S3. SEM images from two areas of the 4-h annealed sample.

Table S1. Chemical compositions (in at%) from spots D1 to D7 and E1 to E3 of the 4-h annealed sample (Figure S3).

Spot	Mg	La	Ni	Ni/(La + Mg)
D1	58.46	8.25	33.29	0.50
D2	86.30	6.04	7.65	0.08
D3	82.10	5.55	12.34	0.14
D4	80.79	5.72	13.49	0.16
D5	75.57	7.31	17.13	0.21
D6	12.88	11.8	75.32	3.05
D7	10.65	12.03	77.32	3.41
E1	81.32	7.78	10.90	0.12
E2	76.02	12.22	11.76	0.13
E3	10.99	12.23	76.78	3.31

MDPI AG

St. Alban-Anlage 66

4052 Basel, Switzerland

Tel. +41 61 683 77 34

Fax +41 61 302 89 18

http://www.mdpi.com

Batteries Editorial Office

E-mail: batteries@mdpi.com

http://www.mdpi.com/journal/batteries

www.ingramcontent.com/pod-product-compliance
Lightning Source LLC
Chambersburg PA
CBHW051709210326
41597CB00032B/5421